WILLIAM DALEY

CERAMIC ARTIST

To Ellen w/L
with Love
William P. Daley
11.1.2013
at TUFANOS

author Ruth Fine

contributors William Daley
Helen W. Drutt English
Janet Koplos
Marian Pritchard
Ezra Shales
Stephen Tarantal

editor Charlotte Daley
editor/designer Barbara Daley Schwab

SCHIFFER PUBLISHING®
4880 Lower Valley Road
Atglen, PA 19310

Library of Congress Control Number: 2013940947

ISBN: 978-0-7643-4523-4
Printed in China

Published by Schiffer Publishing, Ltd.
4880 Lower Valley Road
Atglen, PA 19310
Phone: (610) 593-1777; Fax: (610) 593-2002
E-mail: Info@schifferbooks.com

For our complete selection of fine books on this and related subjects, please visit our website at www.schifferbooks.com. You may also write for a free catalog.

This book may be purchased from the publisher. Please try your bookstore first.

We are always looking for people to write books on new and related subjects. If you have an idea for a book, please contact us at proposals@schifferbooks.com

Schiffer Publishing's titles are available at special discounts for bulk purchases for sales promotions or premiums. Special editions, including personalized covers, corporate imprints, and excerpts can be created in large quantities for special needs. For more information, contact the publisher.

In Europe, Schiffer books are distributed by
Bushwood Books
6 Marksbury Ave.
Kew Gardens
Surrey TW9 4JF England
Phone: 44 (0) 20 8392 8585; Fax: 44 (0) 20 8392 9876
E-mail: info@bushwoodbooks.co.uk
Website: www.bushwoodbooks.co.uk

CONTENTS

A C K N O W L E D G M E N T S

As the subject of this book, I dedicate it to Catherine Stennes Daley and our children, Barbara, Charlotte, and Thomas. On the facing page is a student soldier's 1950 icon to Catherine and our loving art school courtship. I enjoy the conceit that its earnest outcome foretold the beginning of a multi-generational family of emerging artists by combining the gifts of our seminal Celtic, Gallic, and Norse immigrant ancestors.

Insufficient as it is, allow me to offer a list of thank you names of the company above who together inhabit this book's pages. I owe profound thanks to its principal architects, Barbara Daley and George Schwab, Charlotte Daley, Thomas and Donna Weaverling Daley, and our contributing grandchildren, George Andrew and Amanda W. Schwab, Sara and Liam Daley, Edward, Catherine and Ezekiel Giordano.

There is not space enough to cite the specifics of the co-creators of my "in-dwelling" offerings. They are encompassed by the mega-gift of welcoming me and our family to learn with you. You are dear friends, fellow students, teachers, colleagues, mentors, artists, acolyte arts lovers, collectors of loved wonder, designers, craftsmen, heroic master makers, savants, mindful scholar scribes, architects, and engineers. All of you are living gifts in community, inviting mutual trust and nurturing belief. One of my hero mud mentors, Henry Chapman Mercer's mantra was "Plus Ultra;" the best is yet to be!

Thanks to the The Pew Center for Arts & Heritage for their support.

Thanks to Marlin and Ginger Miller for their support of this book.

A special thanks to Alfred and Rica Fenaughty for their lifetime support of my work.
Alfred and I were children together and remain lifetime friends.

— Bill Daley

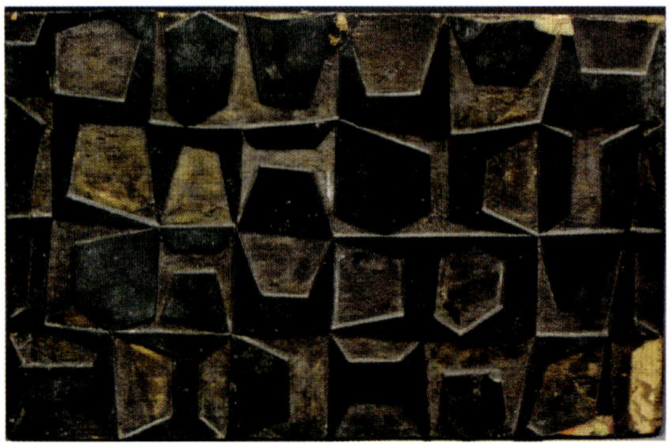

ENDPAPER GRAPHIC: *Block Print* by William Daley, 1949, designed for the *Mass Art Year Book*

Fig. 1, opposite.
Wedding, 1950
Gouache
36 x 28 inches

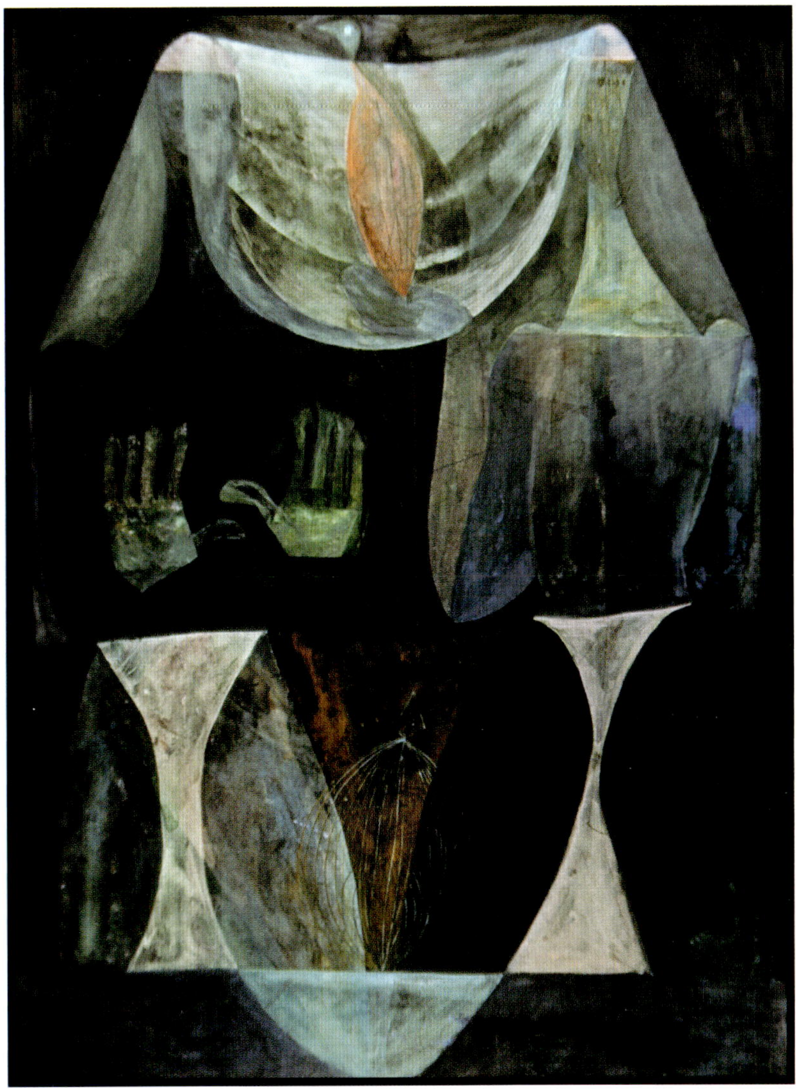

DEDICATED TO

Catherine Stennes Daley

and children
Barbara, Charlotte, Thomas

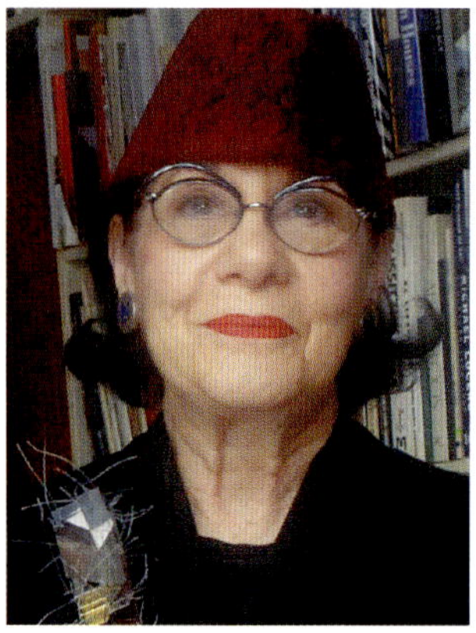

Helen W. Drutt English

Often, we view and judge vessels strictly from the outside; at what moment did Daley invite us to peer over the edge into the interior spaces of the pot? Was it the scale, the absence of a lid, or commissions for architectural projects? Was it his travels that introduced him to unknown territory—mystical trails, caverns, and kivas? The entrance into an open bowl was no longer a hidden or empty space. When did the straight edge of a pot become a rolled edge, or cease to be a constant line or a circle above an octagon? His aesthetic decisions are part of the Celtic consciousness, directing one's visions both backward and forward, like ancient Celtic faces that simultaneously cover past, present, and future.

Gallery 1015, Wyncote, Pennsylvania, in 1962, was among the first in Greater Philadelphia to exhibit the work of potters alongside painters and sculptors. The vision of the founder-director Gladys Meyers eliminated boundaries. It was here that I first met William Daley. He stood erect, proud, and carefully composed, wearing a shirt and tie; above him hung a pot, glazed in gray green. It connected immediately to my personal aesthetic, like a fiery magnet. Organic shapes, carefully articulated with a transferred value of a bulging tortoise shell, had become a container for plants.

As I grew to know him, I was intrigued that Daley's art commissions engaged his entire family in action—Bill, Cate, Barbara, Tom, and Charlotte—not unlike maneuvers; they were, in effect, the classroom on the home front. Trial installations of the completed projects occurred in the backyard of the Daley residence, on Ashbourne Road, making certain that the inception of the work occurred without errors. For instance, images of the 1967 Portuguese National Airlines relief appeared in front of his house. The sheet-bronze sculpture conjured a squadron in vertical flight; it almost reached to the peek of the roof. There is even an image of a motorcyclist stopping in disbelief to stare at what he has just encountered.

Again, in 1979, a trial installation occurred in the backyard. A wonderful photograph exists of Cate and Bill Daley sitting in front of *Cellular Permutation*, commissioned by Raymon Posel for the Ritz Theater, in Center City, Philadelphia, confident that the work will float in the frontal façade between steel frames.

Our friendship followed a definite course: It was connected to his work and our joint mission to empower the crafts as art. Daley became a founding member and central figure in the formation of the Philadelphia Council of Professional Craftsmen (PCPC), an organization founded, in 1967, under my leadership. Among the first major exhibitions sponsored by the fledging organization was Daley's solo exhibition two years later at the Philadelphia Art Alliance. The inaugural exhibition of the Helen Drutt Gallery, in 1974, celebrated Daley's pots among the initial exhibitors. Bill served as the primary instructor in exhibition design and drove the traveling caravan that made the inclusion of nonregional works possible.

During the early years, most works did not bear titles; future pots celebrated regional influences, individuals, and mathematical, historical, and literary references, with an insightful lexicon. Daley's passion in describing the connection between words and work was to become legend, as were his oratorical performances during public lectures or in the classroom. I remember the first Daley pot—an inverted palm constituted the belly with a rolled edge—that was acquired by the Philadelphia Museum of Art, after being exhibited in my eponymous gallery. Darrel Sewell, curator of American art, brought Evan Turner, the director, to examine and assess the work for acquisition approval. Making it a part of the permanent collection gave broader recognition to Daley's work, as well as giving greater credence to the craft community itself.

Further projects continued beyond the gallery itself. In 1976, I invited artists to create contemporary tureens for an exhibition at the Campbell Museum, Camden, New Jersey, in honor of the bicentennial. Can you imagine the response of the trustees—whose passion for seventeenth- and eighteenth-century decorative tureens that once graced royal dining-room tables was well-documented—when presented with an unglazed brown pot reminiscent of healthy female buttocks? Daley's large forms were reduced to the size of a football. The diminutive size irritated Daley, who called me and requested that I never invite him to create a work of that scale again. The museum, however, acquired the Daley tureen for its permanent collection.

The year 1980 saw the organization of "Contemporary Ceramics: A Response to Wedgwood" by both the Helen Drutt Gallery and the Wedgwood Museum, in Bala Cynwyd, Pennsylvania, in honor of the 250th anniversary of the birth of Josiah Wedgwood. It was held at the Museum of the Philadelphia Civic Center. Daley's contribution was entitled *To Josiah W.* The work was a full-bodied inside/outside pot in which a broad band is held over a smooth surface by a rectangular two-dimensional element of diametrically opposed twists.

In 1982, Yvonne Joris and Evert van Straaten traveled throughout the United States making selections for their exhibition "Who's Afraid of American Pottery?" held early the following year at the Dienst Beeldende Kunst, Kruithaus, the Netherlands. This was the first contemporary exhibition of American ceramic vessels held in Europe, although the Stedelijk Museum, Amsterdam, had previously exhibited West Coast ceramic sculpture with an emphasis on the funk aesthetic. *Danced Place* and *Bi-Lateral Invitation* represented Daley's work. We were invited to represent America through a series of alternating lectures. These works helped to redefine the Dutch perception of a contemporary vessel.

Daley has always felt that the spirit in a work is synonymous with its quality. He uses language, selecting descriptive titles to identify works that then move us into unknown territories. Taking advantage of the verbal gift bestowed upon humans, Daley has supported celebratory descriptions expressing his approval as well as derogatory words expressing his disdain. In essence, he is a performer. Can one forget his march through Rittenhouse Square with a colander on his head, pretending he was a German soldier, or his outrageous paper-constructed chapeaux and caps in support of his friends and lectures? Can one forget Daley driving a nail into a pedestal to make certain it would not be moved, or instructing one to remain after midnight to assure that the exhibition installation would not be dismantled by another member of the team? Can one forget his largesse as he celebrates his peers' achievements or guides his students to recognize their potential?

For more than seven decades, Daley's passion and quest for knowledge have been evident in his work. Family maneuvers have continued, as the creation of this publication embraces all aspects of his family's professional expertise. Influences that range from a ruined Spanish church and Pueblo kivas to Frank Furness's architecture, as well as his obsessive reading of texts, past and present, all have supported a creative energy that has transformed clay and, even more important, our understanding of the vessel as a true work of art.

Helen W. Drutt English
Philadelphia, PA

Helen W. Drutt English is world renowned for her significant and continued commitment to the advancement and awareness of the Modern and Contemporary Craft Movement. As a gallerist, author, lecturer, curatorial consultant, and collector, her contributions have been vast, with scholarship and education being at the heart of all of her endeavors.

Whacked Geometry

Ruth Fine

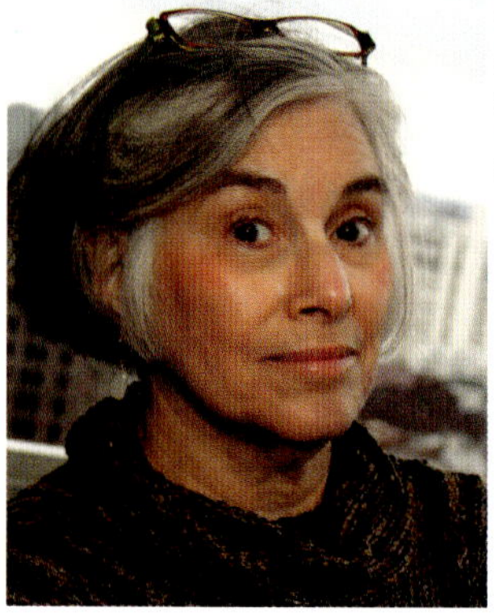

Photo: Charlotte Daley

Ruth Fine

Ceramic artist William P. Daley (Bill, as he is best known) is a brilliant wordsmith. But, he rarely discusses one of the most important experiences of his life. In 1943, at the age of eighteen, Daley was shot down over Germany while on his initial World War II mission with the United States Army Air Corps. He spent ten months as a prisoner of war, initially in solitary confinement in a Budapest state prison. From there, Daley was transferred to Stalag Luft 4A, located some thirty kilometers east of Peenamünde on the Baltic Sea, where the Germans developed and tested their rockets. (Fig. 2)

Daley sustained his spirits by reading aloud with Loyd C. Long, another, more intellectually experienced Stalag prisoner, who readily shared his copies of the *Oxford Guide to Classical Literature* and a volume of William Hazlitt's essays. While ten months is but a moment in Daley's lifetime of more than eight decades, living through this harrowing period had a profound impact on his subsequent experiences: his sense of gratitude for all things good; his immense delight in every aspect of the creative process, which is at the center of his world; and his *"belief in overcoming resistance with verve and belief in possibility becoming."*

Daley and his siblings—an older half-brother, Joseph, and a younger sister, Alice— grew up in a house at 74 Rose Street in the village of Hastings-on-Hudson in Westchester County, New York. (Fig. 3) Their mother, also Alice, was a committed homemaker whose devotion to her family, as well as the family's need for additional funds, led her to care for foster children throughout most of Daley's childhood. This provided him with a "changing family in a small two-story house with more people than actually fit." Daley's *"mega-father," also Bill,* was a colorful character who continues to be his hero, remembered for, among other things, his deep love of language. Daley quotes him continuously in conversation and writings, in one place describing him as a housepainter who acceded to the pressure of providing for his family by giving up a seasonal livelihood to paint year-round in the larger of Hastings-on-Hudson's two factories, Anaconda Copper and Cable. He took fierce pleasure in his capacity for hard work. The elder Daley loved painting walls, reciting A.E. Housman, and talking, particularly about Franklin Delano Roosevelt. As an ardent Catholic and a member of the Admiral Farragut Post 1195, he gave covert testimony to his children on the free use of human energy and the metaphoric possibilities of *"gifts."*

Daley remembers using layered coils to build his first clay pot in elementary school and working on a potter's wheel for the first time during his high school years, both at the Hastings-on-Hudson Public School, "a huge brick horseshoe with the kindergarten on one end, then the grade and junior

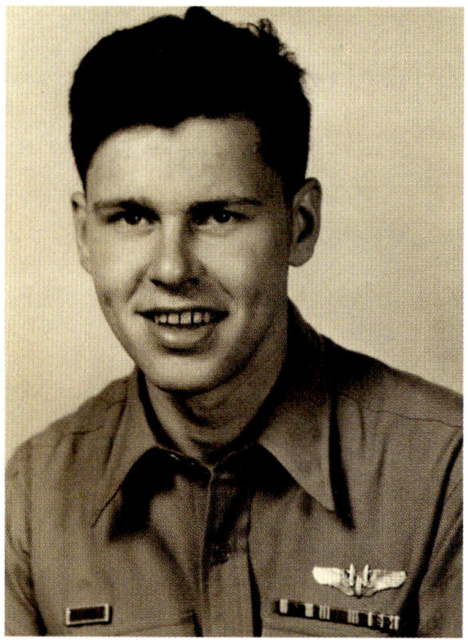

Fig. 2. William Daley, Staff Sergeant, 1945, US Army Air Corps

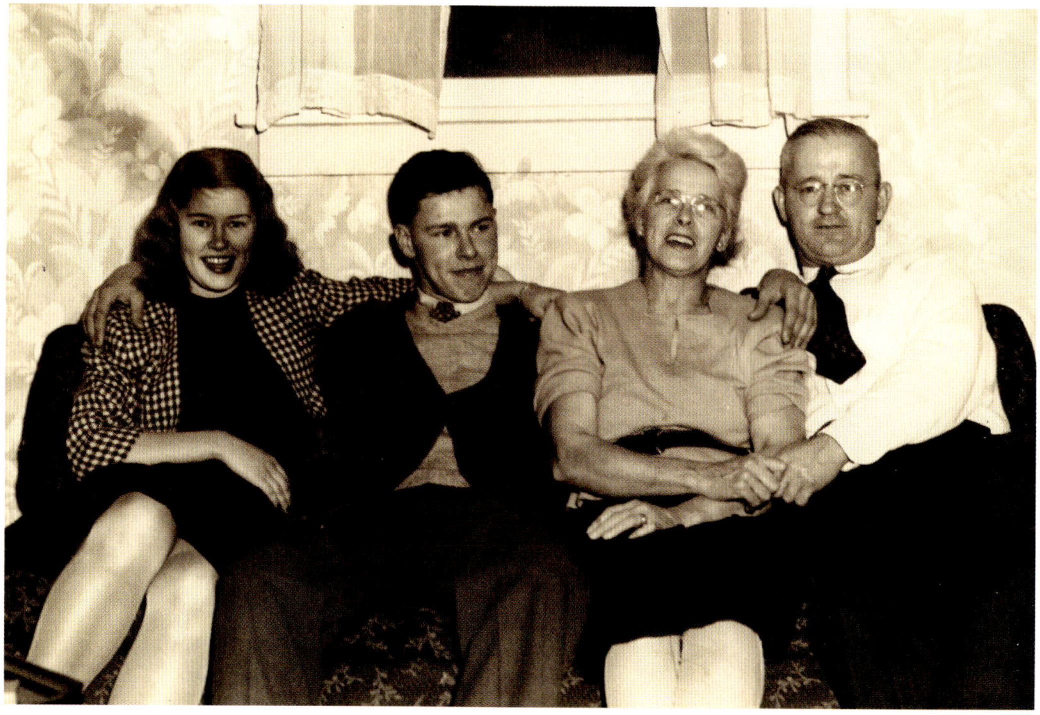

Fig. 3. Sister Alice with Bill and their parents Alice and William Daley, Sr. in the living room at 74 Rose Street, Hastings-on-Hudson, New York, 1948

high [schools], with the high school on the far end.... linking all the children from all the town's sectors in it together—children of mill workers and of commuters to NYC and all in between—in one glorious extraordinary school of students and teachers...part of a multi-ethnic, diverse, economic mix."

During his public school years, Daley became familiar with Helen Gardner's classic art history text, *Art through the Ages*; and Khalil Gibran's book of essays, *The Prophet*, got him off on the mythic romantic direction—"that linked in with my self-induced transports in the 'Mass' and its magical transformations with 'Bells and smells,' and the great splendor of Benediction-transformation and reverberative squinting, down ranks of burning candles after staring at them to make their after-images...turning everything light, black, and everything dark, full of penumbric light—real physical magic."

Daley initially enrolled at St. Matthew's parochial school. As a first-grader, he was chosen to draw seasonal decorations in colored chalk on the class blackboard—pumpkins and turkeys, Santas and saints. A love of drawing carried through high school. Over the years, his Hastings-on-Hudson classes took bus trips to New York City, about twenty miles away. They visited museums and libraries in the five boroughs, institutions devoted to art as well as those focusing on history and on nature, such as the Bronx Zoo, the Van Cortlandt House Museum, also in the Bronx, and the Brooklyn Botanical Gardens. The Museum of Modern Art in Manhattan, however, was Daley's favorite, and he retains strong memories of *"Calder's mobile in the stairwell and Picasso's* Musicians *at the top of the stairs next to Brancusi's* Bird in Flight. " His class also traveled north, to Washington Irving's Sleepy Hollow and other historical properties in New York State, thus providing well-rounded and challenging experiences that jump-started Daley's intellectual life and contributed to his acute sense of the importance of place.

World War II interrupted the flow; but, upon his release from prison camp, the G.I. Bill enabled Daley to attend Boston's Massachusetts College of Art (1946 through 1950), generally referred to as *"Mass Art."*

Mass Art as an Art School then was a wild place. I'm sure the teachers were overcome by the G.I.s and their (crazy) need to believe the carnage was for them (something suffered to learn) that they earned—We were so arrogant—(hungry arrogant)—brattish—arrogant—whatever. We thought we could do anything & everything and do it well—and we tried like Hell. The place was really an Academy—all specialty and fixed pigeon holes and we blew it apart—Art Education was the best program....The most venturesome did it (art ed)—the smartest—most avid—most talented—and the most—visionary. Reading, museum stealthing—play going—making. All-in-all best we could, bathed in the certainty that it was GOOD and so were we—near genius—it was a kind of greedy shared ignorance—going for it full of hope, belief, and juggernaut surety...We were "drawn to" and we "focused on" and we worked hard at walking into the sun—blinded by the "wonder of us." (Fig. 4)

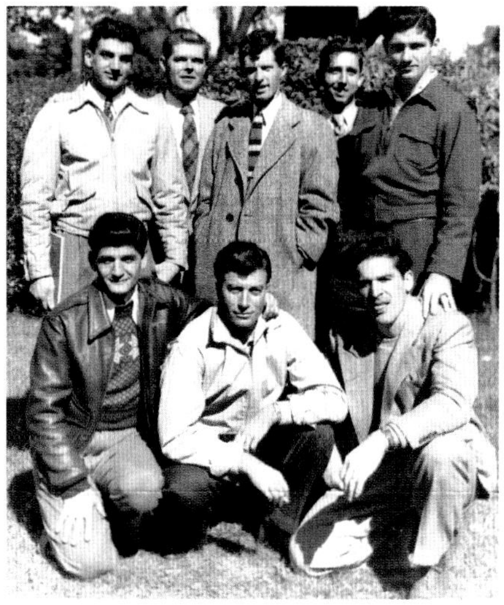
Photo: William Daley

Fig. 4. Veteran classmates of Bill's, Massachusetts School of Art, 1948. Pictured front: Paul Hatgil, Ernest Pelosi, Anthony Carbone; back: Charles Lutheras, Charles Quillan, Milton Sullivan, Anthony Guanea, John Merrill

"The most venturesome did it (art ed)—the smartest—most avid—most talented—and the most—visionary. Reading, museum stealthing—play going—making. All-in-all best we could, bathed in the certainty that it was GOOD and so were we—near genius—it was a kind of greedy shared ignorance—going for it full of hope, belief, and juggernaut surety... We were 'drawn to' and we 'focused on' and we worked hard at walking into the sun—blinded by the 'wonder of us.'"
— William Daley

Will Durant's *Story of Philosophy* served as Daley's *"awakening...to the life of reading beyond story to question belief."* Other books of early importance to him were the writings of Ananda Coomaraswamy, in particular *The Transformation of Nature in Art*; Martin Buber's *I and Thou*; John Dewey's *Art as Experience*; Henri Focillon's *Life of Forms in Art*; Sigfried Giedion's *Space, Time and Architecture*; Emmanuel Kant's *Critique of Beauty*; Herbert Reid's *Education through Art*; and Eliel Saarinen's *Search for Form, A Fundamental Approach to Art.*

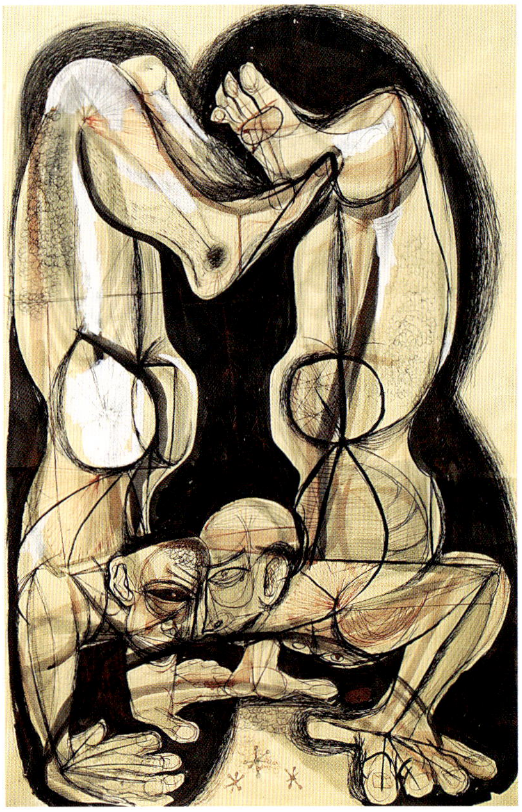

Fig. 5. *Handstand Drawing,* 1948

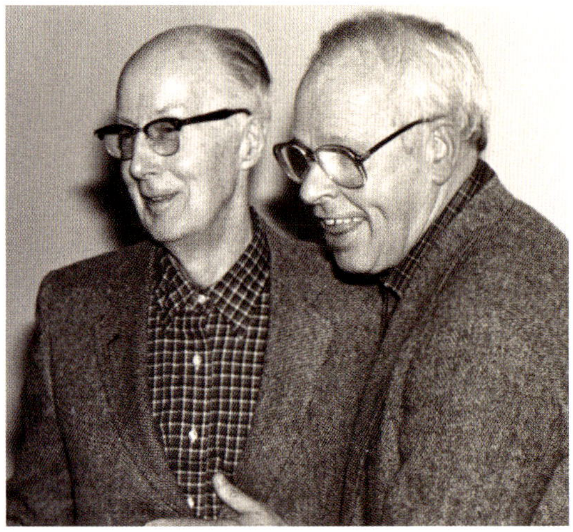

Fig. 6. Charles Abbott with Bill, 1984

Daley immersed himself in literature as well, remembering Thomas Mann's *The Magic Mountain* and Rainer Maria Rilke's *Letters to a Young Poet*, as being among the texts that guided him as a young artist. Over two summers during his undergraduate years, living at home in Hastings-on-Hudson, Daley took literature courses at Columbia University. After working his day-job with the Anaconda Wire and Cable Company, he traveled to Manhattan at night to study *"Restoration Lit., Pope, Dryden, Hazlitt, R. Burns, whom I love, Wordsworth....the teacher was great—all about the 'Structure of the Meaning of Poems.'"*

His sights initially were set on a career in art education with a major in painting, and his two-dimensional works from his formative years such as *Handstand Drawing* (Fig. 5) affirm his talent in this area. Daley's emphasis shifted to clay in his junior year, after he encountered faculty member Charles Abbott (Fig. 6) who had established the ceramics program at Mass Art. Abbott became both teacher and friend to Daley, and their association lasted until Abbott's death in 1986. Abbott's enthusiasm for Asian art, especially pottery, and his commitment to the *"difficult satisfaction of making clay stand up by itself"* were instrumental in Daley's move from painting to ceramics. (Fig. 7) Perhaps most attractive to the young artist was Abbot's open-minded approach to art-making: *"It's about what you're doing while you're doing it"* is how Daley describes his mentor's philosophy, which emphasized the process and practice over the product, an approach to object-making that Daley still observes. Abbott *"never told you what to do—read or think—he just helped you make pots....He thought we ex-soldiers were so after our own answers that we couldn't hear the questions until they came through our hands."*

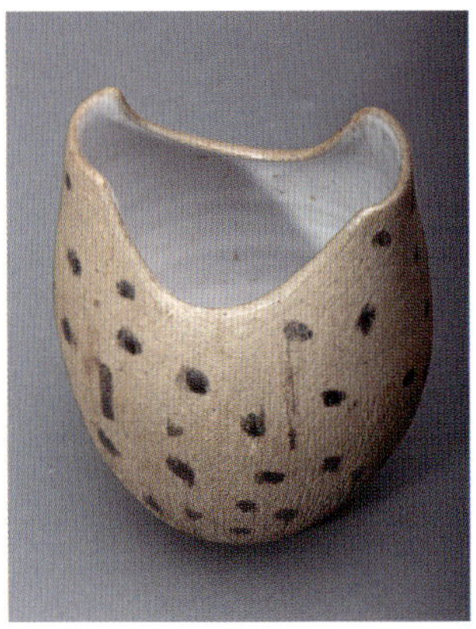

Fig. 7. *Frontal Form,* Vase, 1953
Refer to page 30

Along with the varieties of Asian art to which he was introduced by Abbott, other treasures that repeatedly brought him to the Museum of Fine Arts (MFA), Boston, were the European old master and American modernist paintings. Also at the MFA in 1948, just as he was debating with himself about changing his major, Daley saw a national traveling exhibition of contemporary ceramics organized by the Everson Museum of Art in Syracuse, New York, and sponsored by the Syracuse China Company. It helped clinch his decision to redirect his attention. Yet, despite this commitment to three-dimensional form, Daley's roots in painting and drawing remain essential to his work in *"mud,"* the term by which he refers to the clay material that he has shaped and that has shaped him as an artist for more than fifty years. Even today Daley references the fluidity of paint and the bivalent nature of line when discussing his working methods, stressing the need to keep his clay moist and open if he is to maintain the flexibility essential to the expansive size and complexity of his vessels. (Fig. 8)

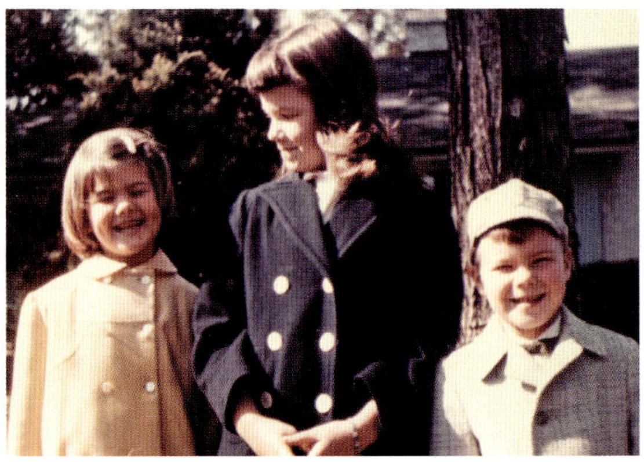

Photo by William Daley

Fig. 10. Daley children: Charlotte, Barbara and Thomas, 1958

Photo: Charlotte Daley

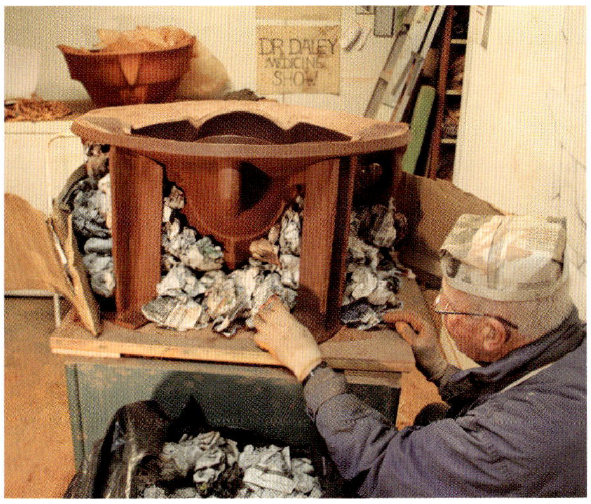

Fig. 8. Bill making a large vessel

The student body at Mass Art was relatively small, and Daley first encountered his beloved Cate (Catherine Stennes at the time), during their freshman year. As juniors, they worked on art education projects together, and as seniors they realized they were in love. The couple married in 1950, shortly after graduation. (Fig. 9) Their family soon included three children, Barbara, Charlotte, and Thomas, all of them now adults with careers in the arts, the direction taken by several Daley grandchildren as well. (Fig. 10)

A virtual dynamo, Daley's extraordinary energy has always been nourished not only by his studio work, but also by family and friends, students and teaching colleagues, church and travel, as well as an intellectual curiosity immersed in a wide range of disciplines, including art criticism and art history, philosophy and religion, science and mathematics, poetry and prose. Essential to all of this, however, is the artist's own critical eye and spirited intuition—his practice of looking closely at every aspect of his surroundings and of considering the broadest visual implications and contextual possibilities of what he sees.

While it may seem obvious and simplistic to mention, all aspects of his life must be taken into consideration to understand the far-reaching meanings of Daley's vessels. He keeps his sights on an unreachable, evolving goal of perfection, extending it with each new formal idea or methodological exploration, always seeking to discover what more might be possible with clay. *Live in your Head: When Attitudes Become Form* was the subtitle for a landmark conceptual art exhibition in 1969. This idea of attitudes becoming form seems equally appropriate to Daley's very physical approach to art-making, one that embraces formal possibilities visible in the world as well as those living in the head.

Eager to share his experience with a community of like-minded associates while developing his own talents, Daley followed his four years in Boston with one in Manhattan, where he continued to work in mud while earning an MA in art education from Columbia University Teacher's College. Swiftly thereafter, he landed his initial job teaching ceramics and sculpture at the Iowa State Teachers College, part of the University of Northern Iowa in Cedar Falls (1951-1954) (Fig. 11). This appointment would call for the first of the Daley family's moves to towns and cities where the most venturesome teaching opportunities were available: Cedar Falls was followed by New Paltz and Fredonia, New York (1954-1957 and 1963-1965 respectively); and Albuquerque, New Mexico (1971-1972). With these exceptions, the Philadelphia, Pennsylvania, region has been Daley's base since 1957.

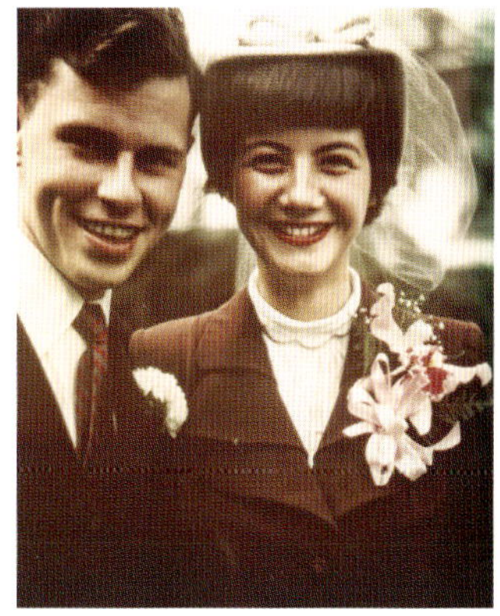

Fig. 9. William and Catherine on their wedding day, September 2, 1950 Belmont, Massachusetts

Photo: William Daley

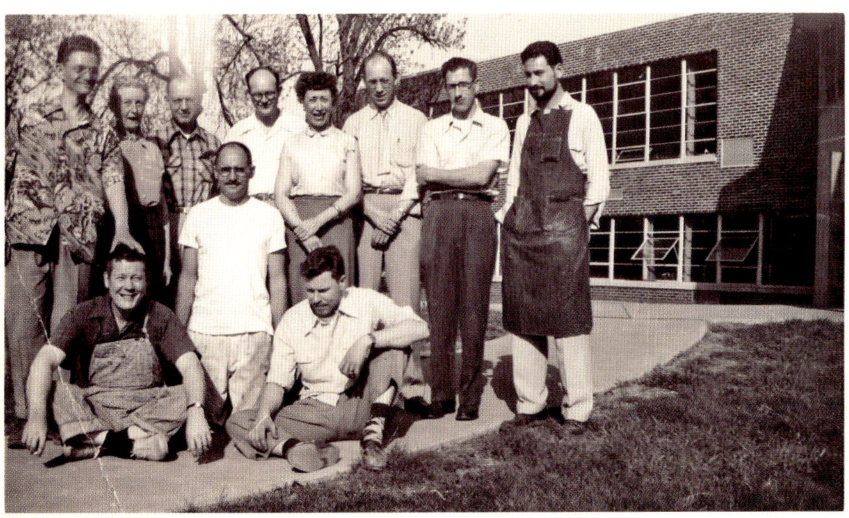

Fig. 11. Iowa State Teachers College, Cedar Falls, Iowa 1952. From left to right, front: Tom Larkin, Jack Berger, Bill Daley; back row: Joseph Bolinsky, Corley Conlon, Clayton Fowler, Harry Guillume, Marjorie Campbell, Paul Smith, David Delefield, Glen Bradshaw

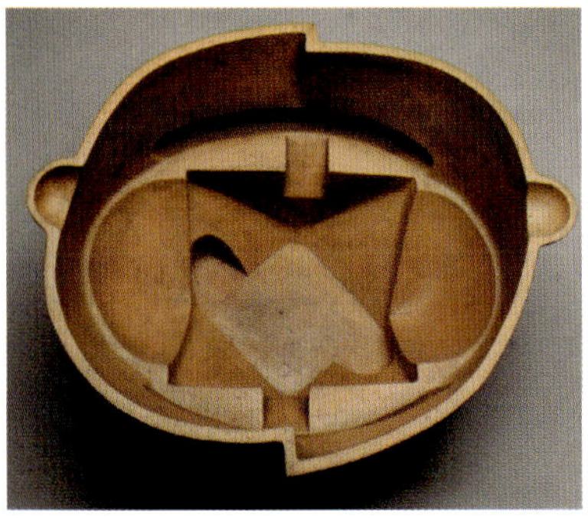

Fig. 12. *Apsed Oval*, 1990
Refer to page 111

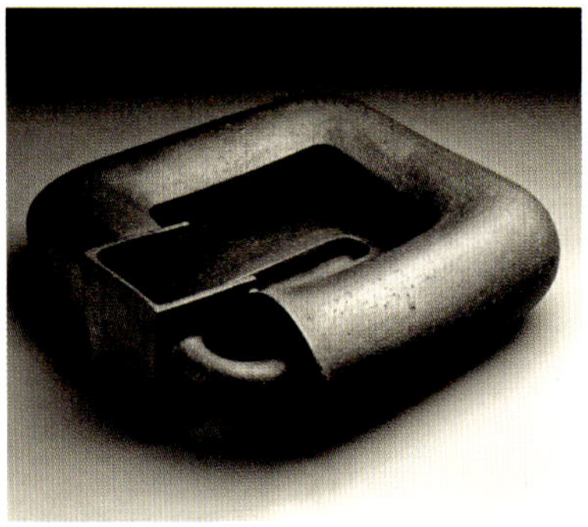

Fig. 13. *Reptilian Chiclet*, 1968
Refer to pages 61, 63

While the logistics of family life were mandated by Daley's teaching career, each move also impacted the materials with which he engaged. In every work place he used a different clay body (which chemically consisted of one part alumina, two parts silica, two parts water). Committed to discovering the variations in color, texture, and density of each region's bounty, Daley sees clay *"as a magic compound—that transforms through heat (oxidation), going from rock to dust and when fired—back to rock (Hardness) so it's the primal alchemical substance of form becoming—witness all the creation stories about it. And it's all true."*

In Iowa, the study of local clays by John Stephenson, a senior at the time, reminded Daley of the mud he gathered from the banks of the Saw Mill River for the pots he made in elementary and high school, thus reinforcing the concern with exploring his surroundings for useful materials. At Mass Art the students had used powdered Fetzer Red Ohio clay mixed with water—an earthenware (low-fired and porous), thus needing to be glazed to hold water. The processed clay they used at Columbia was *"heavily grogged"* (grog being ground-up, pre-fired clay). The clay body Daley used in Iowa was referred to as "Mortar Mix," clay that was ground-up shale, scraped from the cliffs in Fort Dodge, Iowa. Local artists bought it at a lumber yard, where the material was sold to bricklayers for mixing with mortar. *"It was cheap, cheap, cheap—available—and we made earthenware pots with it."*

Following his stint in Iowa, Daley began to invent his own clay bodies. Incorporating more than one kind of clay and adding grog to the mixture, his materials have gotten increasingly complicated over the years.

"The clay at the core of PCA mud on my watch was Jersey stoneware clay—a yellow burning, brick clay that goes to a high temperature without deformation. It doesn't melt. Red clay (earthenware) melts and [becomes] glass at high temperatures. Jersey brick clay is magic; very plastic when wet. With excellent dry strength (shrinking, drying slowly, no cracks). And when high fired is vitreous (water proof). No glaze needed except as cosmetic....

"My clay body [in 2011] has five clays in it—the formula is two parts red clay—earthenware clay from Ohio called Red Art; one part ball clay from Kentucky for plasticity; one part Red Fire clay from California (a high temperature clay like the Jersey clay); one part white fire clay (no iron in it); and 15% coarse grog for temper, to make it strong (to stand up better). I still mix it (with help). Most clays like mine are at workable plasticity with the addition of 30% water by weight. There are other wondrous things about firing it—drying it—coaxing it to stand up while it's becoming rock again—like it was in the beginning. Why beautiful—It's light enough (not chocolate) to show form in light shadow. It's strong as hell. Plastic, just right, and affordable, and I can make it and tweak its make up to say brown a thousand wondrous ways.... "

In addition to precipitating shifts in his clay bodies, Daley's moves to new teaching posts otherwise impacted his work. Over the years at PCA, his schedule included classes in ceramics, sculpture, industrial design, three-dimensional design, two-dimensional design, and descriptive drawing. He also coordinated the graphic design, illustration, and textile design programs and chaired the industrial design department. The courses he taught and the demands of these supervisory responsibilities set his mind in new directions and benefited his own studio-work in diverse ways.

Daley's art continued to embrace strategies, forms, and ideas with origins in artistic predecessors from many times and places, who explore numerous formats. Included are Asian, African, and Western artists, both named and anonymous; paintings, sculpture, drawings, prints, books, and architecture; objects made for ritual and for domestic use; and surfaces that are modified by many kinds of decoration. Important examples of this last are the complex designs on Islamic tiles, as in the fourteenth century AD Alhambra Palace in Granada, Spain. Daley knows the place from book illustrations only, but it has consciously inspired some of the structural patterning essential to his vessels."Information informs format and confirms intuition to be responsive to serendipity and chance play." (Fig. 12)

But Daley's inventions, at times, appear to reference objects and places he has not seen, either in person or in reproduction. These are best understood as forms and ideas with which he recognizes kinship when he discovers them, affinity rather than influence. For example, the 1971-1972 teaching stint in New Mexico brought to his attention a geometric language essential to native art and architecture of the southwestern United States, forms that are closely related to Daley's vocabulary. But, in fact, they were intuitively embedded in his work before he encountered them in the southwest. (Fig. 13) Likewise, Daley employed interlocking linear elements similar to those found in Celtic book illustration (the most

"To me the expressive quality of a vessel form is of first importance. My work is an attempt to experiment with these forms. My interests are sculptural rather than utilitarian. The function of a vessel need not be specific. Work created to be held, touched, and seen is valid to the degree that a person responds to its form. Its presence is the vessel's reason for being. I aspire to be an artist rather than a craftsman." — William Daley

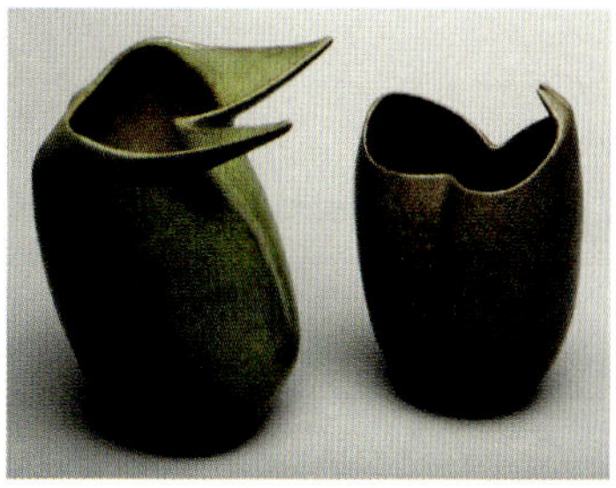

Fig. 14. Glazed pots, 1956
Refer to page 37

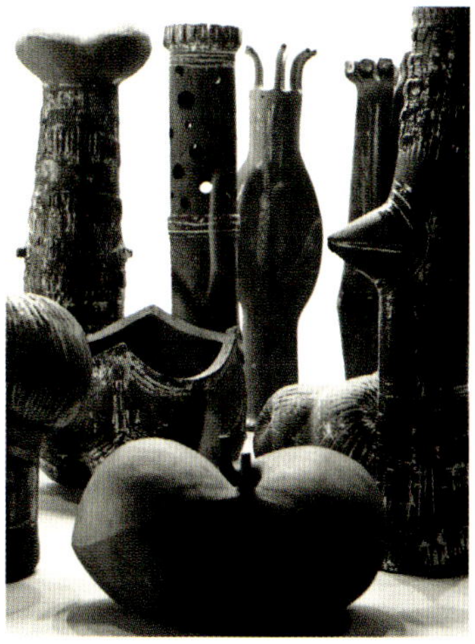

Fig. 15. Composite structures inspired by nature, 1958-59. Refer to page 43

famous example being the *Book of Kells*) prior to his knowledge of this earlier example. The discovery of these kindred spirits, however, encourages him to probe more deeply into the associative and formal possibilities they offer, adding to the visual and metaphorical richness of his art, what he refers to as *"the art of possibility becoming."*

In 1954, when Daley arrived in New Paltz to teach at the State University of New York, the Shang bronzes (1751-1027 BC) introduced to him by Abbott at the MFA remained much on his mind: *"their ellipsoidal cross-section and scooped rims,"* especially a *"three-legged wine vessel stretched in the air with its large bird-like snout and sweeping rim."* He also was using glazes that referenced what he refers to as *"the marvelous green color"* in the oxidized patinas of these objects. (Fig. 14) Within a year, moving beyond the traditions of throwing at a potter's wheel, Daley was slicing into thrown clay, or cutting it into pieces and altering the multiple elements before reuniting them as a new form. Thus he initiated the exploration of what would become his decades-long investigation of inside/outside relationships, with particular emphasis on the wall that divides the two.

In part, Daley's new work was a response to the suggestion to "loosen up" offered by his New Paltz colleague, the painter Jules Olitski. Inspired by a gestural approach in the work of contemporary abstract expressionist painters such as Hans Hofmann and Philip Guston, Daley soon abandoned the wheel completely, to build composite structures from slabs of clay that he formed over pre-determined shapes. (Fig. 15) At times he would use wooden sticks to beat the clay, sprinkling it with coffee grounds which burned out in the kiln; the process altered a work's thickness and enriched the character of its texture: *"Shang shapes and pummeled surfaces."* His molds for forming slabs, at that time, included: tennis balls; rubber balls and condoms filled with plaster; stove pipes and sewer pipes; and oil drums, when larger elements were required. Forms in nature also inspired his art then, as now—trees (both weighty trunk and delicate, gestural branches), seedpods, and summer squash. Suggestions of human anatomy add an erotic element, to which he has referred as the *"breasts and butts"* in his work.

In the late 1960s, a polystyrene building insulation became Daley's medium of choice in creating his library of forms. The material's strength and light weight provide flexibility that enables him to develop a range of personally designed, interchangeable elements. Daley glues or pins together the sheets, which are two-inches thick as manufactured, and then shapes them by cutting, sawing, and shaving. (Fig. 16).

The artist's cogent statement about the work in his first solo exhibition, organized in 1955 by Olitski for the New York University Art Education Gallery, reflects beliefs that essentially remain true today some half-century later: *"To me the expressive quality of a vessel form is of first importance. My work is an attempt to experiment with these forms. My interests are sculptural rather than utilitarian. The function of a vessel need not be specific. Work created to be held, touched, and seen is valid to the degree that a person responds to its form. Its presence is the vessel's reason for being. I aspire to be an artist rather than a craftsman."* (In the 21st century he would say he aspires *"to become a consummate craftsman— i.e., 'maker.'"*)

By 1960, Daley embraced the geometric vocabulary that came to mark his signature vessels, after which the influence of Shang bronzes was less evident in his art. He also

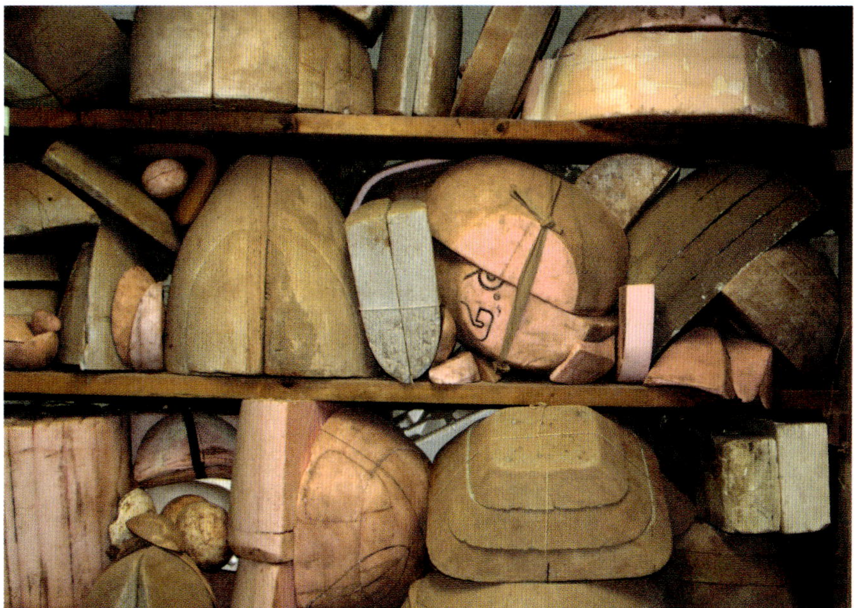

Photo: Charlotte Daley

Fig. 16. Forms in the studio

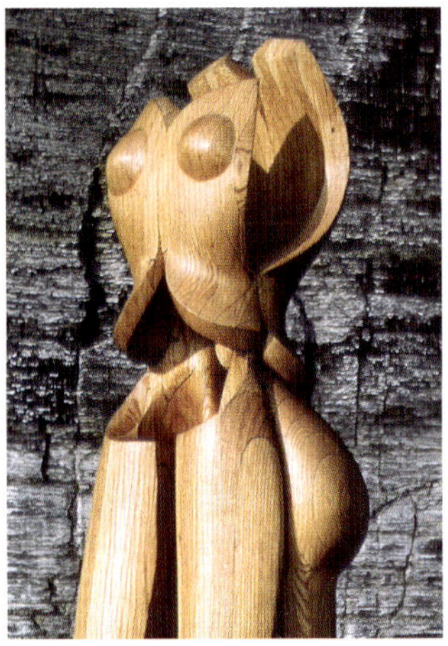

Fig. 17. *Diana,* 1964–65
Refer to page 59

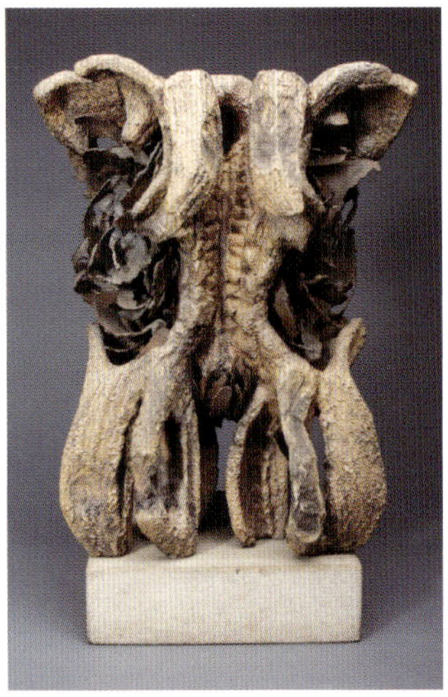

Fig. 18. *Torso,* 1964
Fire clay, table salt and spaghetti combined with welded steel
Refer to page 53

abandoned the use of glazes in 1965, when he left Fredonia to return to PCA, feeling that, for him, their use was more about decoration than form building. His time at Fredonia had been one of extraordinary ferment in his work in many ways. Deciding to *"use every medium,"* he mainly experimented with wood and metal, the latter of which he learned to both shape and weld. He concentrated on making sculpture as well as vessels, in particular works made of what he describes as *"celery stalk-like modules"* cylindrical elements that he placed in two directions, which, in essence, are his earliest play with the issues of inside/outside structure.

Some of these sculptures are more abstract than others, but for the most part they reference the human figure in both size and structure. (Fig. 17) Daley experimented with a variety of woods (carved, laminated, assembled) and metal, primarily bronze (cut, hammered, braised, and reinforced with rods when the size or format required it); he also combined these materials with clay. (Fig. 18)

By this time, Daley had completed several commissions. His association with design and architectural firms in the late 1950s and early 1960s (as detailed in this book's chronology), led him to create several screens, walls, and fountains that were intimately responsive to architectural situations. The screens and walls explored theme and variation within a multiple tile structure, starting with a ceramic screen titled *Progeny* for the downtown Philadelphia home of Max and Mary Raab in 1959. The tiles were marked by a highly personal symbol system by which Daley referenced the idea of language as naming, denoting, in a universal sense, rather than consciously deriving his forms from any preexisting lineal sources.

The Raab screen was followed by a ceramic, wood, and bronze screen installed in the IBM Pavilion at the 1962 Seattle World's Fair. (Fig. 19) Its structure was based on that of an abacus, again incorporating a symbol system, but this time it referenced Babylonian, Pre-Columbian, Hindu, and Hebrew letter forms and numerals. Daley undertook several other wall/screen commissions that are now destroyed or situated in places other than those for which they were completed. Among them was a 1963 assemblage of teak screens for the Singing Water restaurant (now closed) in the newly built Penn Center office complex in center-city Philadelphia and a 1971 wall that was the focal point for the South African Airways offices in Manhattan (see Fig. 252). Built of clay and copper tiles, the spare elegance of their design was inspired by African sculpture.

The last of Daley's wall commissions still stands at its original site. Completed in 1978 for the Ritz Theater, in Philadelphia's Old City historic district, it is built of tiles made of ceramic alone. Their lean geometry clearly reflects the formal interests Daley was pursuing in his ceramic vessels at the time. (Fig. 20)

Having worked with numerous formats and materials in sculpture and commissioned projects, by 1967 Daley realized that the vessel form was his primary interest: ritual object, object of use, work of art, all of which are simultaneously suggested. Daley remembers a sculpture exhibition at the Whitney Museum of American Art that year, in which there

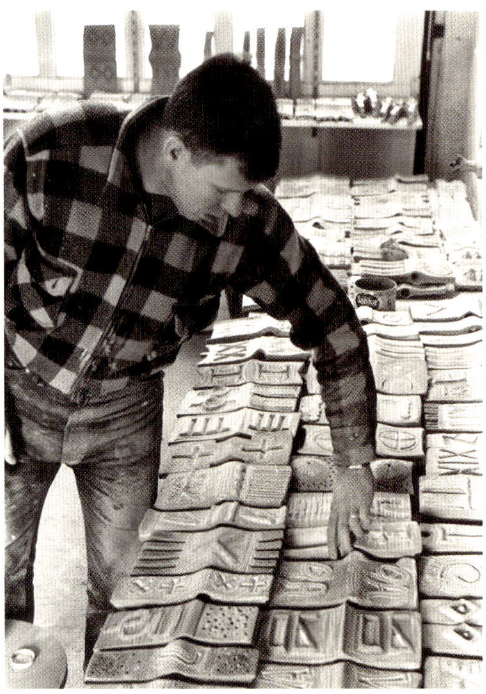

Photo: Murray Weiss

Fig. 19. Test assembly in Bill's Feasterville studio for ceramic wall commissioned by IBM, Seattle World's Fair Pavilion, 1962
Refer to page 168

was a piece that spoke to him quite clearly, as particularly important to his thinking *"about doing more with less through material.... It was a perfect model of containment, protection, exclusion/inclusion all at once."*

That 1967 was critical a critical year for Daley's work is clearly seen in vessels carrying that date, some of which were exhibited in Craftsman '67: First Regional Exhibition at the Museum of the Philadelphia Civic Center, including *Chiclet* (Fig. 13), which reflects his essential concerns at that time:

In [those] pots I put it all together, that creating a pot—a unity—was more than the resolution of opposites. Not "either-or" but "both-and." Not parts arranged, but all of the elements coming together as evocative structure, from the inside out and the outside in—by creating the wall as a membrane (call it a wall that's structure made maximum strength, resisting gravity, standing up with the least amount of material). It is Organic Engineering in Mud, combining Bumps and Holes, utilizing inside and outside by reversal, to make a membrane wall that is strong visually and tactilely engaging.

Decoration in ceramics most often is surface. For me, structure is its own decoration, becoming energized surface through light.... Form becomes Oneness, not just following function but evoking Spirit and Avidity as telling form.

Since that time Daley has fully focused on ideas and forms rooted in the parallel importance of interior/exterior spaces and their dividing wall. This construct, which defines the essence of his mature work, has led him repeatedly to discover new and challenging potential for clay's materiality, continuing into the twenty-first century to push mud in new directions. His emphasis on the vesica form (Fig. 21) is discussed by Janet Koplos in her essay, *"A Complex Harmony: Daley 2000-2010," beginning on page 22.* In general, the work is comprised of rounded and flattened geometric forms with punctures through and impressions into their walls, and rich delineation of the rim that divides inside and outside. There are surprising shifts in the outer shapes that variously reveal themselves depending upon viewers' positioning in relation to the work. A vessel's multiple entrances and exits accelerate our sense of the impact of interior on exterior and vice versa. Viewed as exterior, Daley's vessels suggest several architectural references, including columnar, chiclet, and pyramidal shapes. The interiors generate diverse metaphoric references as well—suggestions

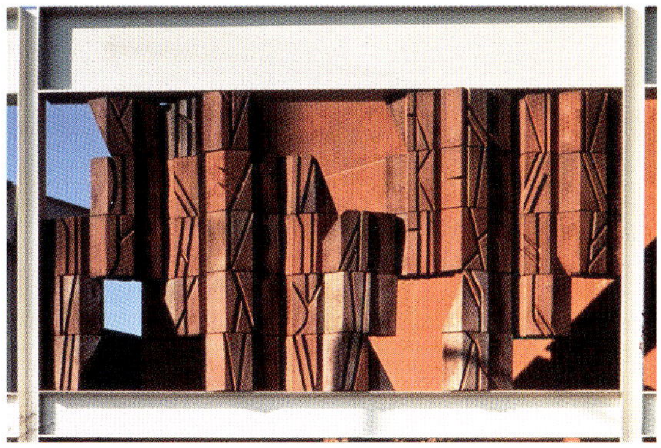

Photo: John Carlano

Fig. 20. Ritz Theater Wall, *Cellular Permentations*, 1978
Refer to page 186

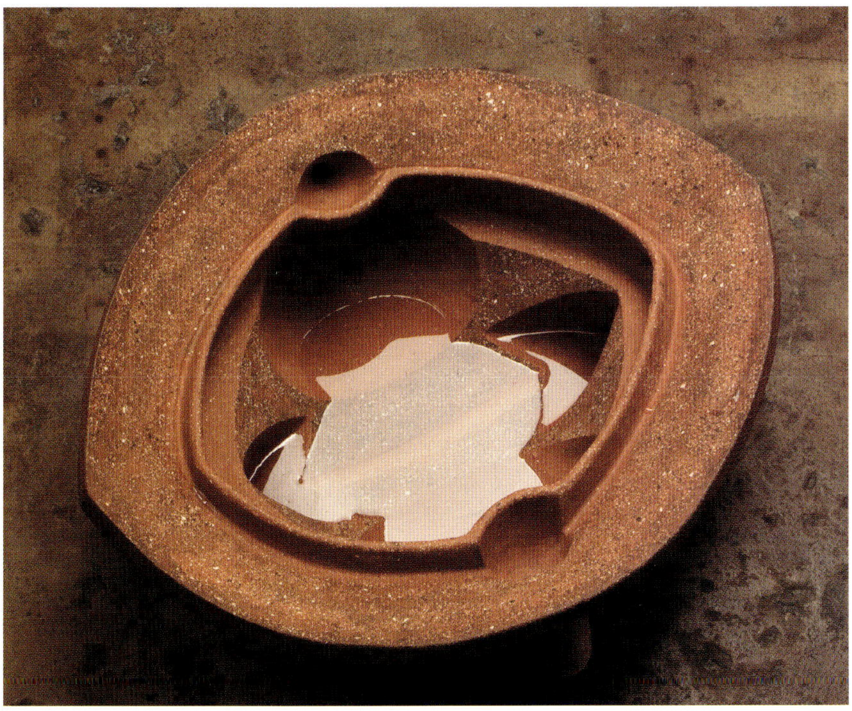

Fig. 21. *Libation Cistern Vesica*, 2010. Refer to page 159

Decoration in ceramics most often is surface. For me, structure is its own decoration, becoming energized surface through light.... Form becomes Oneness, not just following function but evoking Spirit and Avidity as telling form. — William Daley

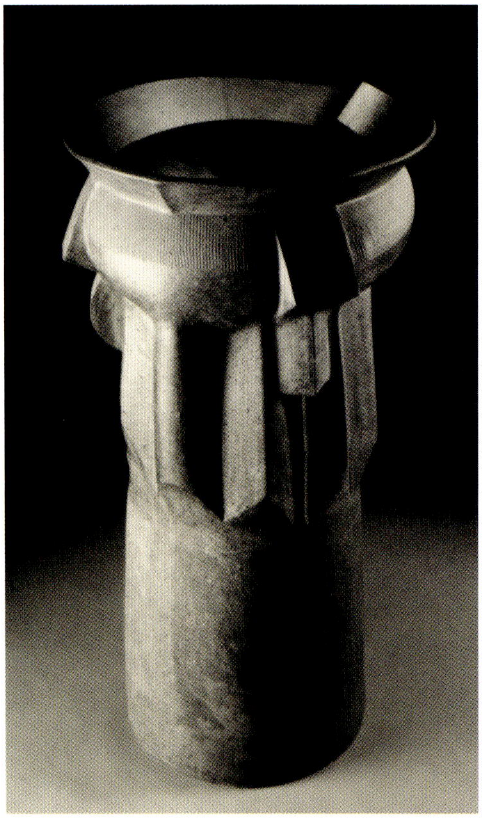

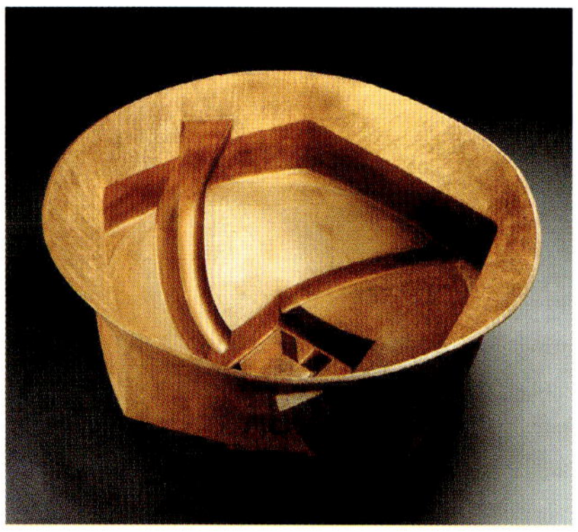

Fig. 23. *Pentagonal Column,* 1986. Refer to page 99

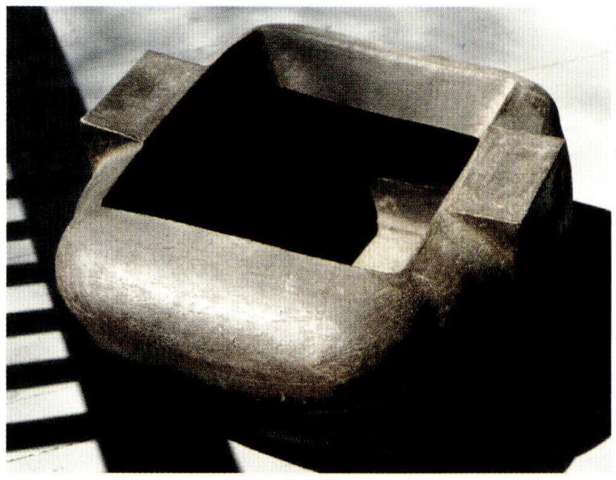

Photo: William Daley

Fig. 24. *Chiclet 3,* 1968. Everson Museum of Art, 1968. Refer to page 63

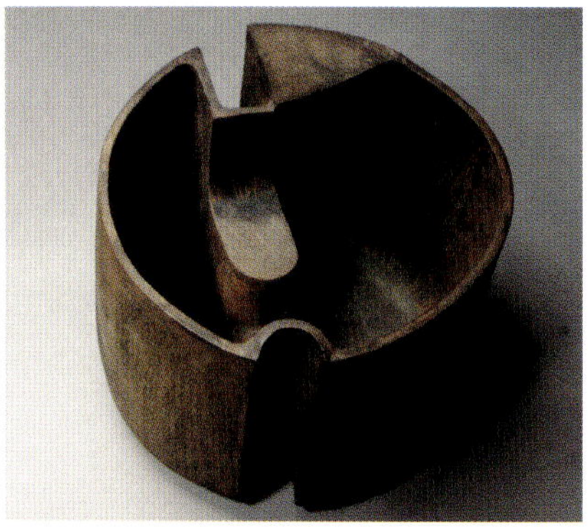

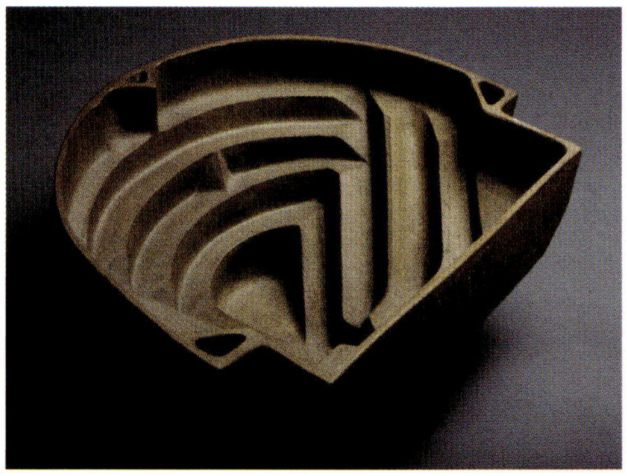

Fig. 22. *Columnar Shang Form* 1977. Refer to page 81

Fig. 25. Untitled, 1968

Fig. 26. *D-Form,* 1976. Refer to page 77

of tubular musical instruments, arenas and stadiums, cisterns and ritual baths. (Fig. 22–26) The work encourages a viewer's imagination to track his or her own experiences, thereby enhancing the multiple, potentially empathetic meanings of Daley's structures.

The derivation of architectural allusions in Daley's vessels likewise have multiple referents. His association with the industrial design field informed his awareness of the form/function relationships that reach their pinnacle in architectural structures. And he has always been interested in the history of architecture per se, an interest later to be extended by his son Thomas's architectural practice and the family collaborations thus generated. Over time, Daley's concerns have embraced not only the Islamic and native American architecture mentioned earlier—the archways and geometric patterning of the former, the round modeled and punctured mud origins of the latter—but structure of other cultures as well. In one of his many unpublished essays, "The Geometry of Residence," 1980, Daley presents *"ideas about location, residence, and pots as habitations"* addressing home and neighborhood:

The pot, vessel, effigy, and tool of clay have become symbols for our communal sense of person. The examples of the pot and figure are symbolic of community in death—the pot is transformation of sustenance. In short the clay object is used to celebrate all forms of the vital in life: success, victory, loss, mourning, new life, hope, and remembrance.

The history of the house used as an analogy for the dwelling as spirit is as ancient as that of using the head or human figure. Han Dynasty tomb furniture incorporating house images and their Egyptian counter parts quickly come to mind. The making of clay transformation models—including shrines, tabernacles, reliquaries, chalices, toys, banks, and dwellings—continues to the present.

As Daley's vessels grew larger, their size and complexity challenged the clay to hold its shape. By necessity, he redeveloped new methods of working in sections, which he attaches to each other and props up with wood T-bars that remain in place as the clay dries. (Fig. 27, 28). In a 1977 letter to Alice Westphal, guest curator of an exhibition at the Evanston Art Center in Illinois in which his work was included, Daley described his process this way: *I keep a way of making pots with slabs over pre-forms and when all goes well it permits me to work with the structure of large vessels. I try to understand structure as it pertains to scale. Most of the time I worry about the mechanical properties of scale, the thickness of clay, its temper and suitability to support an overhang or work as an effective buttress or beam. I try to account*

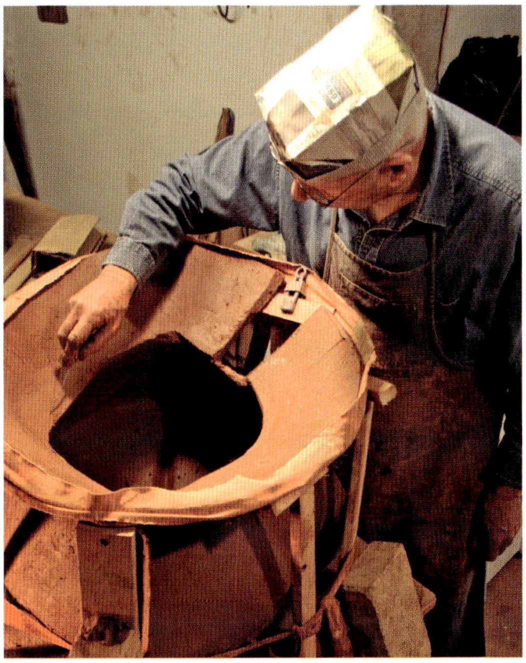
Photo: Charlotte Daley

Fig. 27. Bill constructing the rim of a large pot

"I try to understand structure as it pertains to scale. Most of the time I worry about the mechanical properties of scale, the thickness of clay, its temper and suitability to support an overhang or work as an effective buttress or beam. I try to account for my skill by extending the scale and testing it against the clay."
—William Daley

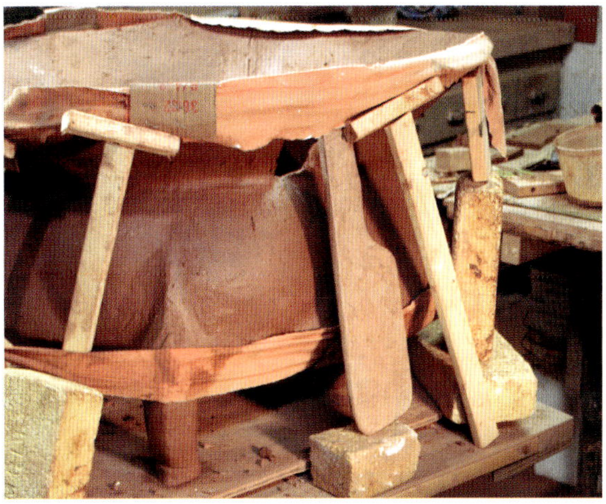
Photo: Charlotte Daley

Fig. 28. Braces used to support clay during construction

for my skill by extending the scale and testing it against the clay.

Sometimes when these matters proceed with assurance, I get canted glimpses of another scale. Scale as the mediator of illusion, a passing invitation to manipulate esoteric events, as forms, which create structures that, qualify the ambiguities of size of inside/outside—vessels as ushers of time aiding passage through edge, plane and volume to present mysteries, volumes, which make gravity, an illusion—pots in which logic and caprice are enfolded to extend illusion.

With each beginning, Daley considers anew the walls to be built, the base, the interior spaces and exterior surfaces and their impact on each other's structure, and, perhaps most importantly, the transitional rim that joins interior and exterior. In his world, there is no *"blank canvas"*—to use a painter's metaphor—cognizant as he always is of the work just finished, which, at least in part, establishes the next beginning. High spirited as he is in public, Daley is a brooder in the studio, concerned with issues of repetition and opposition, echoes and parallels; cruciform-based *"whacked geometry"* (a diamond becoming a hexagon becoming a circle); an interlaced caprice. He also is concerned with parts and whole—one and many—and with enveloping history while imagining the future but remaining rooted in the present.

Think about his process. For example, consider the rim of a vessel: should it be straight or slanted? If slanted, how much? The same thickness as the walls? If not, thinner or thicker? Solid or hollow? Sharp or rounded? Consistent or varied in all of these respects? How should its surface compare to the exterior of the vessel? To the interior? Take these questions and dozens more and apply them to every aspect of the vessel, and then add the element of intuitive response that is always at the essence of Daley's work. Consider also the vessels' surfaces. Daley has eschewed glazes since 1965, shifting, those many years ago, to coloration that is dependent upon modifications to the clay body itself, to firing methods, and to alterations in the kiln temperatures used. He has commented that *"As a potter, I should get used to the fact that the kiln door makes all victories provisional and all intentions subject to revision,"* an idea he credits to ceramic artist Wayne Higby. Daley also devotes significant time and attention to grinding, sanding, and otherwise working the fired surfaces after the kiln has completed its magic of turning mud to earthenware or stoneware *"to reveal the 'bones,' the inside core as surface."*

Daley generally fired his own work using kilns where he taught. At PCA this eventually was a walk-in gas-fired kiln affectionately named *"Bertha."* Upon his retirement he acquired an electric stacking ring kiln, which he assembles ring by ring around the

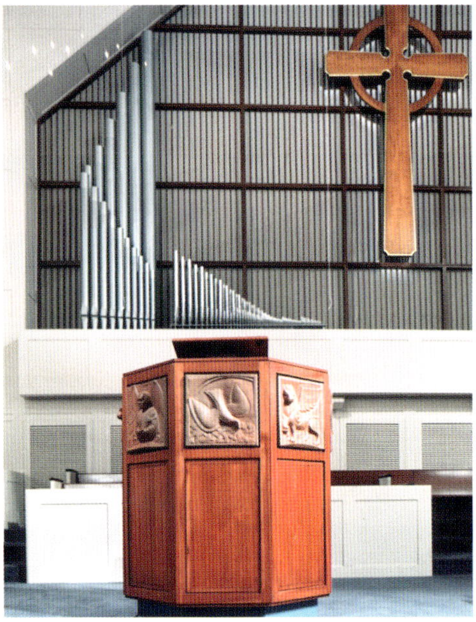

Fig. 30. Samuel Crothers Associates Architects (now Daley + Jalboot Architects) and William Daley collaboration on Westminster Presbyterian Church, 1994. Refer to page 190

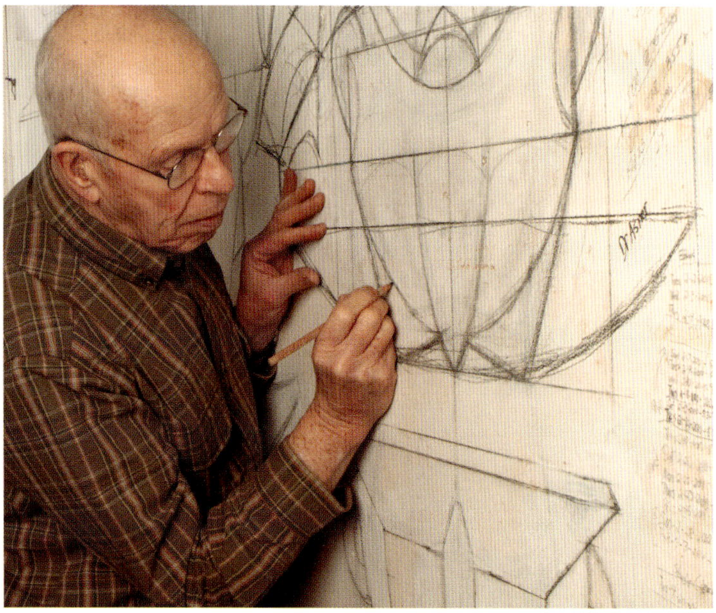

Photo: John Carlano

Fig. 31. Bill at the "Drawing Wall" in his basement studio, 2011

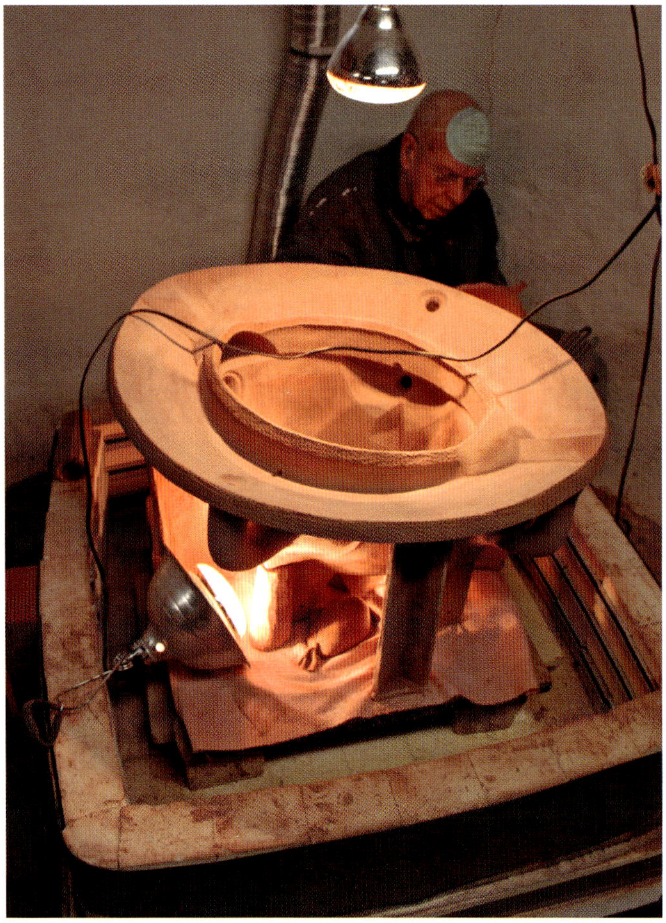

Photo: Charlotte Daley

Fig. 29. Bill drying a pot with a heat lamp, one ring is placed on the kiln, 2010

"As a potter, I should get used to the fact that the kiln door makes all victories provisional and all intentions subject to revision" — William Daley

vessel each time he fires a new one. (Fig. 29) Among the most recent have been several commissioned projects for ceremonial religious furniture, made primarily for projects Daley has undertaken in collaboration with his architect son, Thomas. (Fig. 30)

For the most part, Daley's work comprises unspecified series that are neither sequentially created nor marked by a defined beginning or end. Rather, an idea, a combination of forms, a pattern, a volume, provokes repeated examination within the context of multiple works. The vesica shape on which he has focused for the past two decades has origins in works from the 1950s, for example, a boat shaped planter. The vesica results from overlapping two circles, thereby creating an ovoid form. As curator Matthew Drutt pointed out, since the Middle Ages the structure has carried multiple symbolic meanings, its fish shape, for example, associating it with metaphors of Christ. Drutt also outlined its use in Gothic architecture. Likewise the vesica has anatomical associations to the human bladder.

With clay as his primary working material, the drawing practice Daley cultivated during childhood has remained a mainstay of his art. Hundreds of

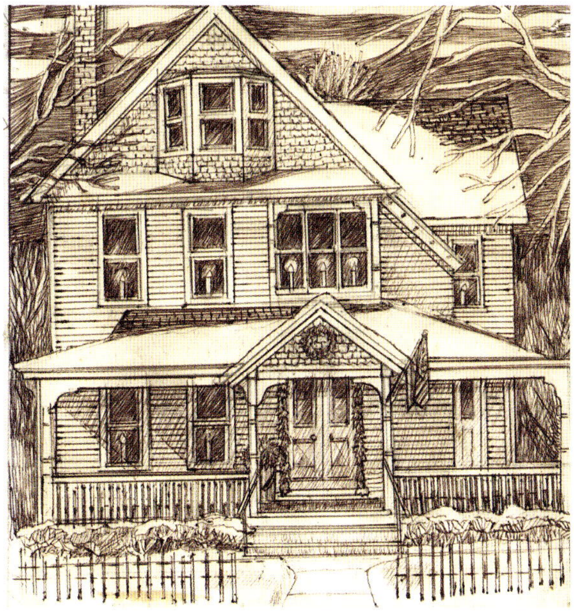

Fig. 32. Drawing of Daley home in Elkins Park, Pennsylvania

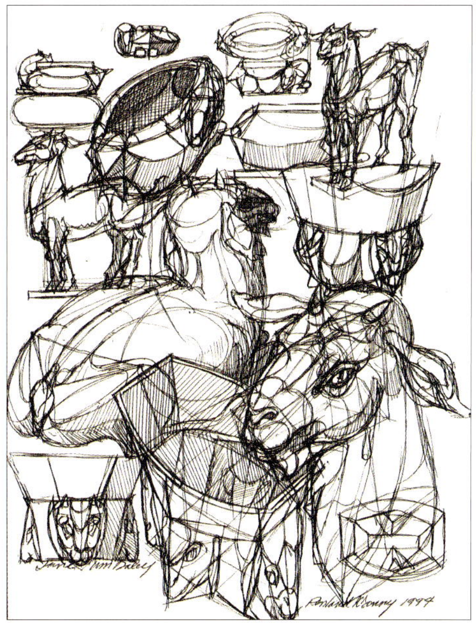

Fig. 33. *Penland Nanny,* Drawing, 1994

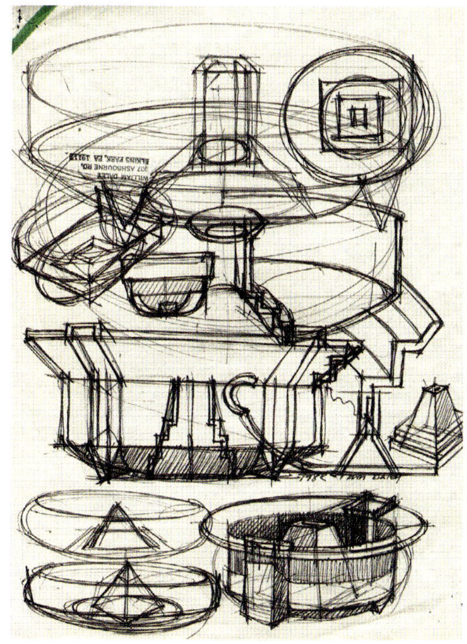

Fig. 34. *Fountain Sketch,* 1982

unknown drawings made throughout the decades represent a substantial aspect of his practice. He essentially considered this as a private activity until the early 1970s, when he allowed drawings out of the studio as part of exhibitions of his vessels and also as independent works in group shows of drawings. Moreover an ex-ray excavation of Daley's studio wall would reveal hundreds of layered, painted-over drawings of vessels in progress, drawings that are repeatedly destroyed when a clean wall is needed for thinking through the next project. (Fig. 31)

Drutt referenced both the calligraphic exercises in which Daley's father engaged and Daley's own spiritual escape through drawing from *"study by demand"* school work. His studies of imagined rounded and tubular elements of the sort that are essential to his art today are not so divorced from his childhood renditions of As, Bs, and Cs. Responding also to elements in nature that have inspired his dimensional concerns, Daley's drawings likewise reference seed pods, tree trunks, and other landscape and architectonic elements such as his house, as well as images of animals and birds and studies for pots. (Figs. 32–34)

As apparent from his writings, Daley believes that through dialogue with colleagues, as well as through teaching, he has learned as much as he has imparted, that in a dialogue about the making of art everyone is a student, some more advanced than others, but students nonetheless. Always an advocate for younger artists, the expansive collection of ceramic art that graces the Daley home is equally comprised of works by esteemed and renowned colleagues, as well as works purchased from students and young artists entering the field. (Fig. 35)

Philadelphia art schools long have been strong in the craft subjects, so the city has been a magnet for dozens of talented faculty (and students) with whom Daley has formed important friendships. Growing from that situation, the Philadelphia Council of Professional Craftsmen, started in 1967 under founding director Helen Drutt, fostered a strong community. Those closest to Daley have been many PCA/University of the Arts colleagues—Richard Reinhardt (metal); Ruben Eshkanion (fibers); Aurelius Renzetti, William Parry, Petrys Vaskys, and Wayne Bates (clay); and Daniel Jackson (wood). Temple University's Tyler School of Art has generated many close associations as well, including Albert Paley, Rudolph Staffel, Paula Winokur, and Robert Winokur.

Daley's public lectures are always invigorating and challenging. The spirit he invests in them is suggested by previously unpublished excerpts from his 1991 graduation address to students at the Nova Scotia College of Art and Design.

Since he *"retired"* from teaching, the artist has continued to conduct ceramics workshops throughout the United States on a frequent basis, He brings to these sessions his unique, sometimes whacky, approach, one aspect of which consistently focuses on making (and wearing) hats, which in this context

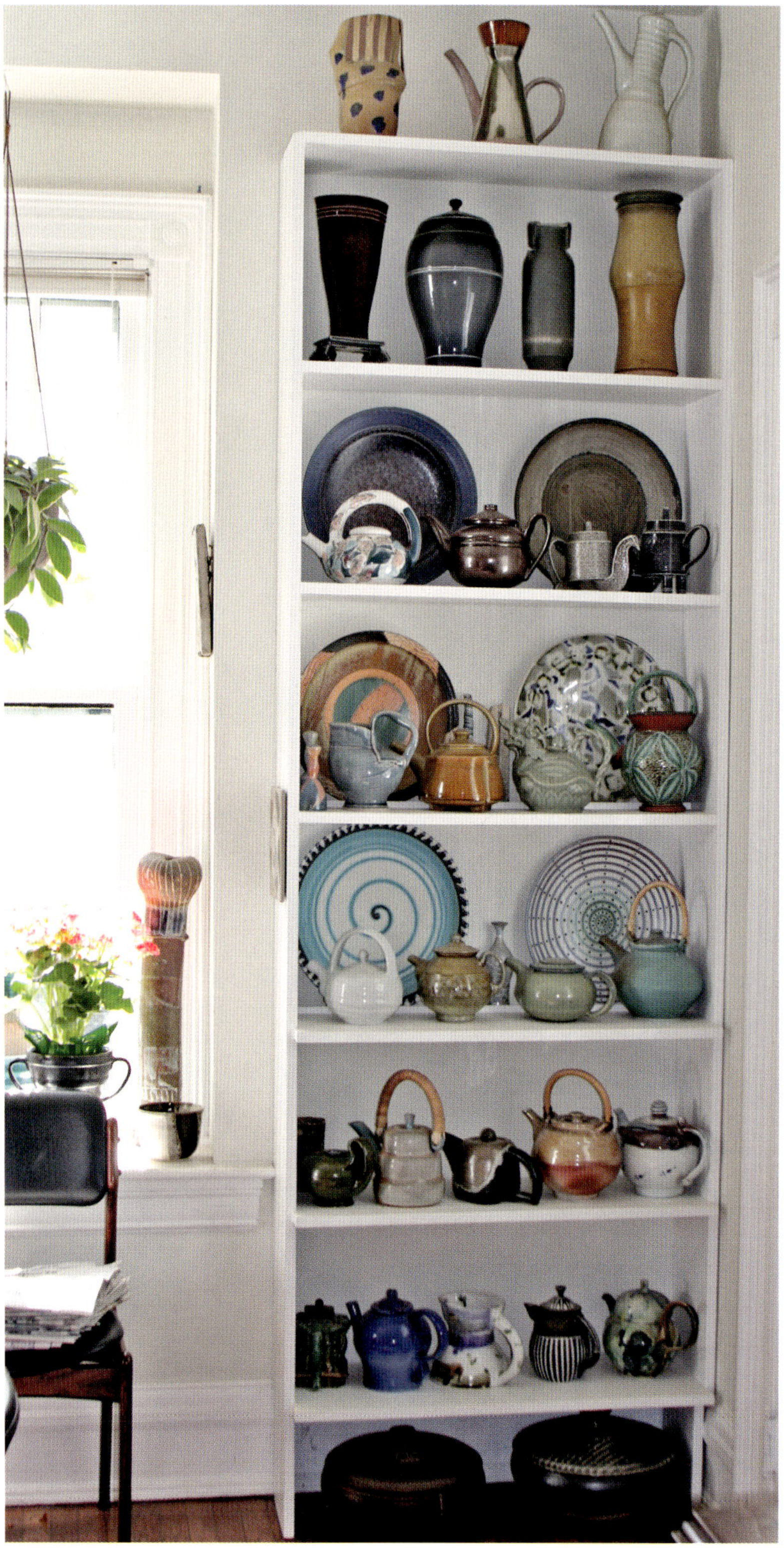

Photo: Charlotte Daley

Fig. 35. Shelves in the dining room of the Daley's home in Elkins Park, PA, 2010

Notes: This essay owes much to the works listed in this volume's bibliography, especially Matthew Drutt's essay in the catalogue for the exhibition William Daley: Ceramic Works and Drawings, that he organized, for the Levy Gallery for the Arts in Philadelphia at Moore College of Art and Design, which circulated to the Renwick Gallery and The Arkansas Art Center from November 1993 through May 1994. The essay is more firmly rooted, however, in hundreds of undocumented conversations with William, Catherine, Charlotte, Thomas Daley, and Barbara Daley Schwab from approximately 1969 to the present.

Quotations are from Daley's writings, recorded conversations and notes were taken during telephone conversations between artist and author in August 2009 and January 2010.

"I have come to collectively call the distillation of my preferences over time the place where my temperament and possibility meet to locate my voice, my lexicon of "Whacked Geometry."

— William Daley

suggest upside-down hanging pots. Documented by photographs taken at these sessions, (Fig. 36) the hats seem to function as brain-containers, developed by a brain that is ever ready to absorb new data from previously finished vessels, books read, places visited, dreams, to say nothing of the quotidian rituals of daily life.

Now in his mid-eighties, Daley is a unique phenomenon, his energy and curiosity seemingly boundless, his health maintained by a strict non-fat diet and a daily regimen in which the practice of yoga alternates with rapid walking sessions around a high-school track near his home. His studio is reached by walking down a staircase off the kitchen of the Victorian home he has shared for forty years with Cate, whose participation in the studio, working on pots as they are developed has been essential to Daley's art over these decades. Cate's contributions have included critical comments about works in progress to which Daley has responded with substantive changes; assisting with the assembly of many of Daley's vessels; and *"her refined touch set the standard for burnishing, finishing"* their surfaces. For a period of time in the 1970s and 1980s Daley signed his vessels "W and C Daley," but Cate objected, believing her contribution to be solely manual. "W" resisted; "C" insisted, and so he stopped including her initial in the signature. (Fig. 37) From the studio a visitor can walk out to the garden, the furthest reaches of which are bounded by a stream. This proximity of nature is important to Daley, whose *"whacked geometry,"* in fact, is as dependent upon the living world and the reverie of meditation and dreams, as it is upon mathematically-based forms.

Like Alfred Stieglitz's life-long battle to achieve for photography the respect accorded painting and sculpture, Daley has strived to break down similar hierarchical considerations surrounding traditionally craft (use) endeavors in relational to what are considered fine arts undertakings (paintings, and sculpture, drawings and prints, expanded in recent decades to include photography, video, and installation art). Daley's battle may be more difficult than Stieglitz's, however, given that photography was a relatively new discipline when Stieglitz argued for its importance, whereas attitudes toward craft objects have deeply rooted historical precedent as well as what Daley views as autocratic prescriptions for right practice.

Daley seeks to eliminate the notion that parallel universes necessarily exist among artist-practitioners

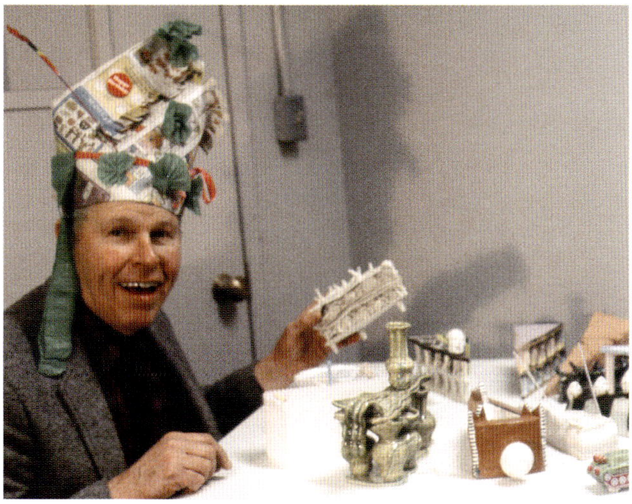

Fig. 36. Bill's retirement party, 1990

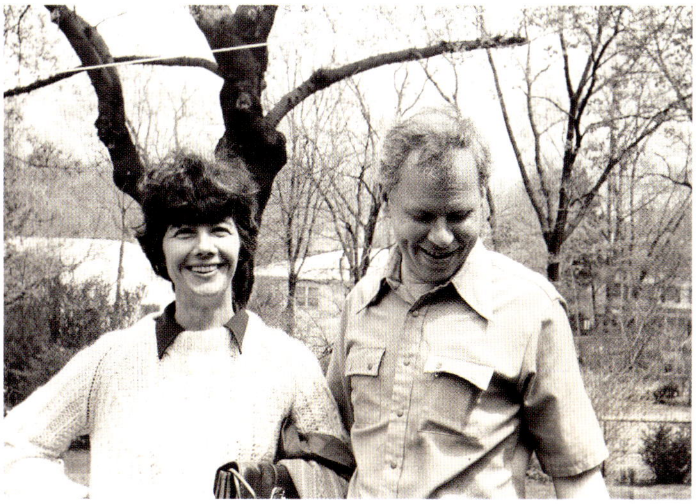

Fig. 37. Bill and Catherine Daley, 1974

today, to negate the separations based on the materials used and the nature of the resulting forms. His argument would probably be that in the main floor galleries of New York's Metropolitan Museum of Art, sculpture and vessels of ancient Greek origin are accorded equal status, installed in adjacent galleries. And they generally are equally admired by students of this culture. Greek painting essentially is known only by its decorative presence on clay pots. A twentieth century parallel might be found in the pottery makers of Japan, who are accorded the status of National Treasures in equal measure with artists in other media. Yet in the culture for which Daley has created an extraordinary body of ceramic art, the maker of mud vessels is separated from sculptors and labeled a craftsperson, a designation Daley believes is seen by some as less intellectually acute.

Daley continues to investigate in clay the possibilities embedded in the challenge presented by his teacher Charles Abbott more than a half century ago, to make a pot that, like a Sung bowl, demands to be understood so deeply as to transcend itself. In committing himself to the vessel, Daley committed himself to making the form carry the philosophical and artistic weight that is accorded paintings, the discipline he had originally intended to pursue. He summarized his journey in a 2005 statement for the International Architecture/Ceramic Exhibition in Seoul:

Over fifty years I have made many moves and turns. A few of many examples that expanded the boundaries beyond my loving study of my form heroes are or were playing "Inside/Outside" passage games with bumps and holes to make bivalent walls entries and exits that mediate the scale of volumes. The uses of Infinity Diagrams and world sacred geometries such as Spirals, Pentagons, and Whirling Geometries to extend the perception of passage to residence. Stacking primary geometries vertically to assure stability and enhance a sense of enclosure. Evolving tubular wall sections as columns to carry weight. The use of skewed geometries, like the diamond, the oval, and the vesica change the rates of passage around and through a pot. Over the decades I have come to collectively call the distillation of my preferences over time the place where my temperament and possibility meet to locate my voice, my lexicon of "Whacked Geometry."

Daley's ability to prosper is owing to his commitment to engage every aspect of his experience, such as describing how *"much of prison camp—particularly the long march—had edges of amazing aliveness—clarity—vividness—which were the dual of anxiety—fear and dread—a kind of existential growth/and that art offers a window into consciousness witnessed."* Duality is essential to his thinking, not in any sense setting up a conflict situation, but an enriching intellectual tug-of-war. Impulsive actions are balanced by carefully worked meditations on a specific combination of forms. The spirit of the work is embedded in the form as the form is generated by the spirit of this committed church-goer who is always questioning the validity of scripture. Memories of childhood mornings roaming the cliffs above the Hudson River in Hastings-on-Hudson are as vivid to Daley as a recent walk in his small, suburban Philadelphia town. He still periodically paints his and Cates's home in homage, no doubt, to his father's profession. Memory, after images, reflection: *I think I have become a pantheistic agnostic that believes deeply in how great it is to be part of consciousness in a "telling" way. It all makes a continuous set of interweaving spirals.*

Ruth Fine

Now based in Philadelphia, Ruth Fine was curator of special projects in modern art at the National Gallery of Art, Washington, D.C., where she organized numerous exhibitions devoted to early twentieth century and contemporary American artists, from John Marin and Georgia O'Keeffe to Romare Bearden, Helen Frankenthaler, Roy Lichtenstein and Jasper Johns.

A Complex Harmony: Daley 2000-2010

Janet Koplos

Photo: Charlotte Daley

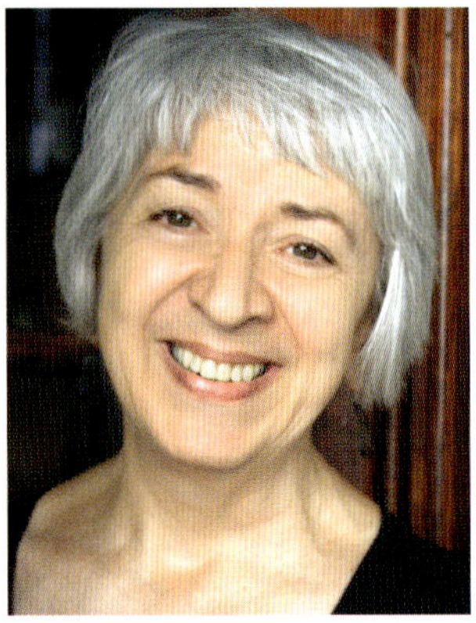

Janet Koplos

Bill Daley was always a rarity—one of the small number of college teachers with enough talent and drive to avoid being swallowed up by that profession and to simultaneously achieve a major career as a studio artist. So it is not surprising that he has produced a substantial and compelling body of work since retirement from the University of the Arts in 1990 allowed him to work full-time in the studio.

This essay, which looks at the work of the last decade, describes not so much a changed path as a honed and clarified one in which he retains the signature aspects of his work without simply repeating. While he continues to create architectonic ceramic vessels as well as drawings related to them (executed both before and after his hand-building process), he has found new elements and new expressions. He may be moving toward a slightly simplified vessel structure, but at the same time, he is giving increased attention to densely linear drawings. He has added to his repertoire forms he calls libation vessels and cisterns. He has also introduced feet to an oeuvre previously marked by flat-bottomed stability.

Daley has been influenced by primal forms and ancient structures and objects throughout his career. He had discovered the eye- or boat- or cupped-hands-shaped vesica by the 1970s, but he has concentrated on that form only in the last decade or so. His sculptural career is of a piece; he is moved by, and moved to create, elements found in many cultures and many times, carrying various meanings. He embraces the purity of geometric solids (cube, cylinder, sphere, and pyramid), their realization in human constructs (from kivas to stair steps) and softer organic incarnations (such as pods). His recent works favor stairs, meanders, and crosses of various sorts. The imperfectly symmetrical forms have a central axis (often marked by a cross on the interior base), which in his mind refers to the *axis mundi* connecting heaven and earth. The vesica—complete, truncated, or just implied—appears almost everywhere in his work: as the shape of the base, marking the walls or, increasingly, defining the rim.

Daley has always explored topological transformations: what is convex on the outside is concave on the inside, or a protrusion that serves as a foot or as a rim "platform" continues as a cylinder or vesica inside. The sculptures insist that viewers see both aspects. The works cannot be grasped in the silhouette/profile that traditionally summarizes a pot, nor is their appearance defined by surface

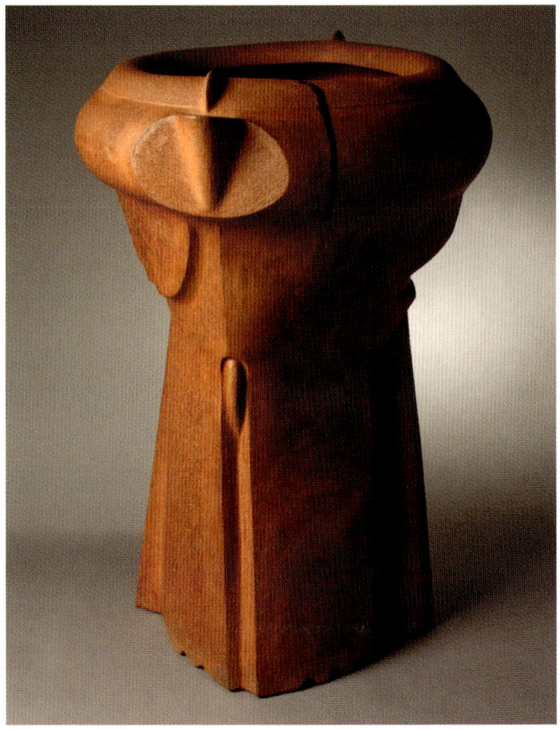

Fig. 38. *Guardian Vesica,* 2005. Refer to pages 138 and 139

treatment. Daley sees the symbolic implications and the formal sensuality of various familiar shapes. By selecting, simplifying, focusing, and repeating, he calls attention to them.

The vesica is an ancient symbol that relates, among other meanings, to sexuality. "I've been interested in bumps and holes since puberty," he jokes. Although he is a funny, lively individual, that personal style is not reflected in the works, which are unrelentingly serious. In his art, he connects himself to big, timeless themes.

Daley is committed to the emotional and intellectual suggestions of abstraction. Today classic abstract sculpture—in the style of, say, Brancusi or Noguchi—may be much admired but is not so much practiced. Figurative sculpture is strong now, along with installations that deal with social or political notions. However, Daley shares with younger sculptors such as Tara Donovan a fascination with material and with methodical construction. And even more, he shares the attitude of potters, who exult in the abstract beauty of vessel curves or angles, for it is in functional pottery that abstraction is most prominently expressed in clay today, free of illustration or external subject matter. The floating geometric decoration of ceramists such as Maren Kloppmann and Nicholas Bivins, which sets rectilinear against contoured, or the sculpted or articulated curves of pots by others, such as Christa Assad, Deborah Schwartzkopf, or Masamichi Taguchi, recall the configurations that

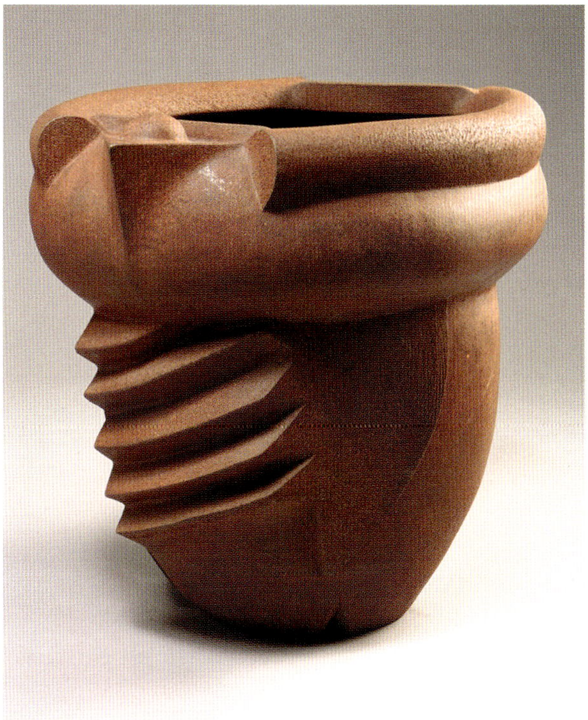
Fig. 39. *Coupled Vesica,* 2001. Refer to page 123

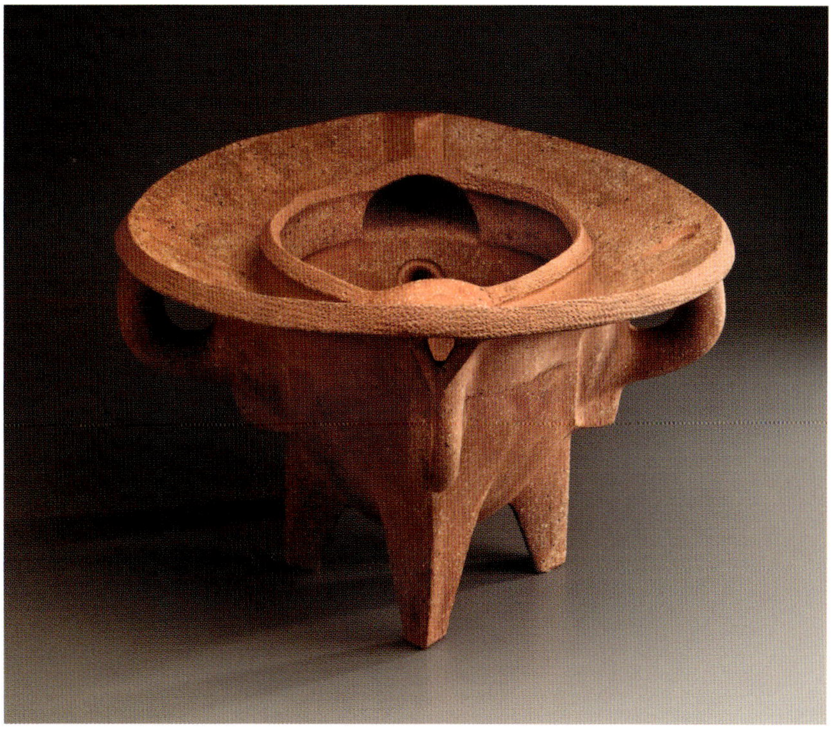
Fig. 40. *Joy-Full Libation Vesica,* 2010. Refer to page 158

Daley has long explored without concern for function or hand scale.

There is no causal relationship here. It's not that Daley is enlarging pottery elements into sculptural scale, but rather that his formal discoveries reveal the kinship between small and large structures, namely pots and architecture. A pot can offer to the receptive individual a sense of naturally evolved landscape or of constructed space; one can mentally project oneself into its recesses or onto its shoulders to imagine climbing or sliding—as if on eroded mesas or slot canyons (when rounded) or narrow cornices or stairwells (when angular). Architecture, of course, offers such shapes and spaces in life-size or larger-than-life dimensions. Daley's scale specifies a space of its own, neither utilitarian nor architectural but partaking of both. This can be understood by visual means when his sculptures are presented in the controlled environments of galleries or museums, but Daley often urges viewers to touch for another experience of the work.

The sculptor Tony Smith, whose integral single objects alter viewers' sense of space, once famously explained why his six-foot cube *Die* (1962/68) is the size it is: if it were smaller it would be an object but if it were larger it would be a monument. Thinking of that statement induces a provocative reconsideration of Daley's work. It is definitely an object, a vessel of generous size. In addition, there is a hint of the figurative—especially in the works, low or tall equally, that have a tubular ring at the rim. In *Coupled Vesica,* 2001 (Fig. 39) and *Guardian Vesica,* 2005 (Fig. 38), for example, that tube feels like an embracing arm. Daley's tall works relate more in scale to the body than to a pot. Yet all are equally clearly monuments, meant to ruminate upon, meant to take viewers to a different frame of reference than the minutiae of everyday. Daley's works of the 1970s imply the overwhelming proportions of a building (or the spaces between buildings) even if their reality remains grounded in understandable and familiar human scale. But during this last decade his work seems to show a drift away from explicit architectural reference and a subtle shift toward container forms (although, of course, buildings are containers, too).

The spatial relationships make one think of other relationships: everything connects. Daley's most recent works such as *Joy-Full Libation Vesica,* 2010 (Fig. 40), with their allusion to cisterns, evoke water-collection systems, drains, and catchments. One might link this to the new environmental interest in urban water reclamation and in domestic gardeners' storing rainwater rather than drawing on city water. But many houses a century or more old have cisterns for that purpose built into their basements. It is, in fact, an ancient technique and thus symbolic of human problem-solving, like other architecture. In a more abstract sense of the word relationships, Daley's works are typically characterized as comparisons or extensions of parts, or as arrangements, or as contrasting or evolving plural views of inside and

Photo: Amanda Schwab

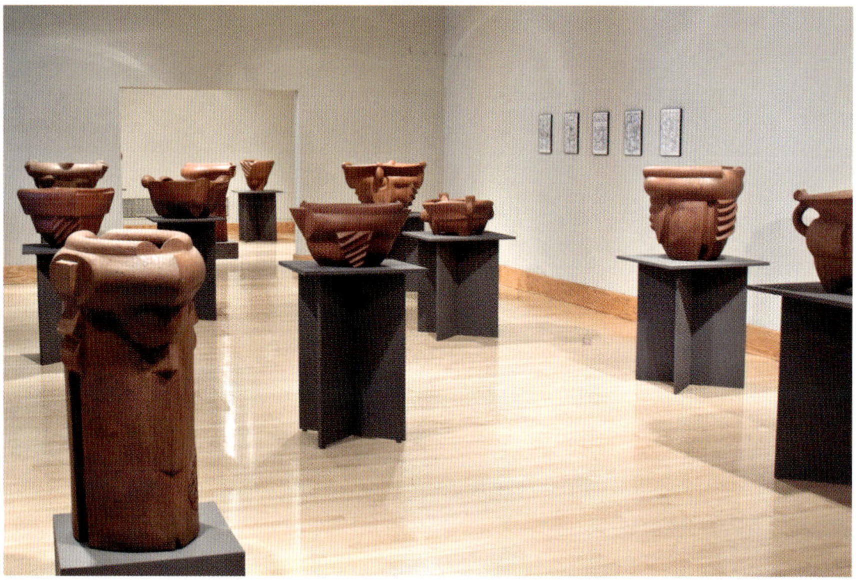

Fig. 41. *Vesica Explorations,* exhibit at the List Gallery, Swarthmore College, Swarthmore, PA, 2009. Refer to page 150

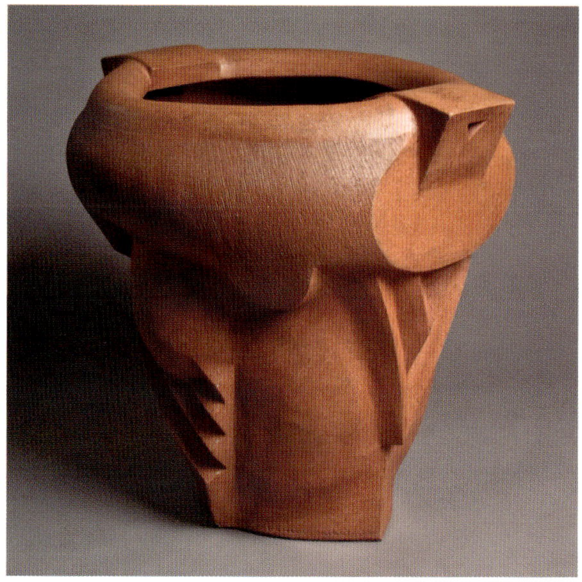

Fig. 43. *Apollonian Vesica,* 2003. Refer to page 128

outside. Viewers study the forms and lines with the feeling that they exist in a complex harmony rather than in aesthetic—or ascetic—isolation.

This feeling comes across even in pictures. One photograph of his 2009 exhibition *Vesica Explorations* exemplifies such harmony. The view shows 11 vessels, all on X-base pedestals that are painted gray (Fig. 41). The size and mass of the multiple works recall Peter Voulkos's 1995 retrospective exhibition as seen at the American Craft Museum—an expansive field of visually related objects—but these are not as formally repetitive as Voulkos's *Stacks*. They range from low bowls to columns, and the pedestals vary in height, too, so all the vessels are not at the same eye-level. The forms speak to each other as relatives, with motifs passed among them like inherited traits.

To look across the last decade through a succession of works is to see those family resemblances in new generations. *Vesica Vessel*, 2001 (Fig. 42) has the "embracing arm" tubes with a gently curving stand-up rim above them. At both ends of the vesica-shaped rim are "platforms"—flat-topped, rounded-edge protrusions that echo the vesica shape as they taper to nothing down the sides. Below the tube area is a section of sharply angled stair steps that creates a sense of motion. Overall the sculpture has a bowl-like profile with limited architectural allusion although the familiar devices are present, and it has just as strong an interior definition as exterior character. The base is flat.

Two years later the *Appollonian Vesica* (Fig. 43), at 23 inches, is just over twice as tall as the *Vesica Vessel*. Likewise it has the tubular embracing arm near the vesica-shaped rim, the ends of which are flattened to a plane. A block structure appears to be embedded diagonally in this plane, creating a platform in the rim. Below that is a V-shaped contour in the vessel wall. On opposite quadrants, near the base, is the stair-step motif, here inverted so that it looks saw-toothed.

Another two years later, from 2005, *Guardian Vesica* (Fig. 38) is columnar and 40 inches tall. It has darker shading than typical Daley works, a result, he explains, of firing it wrapped in cloth to create a look like reduction firing. Again the ends of the vesica-shaped opening are truncated, leaving a flat oval; and again (more strikingly) the rim rolls inward above the embracing-arm tubes. Lined up vertically on each flat oval is a thick disk on edge above a cone; both are embedded in the wall and so seem incomplete, leading the viewer to look inside for continuation. Below that is a long slot filled at the top with a narrow bulge that might suggest female genitalia. The continuation of this slot to the base, plus the stair-step zigzag that serves, this time, as the base,

Photo: Amanda Schwab

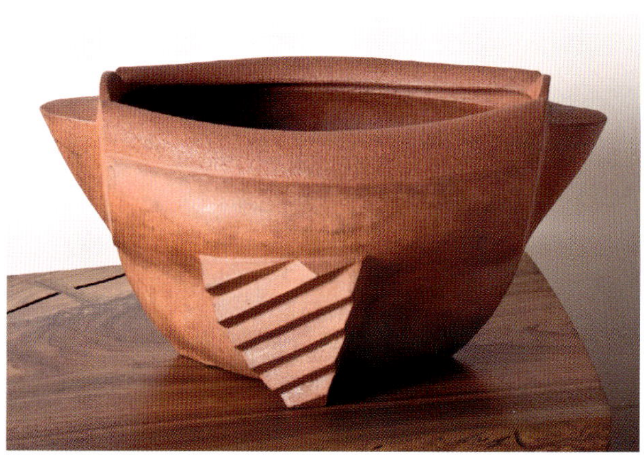

Fig. 42. *Vesica Vessel,* 2001
Unglazed stoneware
11.5 x 16 x 24 inches
Collection of Marlin and Ginger Miller

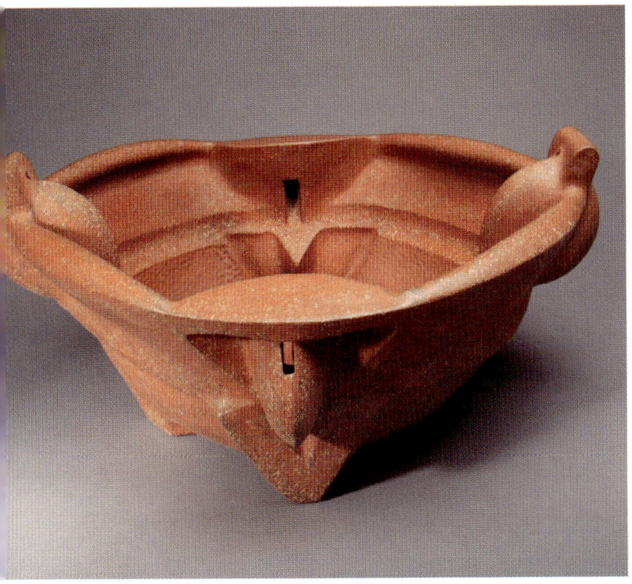

Fig. 44. *Lovers Vesica,* 2008. Refer to page 148

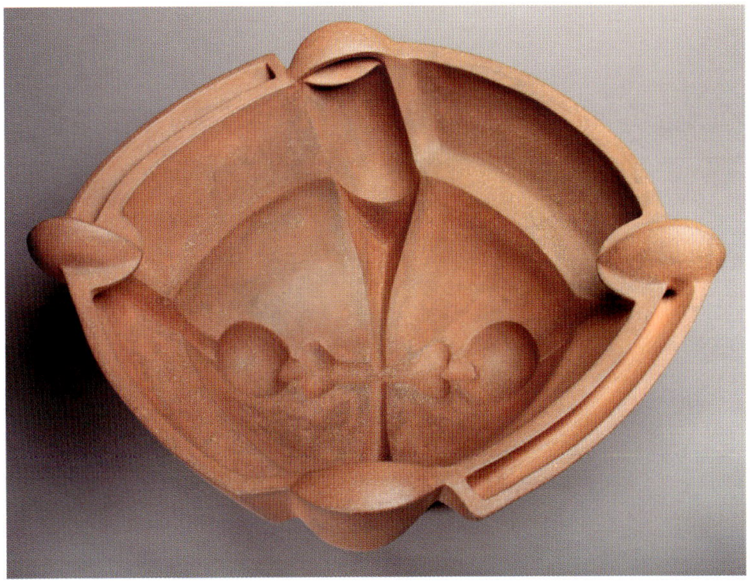

Fig. 45. *Azoic Vesica,* 2008. Refer to page 144

lifts the sculpture and gives it a vaguely skirt-like lower edge. The overall profile is like a chainmail dress—big shoulders and a kilt hanging straight from the waist.

Lovers Vesica, 2008 (Fig. 44) is an inverted helmet shape, with very short feet in a base much smaller than the rim. The feet are formed of blunt cutout V shapes, the familiar detached ends of vesica but here a thick outline, not a solid. The V is the base of an indented triangle, in the middle of which is a "bagel" form that is punctured by a tiny rectangular hole. The rim is unusual: most of it has a kind of rolled-over edge, softly thickened. It terminates in another bagel shape with a tab on the top. The name of this work, along with *Couples Vesica* and *Nuptual [Nuptial] Vesica,* implies that something familial is on Daley's mind now, although, as always, he addresses universal ideas of pairs and connections.

The rectangular hole in *Lovers Vesica* is like a lookout window, but by the next year, in the open bowl of *Azoic Vesica,* 2008 (Fig. 45) the opening has clearly become a drain. Here Daley introduces the notion of "collection," leading to his strikingly different libation vessels and cisterns of the last year or so. Today the word libation is a tongue-in-cheek term for alcoholic beverages, but formerly it alluded to ceremonial events, and even longer ago libations were associated with sacrifice or with honoring the dead. *Joy-Full Libation Vesica,* 2010 (Fig. 40), which is made of a flecked clay body that Daley makes with a self-grog, sits firmly on four short legs. They grow smoothly from the pendulous body of the vessel, which hangs low, almost touching the ground. Above each leg a thick loop handle emerges from the wall and returns just below the wide flaring rim, which has a thick edge so heavily textured that it looks like corduroy or like it was beaten with a meat-tenderizing hammer. The stubby legs, thick loops and broad rim create the suggestion of a person bent under a load with arms raised to support it. Confirming details are not there, of course—this is not figurative sculpture—but still this faintly anthropomorphic quality seems quite different from the wall-like impassiveness of Daley's architectonic vessels. The inward sloping rim has an inner ring to direct the flow of some unknown liquid. An interior view shows that the handles are also collection devices, having holes to channel and drain.

All Daley's works are, it must be admitted, hard to describe, although they feature recognizable and repeated elements. They recall what the critic Dave Hickey once said of Ken Price's ceramic objects, that words don't stick to them. It's easy to talk all around Daley's sculptures, though, since they connect with fundamental forms and practices from prehistoric to pre-Columbian, from Ancient Greek or Chinese through the Roman Catholic ritual in which Daley was raised. The solemnity of the forms seems rooted in spiritual experiences, elemental human drives or humanist confidence.

Even if the sculptures elude description, they are readily accessible to physical encounter. Viewers can kinesthetically understand them in relation to their own body and environment. Daley's works give an appearance of symmetry, which is stabilizing and comforting. But he constantly denies expectations, so the forms surprise the eye and hand. The vessels, hinting at rituals and ancientness and memory, are studded with points of intimacy—holes, crossing lines, textures—that give them an unexpected immediacy.

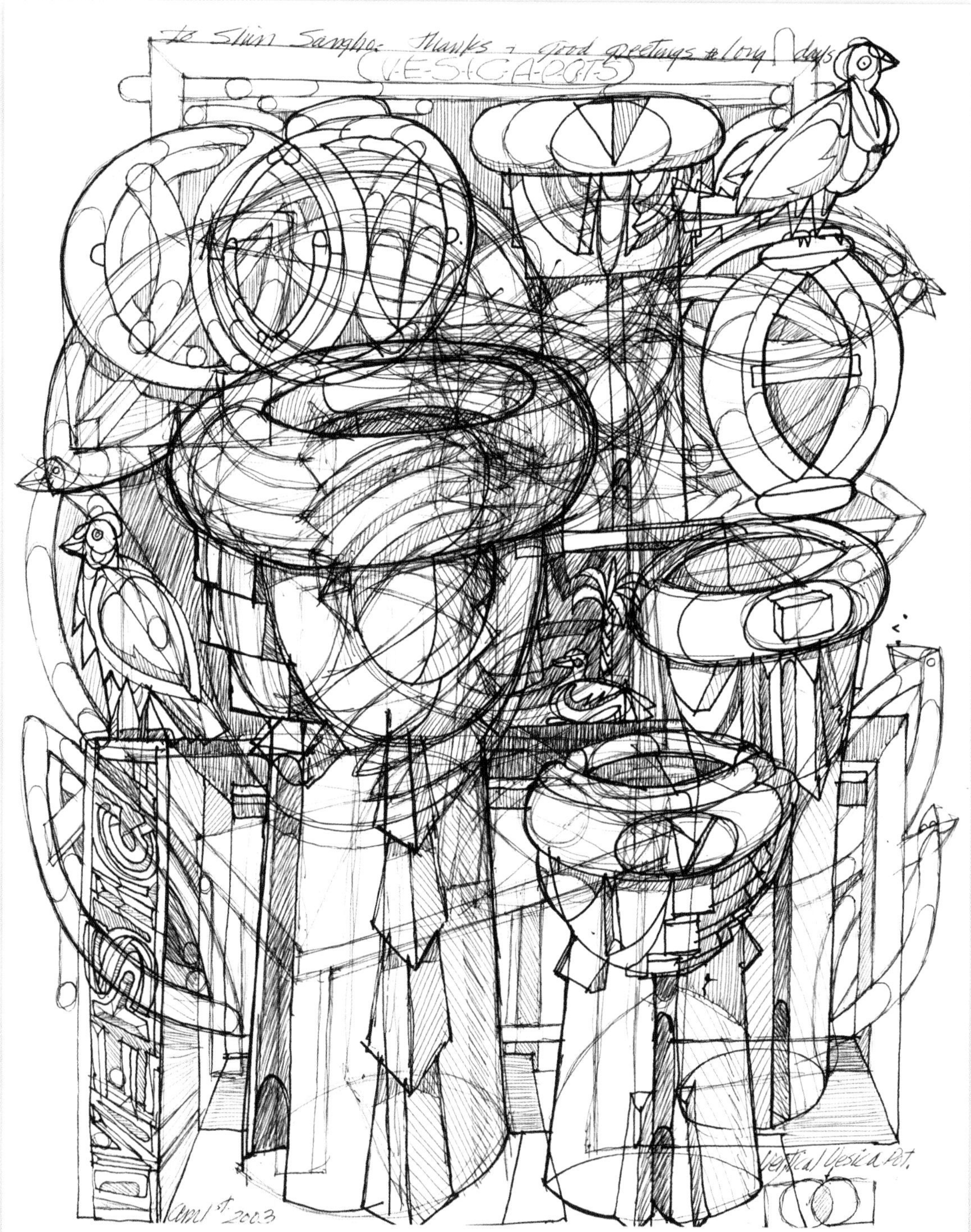

Fig. 46. Drawing, *Vertical Vesica,* 2003

The recent drawings (Figs. 46, 47)—now archival ink on paper rather than ballpoint on notebook pages—are dominated by overwriting. Daley draws and draws again, so that the recognizable individual vessels he maps out are tangled in webs. These look like studies: here's a vessel with one cube embedded in its tubular rim accompanying one with a cascade of cubes down the side. He tries out assemblies not yet seen in the ceramic works, such as one complete vesica outline perpendicular to another, to make a kind of star shape. The drawings also look simply like doodles, like direct connections from Daley's mind through his arm and hand to the paper, thus providing him with the quick, spontaneous record that his slow, deliberately constructed vessels (which require measured drawings) cannot. He plays out his fascination with repetition in congeries of hatch marks indicating shading or by echoing the forms of letters, such as the word vesica. (There's also a place for his wit, when city pigeons coo their way into a drawing.)

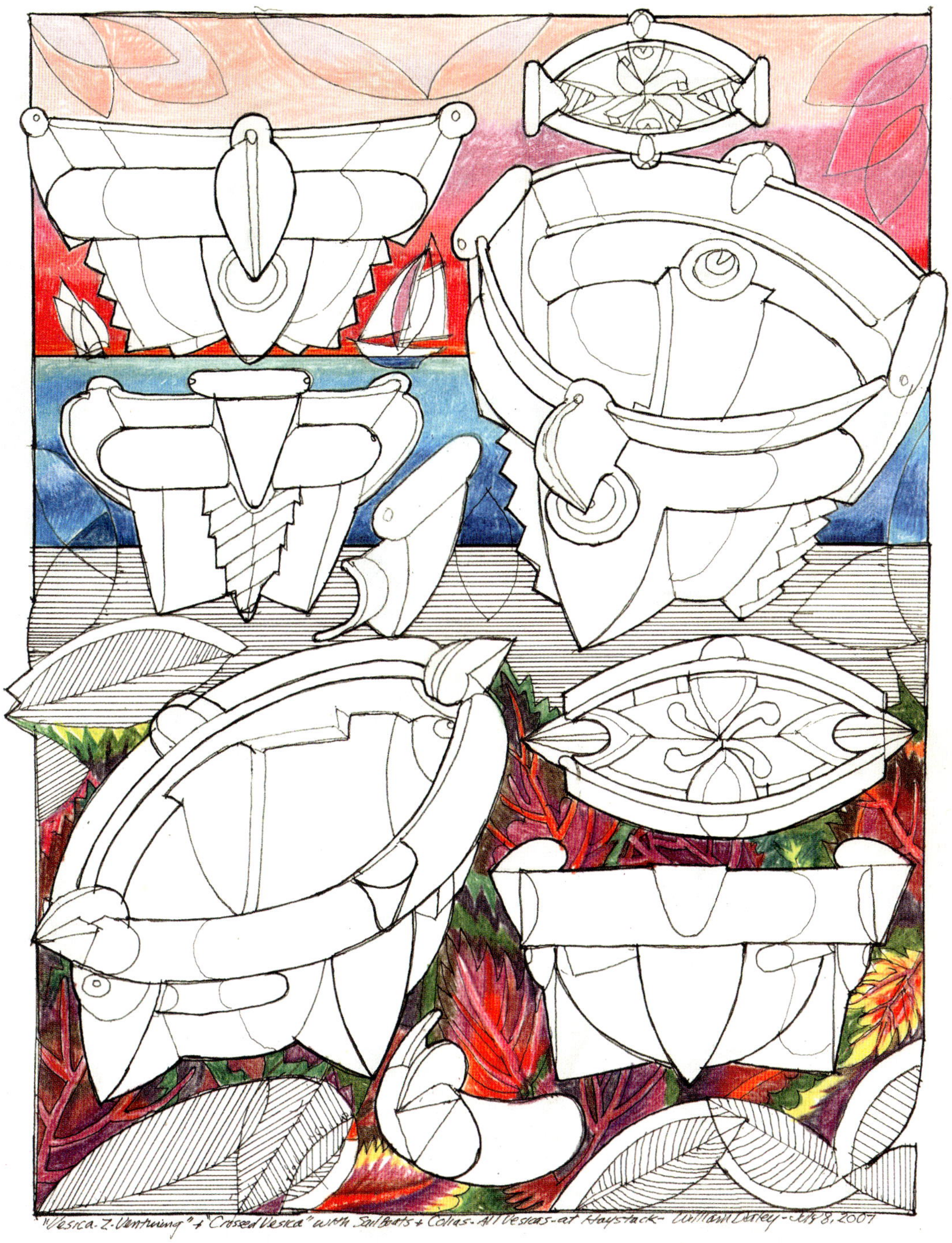

Fig. 47. Drawing, *Vesica Z Venturing + Crossed Vesica* 2007

The drawings are all those things and more: they are a way of thinking and experimenting. Daley has introduced color in them, starting with a mix of soft tans or an appealing blue as painted-in background that makes the sculptural depictions stand out. In any case, the overwriting also implies history, palimpsests, the multitudes of ceramic forms—or any human endeavor—repeated over the ages, and above all, symbolic lines of continuity.

Janet Koplos

Janet Koplos, a New York-based art critic, is a contributing editor to *Art in America* magazine. She is co-author of *Makers: A History of American Studio Craft* (University of North Carolina Press, 2010) and has published about 2,500 essays, articles and reviews since 1976.

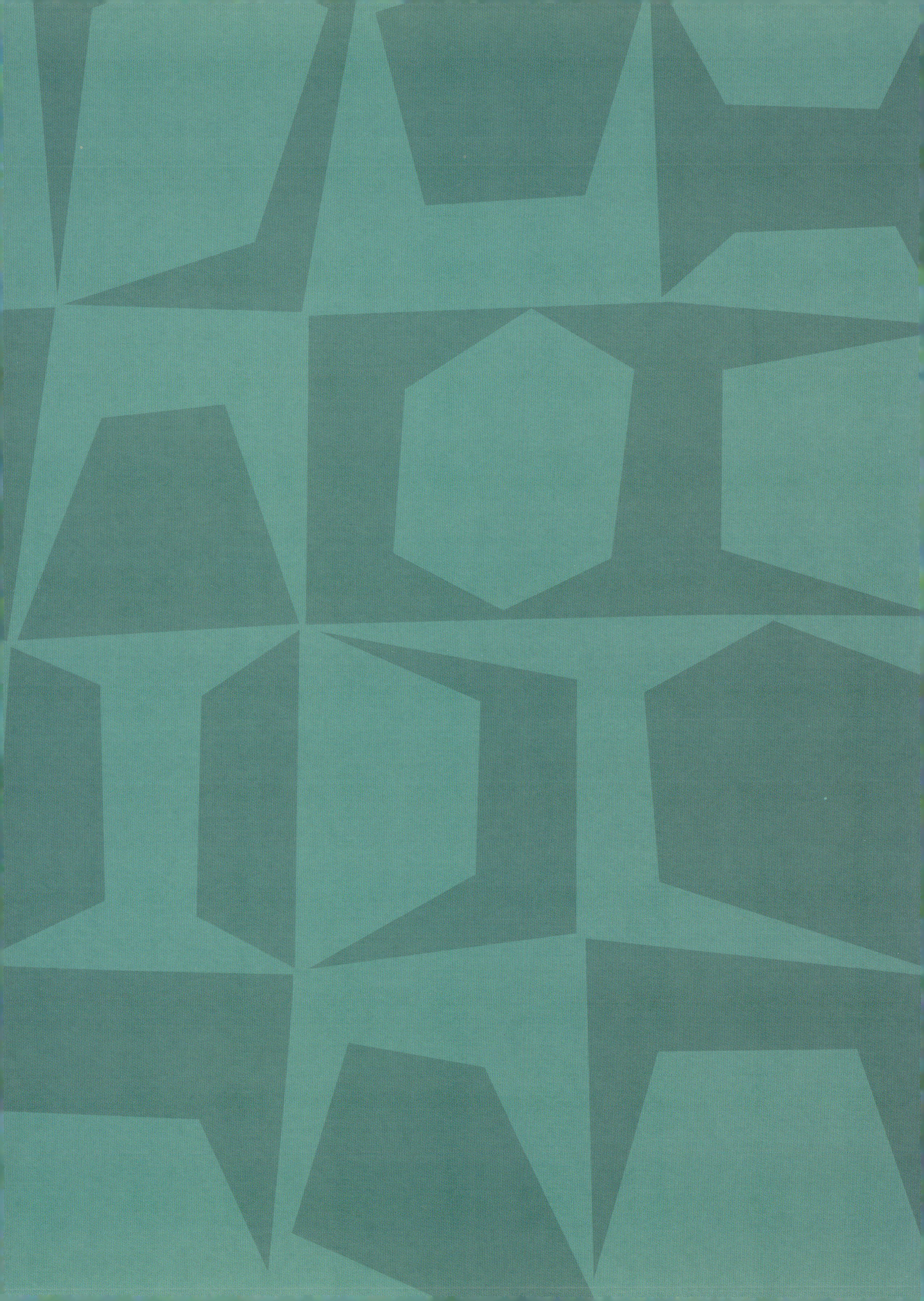

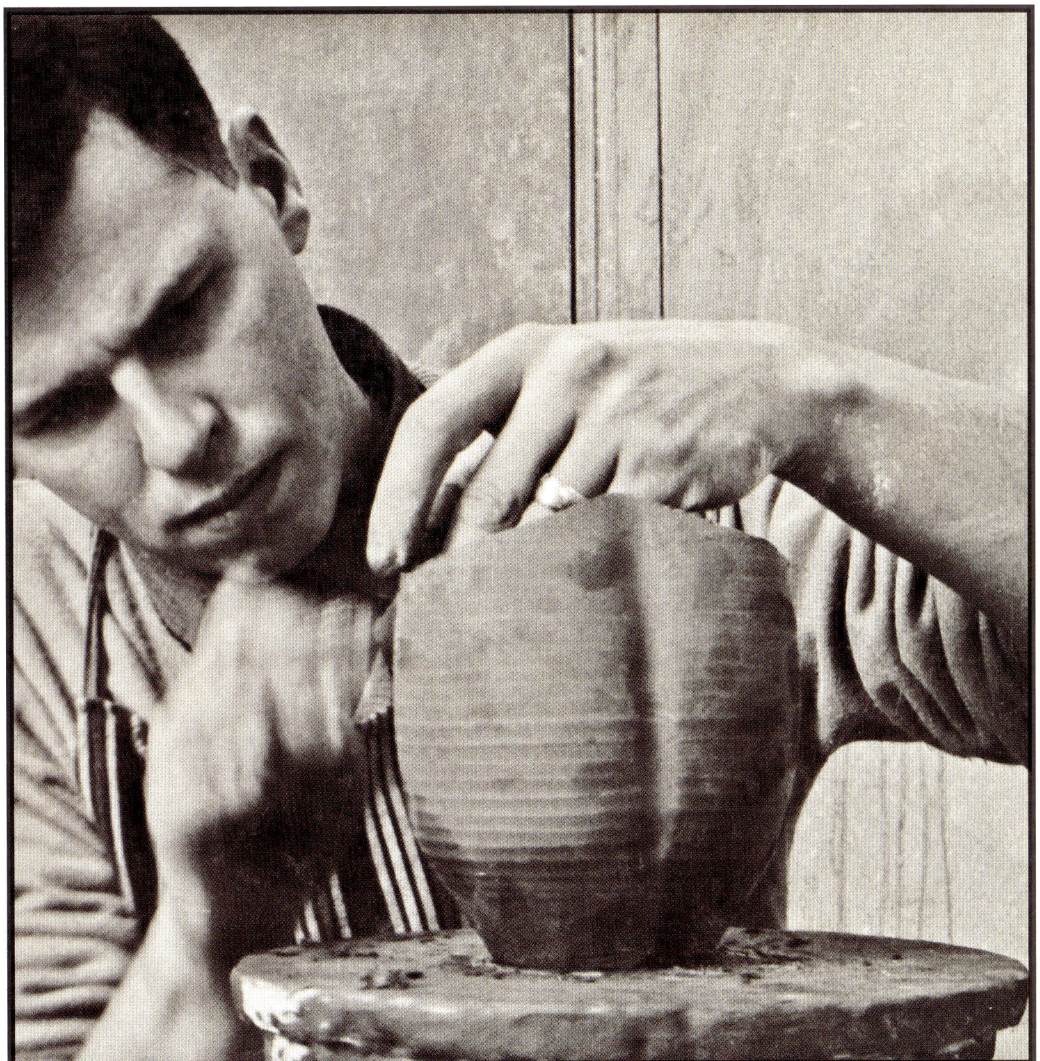

SEVEN DECADES

Fig. 48. Bill Daley, 1956
New Paltz, PA

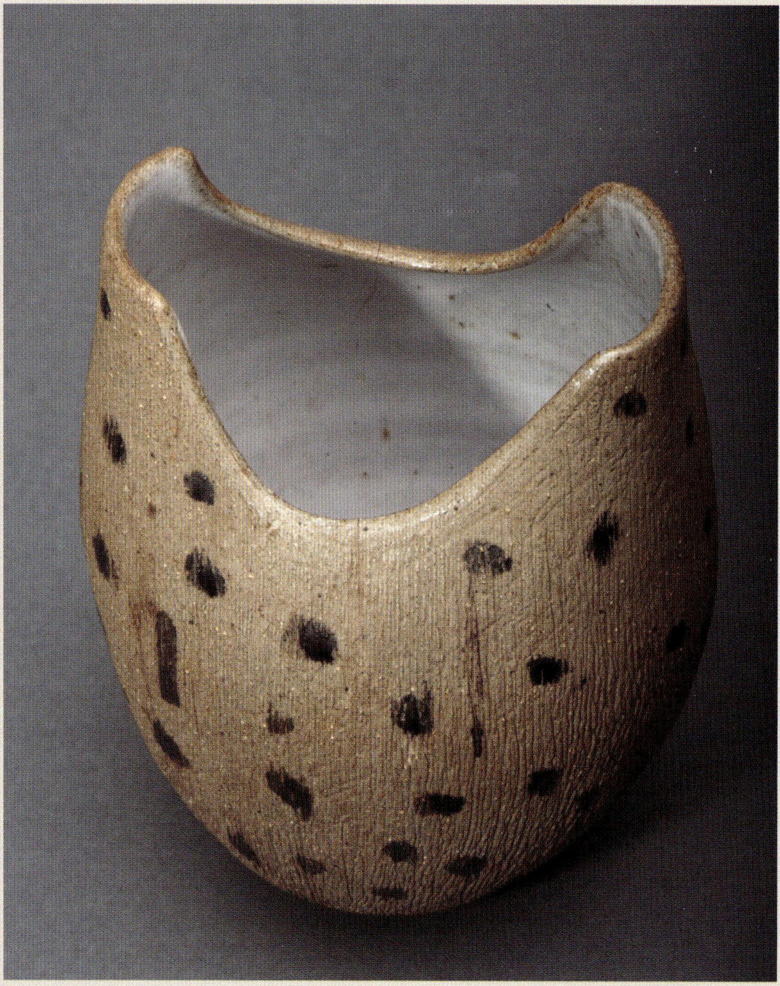

Photo: Amanda Schwab

Fig. 49. *First Frontal Form,* 1953
Exterior: unglazed stoneware,
wire brushed with slip spots
Interior: glazed
5.5 x 3.75 x 4.25 inches
Alfred, New York
Collection of William and Catherine Daley

Fig. 50, opposite.
Cyclops Eyecup, 1952
20 x 12 x 17 inches
Stoneware
Made at Columbia University
Collection of William and Catherine Daley

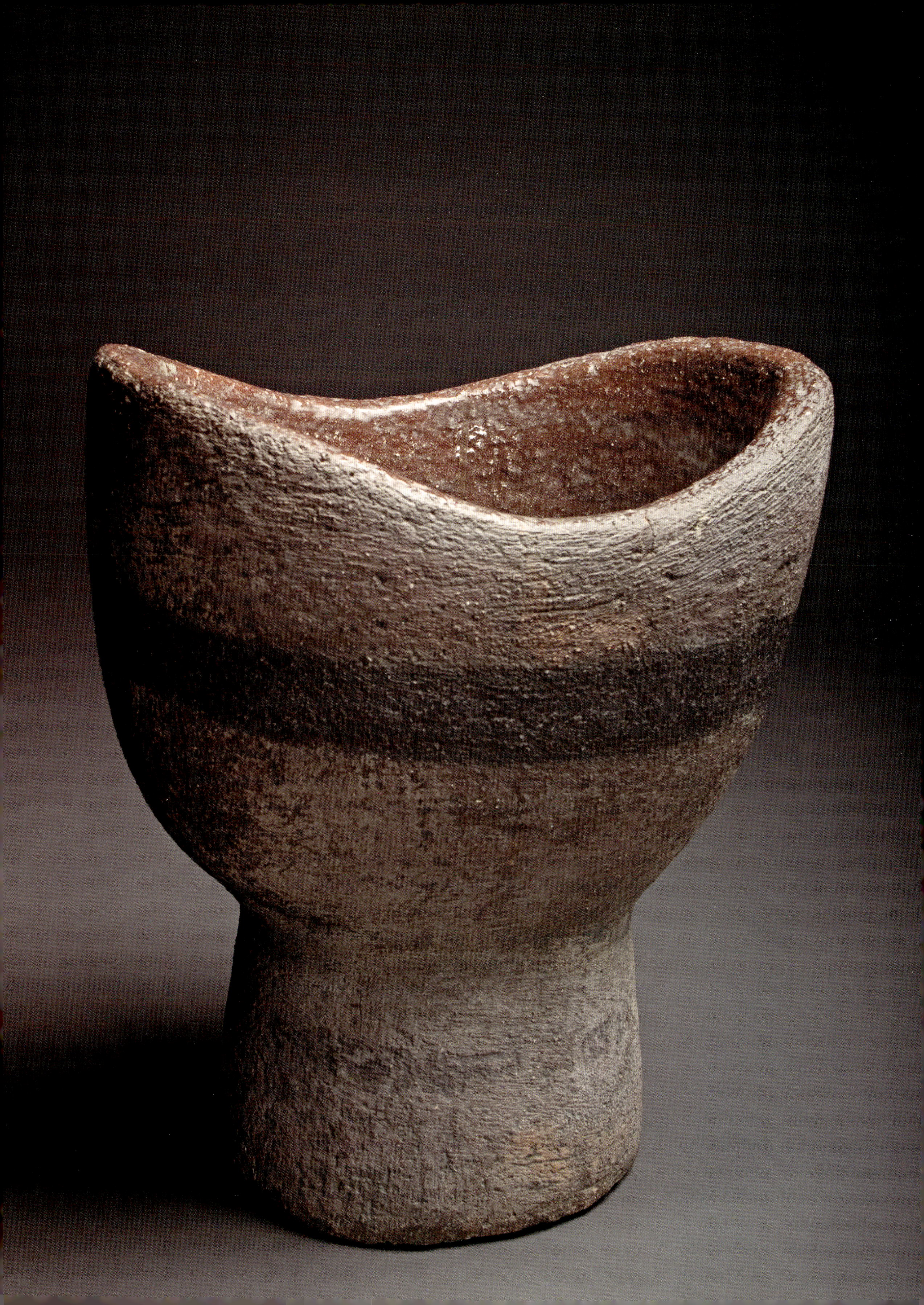

Photo: William Daley

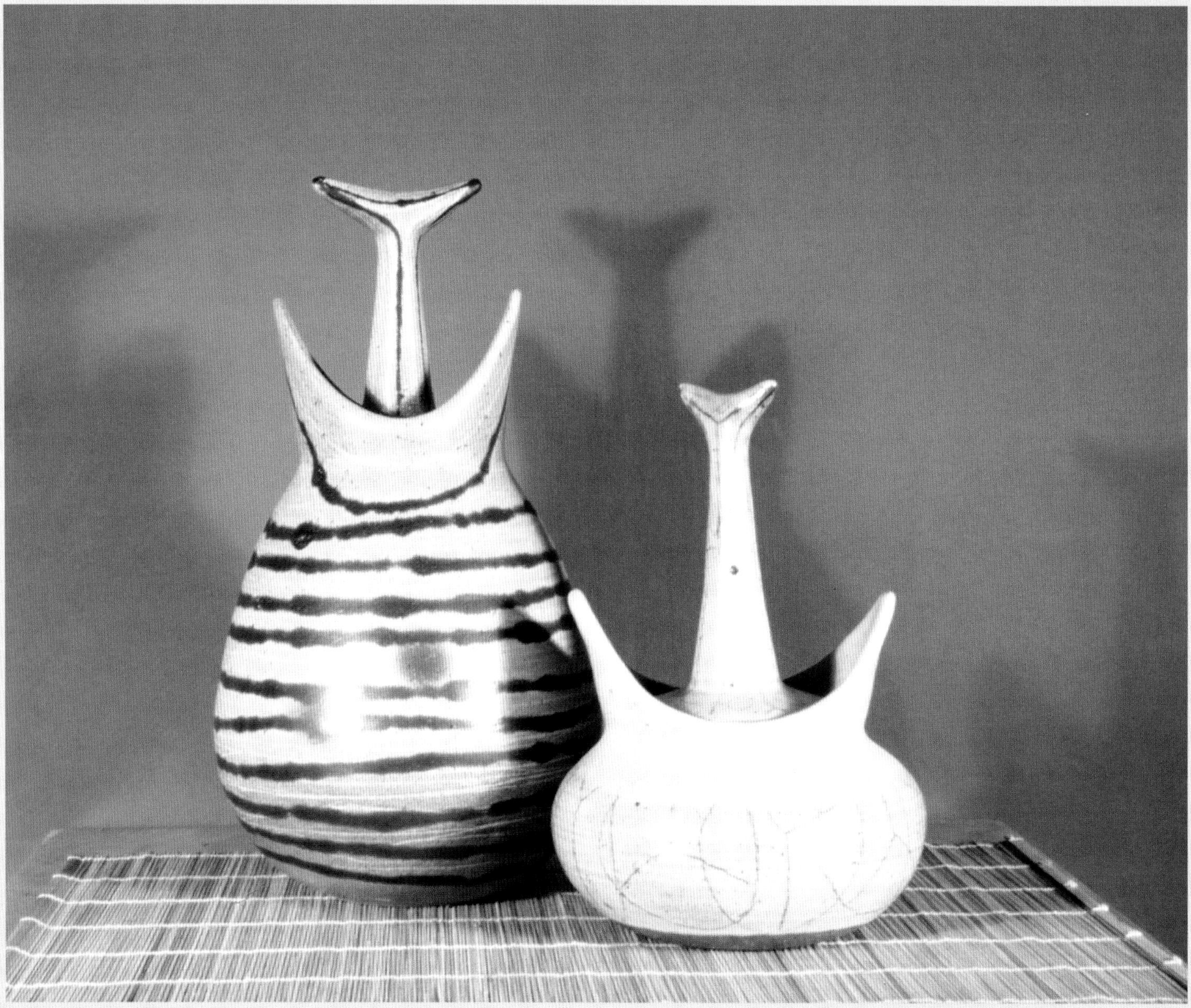

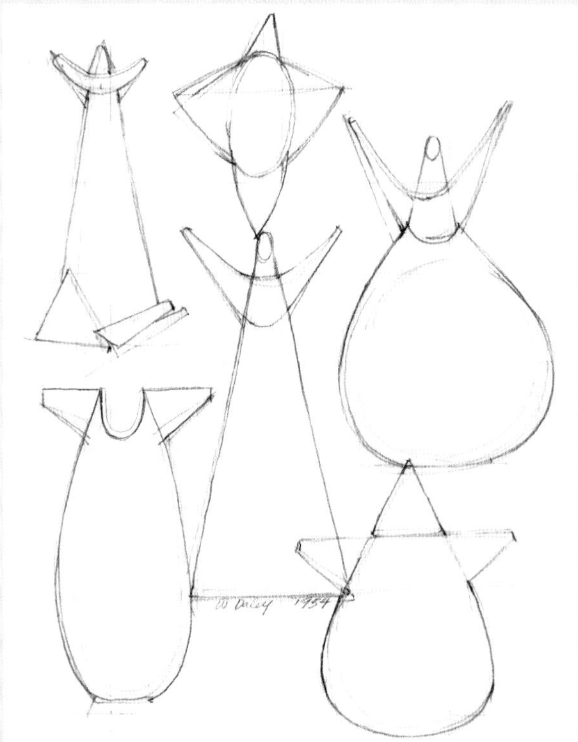

Fig. 51, above.
Covered jars, 1953
Left: glazed earthenware
12 x 6 x 6 inches
Cedar Falls, Iowa
Right: stoneware glazed
9 x 5 x 5 inches
Cedar Falls, Iowa

Fig. 52, left.
Drawing, 1954
Pencil on vellum paper
11 x 8.5 inches

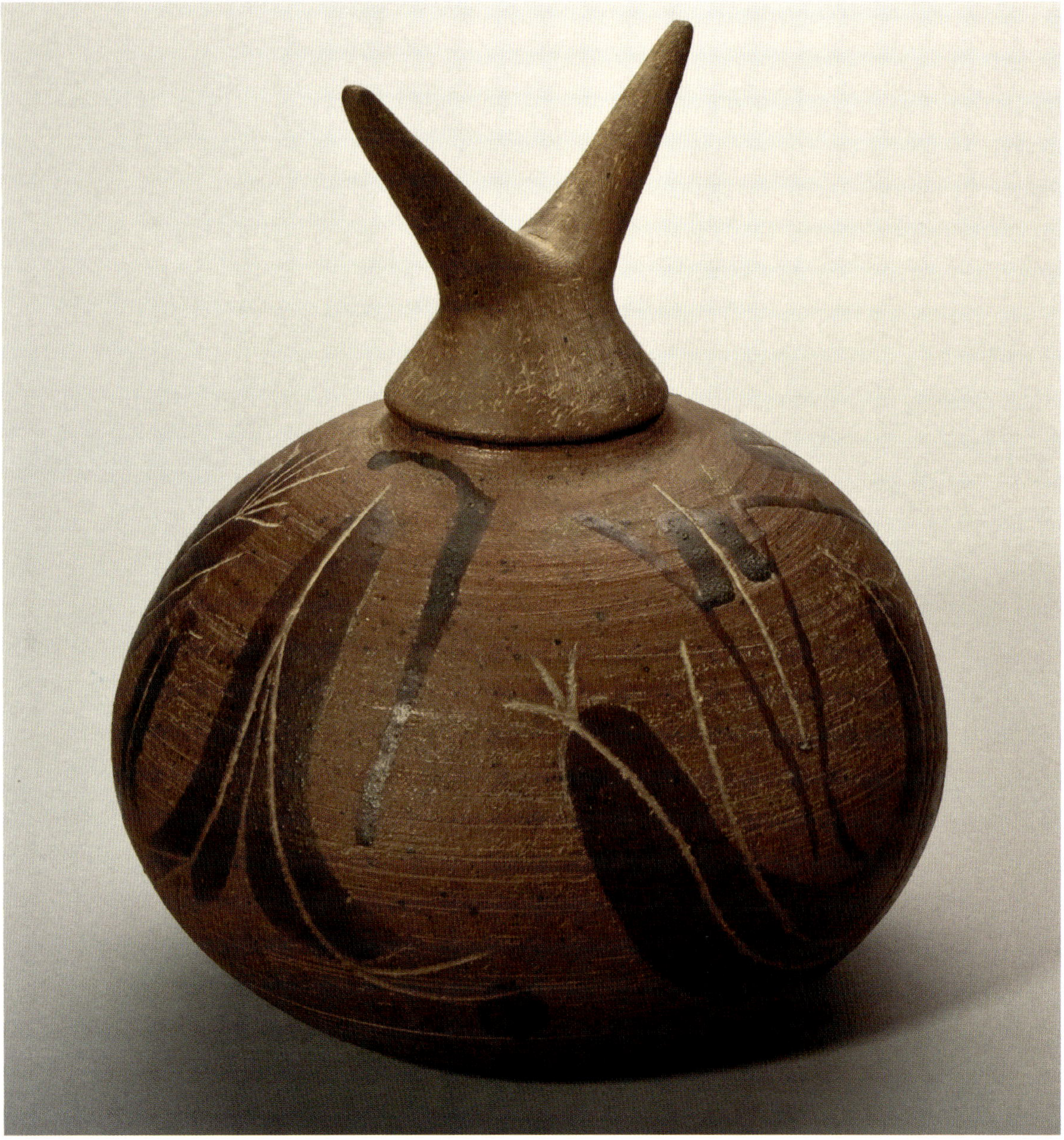

Photo: Amanda Schwab

Fig. 53, left.
Untitled, 1953
Stoneware with slip and sgraffito decoration
9 x 8 x 8 inches
Alfred, New York

Fig. 54, below left.
Hanging pot, 1954
Glazed earthenware
6 x 8.5 x 11 inches
Cedar Falls, Iowa

Fig. 55, below.
Hanging planter, 1953
Glazed earthenware
3 x 9.25 x 9 inches
Cedar Falls, Iowa
Collection of Dean and Geraldine Schwarz

Photos: Amanda Schwab

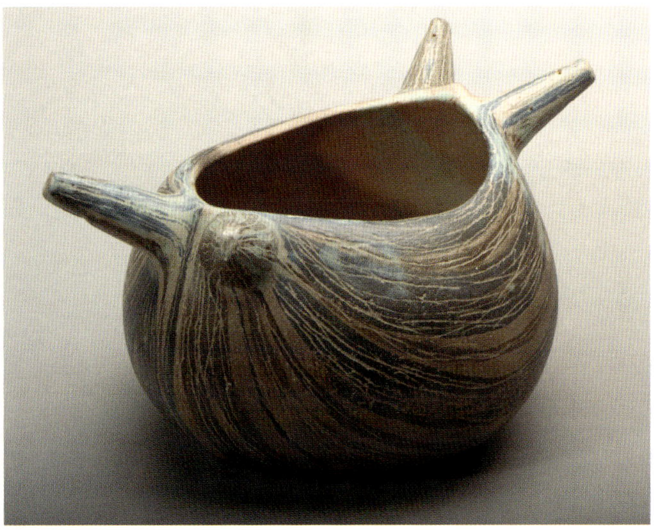

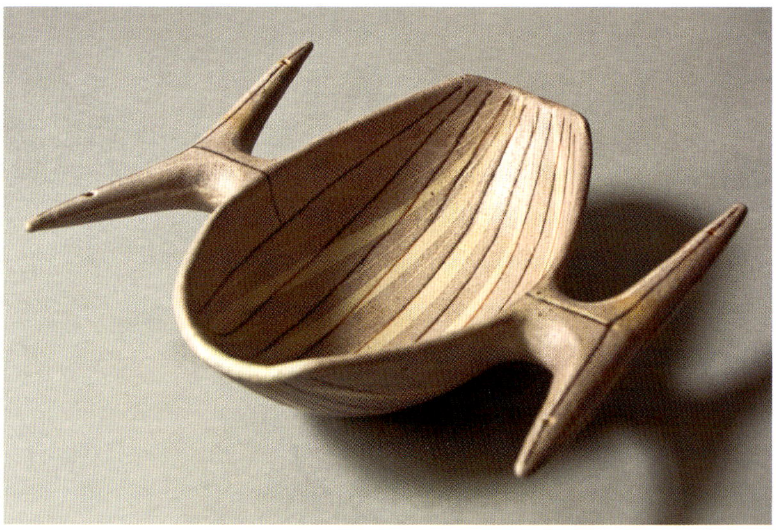

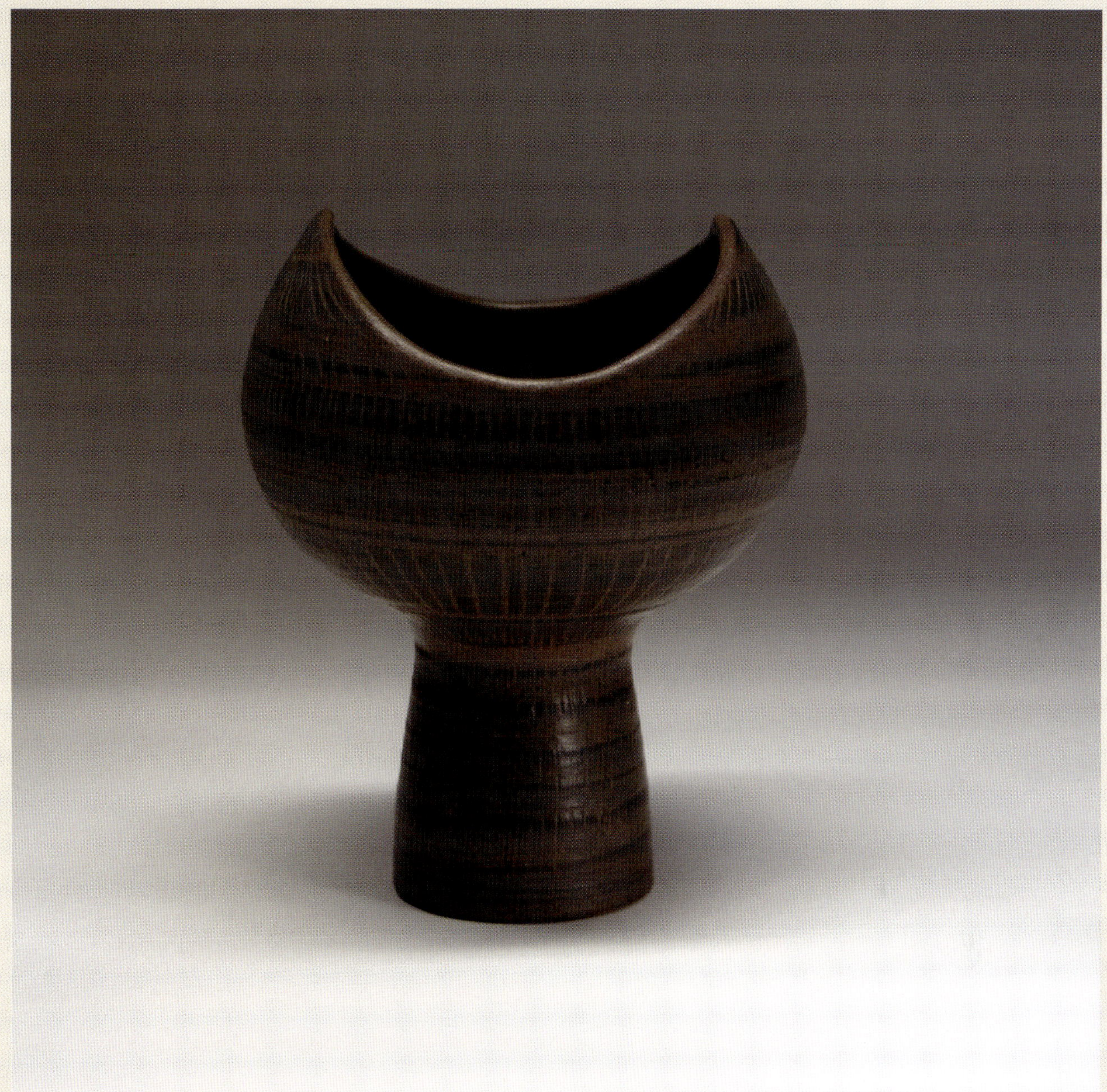

Photo: Christine Carreiro

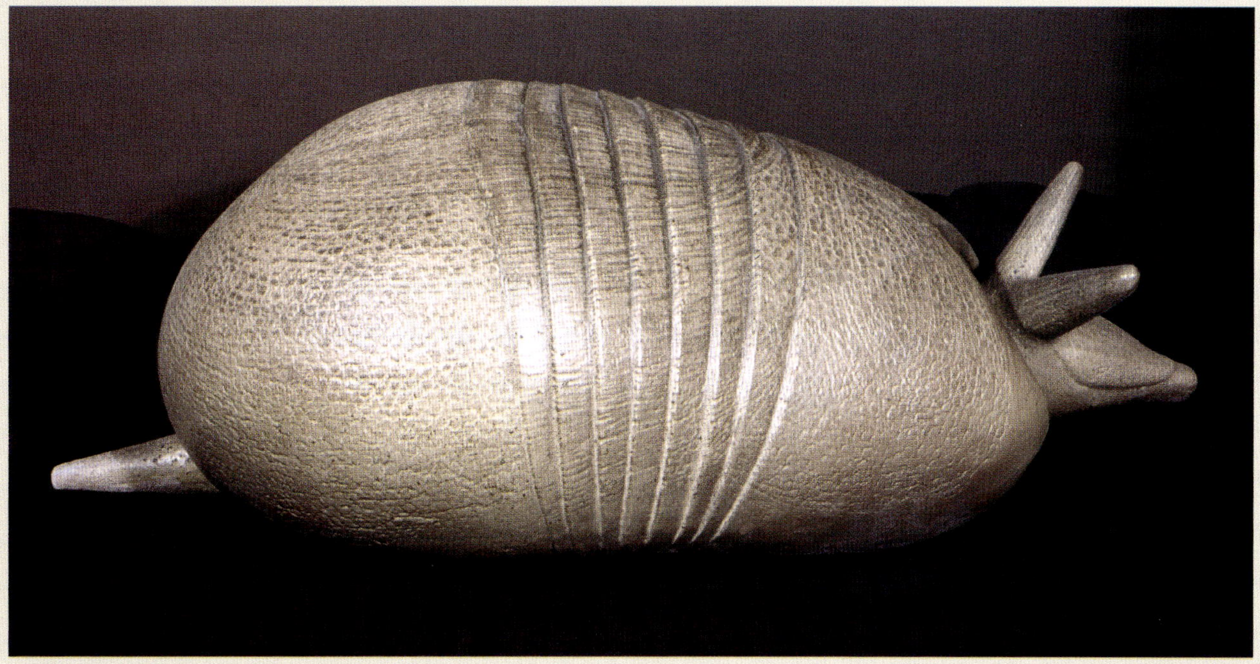

Photo: William Daley

Pots for Show 1956
Young Ceramist Aims at Works of Art Not Useful Objects

Jules Olitsky

Unlike the potter of bygone days who was primarily concerned with creating useful and necessary ware, the contemporary ceramist presented here eschews utility in favor of the work of art—and his is not alone in his point of view. On the other hand, a considerable section of today's potters maintain that to be really good a pot must "give assurance of practical service." This was the view presented in "Pottery today—for Show or Us?"
— Carlton Atherton, *Ceramics Monthly,* July 1954. (Ed.)

When the Art Education Gallery of New York University presented an exhibition of pottery by William Daley (last spring,) the twenty-seven pieces displaying aroused such intense interest and appreciation that the show was held over for two weeks beyond the original days.

Why this unusual response—from faculty, students and the public? The answer is to be found in the attitude of Mr. Daley toward his art. This attitude is directly reflected in his work. In short, Mr. Daley is one of those potters who regard themselves as artists, and insist that their pots be judged as works of art, deserving the serious consideration we accord without question to paintings or sculpture.

The potter who looks upon himself as a creative artist is inescapably involved with innovation and experiment, with creation and the discovery of new forms. He is not interested in the merely utilitarian function of pottery. Concerning his objectives Mr. Daley has said, *"I am not interested in utility...but in the expressive. For me pottery is not a craft. I do not care about making 'useful objects for better living' nor do I care about producing numbers of any one pot. I am interested in the sculptural and should like my pots to be viewed as ends in themselves...What I am trying to do is to make works of art."*

In spite of his disavowal of craft, Mr. Daley is a first rate craftsman, disciplined and conscientious in his awareness of the natural limitations of his medium. It is instructive to see him at work. First the basic form is thrown on the wheel. Then, with the utmost freedom and yet startling accuracy, typical of the skilled craftsman, the clay is pierced, incisions are made, and parts of the vessel cut completely away. At this point the potter stops and surveys his work, slowly circling the pot from a distance. Soon he is back, re-shaping a contour, pressing together or pulling further apart the edges where an incision has been made. Manipulating the clay and making changes he is getting closer all the time to that ultimate form which hovers in his imagination. When the pot is leather hard, such appendages as are indicated are added. And then the final decisions concerning the form are made.

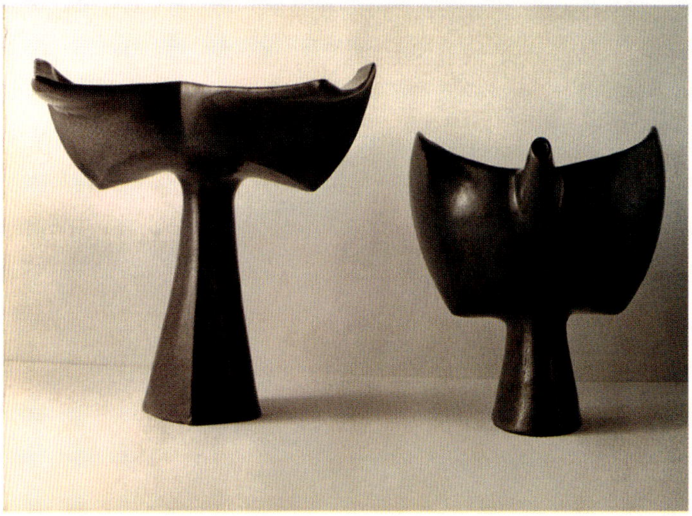

Photo: William Daley

In the New York show, all of Mr. Daley's pots bore the same lead matt glaze fired at cone 04. The color effects varied with each pot although the colors were severely limited to green, dark green and tan. The muted surface colors enhanced, and quietly emphasized the bold expressiveness of form.

Fig. 56, opposite above.
Small Pot, 1954
Earthenware with slips
8.5 x 5.5 x 4 inches
Private collection

Fig. 57, opposite below.
Armadillo, 1954
Glazed earthenware
7.5 x 21 x 10 inches
Cedar Falls, Iowa
Collection of Joseph and Karen O'Mara

Fig. 58, above.
Left: *Pedestal Pot,* 1956
Right: *Pouring Pot,* 1956
Glazed earthenware
New Paltz, New York
Included in New York University show, New York, NY, 1956

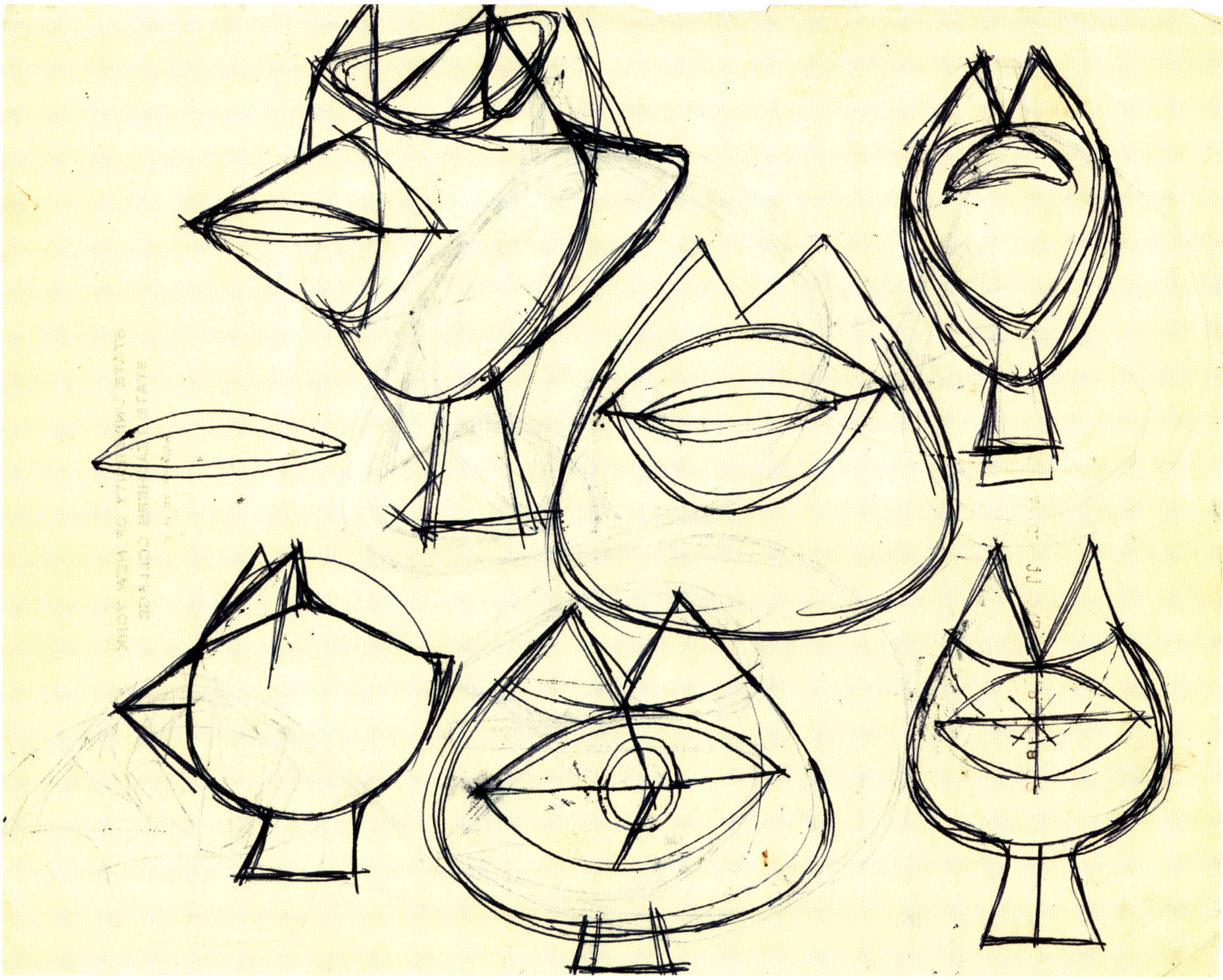

Fig. 59. Drawing, 1956
Pen on paper
8 x 10 inches

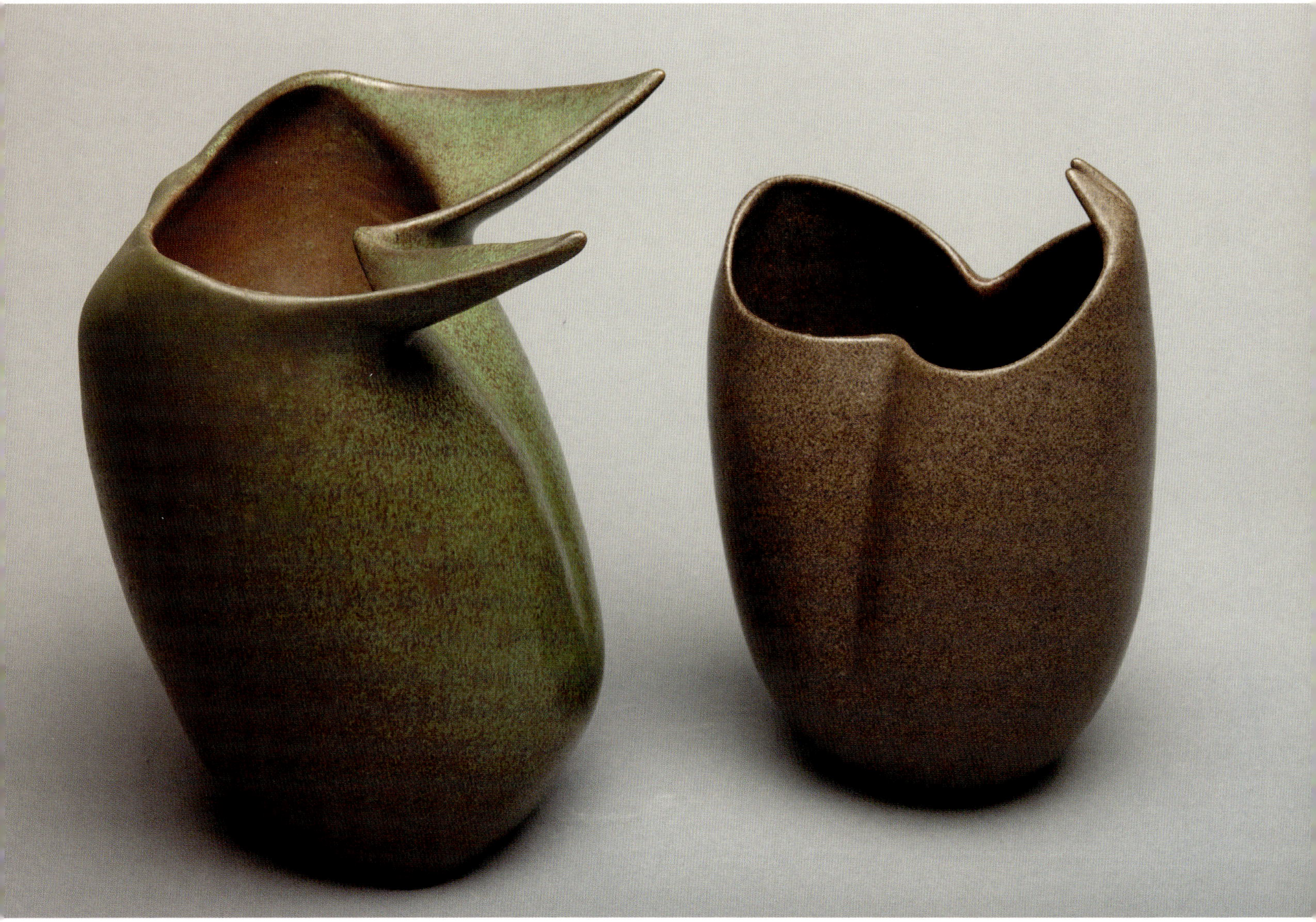

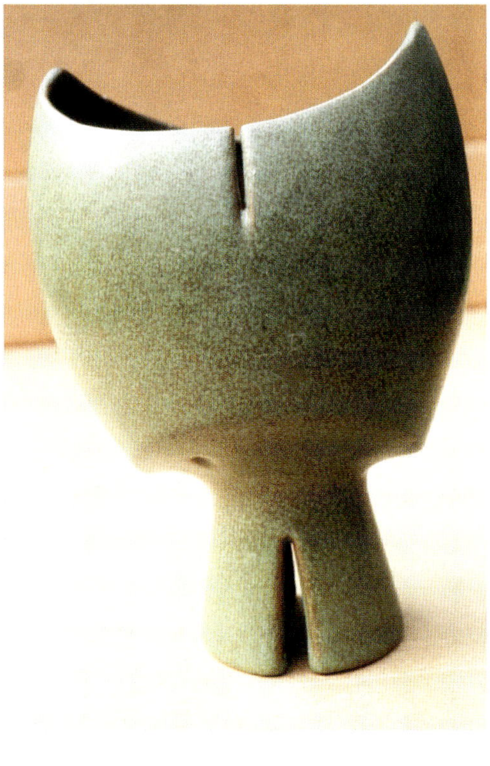

Photo: William Daley

Fig. 60, above.
Left: *Pregnant Cate Pot,* 1956
From NYU Show, 1956
Earthenware with copper matte glaze
8.25 x 8 x 6.5 inches
Collection of William and Catherine Daley

Right: Untitled, pot from NYU Show, 1956
Earthenware with manganese matte glaze
7.5 x 4 x 5 inches
Collection of William and Catherine Daley

Fig. 61, right.
Shang Variations, 1954
Glazed earthenware
approx size 10 inches height
Included in New York University show,
New York, NY, 1956
Collection of Dorothy Carreiro

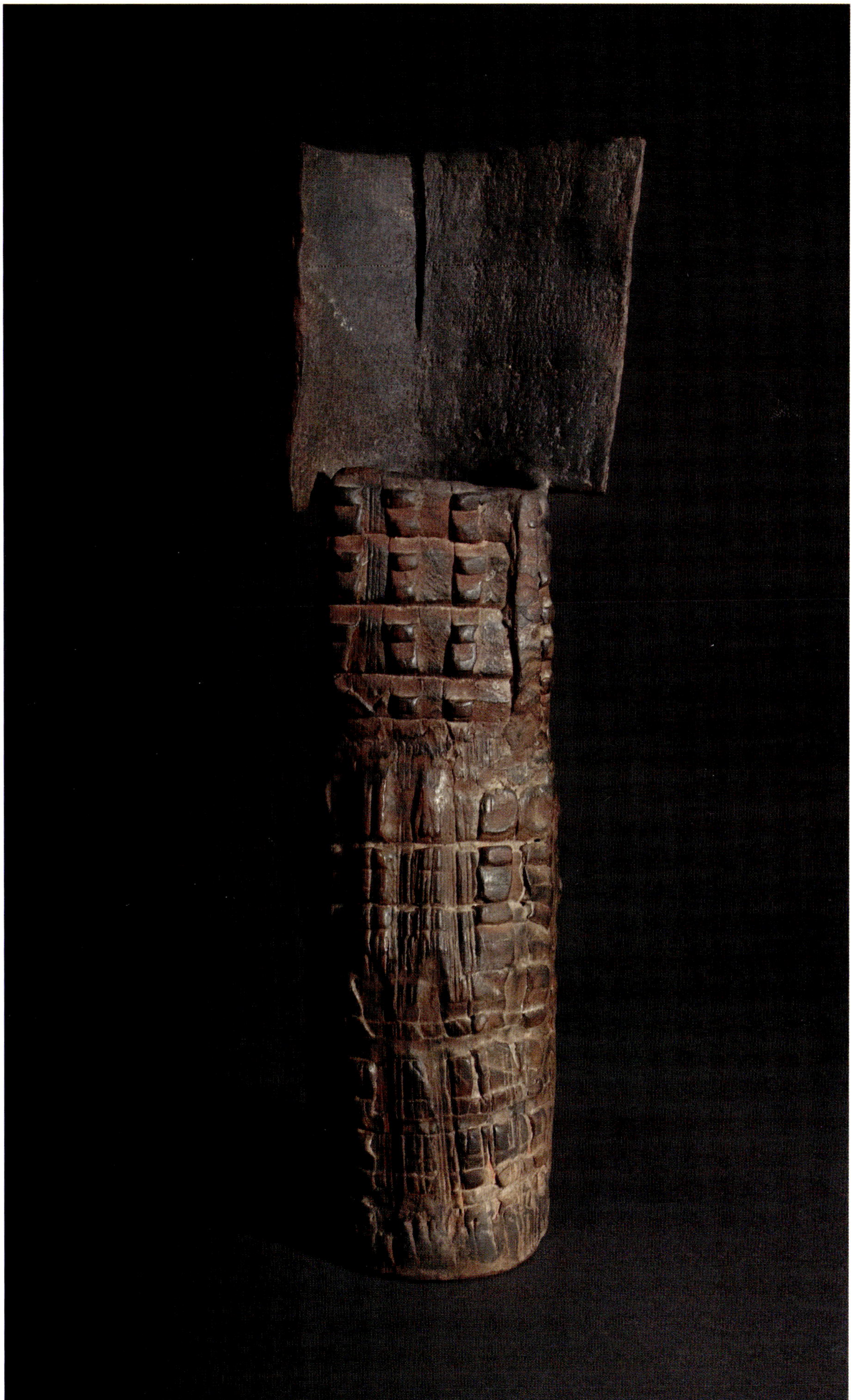

Photo: Amanda Schwab

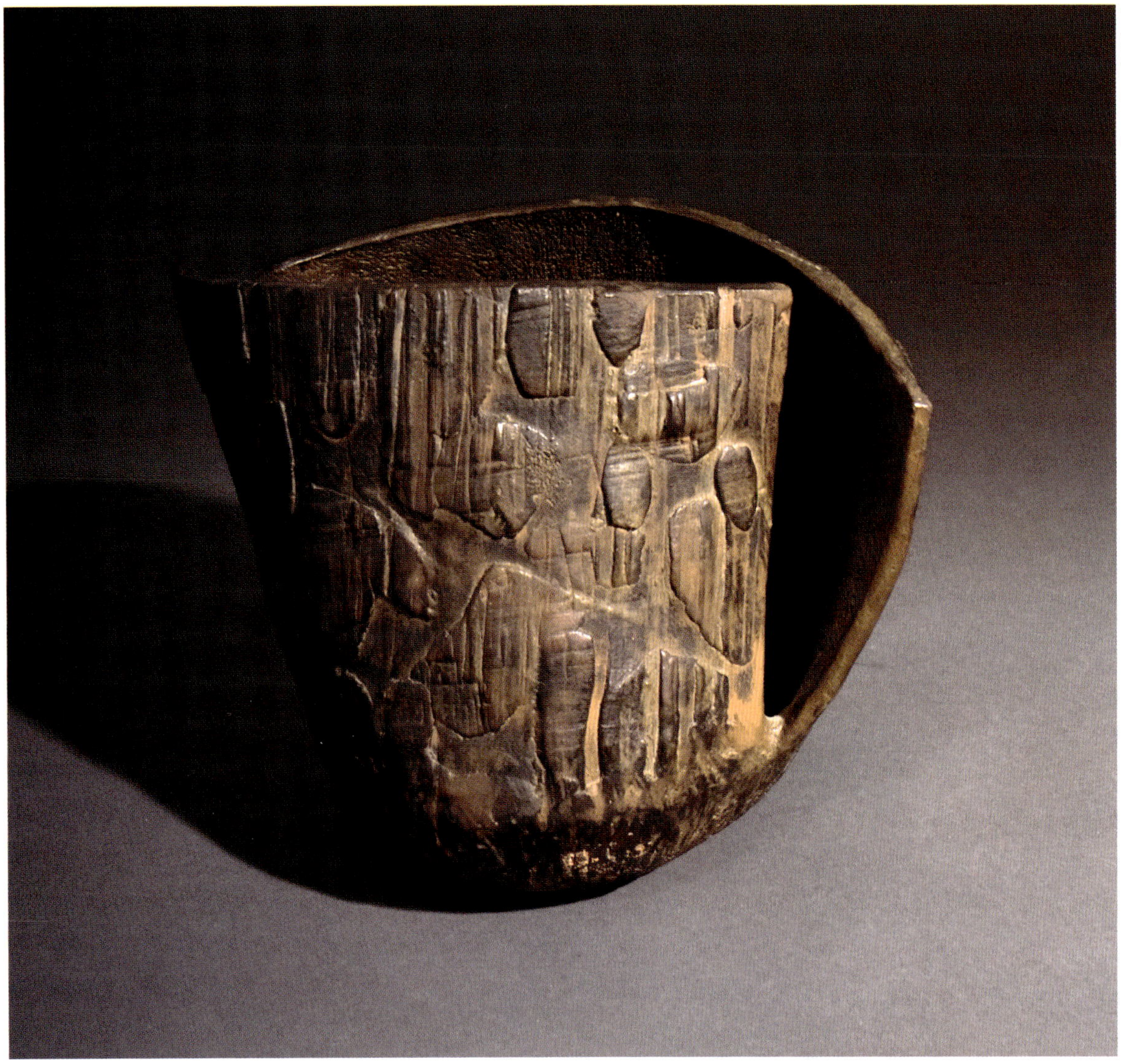

Photos: Amanda Schwab

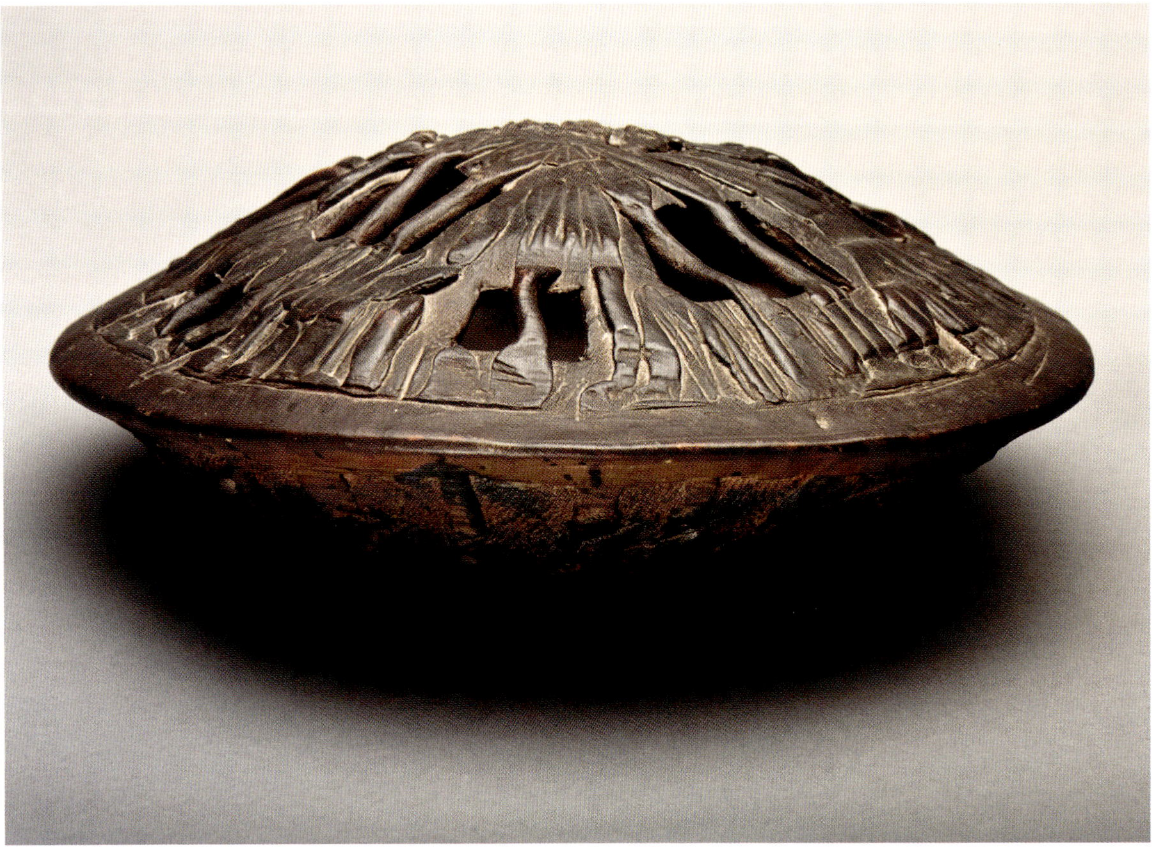

Fig. 62, opposite.
Tall Split Form, 1956
Glazed earthenware
16 x 3.5 x 3.5 inches
Collection of Charlotte Daley

Fig. 63, above.
Split Form, 1956
Glazed stoneware
11 x 4 x 6 inches
Collection of George and
Barbara Daley Schwab

Fig. 64, left.
Petrified Pizza, 1956
Glazed earthenware
4 x 11.5 x 11.5 inches
Collection of Thomas and
Donna Weaverling Daley

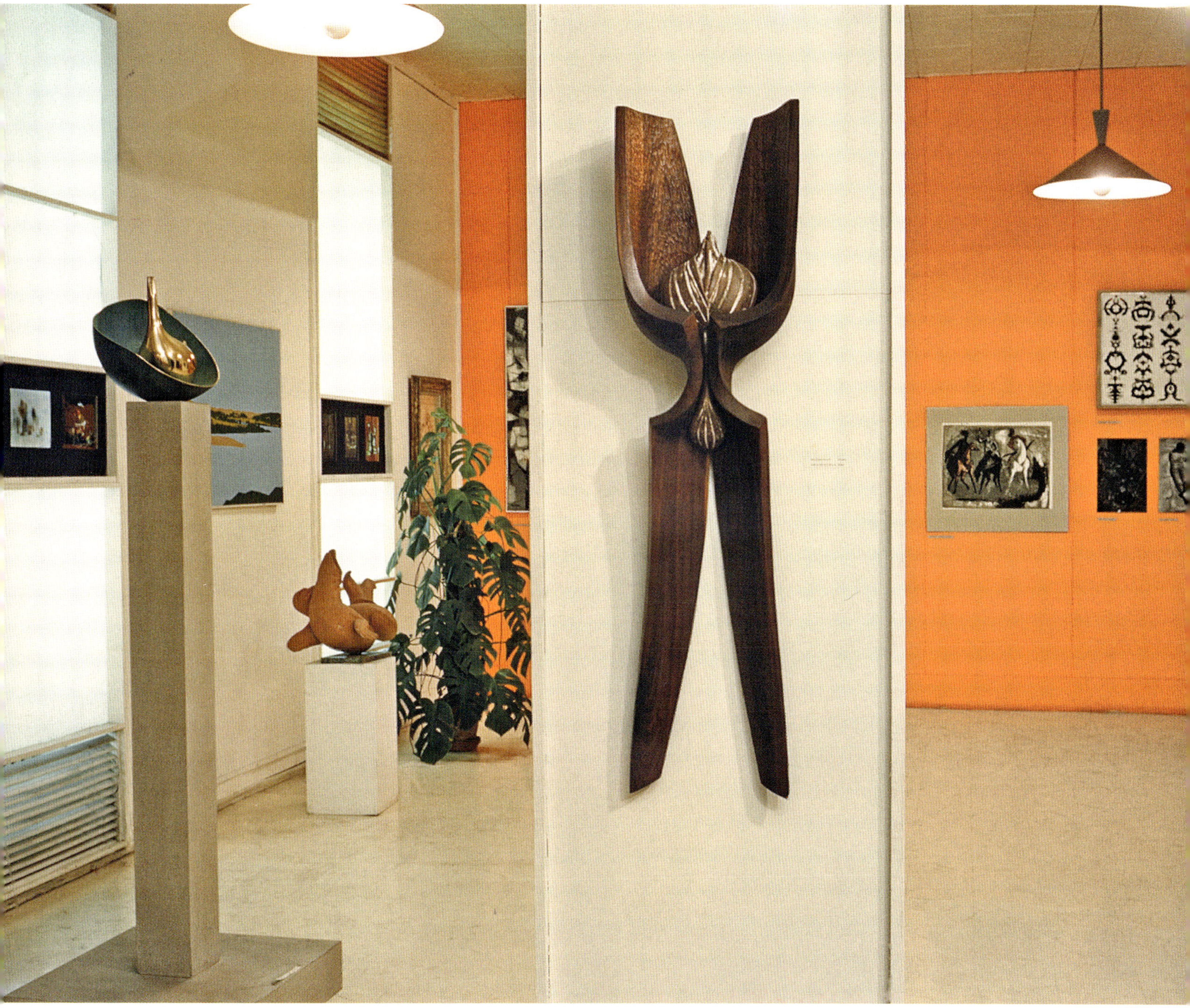

Fig. 65, above.
Center: *Columbine,* 1960
William Daley
Faculty exhibition Philadelphia Museum School of Art
60 x 15 x 8 inches
Ceramic and carved walnut

Left: Sculpture on tall limestone pedestal made with cast bronze and spun aluminum by artist Aurilius Renzetti

Fig. 66, opposite.
Phallic Pot, 1958
Glazed stoneware
23 x 5.5 x 5 inches
Exhibited Pace Gallery, Boston, MA 1959
Collection of William and Catherine Daley

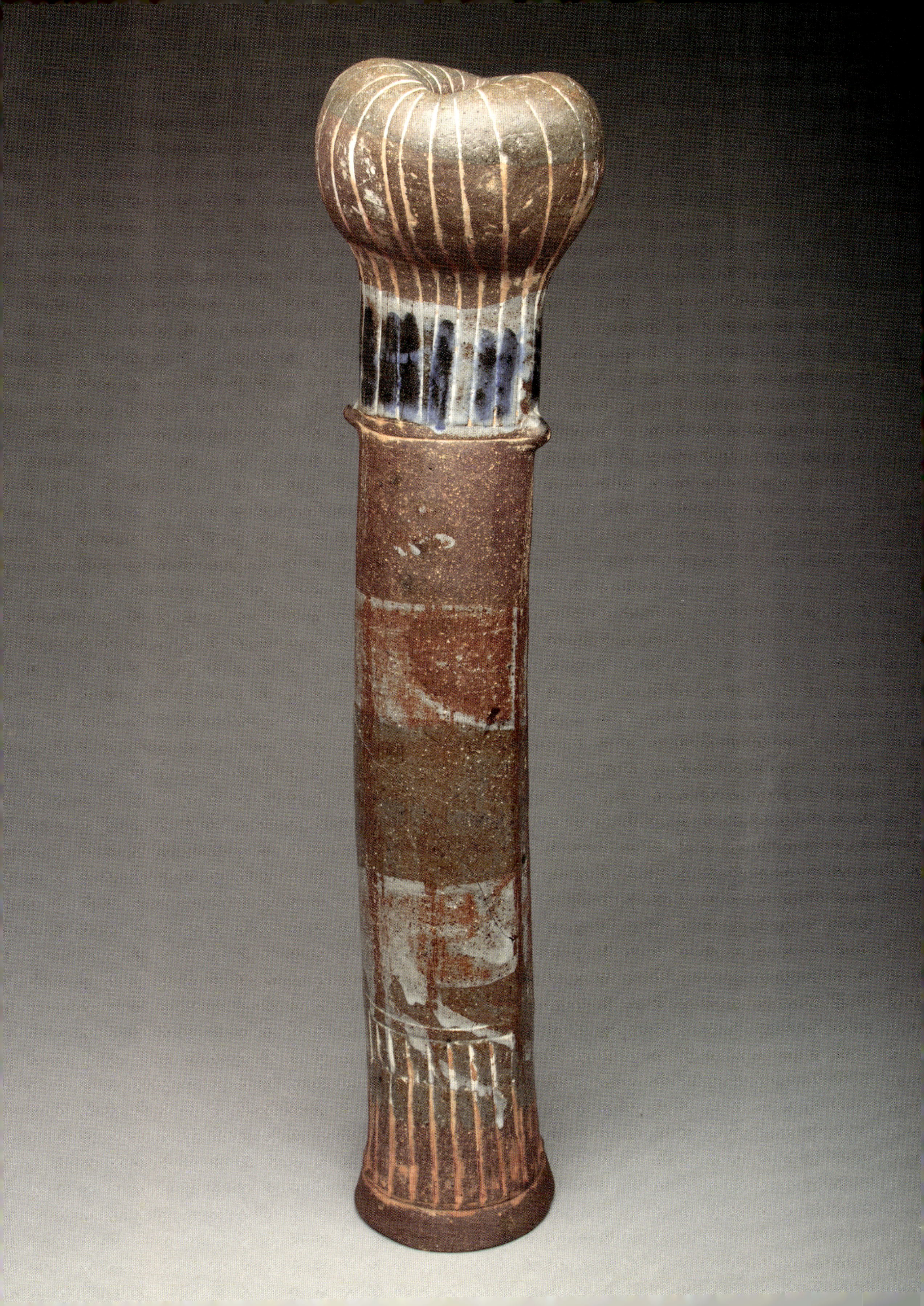

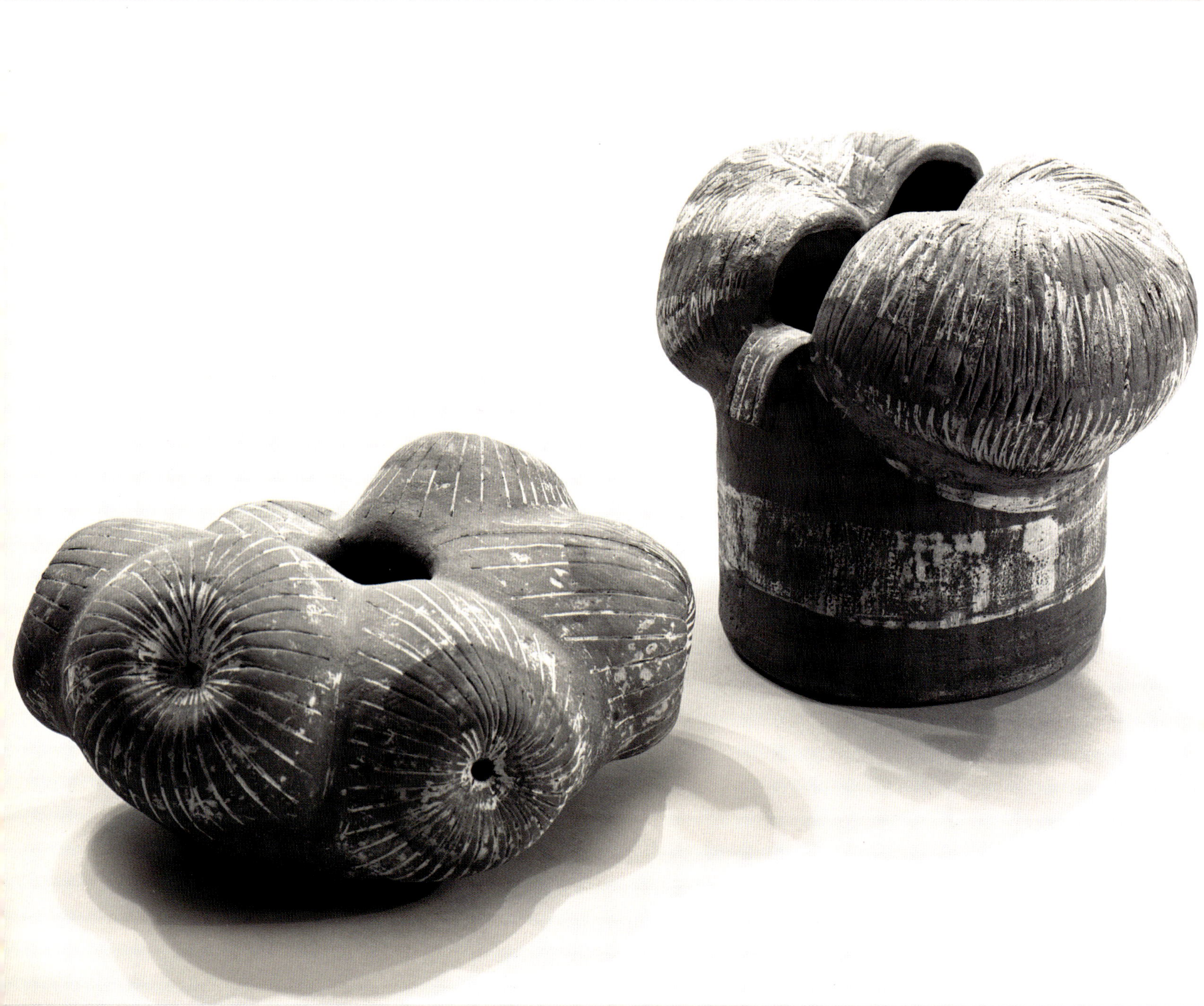

Fig. 67, above.
Left: *Summer Squash,* 1958
Stoneware with white slip
9 x 16 x 16 inches
Collection of Jack Lenor Larsen

Right: Untitled, 1958
Stoneware with white slip
14 x 14 x 9 inches

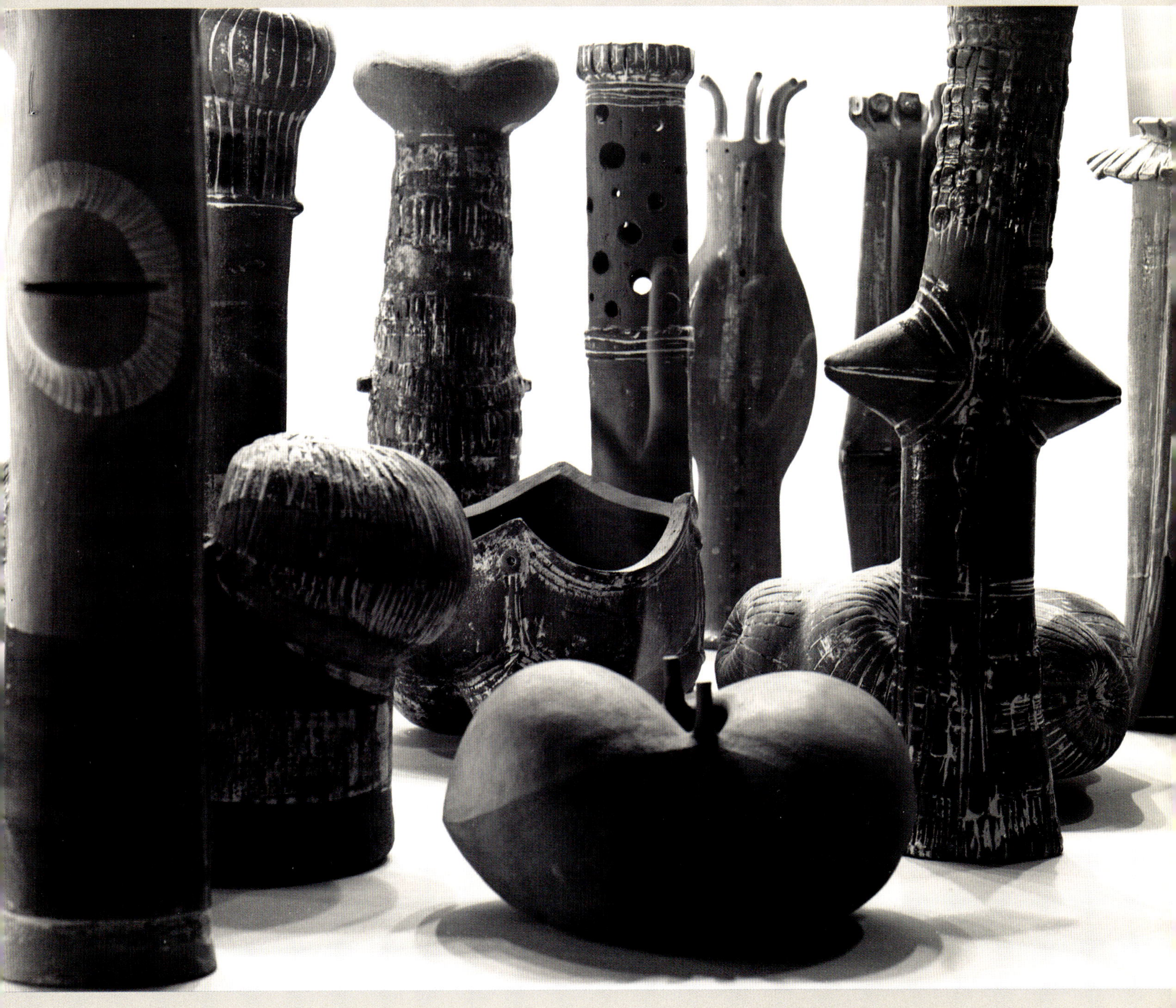

Photos: Robert Robertson

Fig. 68, above.
Various pots, 1958-59
Glazed stoneware
Shown at Pace Gallery, Boston, MA, 1959

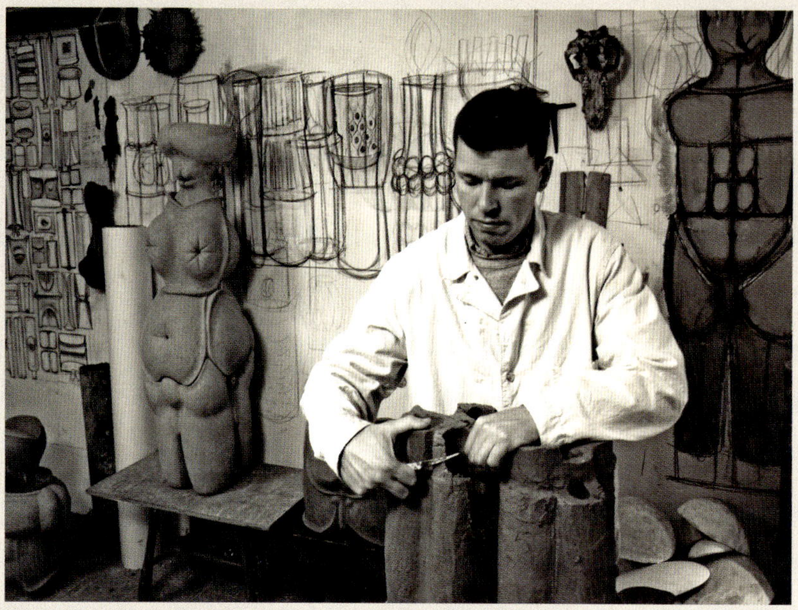

Photo: Murray Weiss

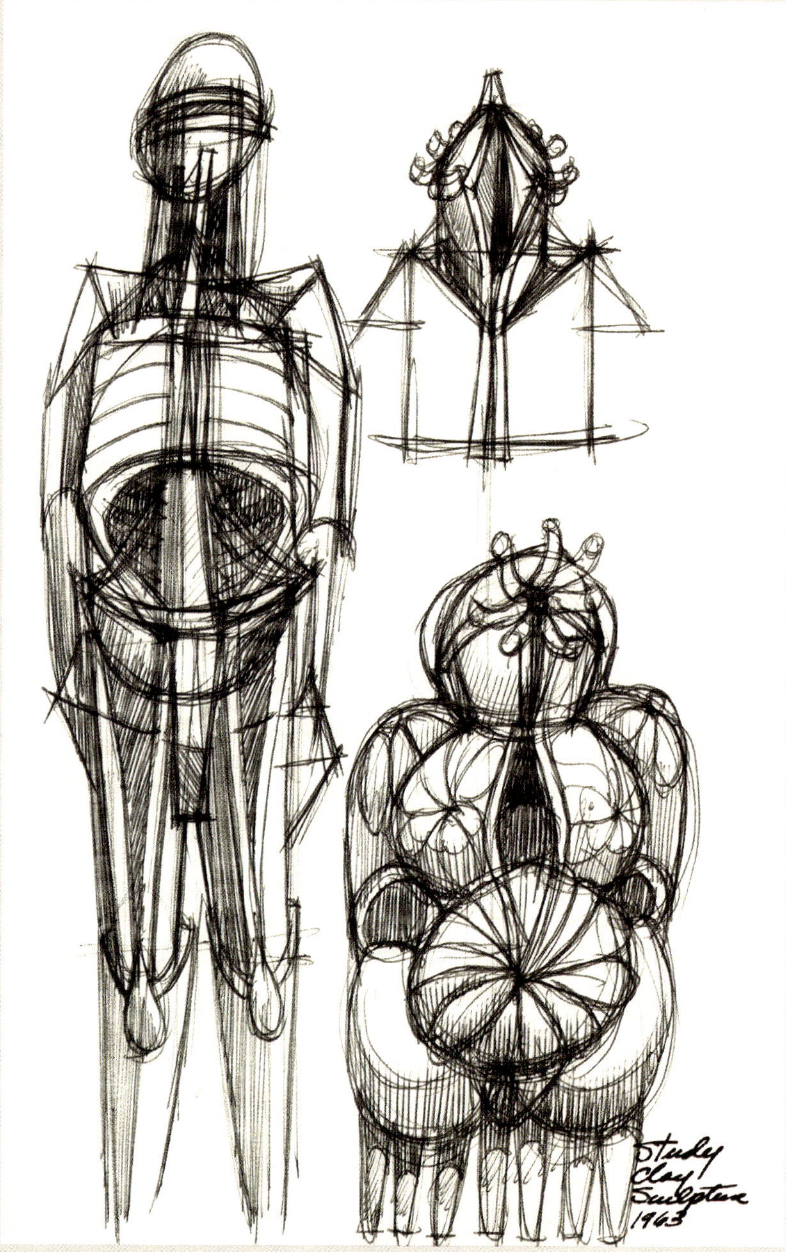

Fig. 69, above.
William Daley constructing Adam's leg section in his studio in converted garage, Feasterville, PA, 1962

Fig. 70, right.
Drawing, *Study for clay sculpture*, 1963
Ink on paper

Fig. 71, opposite.
Adam and Eve, 1962
Unglazed stoneware
Adam–52 x 14 inches
Collection of William and Catherine Daley
Eve–49 x 14 inches
Collection of Robert L. Pfannebecker
Gift of the artist

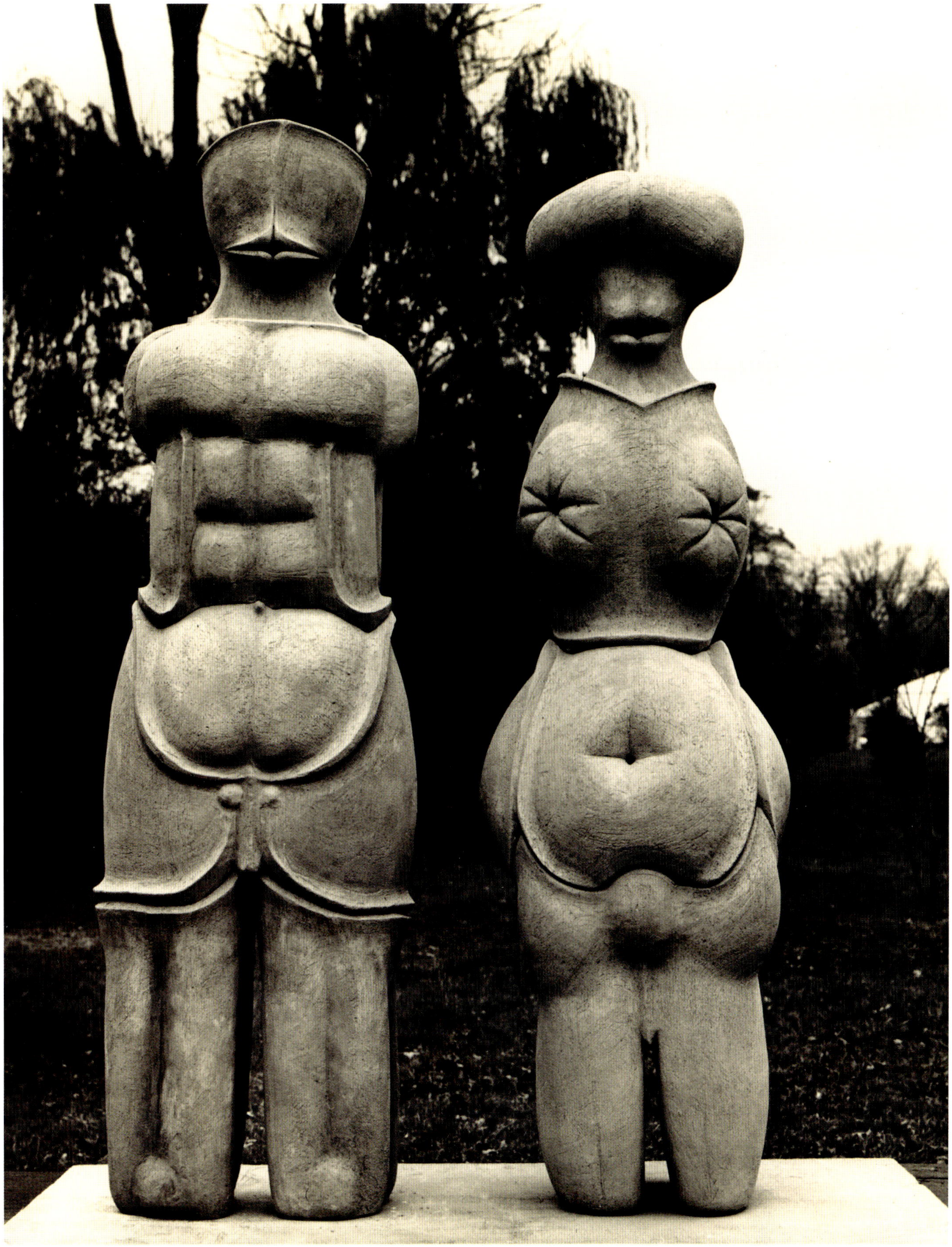

Photo: William Daley

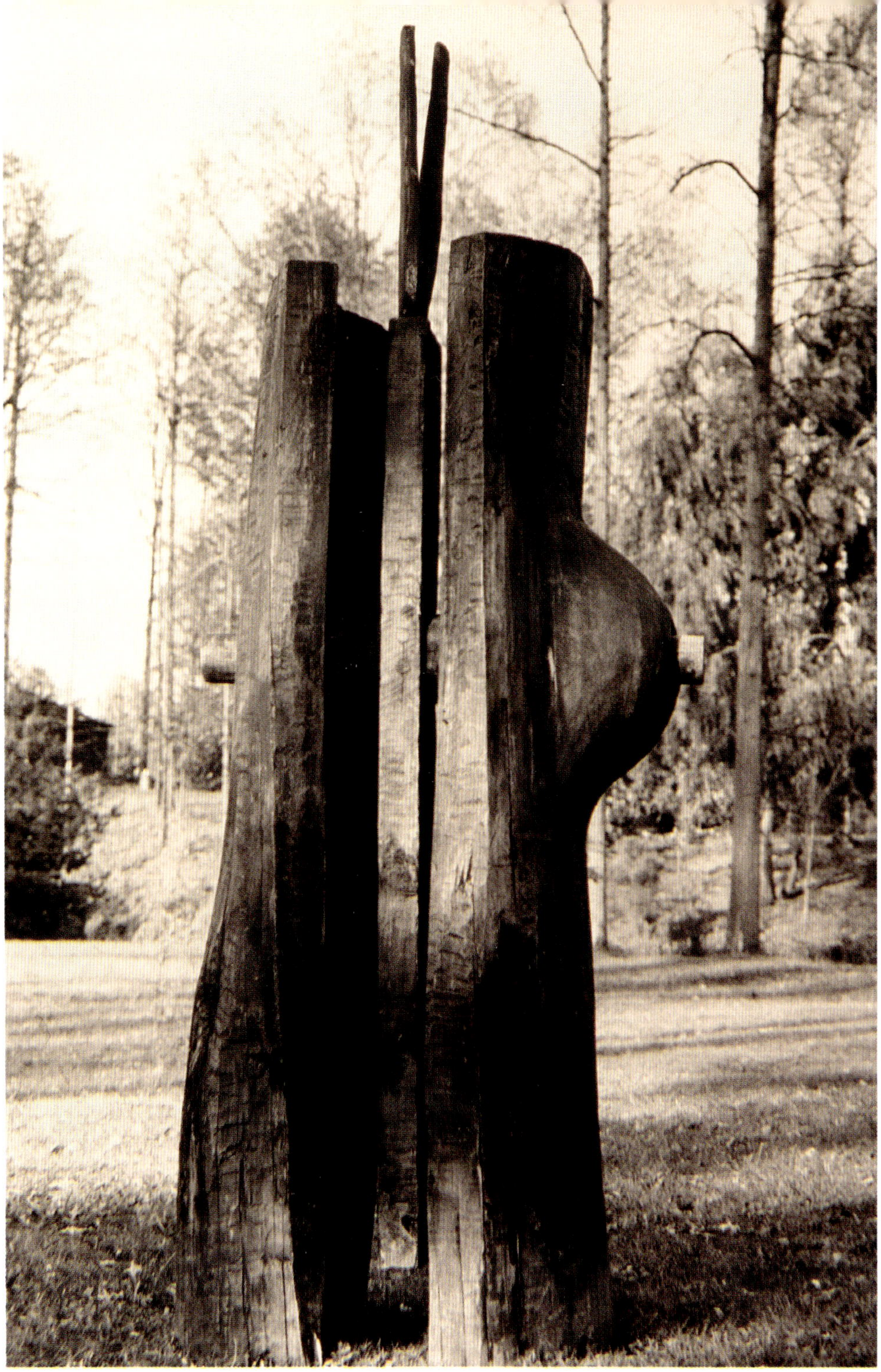

Photo: William Daley

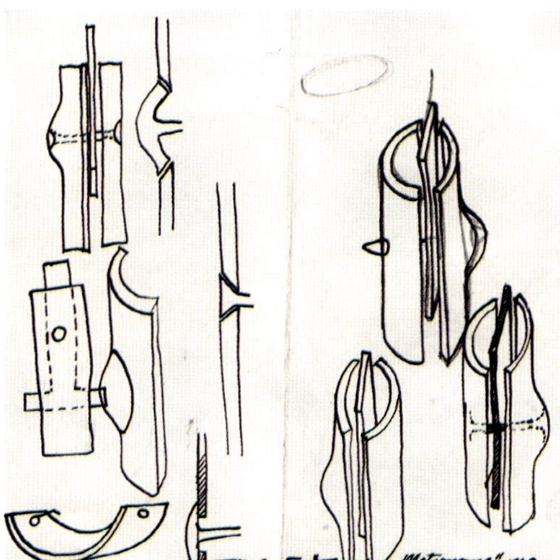

Fig. 72, above.
Timeless Metronome, 1960
Adzed split poplar log
88 x 54 inches

Fig. 73, right.
Drawing *Study for "Timeless Metronome,"* 1960
Pen on paper

Fig. 74, opposite.
Ritual Carcass, 1961
Wood and clay
80 x 22 x 5 inches
Collection of William and Catherine Daley

Photo: William Daley

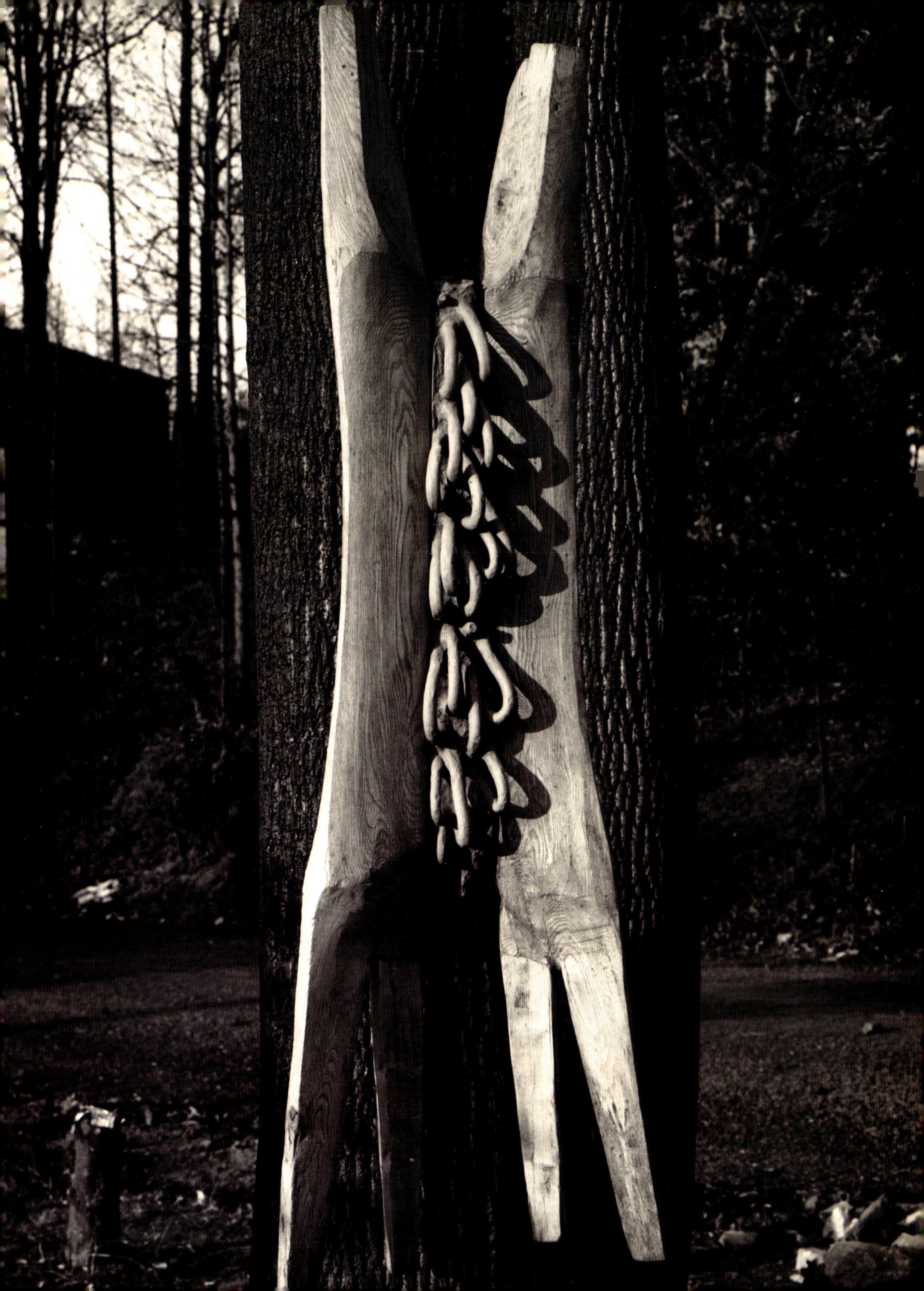

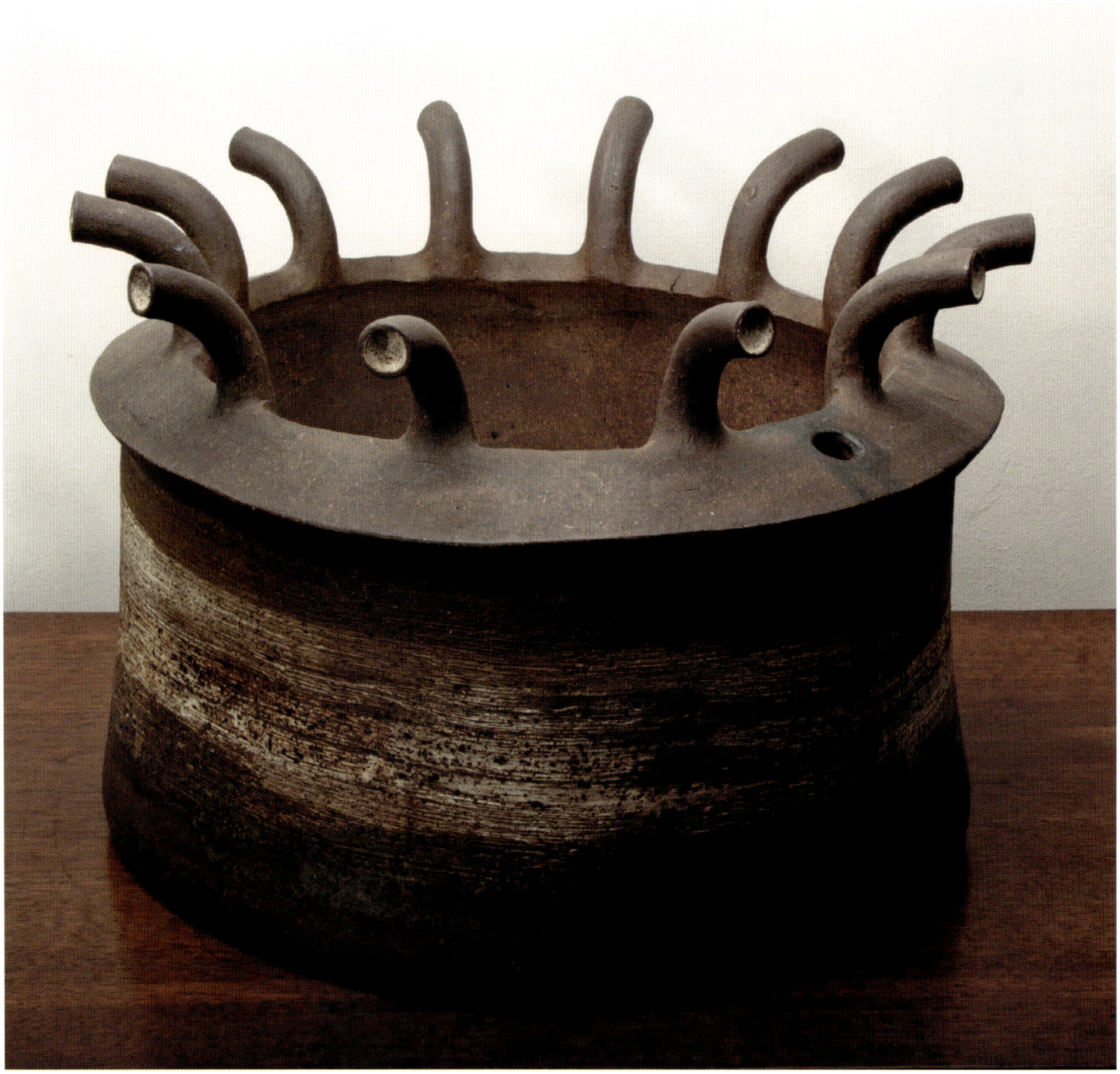

Photo: Amanda Schwab

Fig. 75.
Oval Tank Pot, 1962
Stoneware with slip
10 x 14 x 16 inches
Collection of William and Catherine Daley

Fig. 76.
Three Floor Vases, 1962
Stoneware with slip
approx. 24 x 9 inches
From left to right: In the collection
of Noel Mayo, Ruth Fine, and unknown

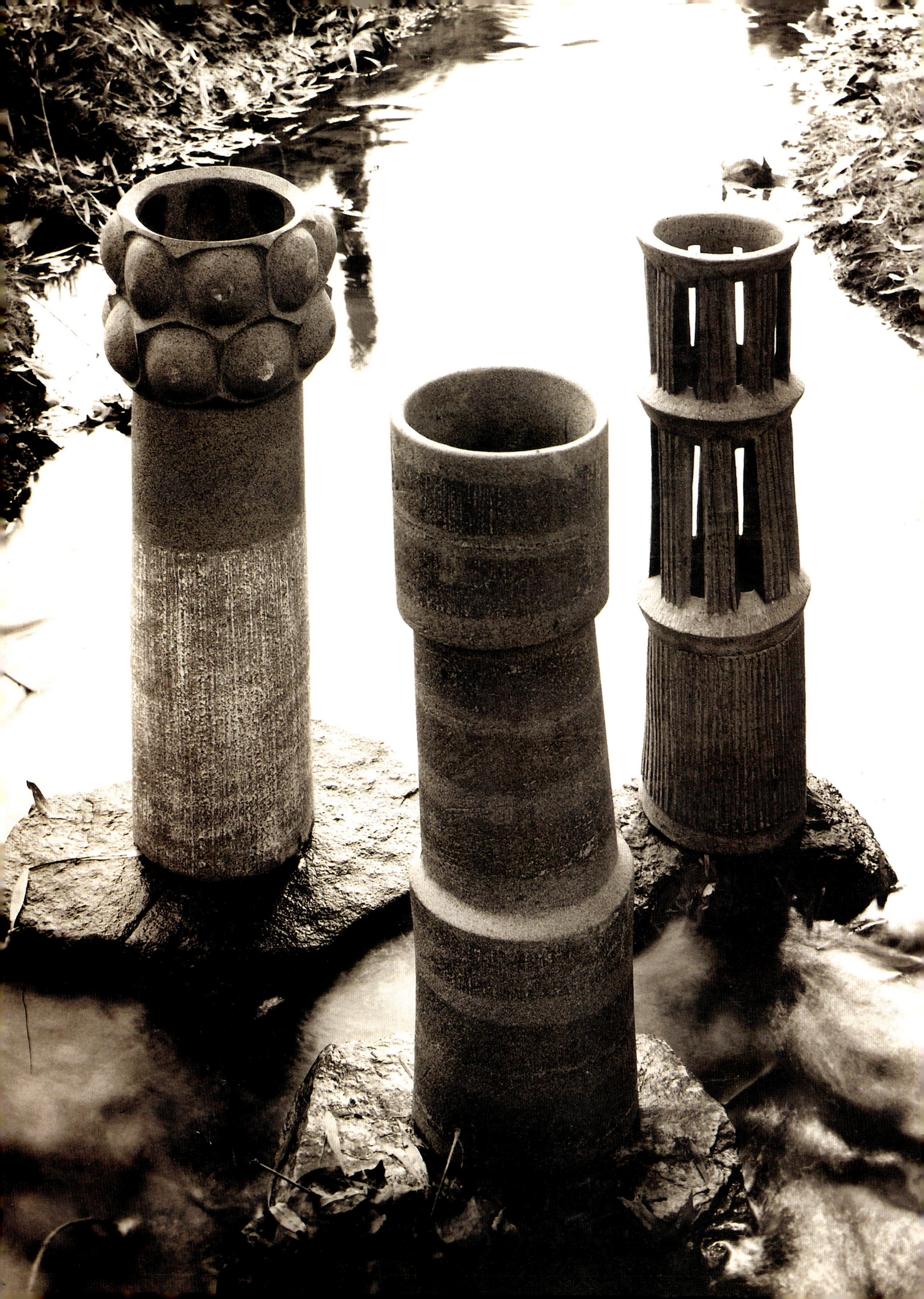

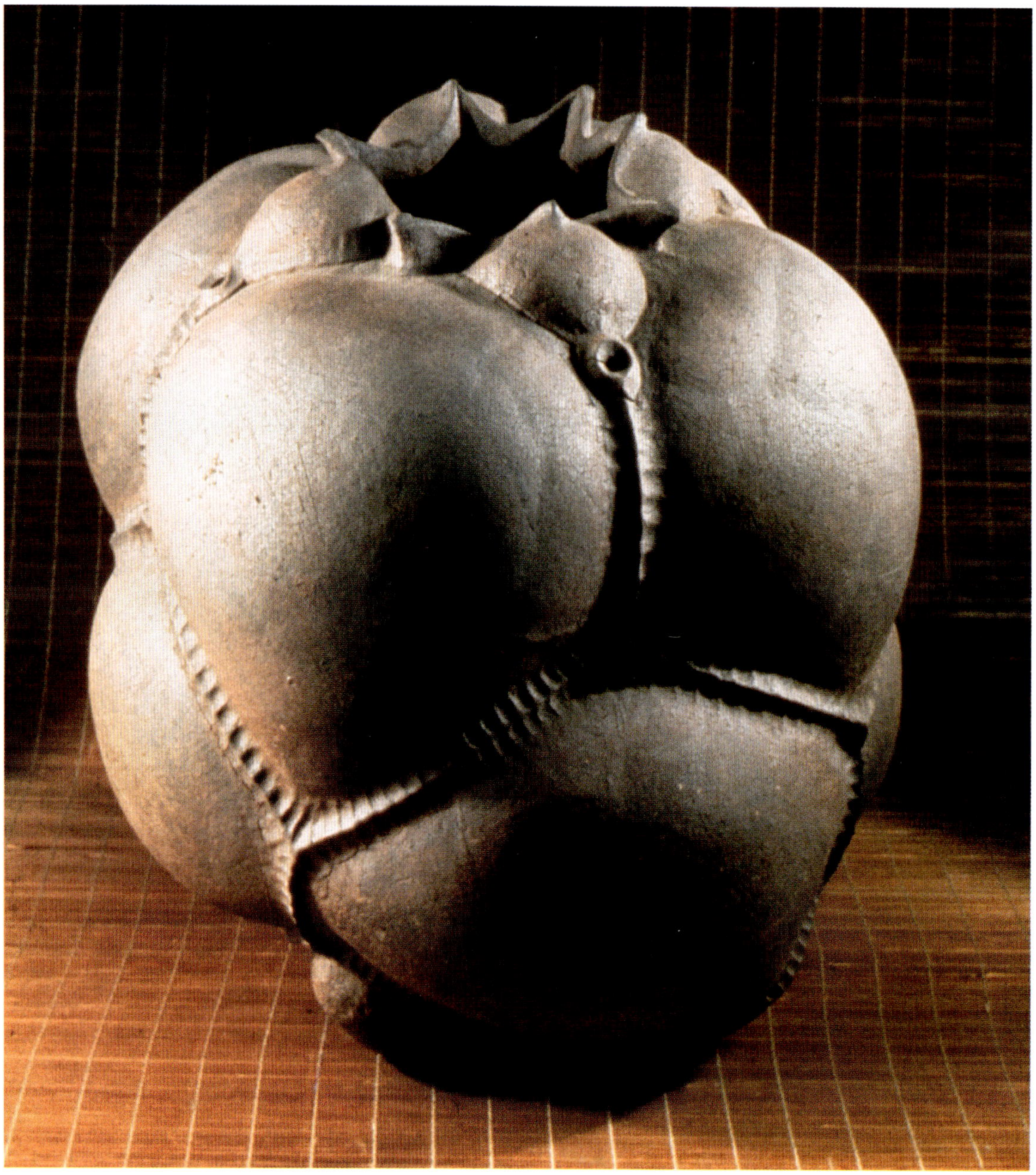

Photo: A Renzetti

Photo: Amanda Schwab

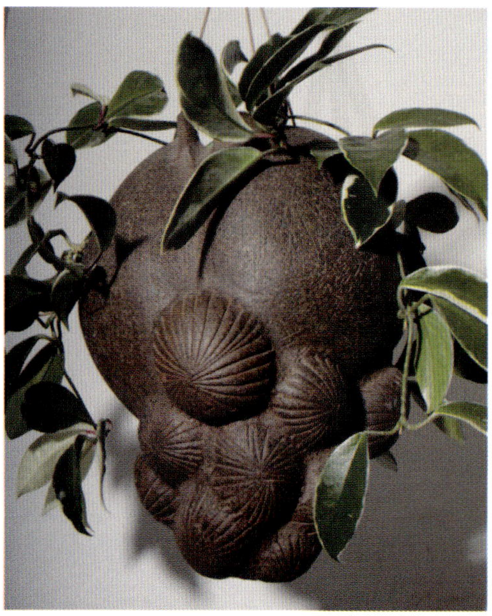

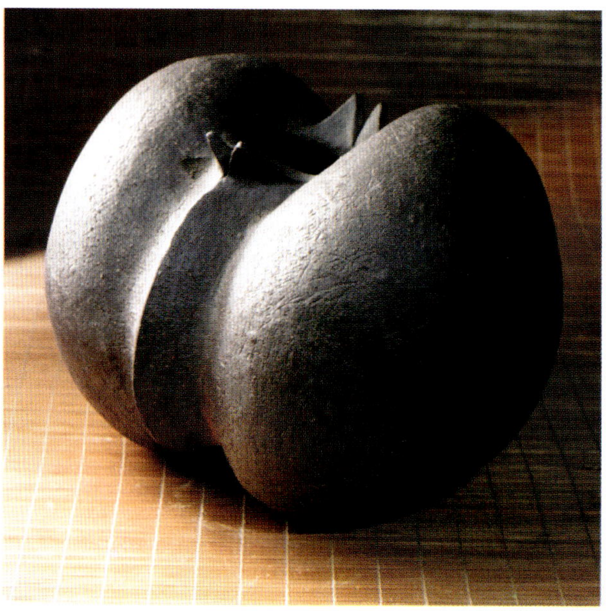

Photo: A Renzetti

Photo: A Renzetti

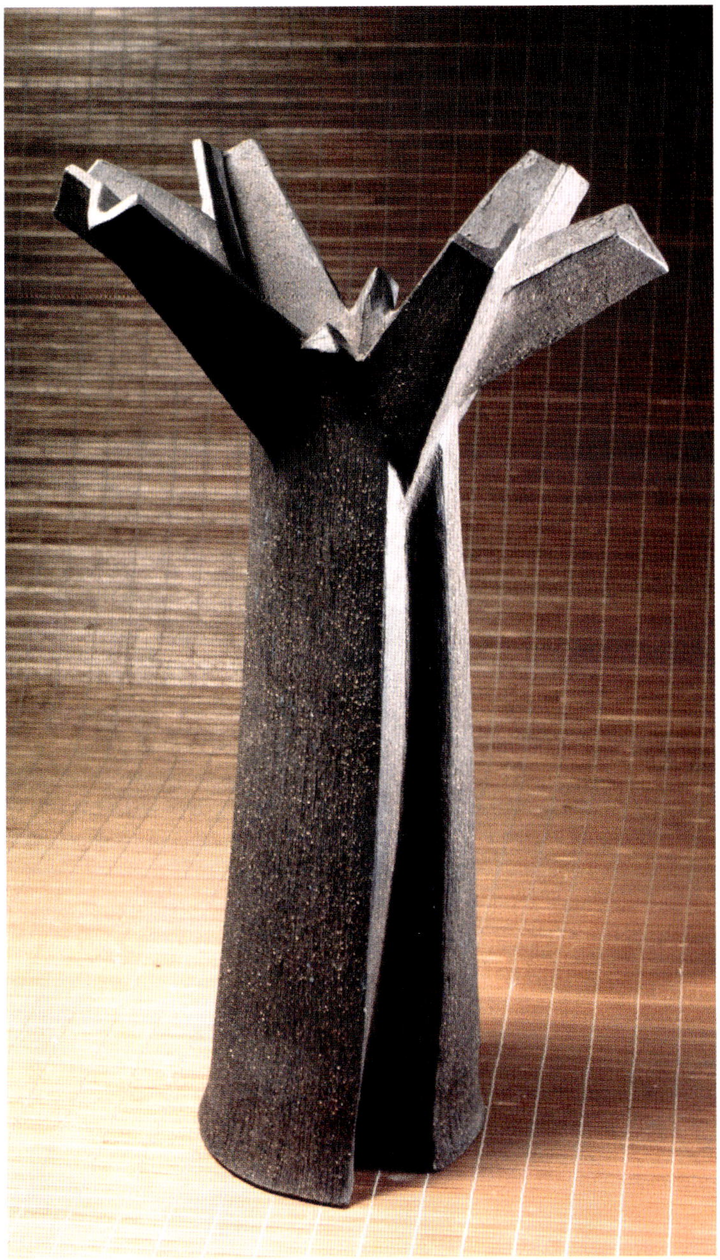

Photo: Alfred Sending

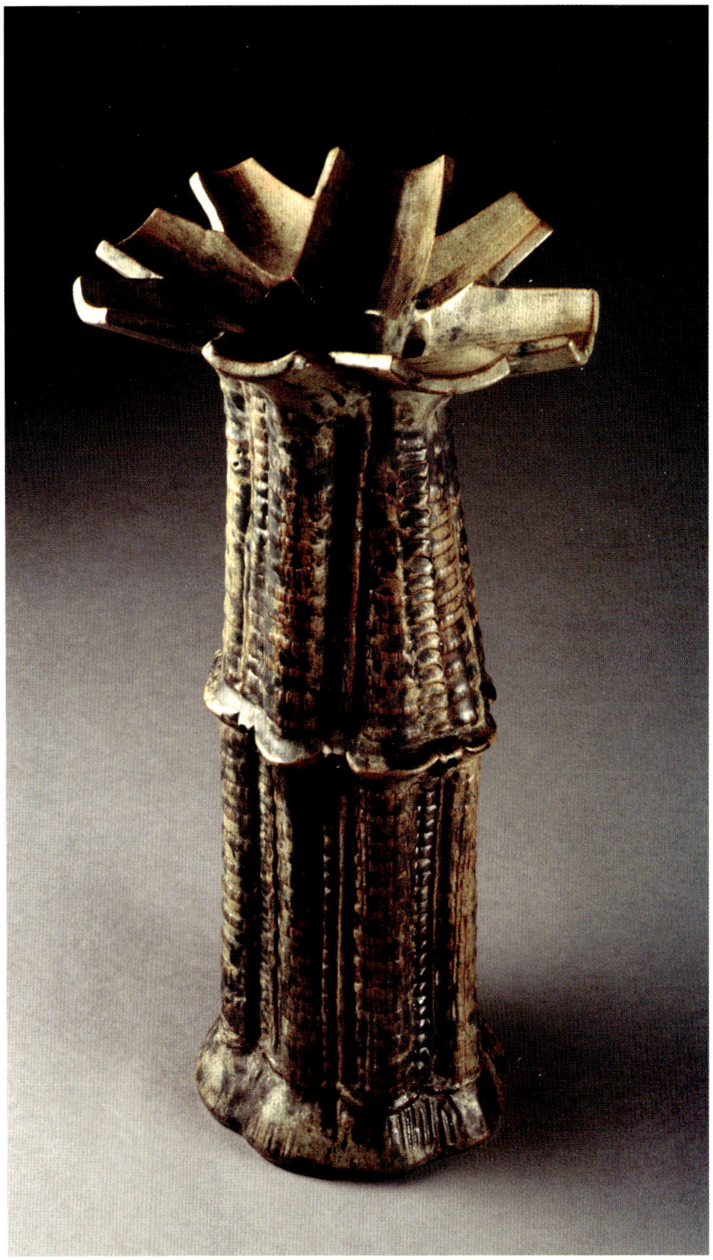

opposite:

Fig. 77, top. *Pod Form,* 1962
Stoneware, valentine clay
16 x 14 x 14 inches
Collection of William and
Catherine Daley

Fig. 78, left. *Hanging Pot,* 1966
Unglazed stoneware
14.5 x 9.6 inches
Collection of William and
Catherine Daley

Fig. 79, right. *Nut Pot,* 1967
Unglazed stoneware
10 x 12 inches

Fig. 80, above left. *Tree Pot,* 1967
Unglazed stoneware
18 x 9 x 9 inches
Private collection

Fig. 81, above right. *Tree Pot,* 1964
18 x 7 x 7 inches
Earthenware, matte glaze
Fredonia, NY
Collection of Alfred and Rica Fenaughty

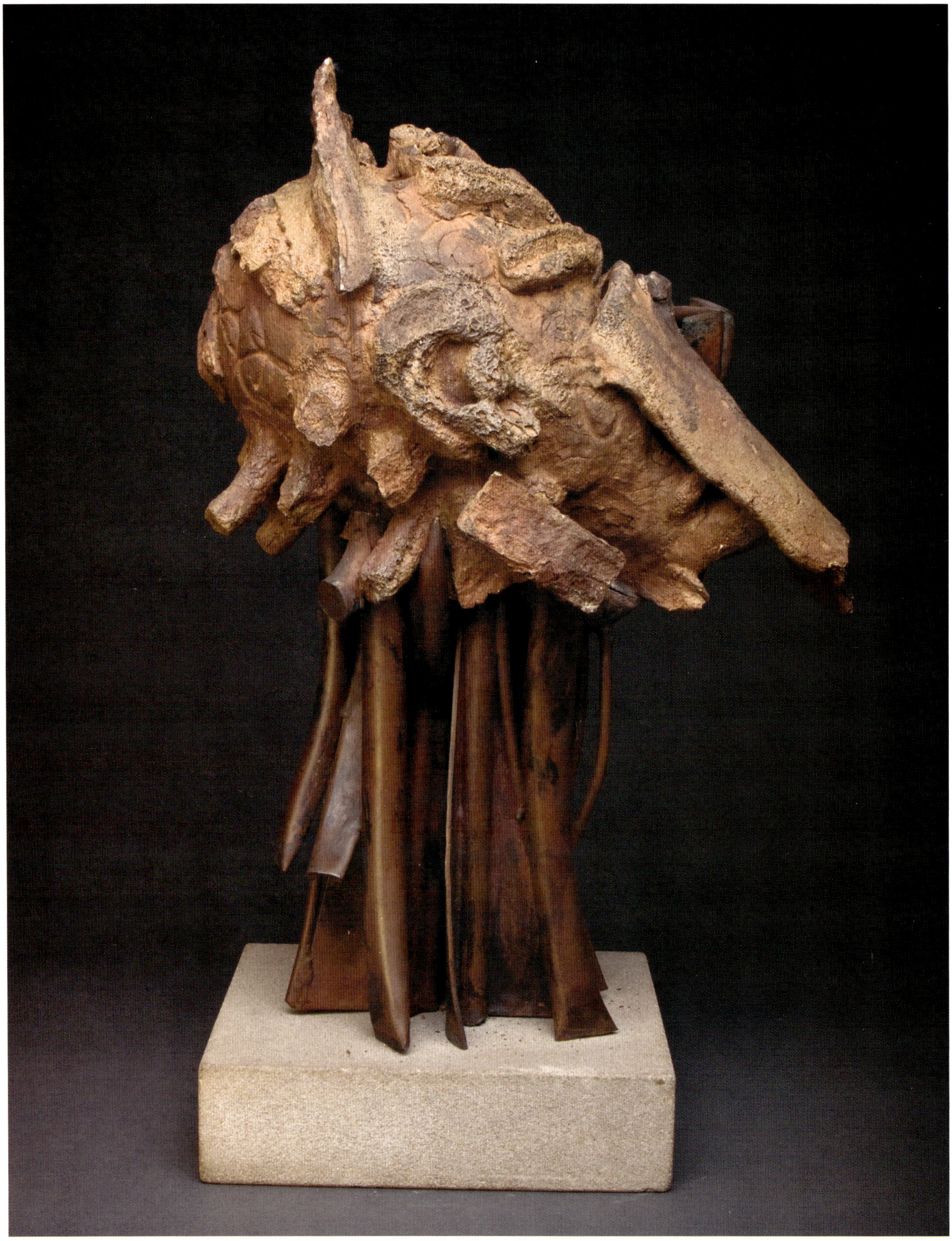

Photo: Amanda Schwab

Fig. 82. *Mantis Head*, 1964
Fire clay, table salt, and spaghetti
combined with brazed copper
18 x 10 x 13 inches
Collection of William and Catherine Daley

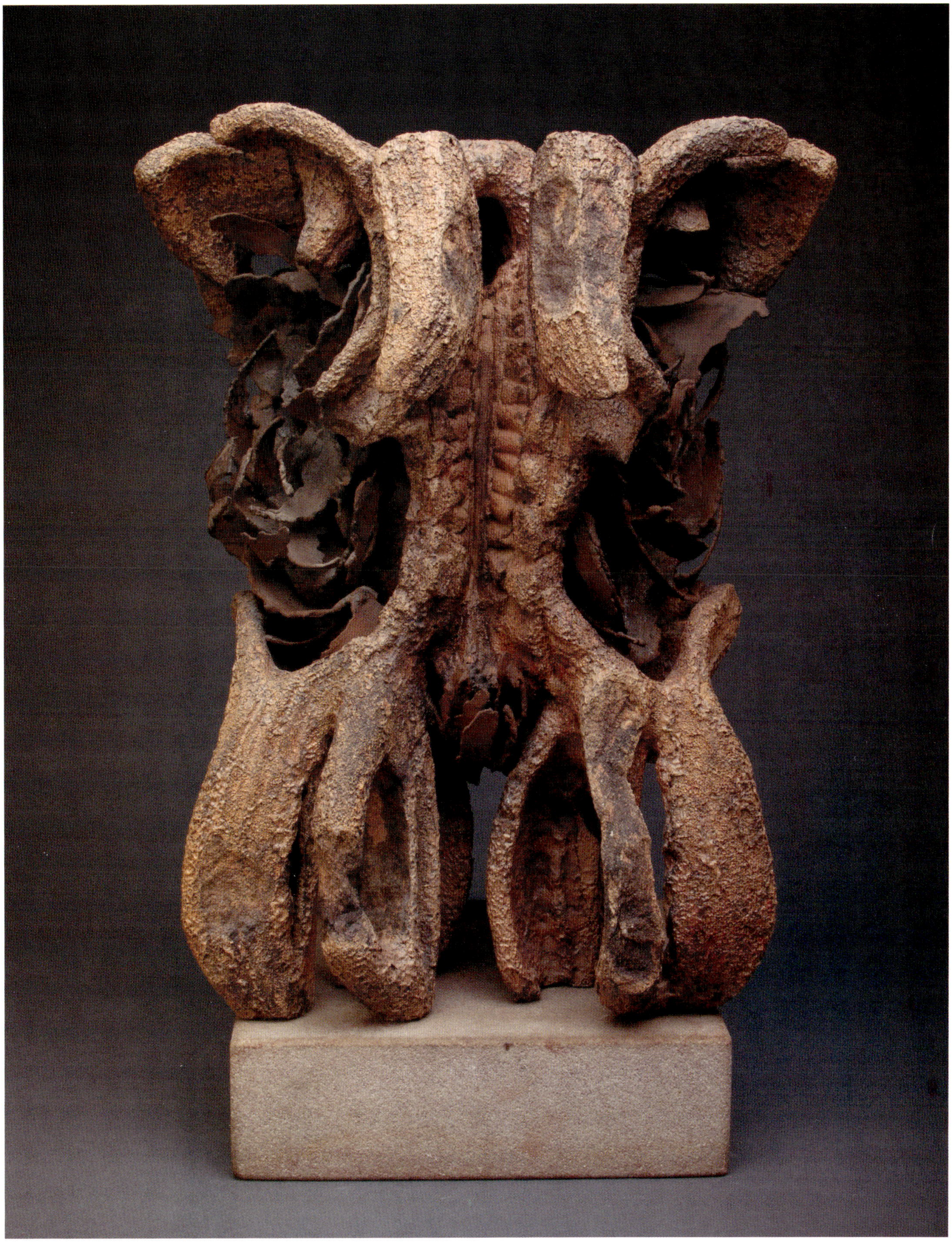

Photo: Amanda Schwab

Fig. 83. *Torso,* 1964
Fire clay, table salt, and spaghetti
combined with welded steel
15 x 12 x 9 inches
Collection of William and Catherine Daley

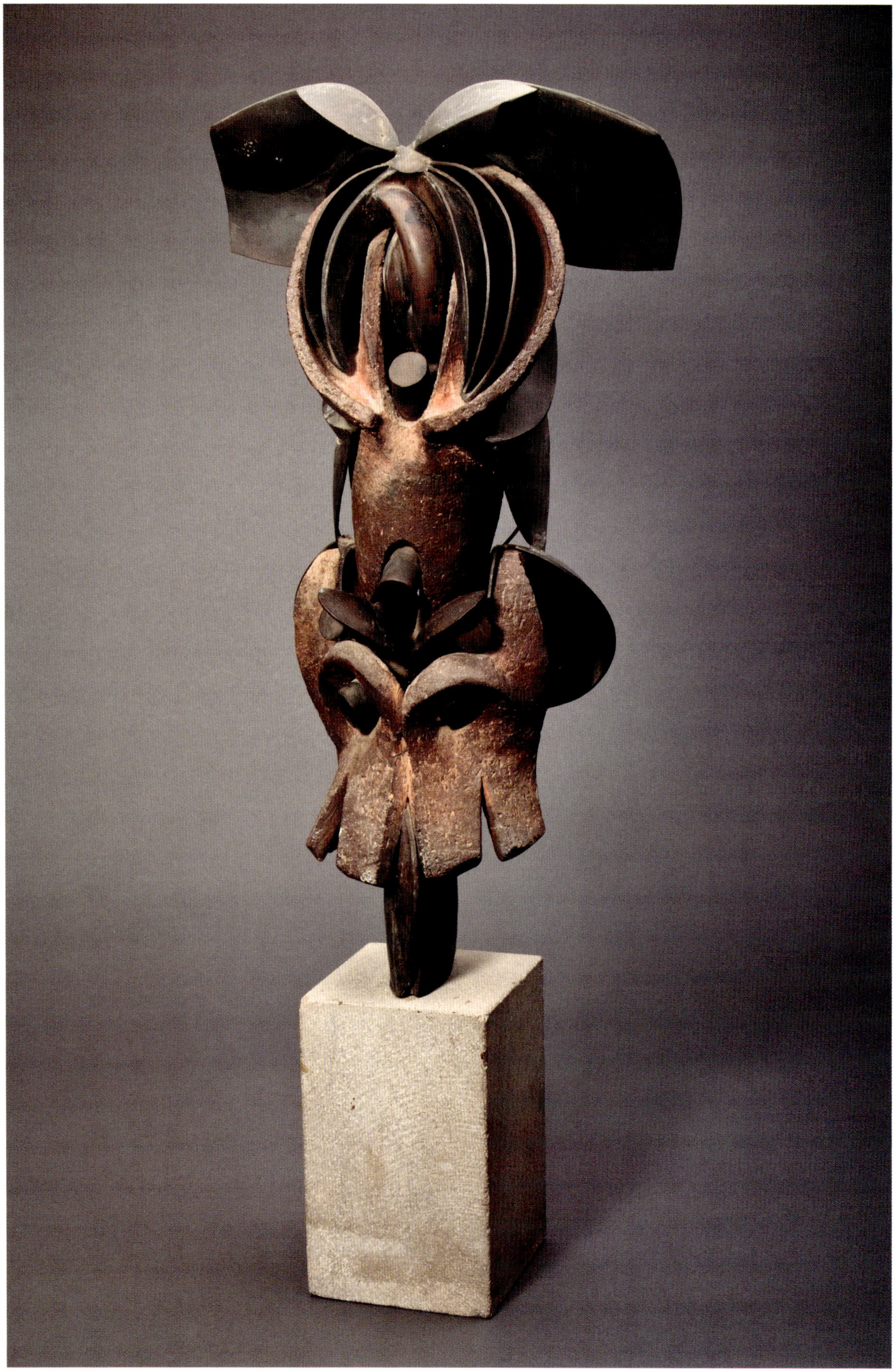
Photo: Amanda Schwab

Fig. 84. *Totemic Figure,* 1966
Sheet bronze and clay
10.75 x 9 x 5 inches
Collection of William and Catherine Daley

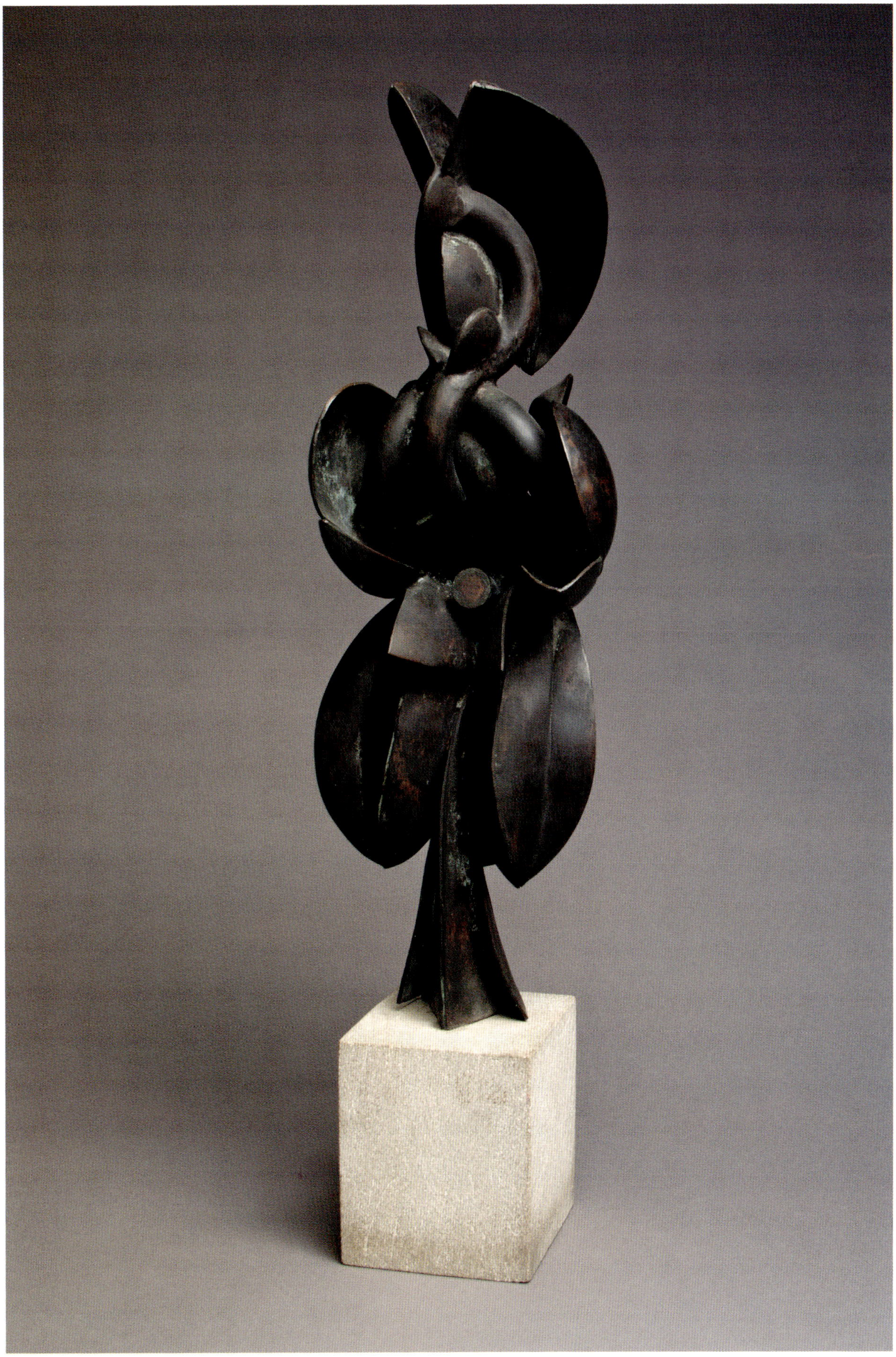

Photo: Amanda Schwab

Fig. 85. *Figure,* 1967
Sheet bronze, brazed
16 x 5.5 x 5 inches
Private Collection

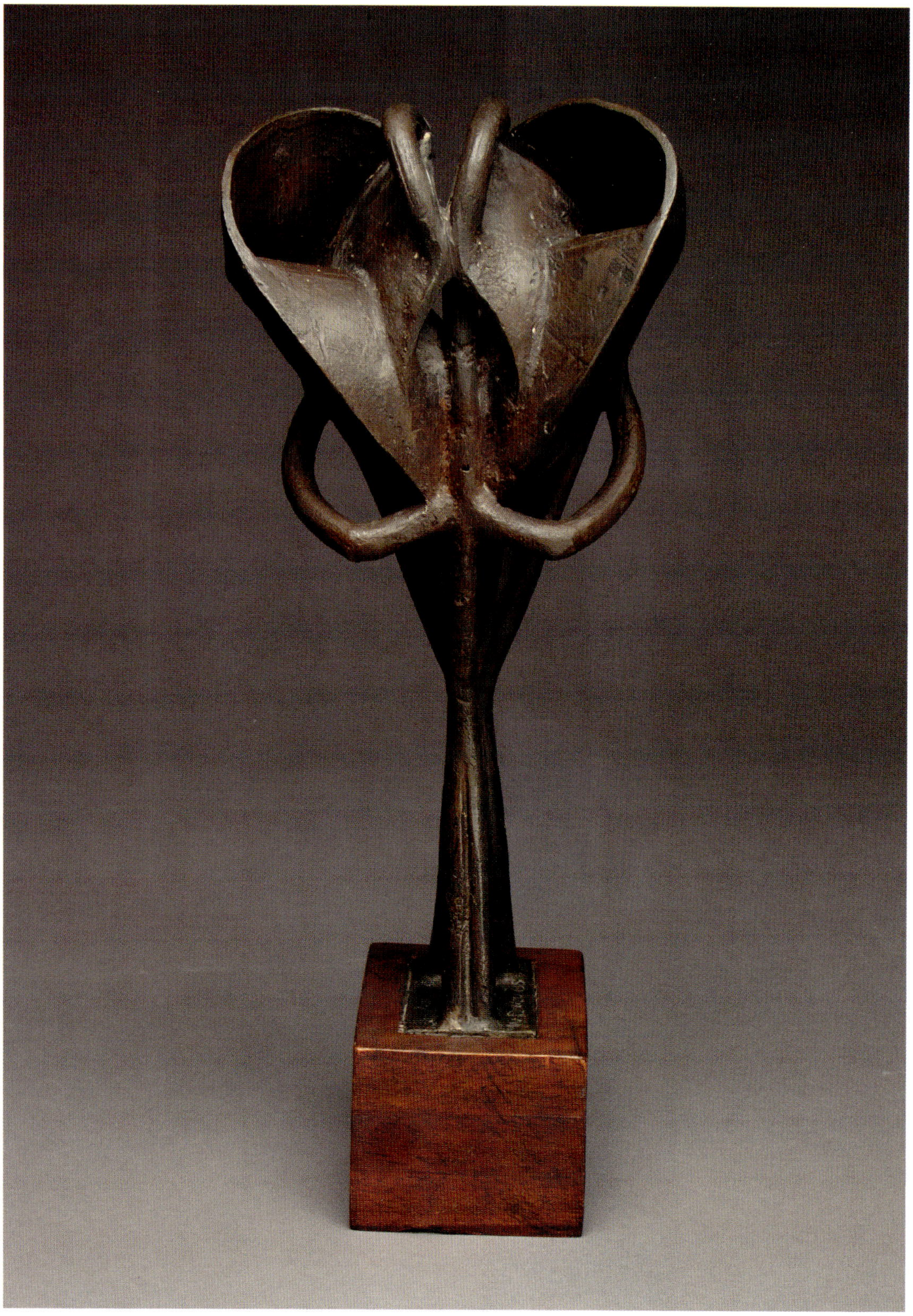

Photo: Amanda Schwab

opposite:

Fig. 87, top left.
Figure, 1966
Tomisil bronze, cast
17 x 7 x 3.25 inches
Collection of Barbara Forth

Fig. 88, top right.
Figure, 1966
Tomisil bronze, cast
9 x 2.5 x 2.5 inches
Cast in PCA foundry
Collection of Dean and
Geraldine Schwarz

Fig. 89, bottom left.
Figure, 1966
Tomisil bronze, cast
9.5 x 3.5 x 3.5 inches
Reinhardt family collection

Fig. 90, bottom right.
Head, 1966
Tomisil bronze, cast
9.25 x 8 x 6 inches
Collection of William and
Catherine Daley

Fig. 86, above. *Figure,* 1966
Tomisil Bronze, cast
10.5 x 5.25 x 3.5 inches
Private collection

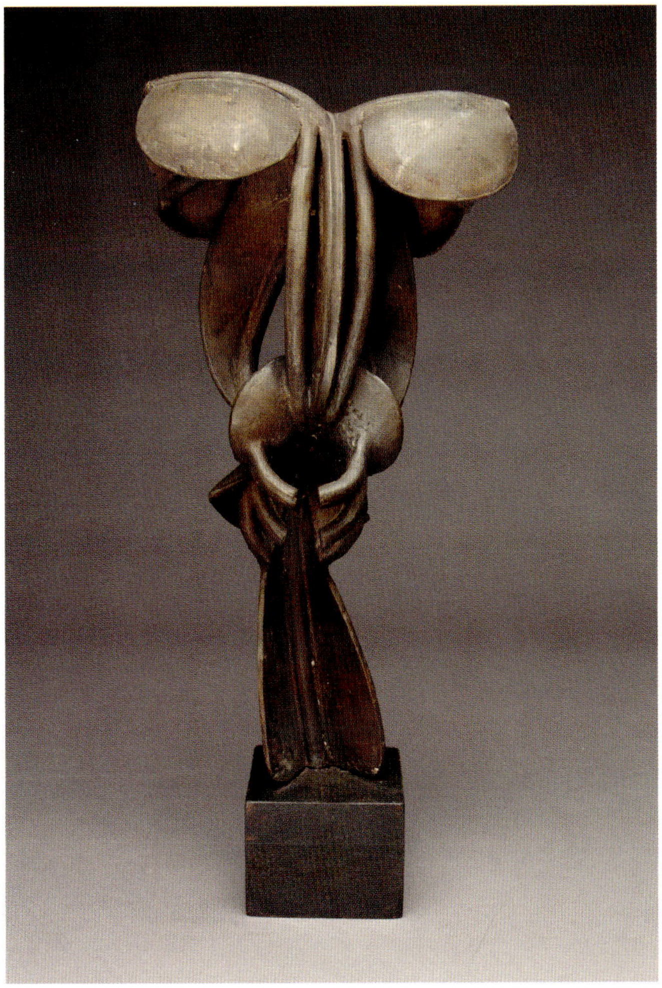
Photo: Amanda Schwab

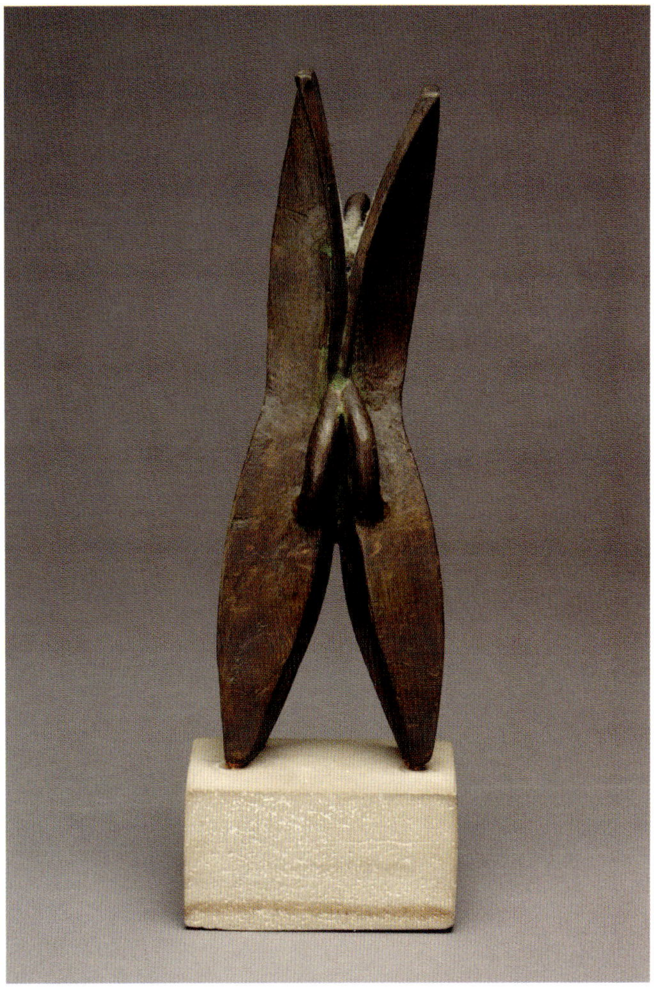
Photo: Amanda Schwab

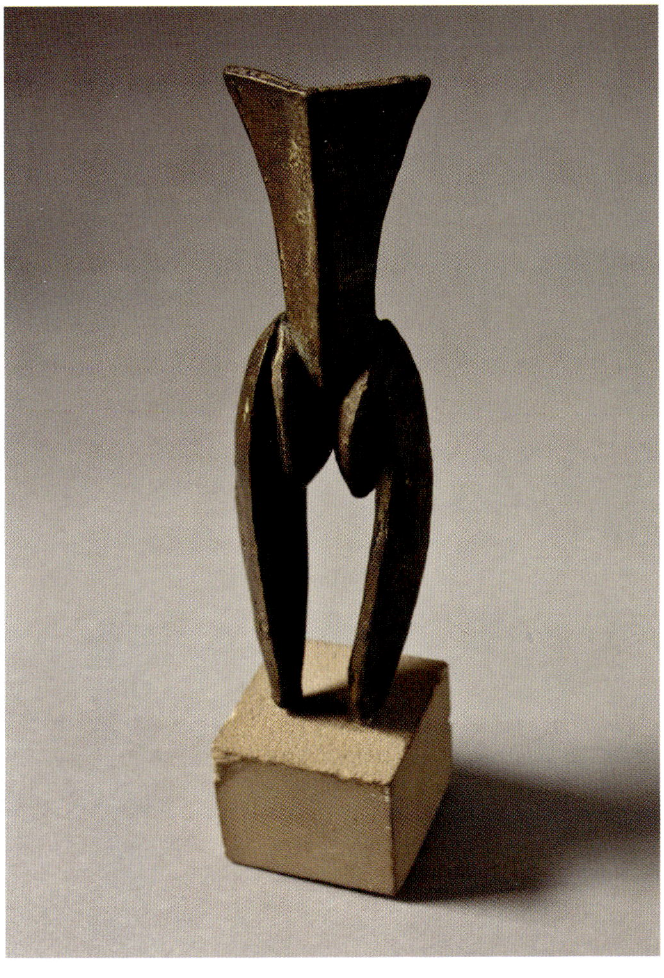
Photo: Amanda Schwab

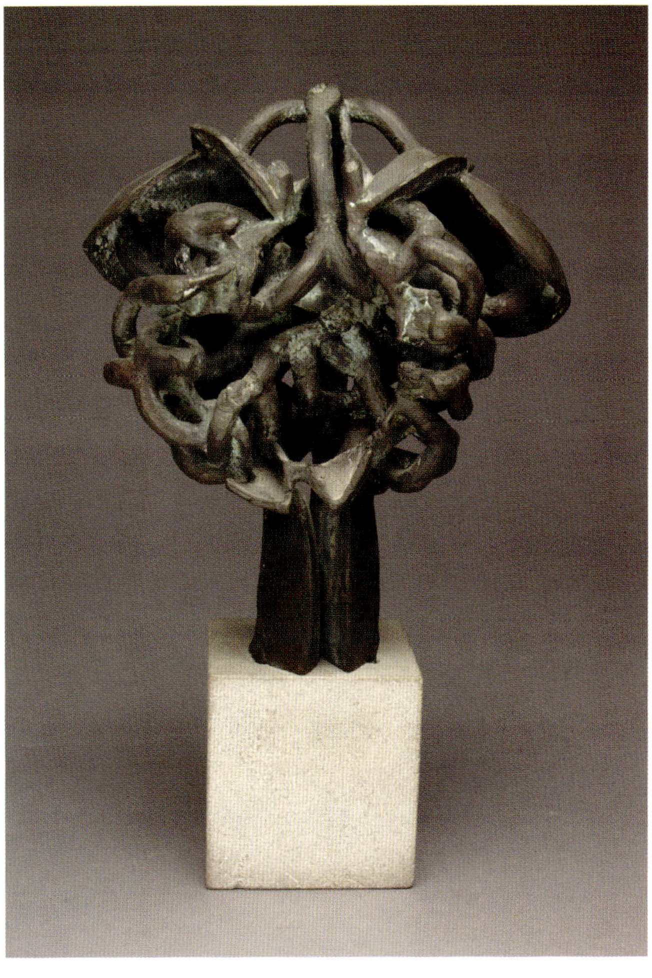
Photo: Amanda Schwab

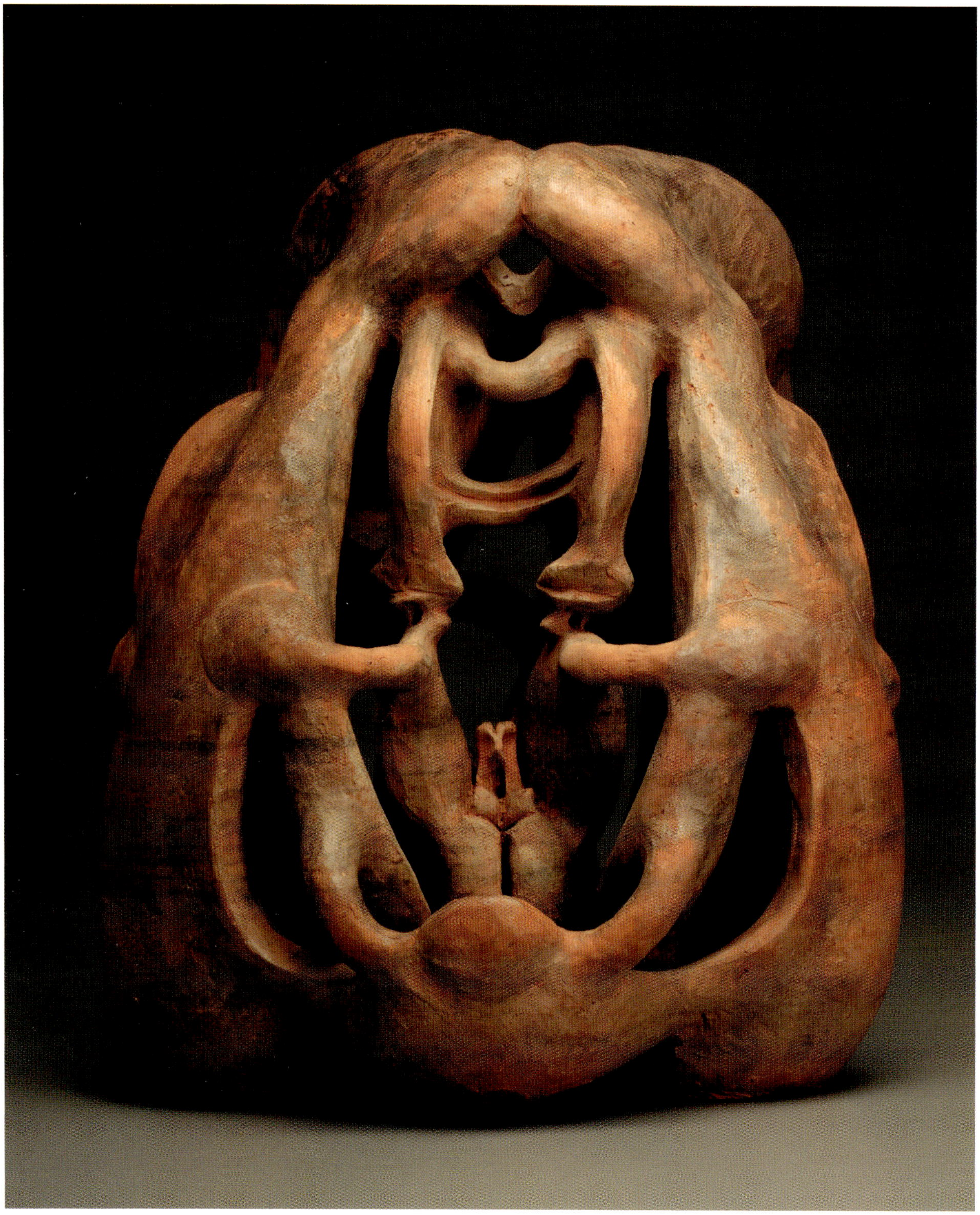

Fig. 91. *Vascular Professor,* 1964
16.5 x 13.25 x 14 inches
Fredonia, New York
Collection of William and Catherine Daley

Fig. 92, opposite. *Diana,* 1964 – 65
Laminate Red Oak carved and constructed
60 inches height
Fredonia, NY
Permanent collection of
St. Lawrence University, Canton, NY

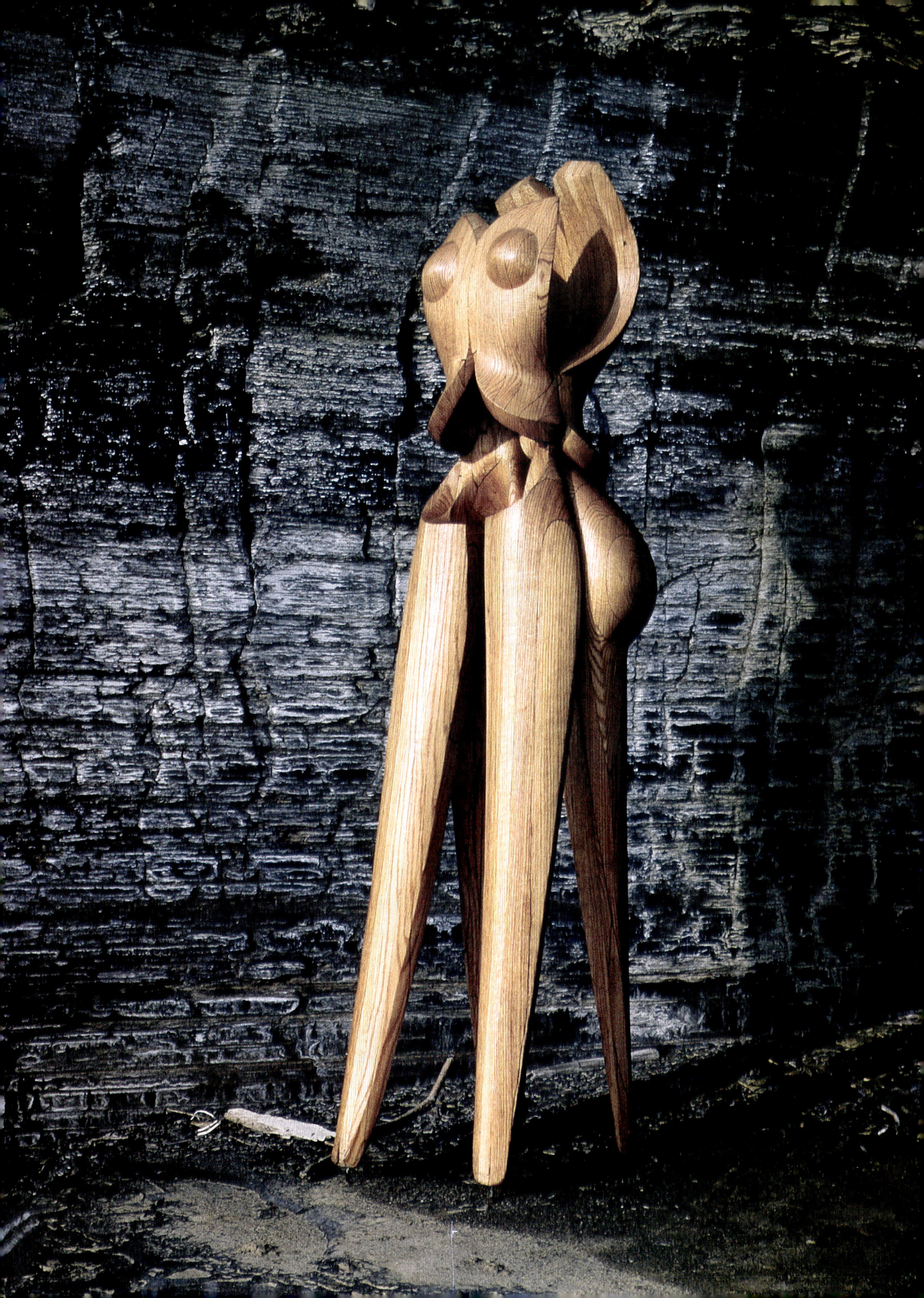

Photo: James Quaile

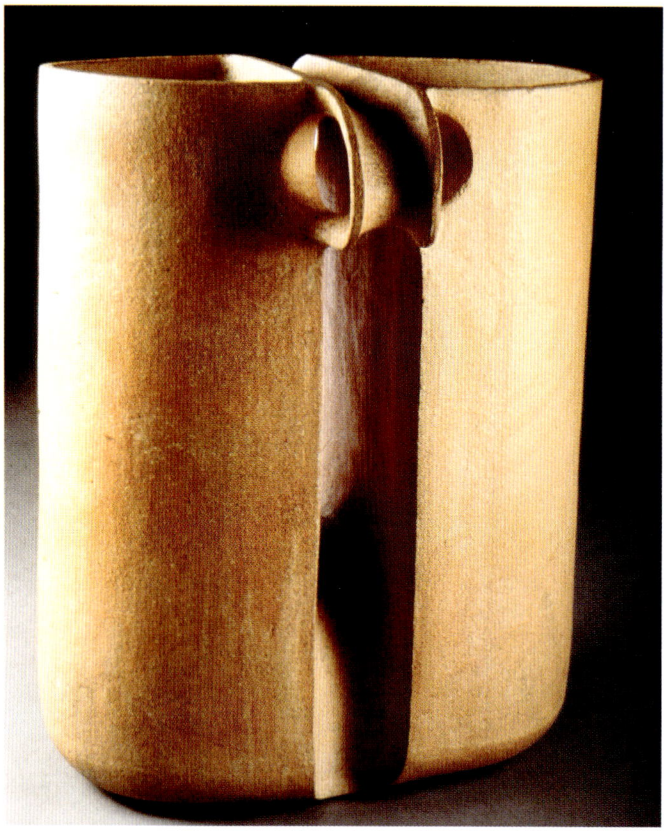

Photo: Ave Pildis

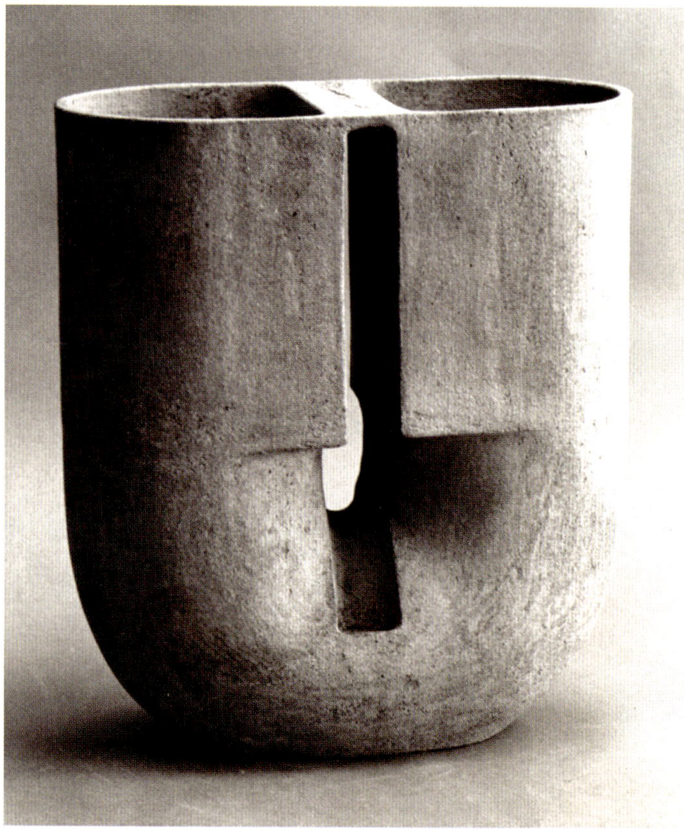

Photo: Amanda Schwab

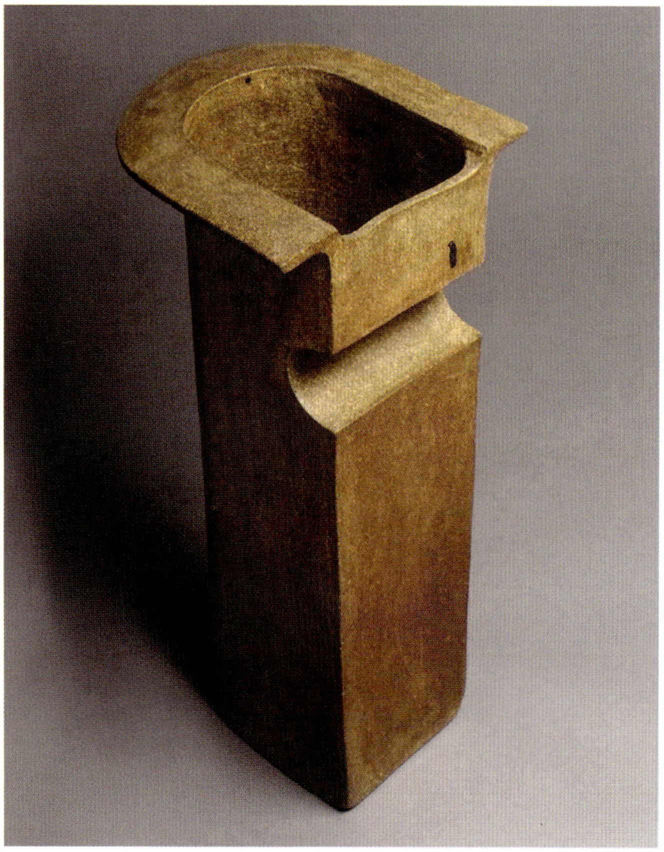

Photo: Amanda Schwab

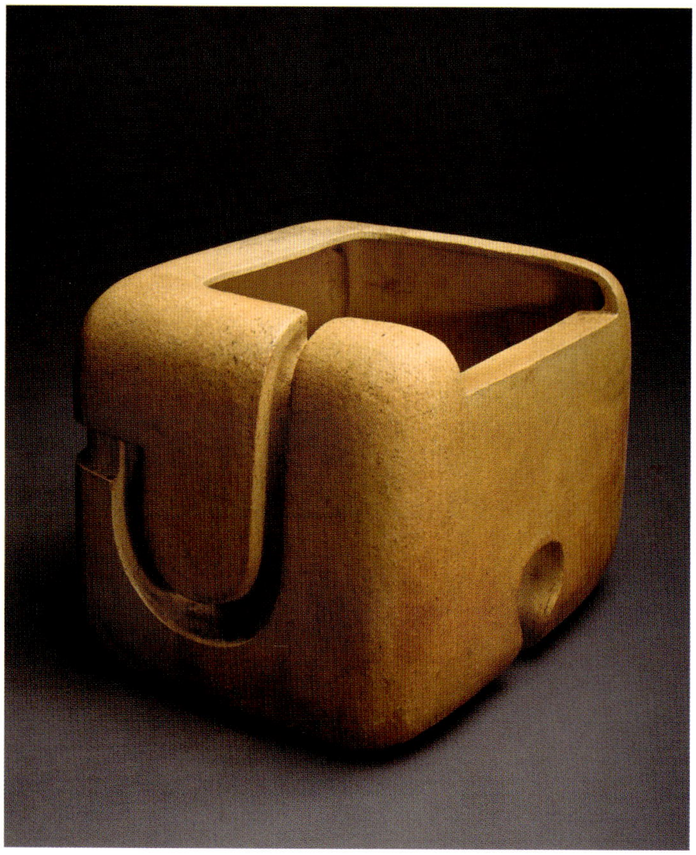

Fig. 93, top left. *U Form,* 1968
Unglazed stoneware
18 x 14.5 x 14.5 inches
Reinhardt family collection

Fig. 94, bottom left. *Floor Pot,* 1968
Unglazed stoneware
20.5 x 9 x 9 inches
Collection of Helen Drutt English
Pot was part of the PCPC show

Fig. 95, top right. *U Form*, 1968
Unglazed stoneware
16 x 13 x 8 inches approximately
No longer extant

Fig. 96, bottom right. Untitled, 1968
12.5 x 14 x 14.5 inches
Unglazed stoneware, Jersey brick clay
Collection of William and Catherine Daley

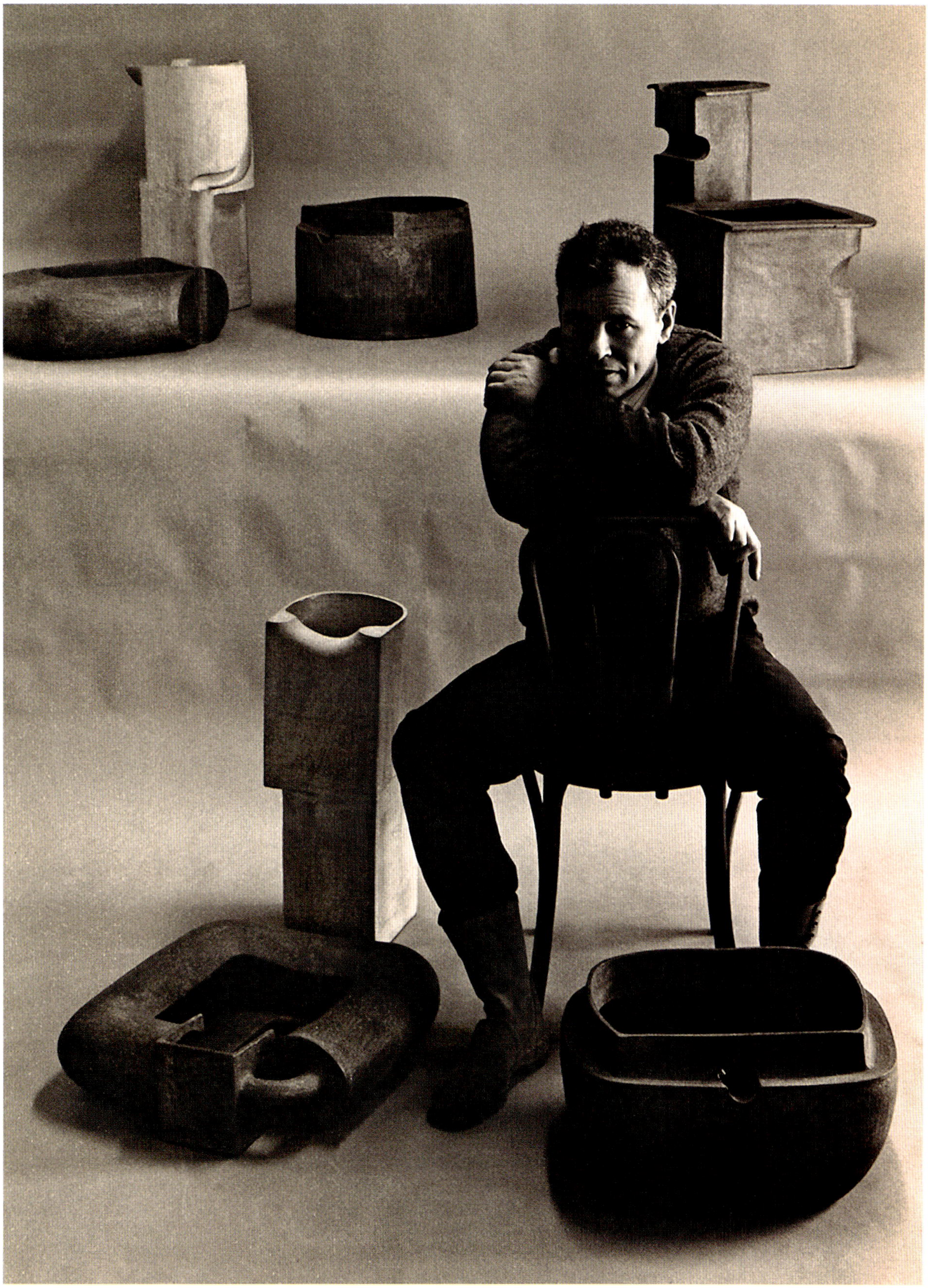

Photo: Ave Pildis

Fig. 97. Section of poster for *William Daley* show at the Philadelphia Art Alliance, sponsored by Philadelphia Council of Professional Craftsmen (PCPC), 1969

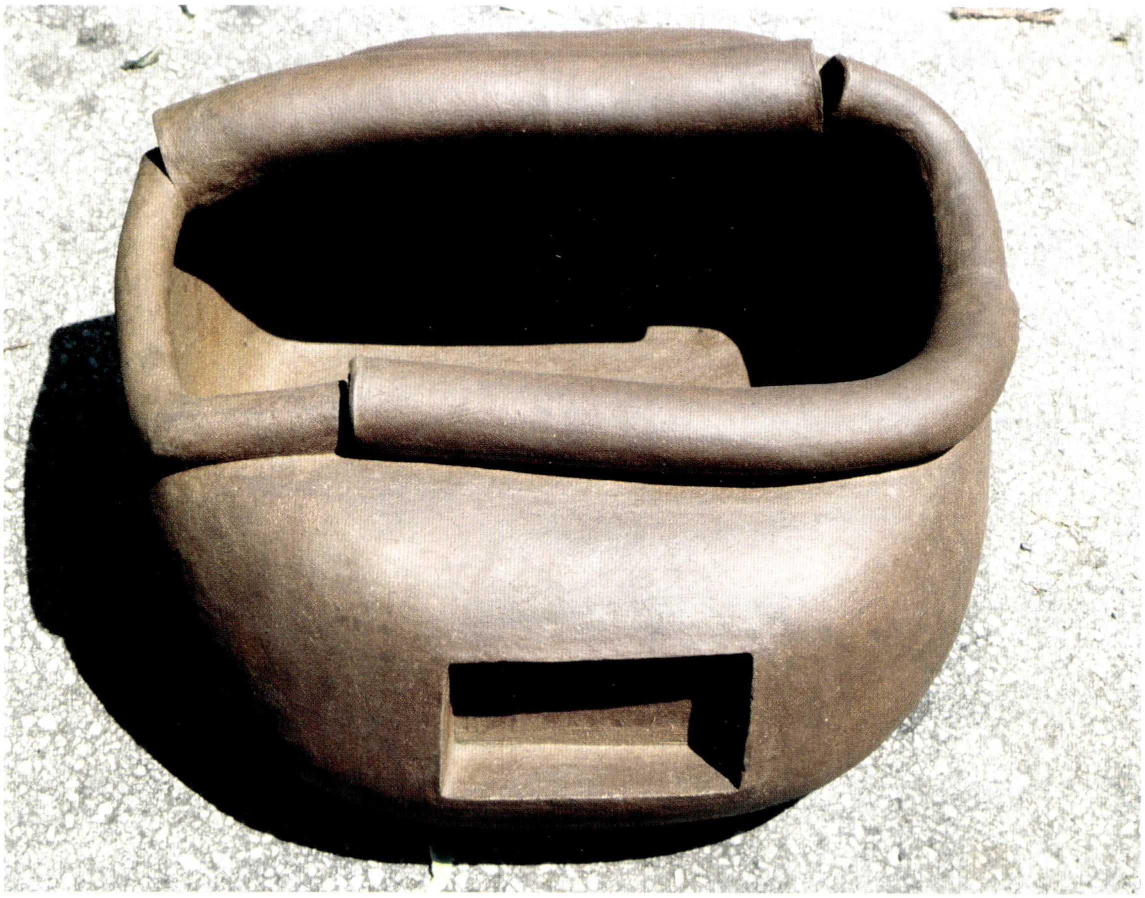

Photos: William Daley

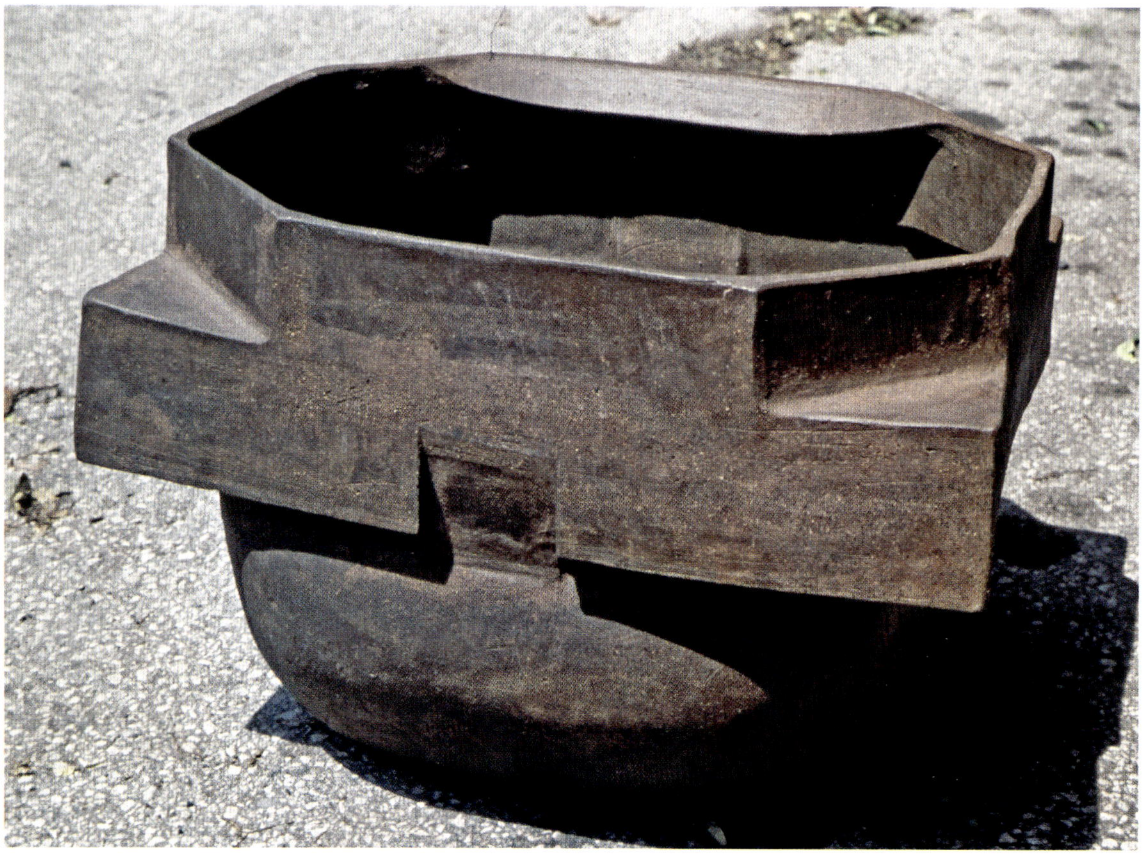

Fig. 98, top. *Relaxed Chiclet,* 1967
Unglazed stoneware
13 x 24 x 24 inches
No longer extant

Fig. 99, above. *Composite Form*, 1967
Unglazed stoneware
15.5 x 21 x 19 inches
Formally in collection of the Philadelphia Civic Center Museum
Currently in the collection of William and Catherine Daley

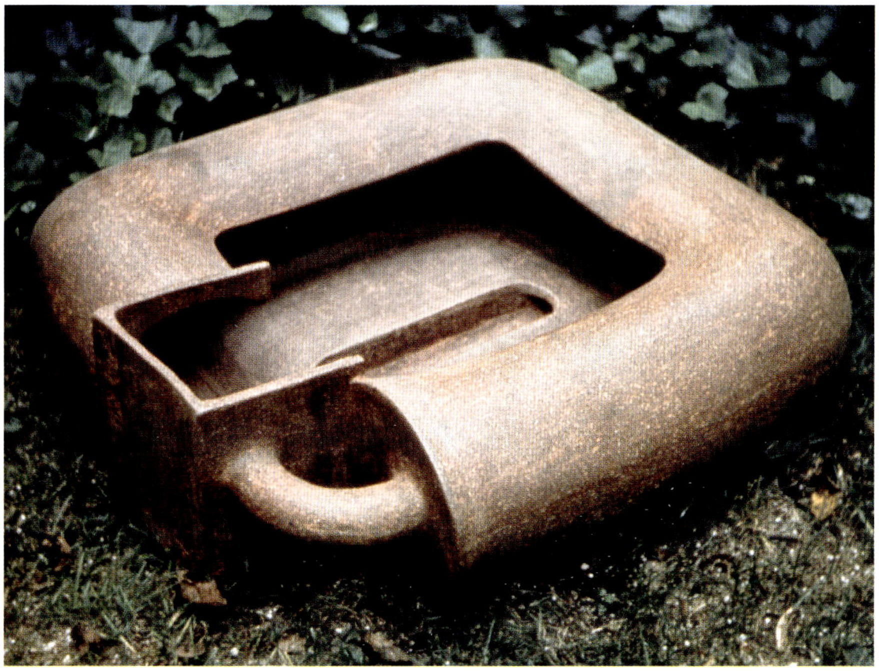

Photos: William Daley

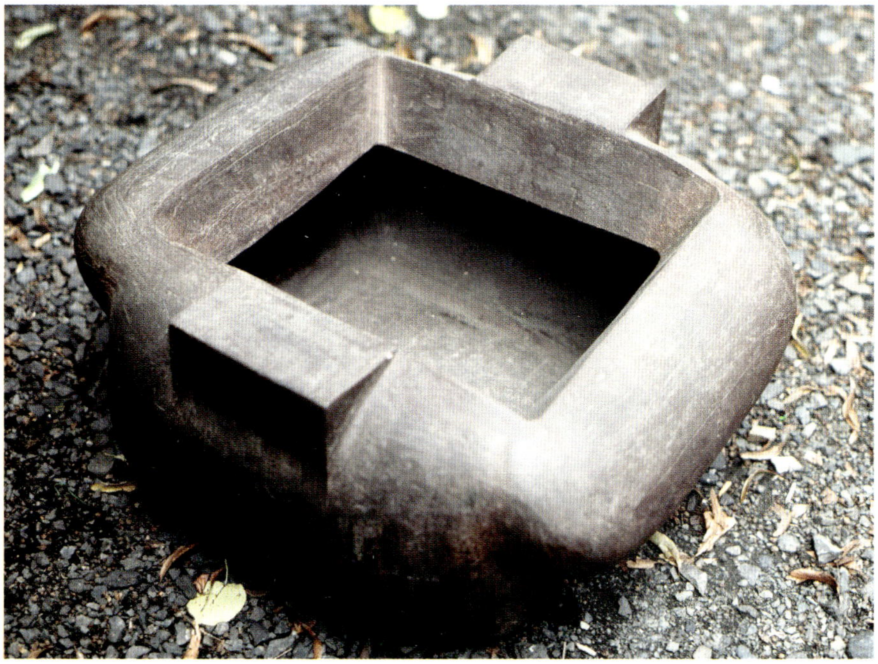

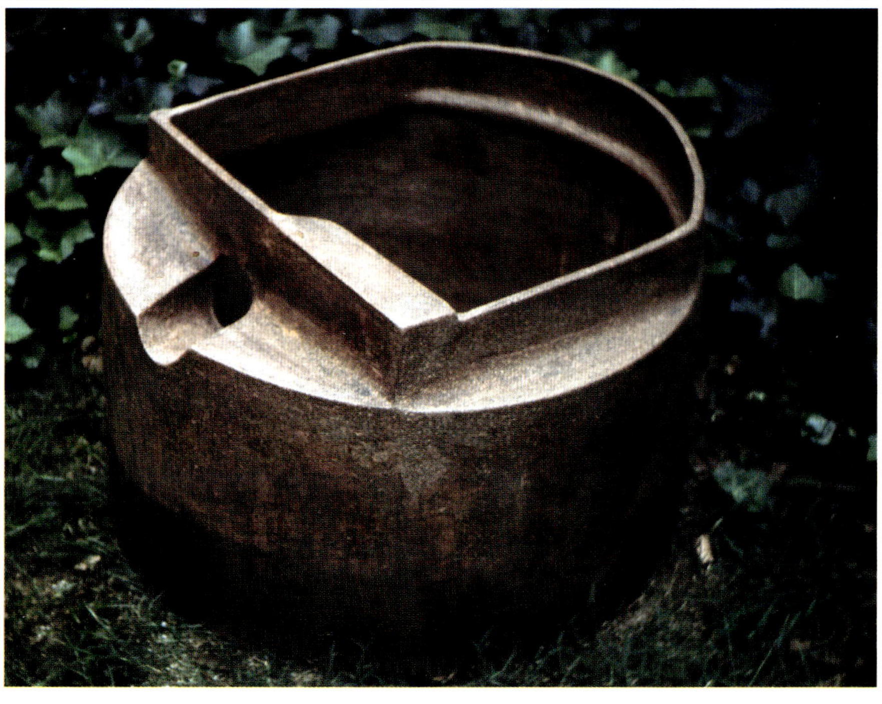

Fig. 100, top. *Reptilian Chiclet*, 1968
Unglazed stoneware
10 x 21.5 x 20.5 inches
William Daley exhibition Philadelphia Art Alliance
Philadelphia Council of Professional Craftsmen
Collection of William and Catherine Daley

Fig. 101, middle. *Chiclet 3*, 1968
Unglazed stoneware
11 x 21 x 24 inches
Collection of Everson Museum of Art
Syracuse, NY
25th Ceramic National Exhibition
Given by Thomas and Donna Weaverling Daley

Fig. 102, bottom. *Tank Pot*, 1968
Unglazed stoneware
11 x 14 x 12 inches
Solo Exhibition: *William Daley*
at the Art Alliance, sponsored by the Philadelphia
Council of Professional Craftsmen
Collection of Jennifer Pickett Duffy
from the John Pickett collection

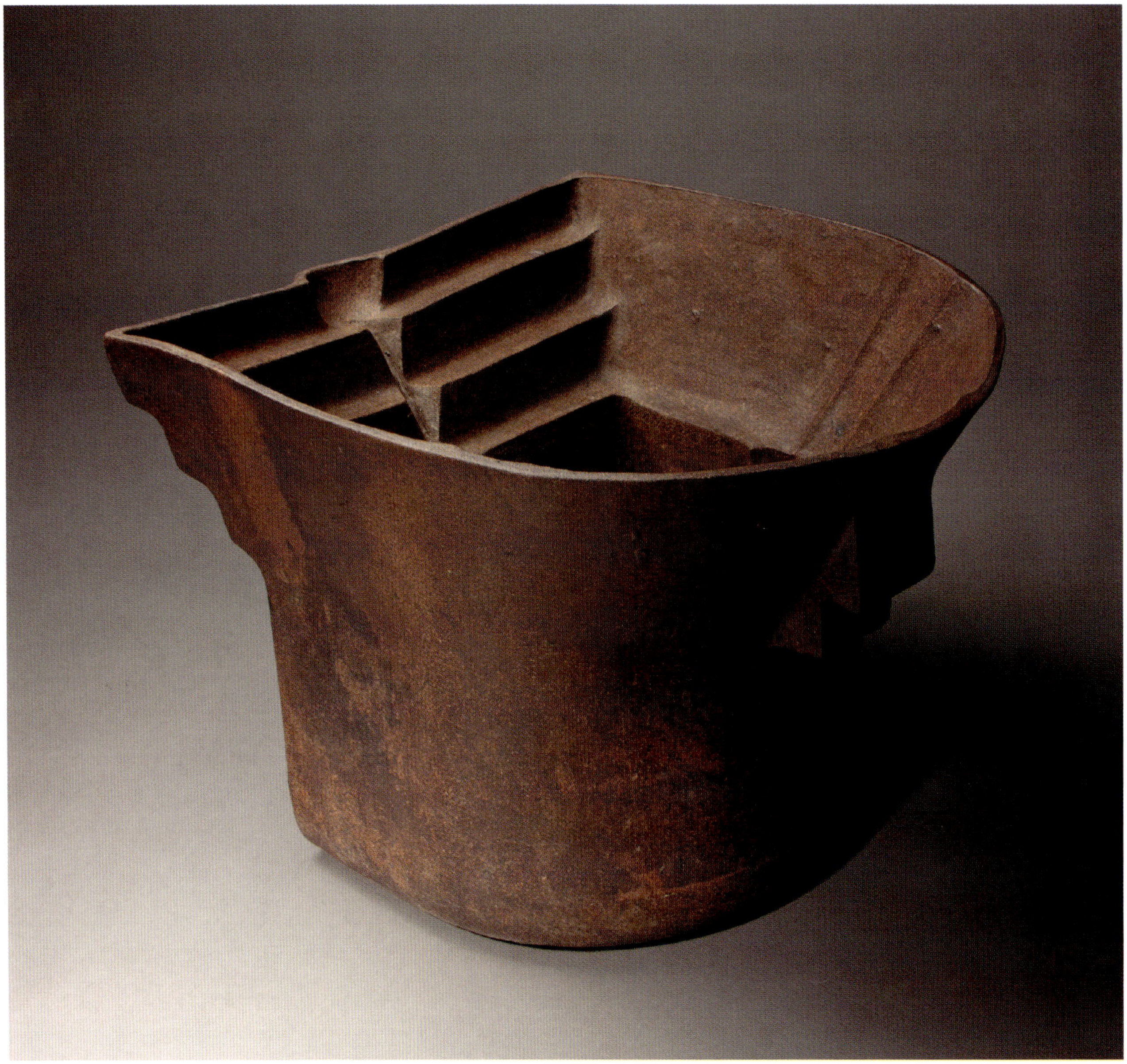

Photos: Amanda Schwab

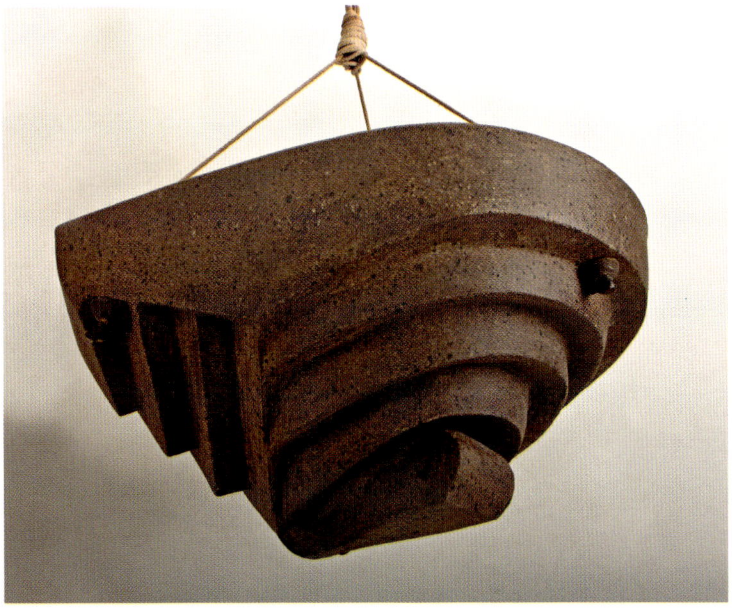

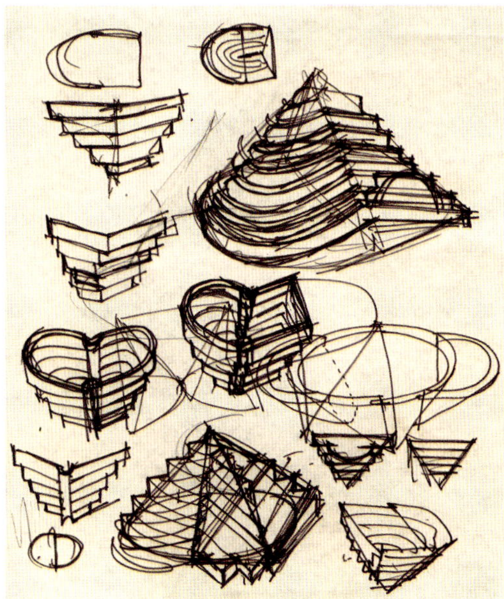

Fig. 103, top. *Floor Pot,* 1969
Unglazed stoneware
17 x 20 x 24 inches
Collection of Alice Daley LaPier

Fig. 104, above. *Hanging Pot*
Unglazed stoneware, 1970
9 x 17.5 x 19 inches
Collection of Drs. Harry and Betty Gotlieb

Fig. 105, above. Drawing hanging pot study
Pen on paper
11 x 8.5 inches
Collection of William and Catherine Daley

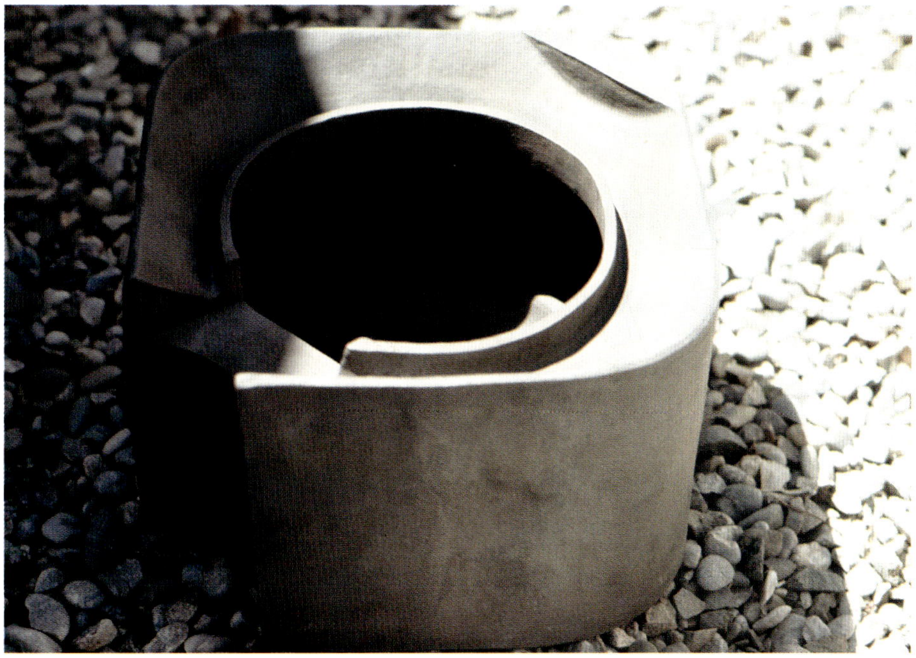

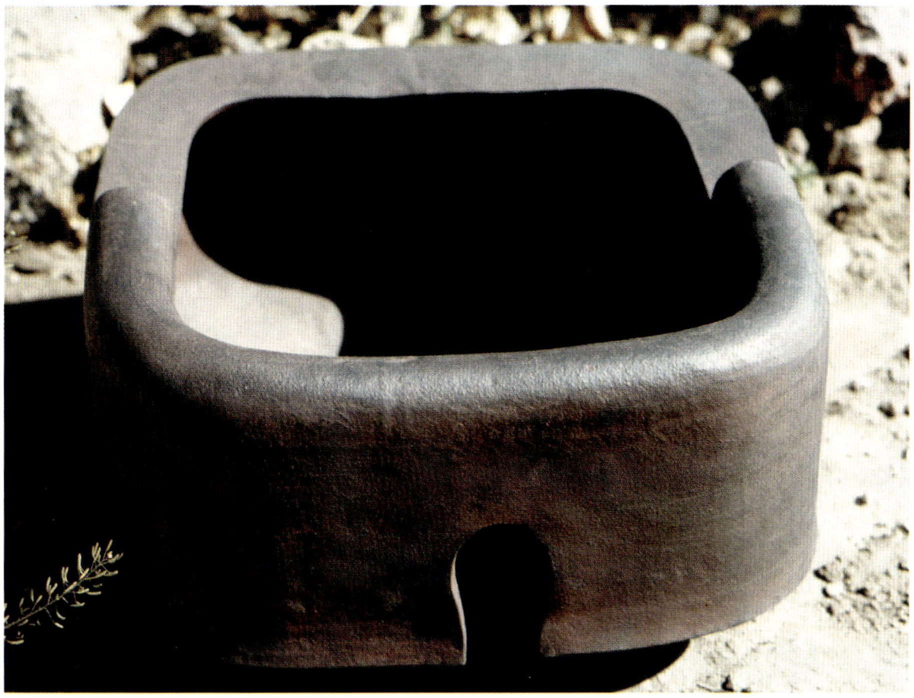

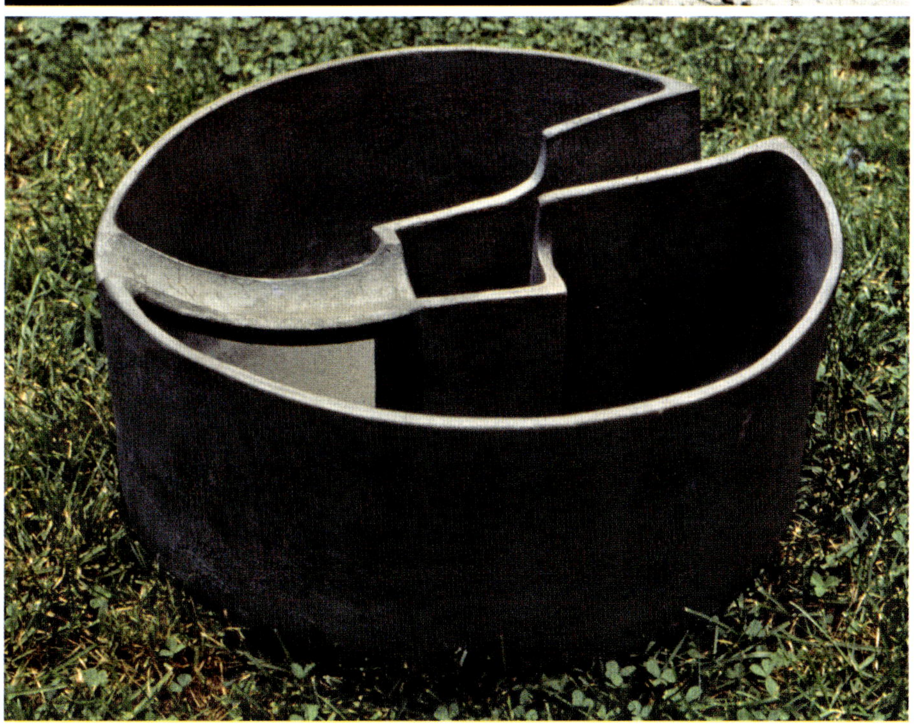

Photos: William Daley

Fig. 106, top. *Albuquerque Pot,* 1972
Unglazed stoneware
Graphite burnished
10 x 16 x 16 inches
Private collection

Fig. 107, middle. *Albuquerque Mesa Pot Chiclet,* 1972
Unglazed stoneware
Graphite burnished
10 x 16 x 16 inches

Fig. 108, bottom. *Canyon Pot,* 1970
Unglazed stoneware
7 x 15 inch diameter
Peace Exhibition
Philadelphia Museum of Art
Collection of David Pease

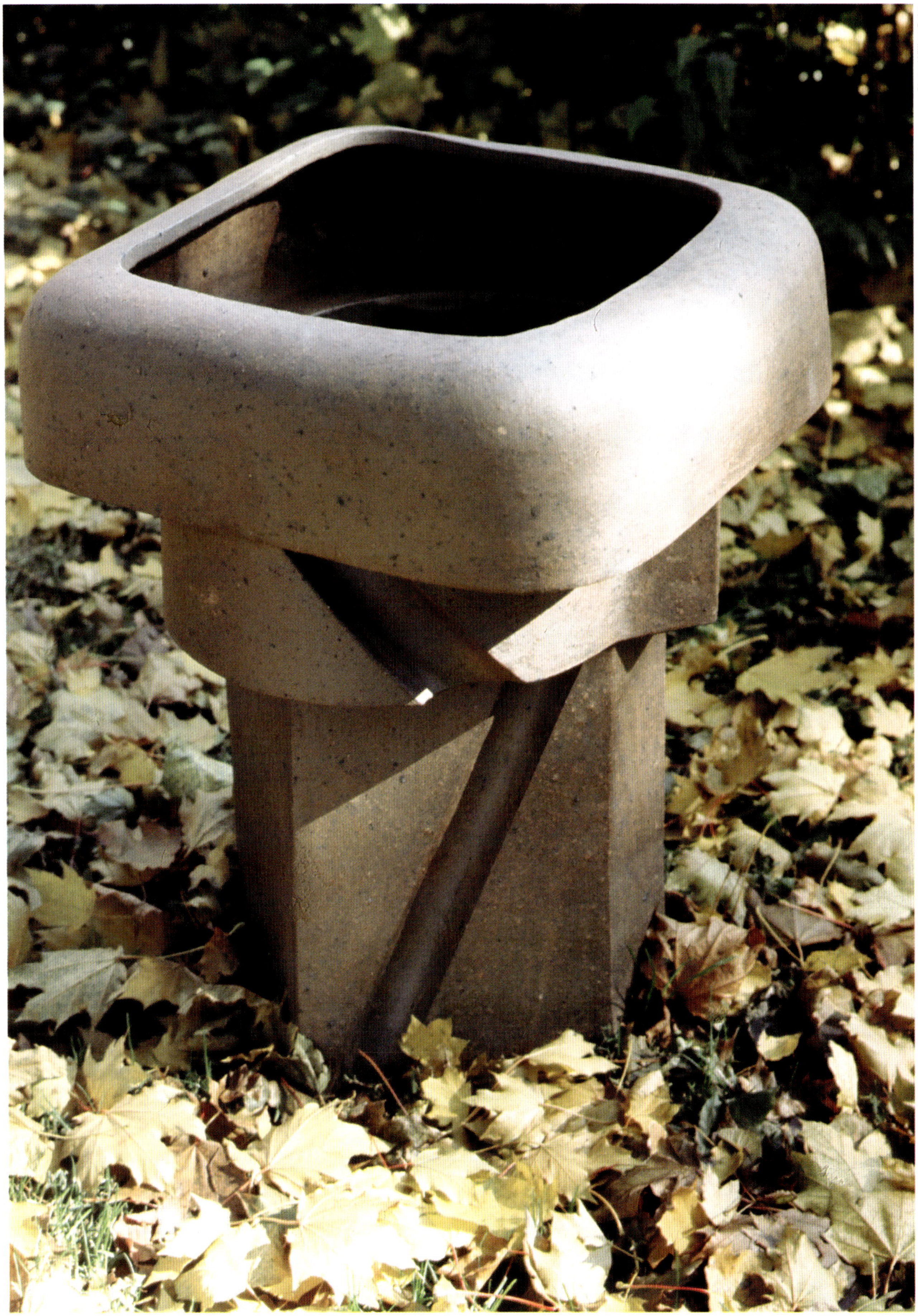

Photo: William Daley

Fig. 109. *Floor Pot,* 1973
Unglazed stoneware
18 x 20 x 20 inches
Exhibited at the Helen Drutt Gallery 1974

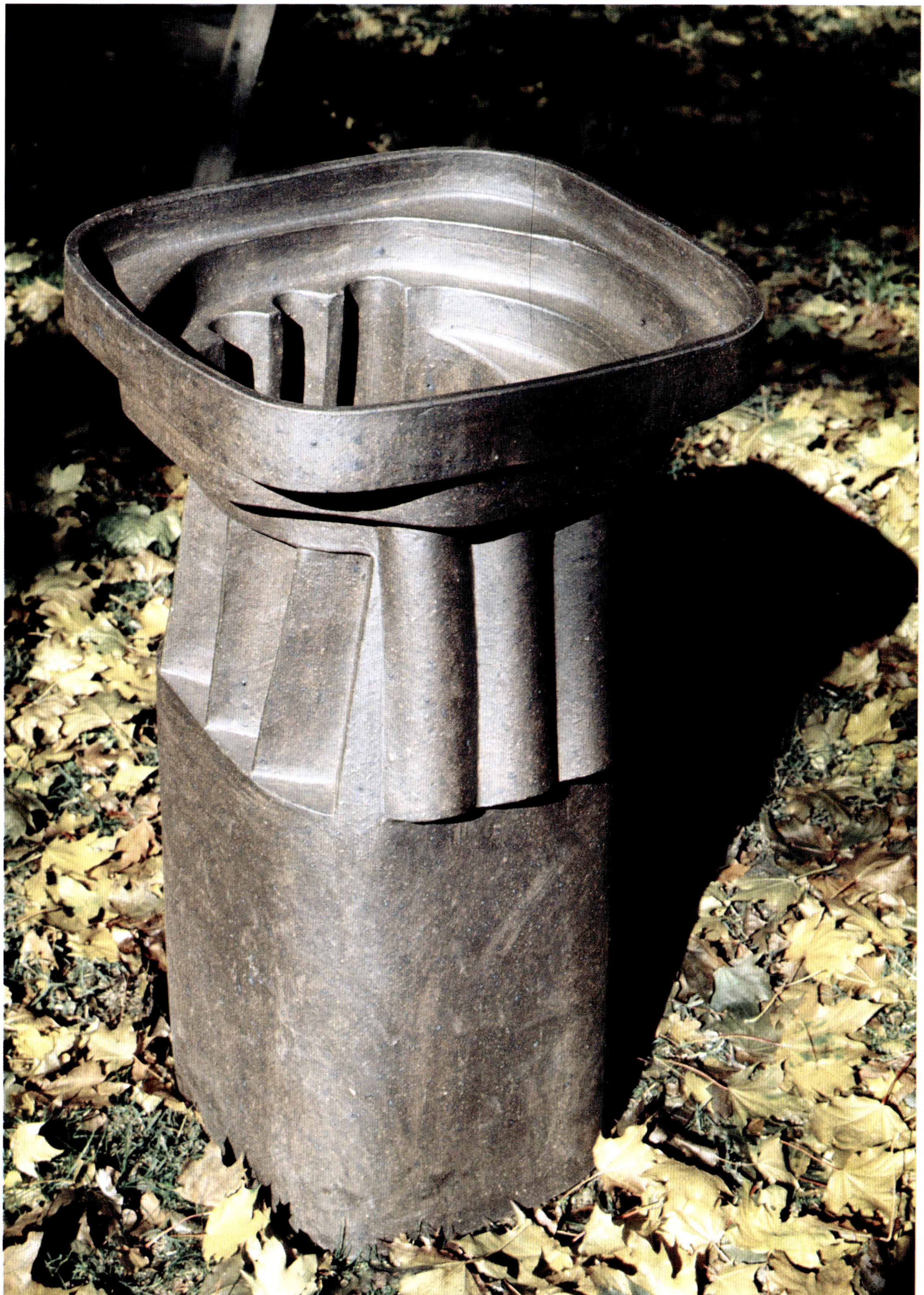

Photo: William Daley

Fig. 110. *Floor Pot,* 1973
Unglazed stoneware
30 x 15 x 15 inches
Exhibited at the Helen Drutt Gallery 1974
Collection of Helen and Jack Bershad

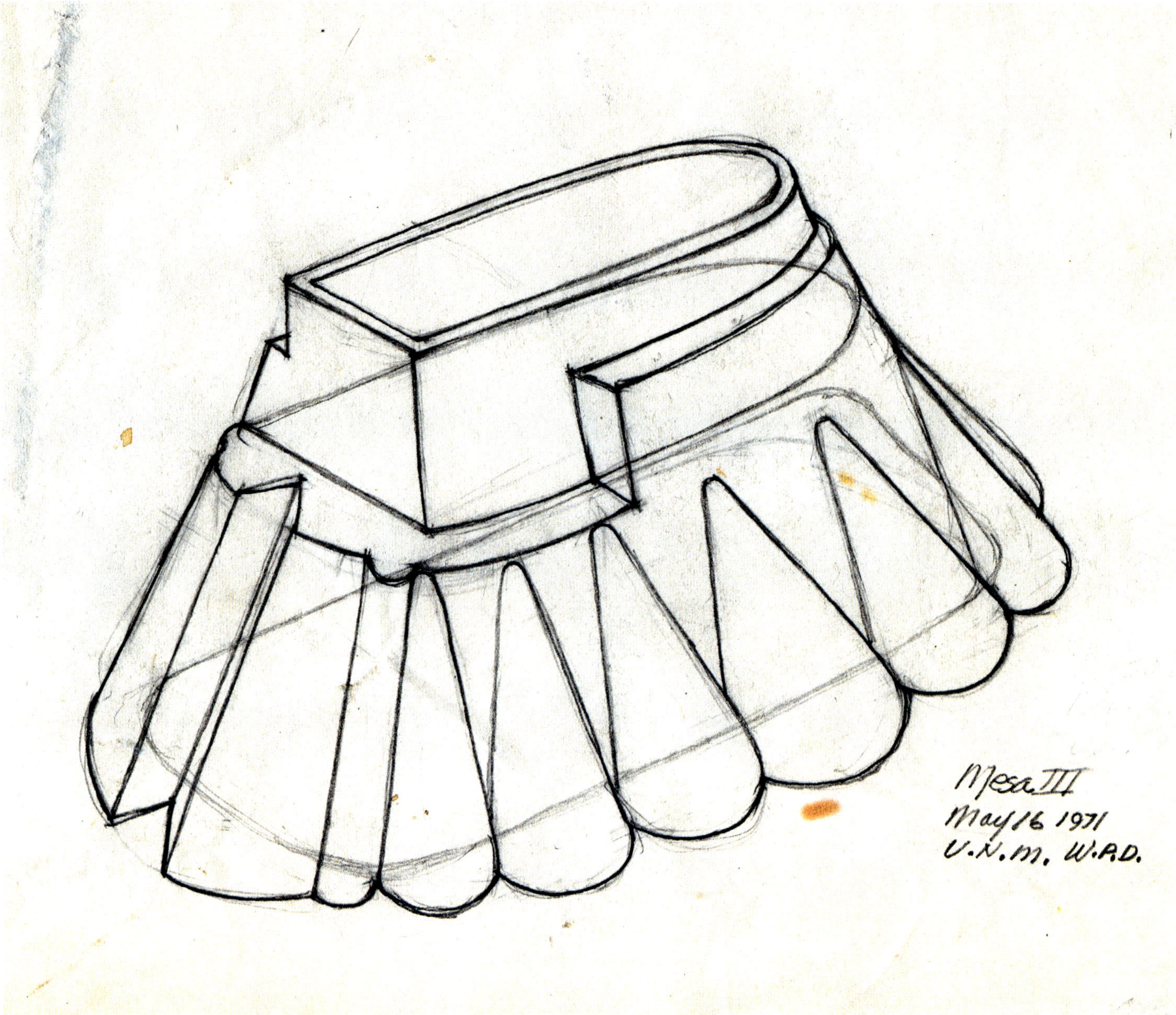

Fig. 111. Drawing, *Mesa III,* 1971
Pencil on paper
8 x 11 inches
Collection of William and Catherine Daley

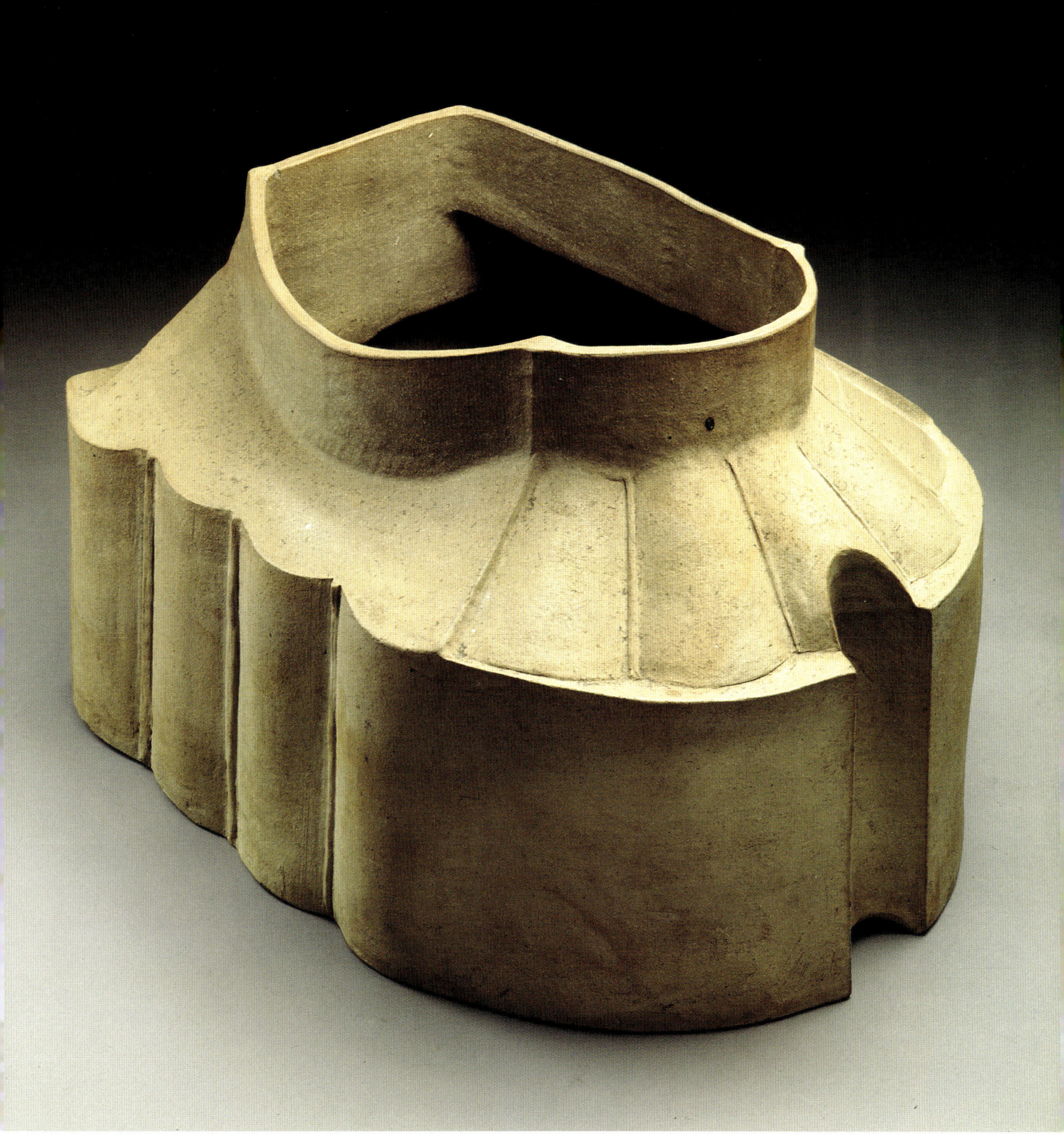

Fig. 112. *Unrooted Mission Church Pot*, 1972
Unglazed stoneware
13 x 16 x 20 inches
Collection of William and Catherine Daley

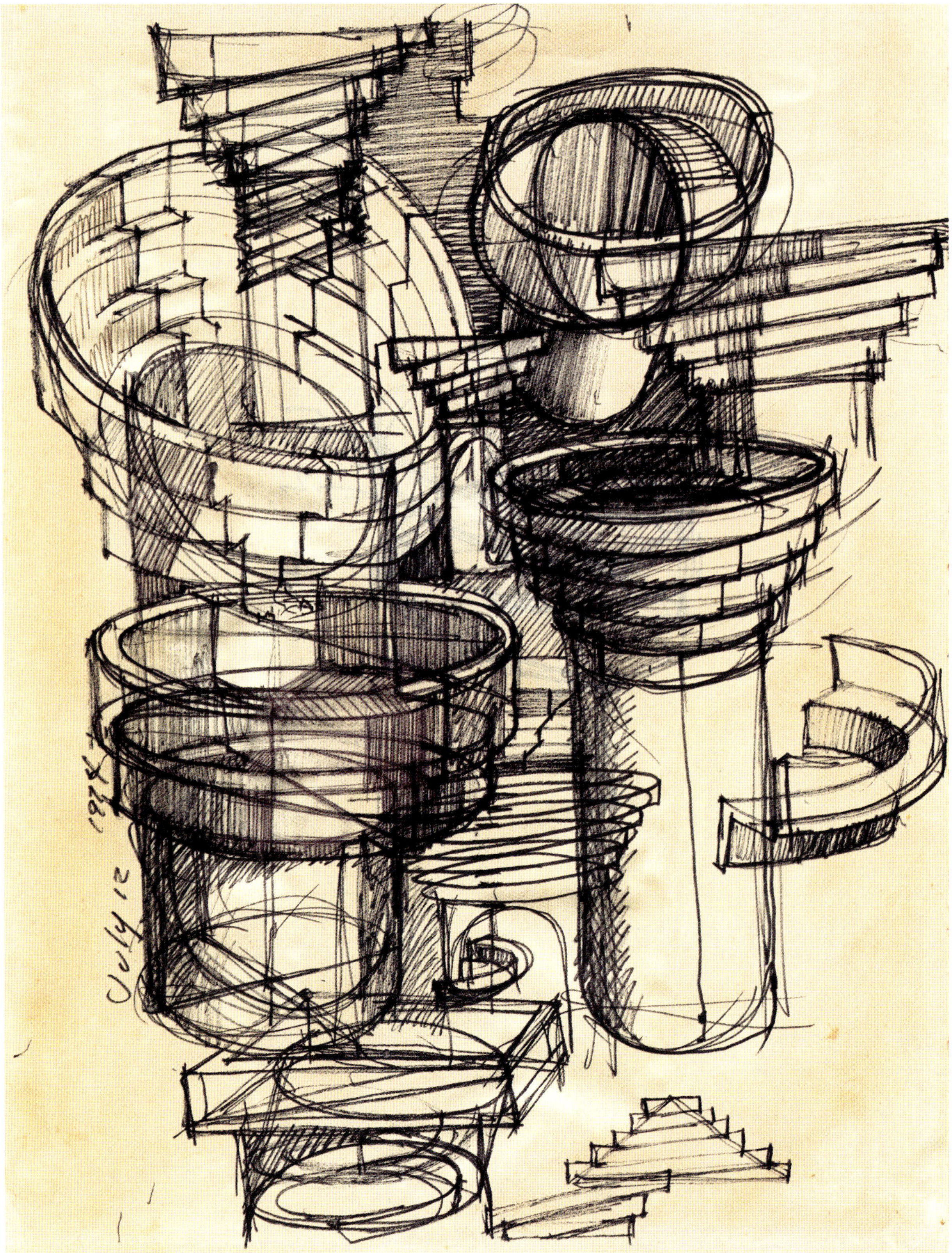

Fig. 113. Drawing of *Columnar Floor Pot,* 1974
Ink on paper
11 x 8.5 inches
Collection of Charlotte Daley

Fig. 114, opposite. *Columnar Floor Pot,* 1974
Unglazed stoneware
27 x 16 x 16 inches
Collection of William and Catherine Daley

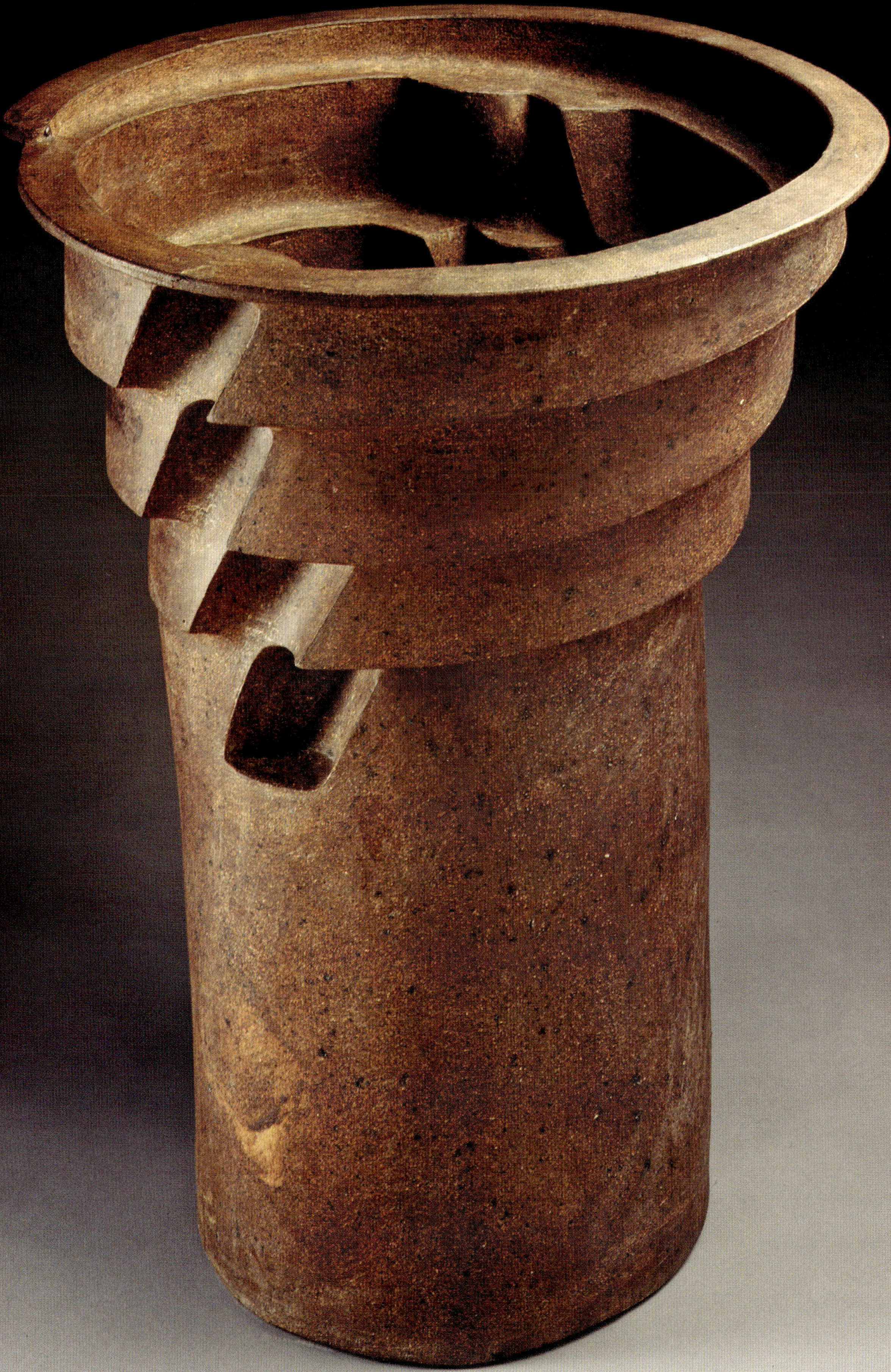

Photo: William Daley

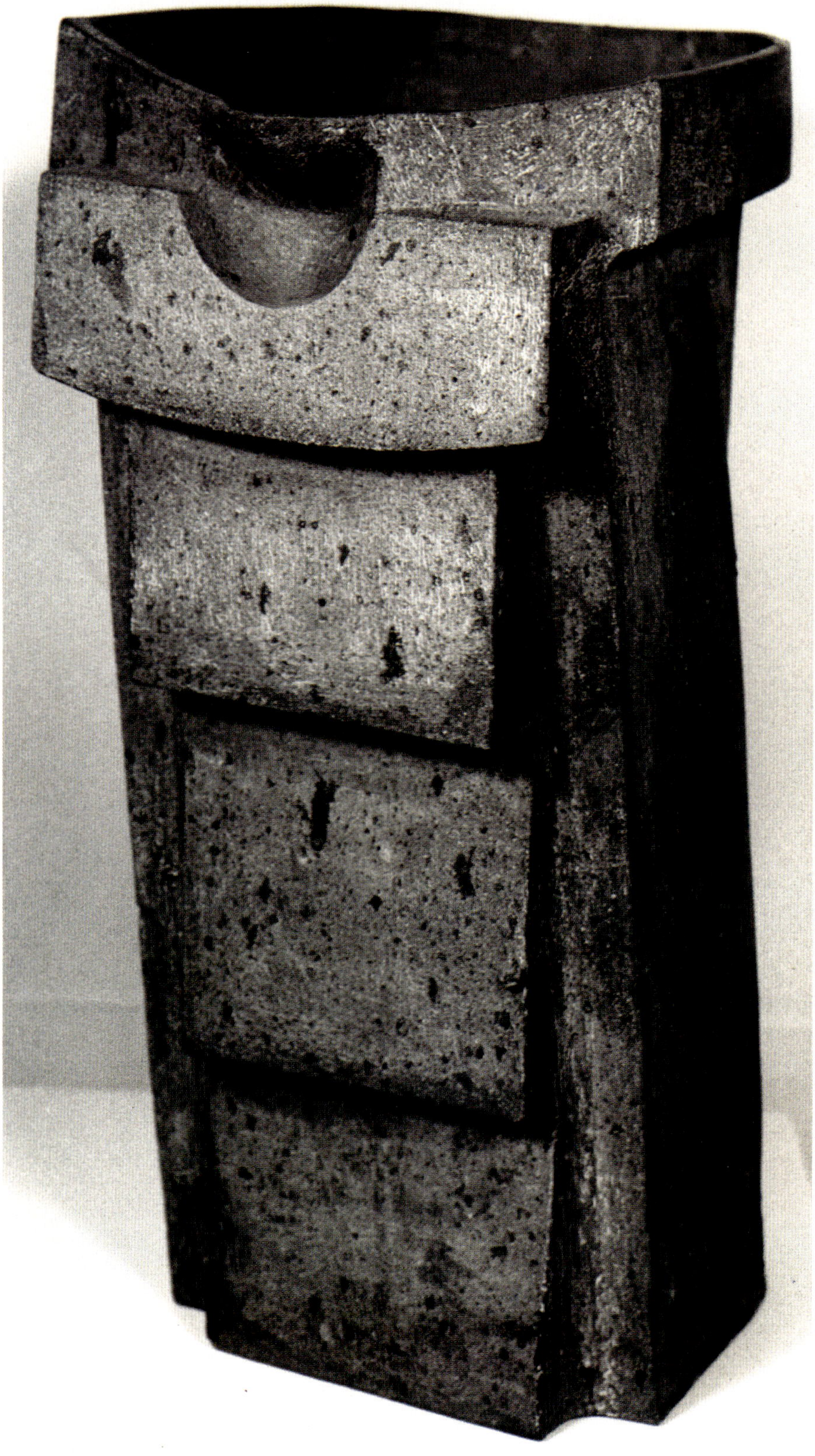

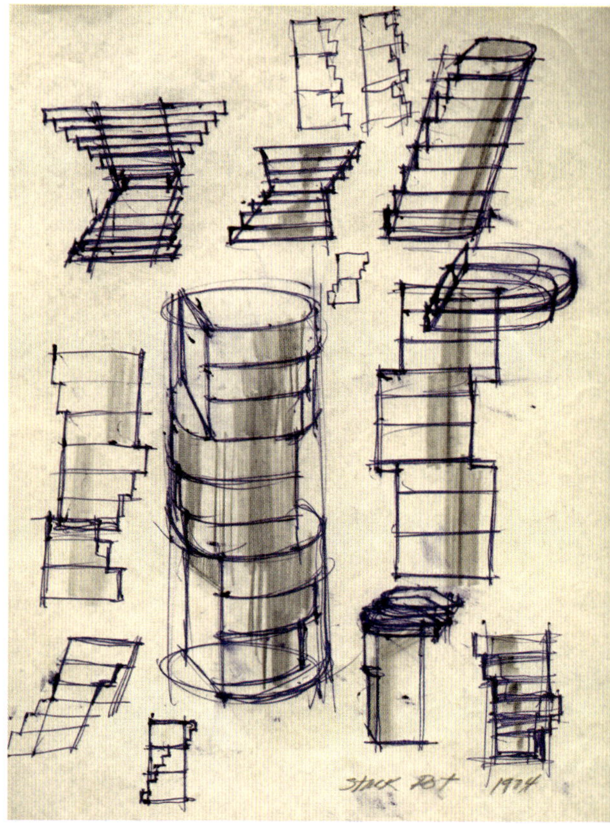

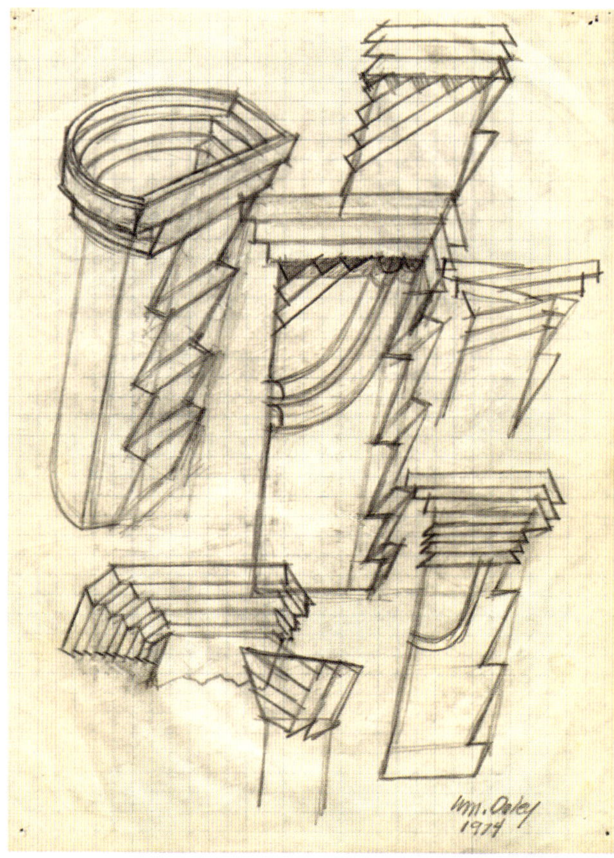

Fig. 115. *Stacked Pot,* 1974
Unglazed stoneware
32 inches height
Exhibited at the Faculty Exhibition
Philadelphia College of Art

Fig. 116, top. *Stack Pot* sketch, 1974
Pen and pencil on paper
11 x 8.5 inches

Fig. 117. Sketch, 1974
Pencil on paper
11 x 8.5 inches

Photo: William Daley

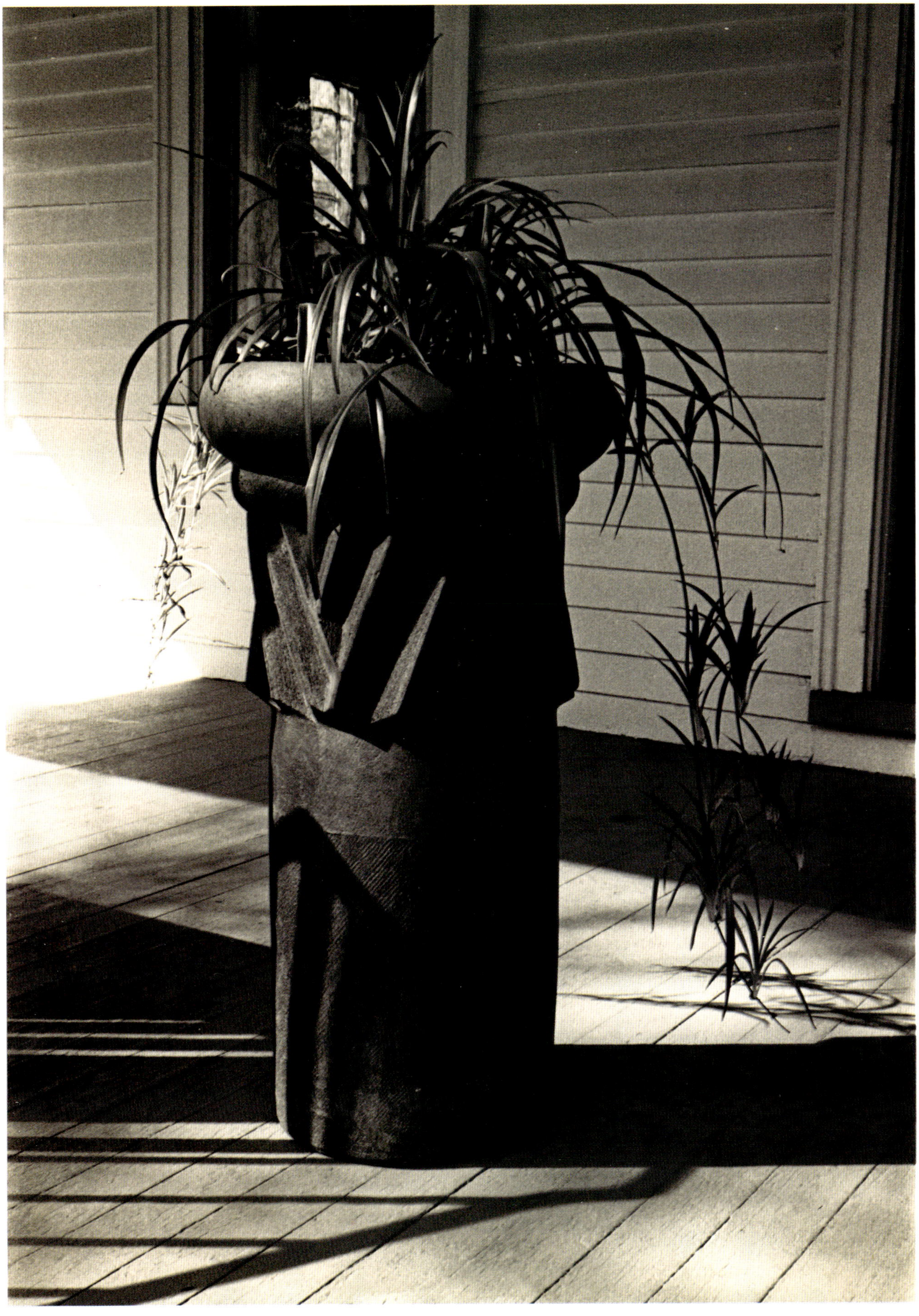

Fig. 118. *Floor Pot,* 1974
Unglazed stoneware
33.25 x 19 inch diameter
Photographed on the front porch of the Daley home
Collection of the Philadelphia Museum of Art

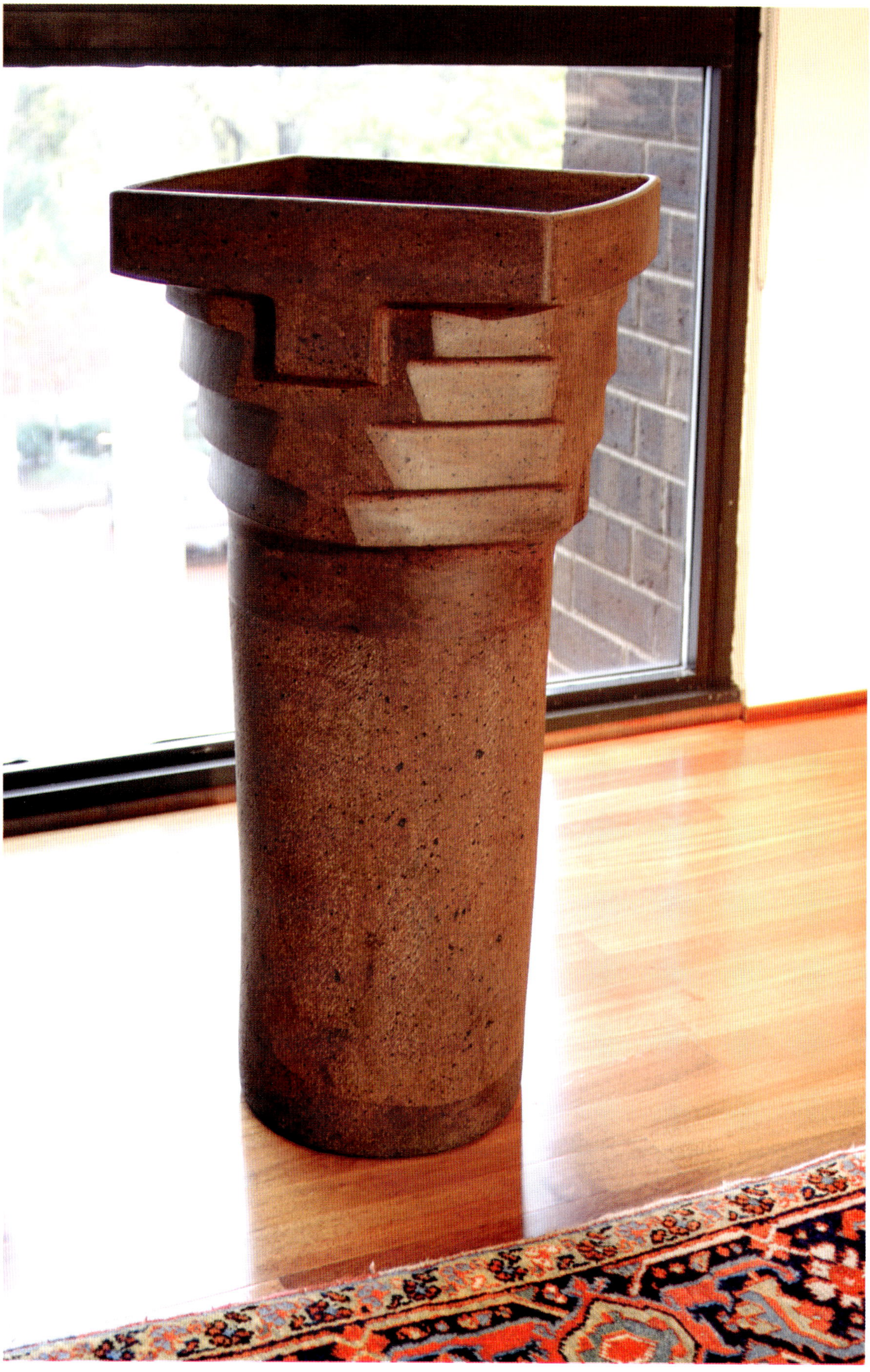

Photo: Charlotte Daley

Fig. 119. *To Frank Furness,* 1977
Unglazed stoneware
36 x 18 x 18 inches
Collection of Adele Fine

Fig. 120, opposite. Various pots, 1974
Unglazed stoneware
Preparation for show at the Helen Drutt Galley
Photographed in the backyard of the Daley home

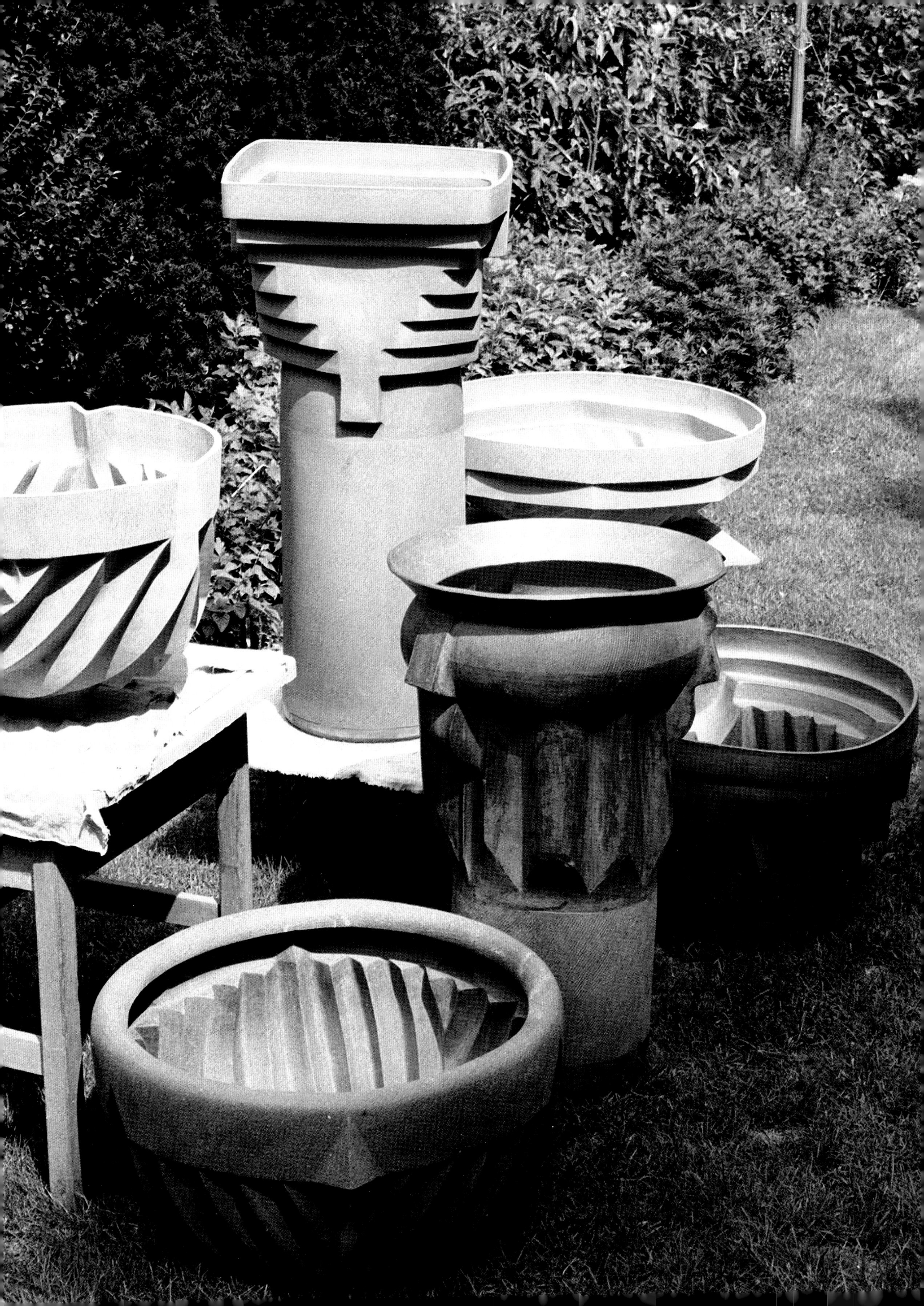

Photos: Amanda Schwab

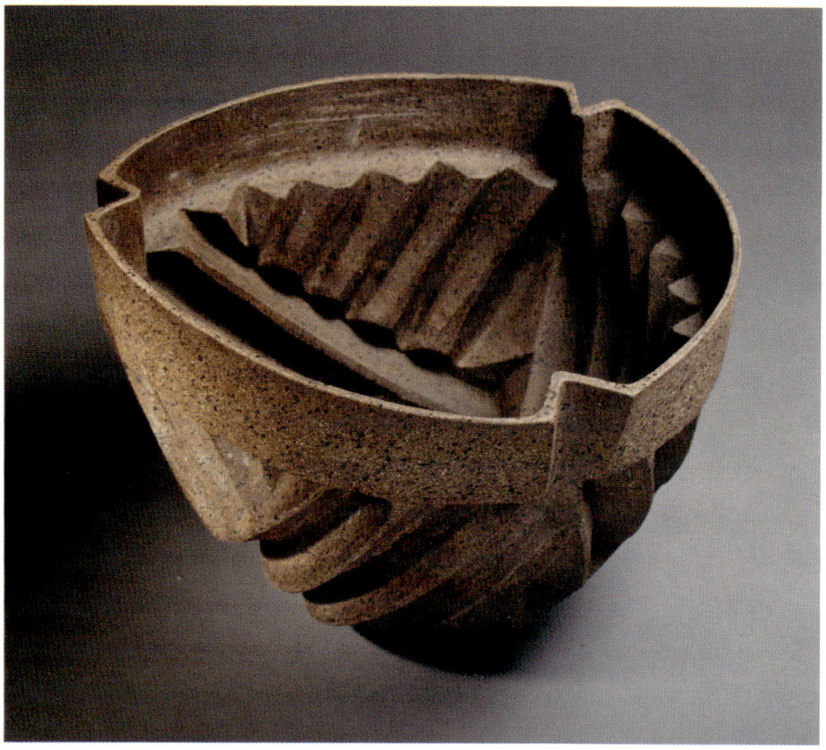

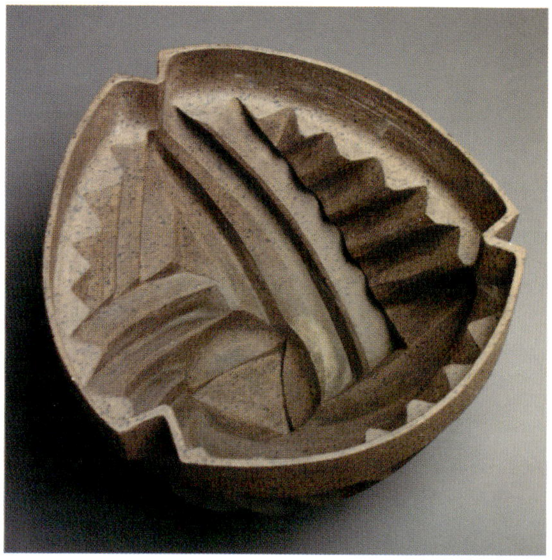

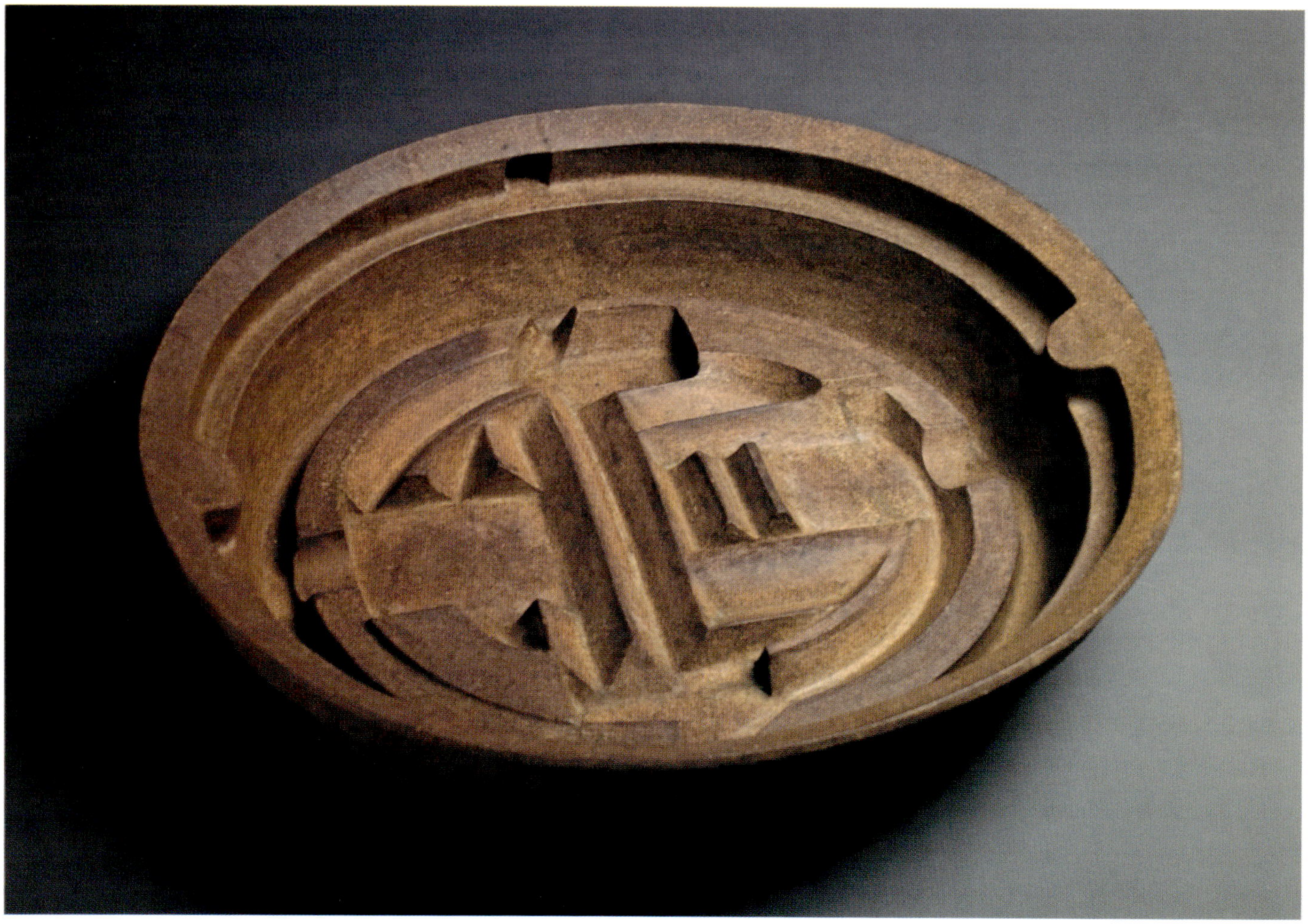

Figs. 121 & 122, top.
Triangular swirling hanging floor pot, 1976
Unglazed stoneware
14.75 x 18 x 18 inches
Collection of Helen W. Drutt English

Fig. 123. *Labyrinth,* 1974
Unglazed stoneware
7.5 x 27 x 29.75 inches
Collection of Alain and Françoise Schremmer

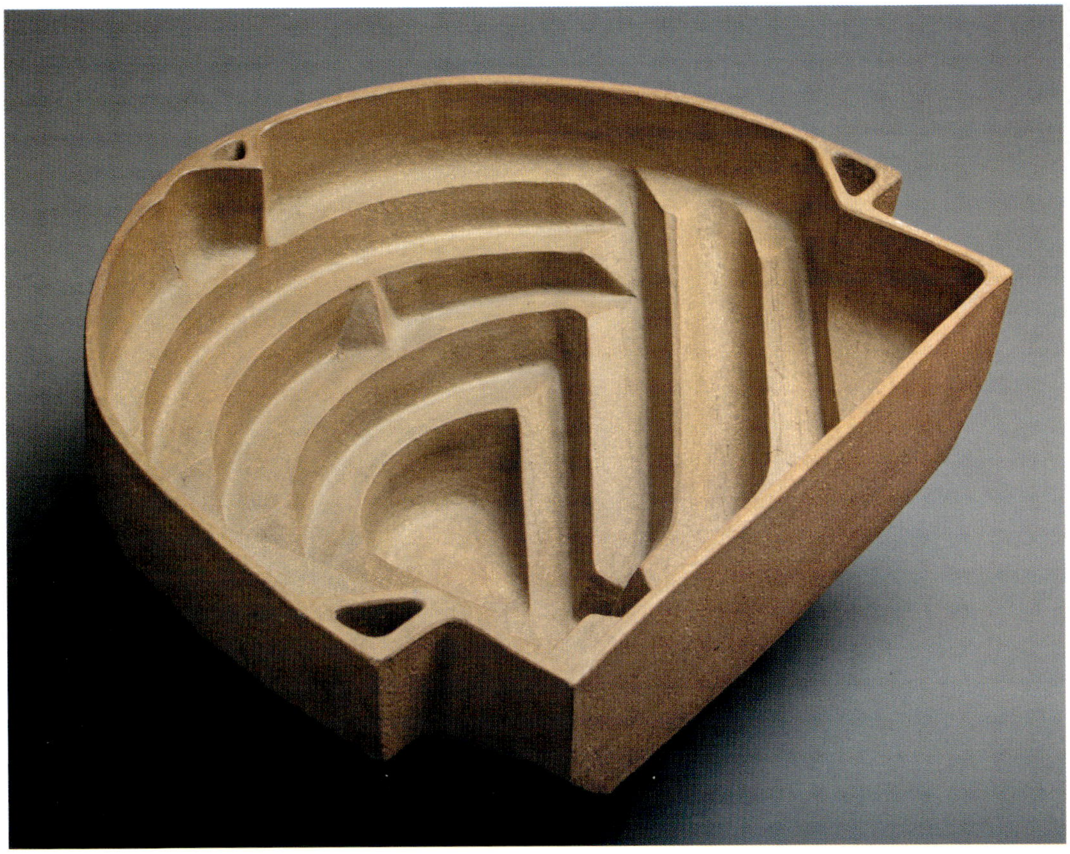

Photos: Amanda Schwab

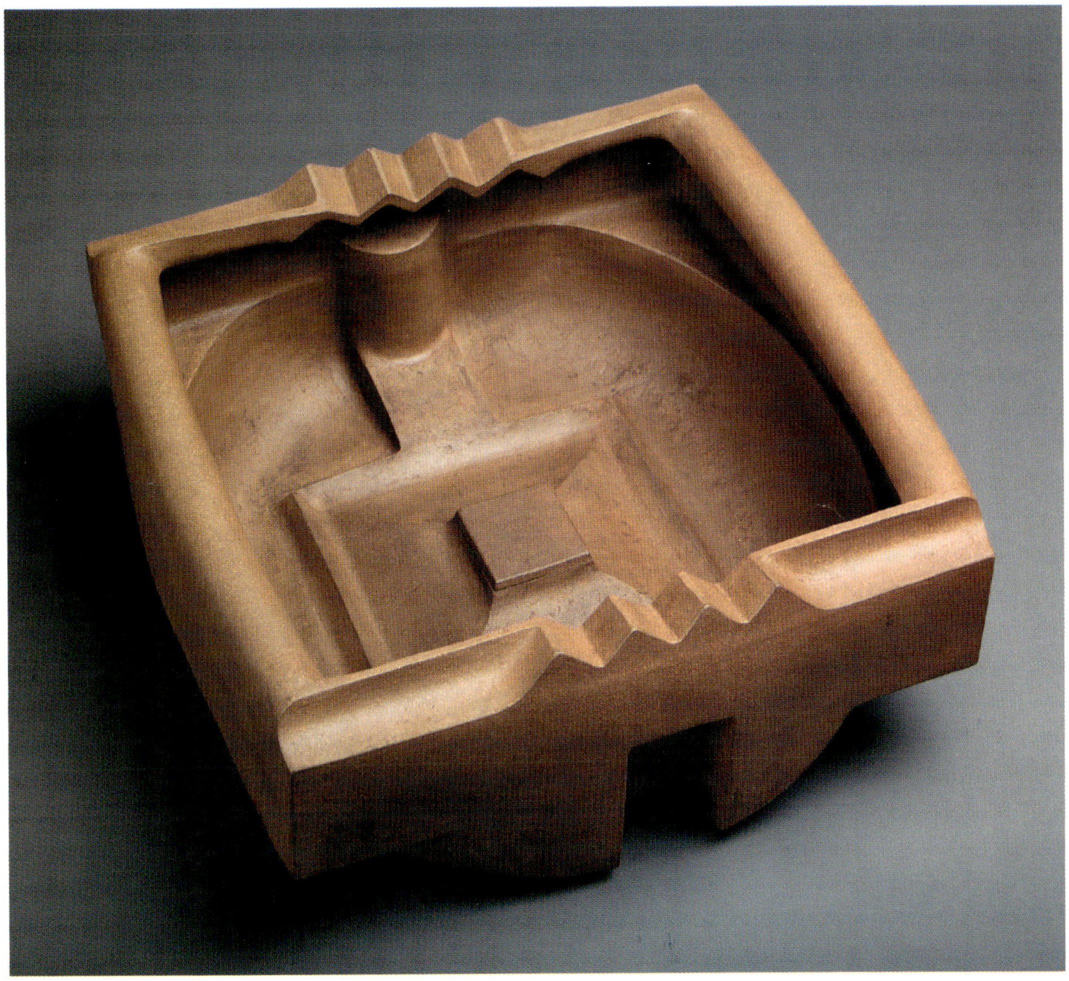

Fig. 124, top. *D-Form* hanging planter, 1976
Unglazed stoneware
9 x 19 x 17.25 inches
Collection of George and Barbara Daley Schwab

Fig. 125, above. Untitled, 1978
Unglazed stoneware
11 x 22 x 23 inches
Collection of William and Catherine Daley
Published in *American Ceramics, The work of Twenty Modern Masters,*
by Michael Mctwiggen and Garth Clark

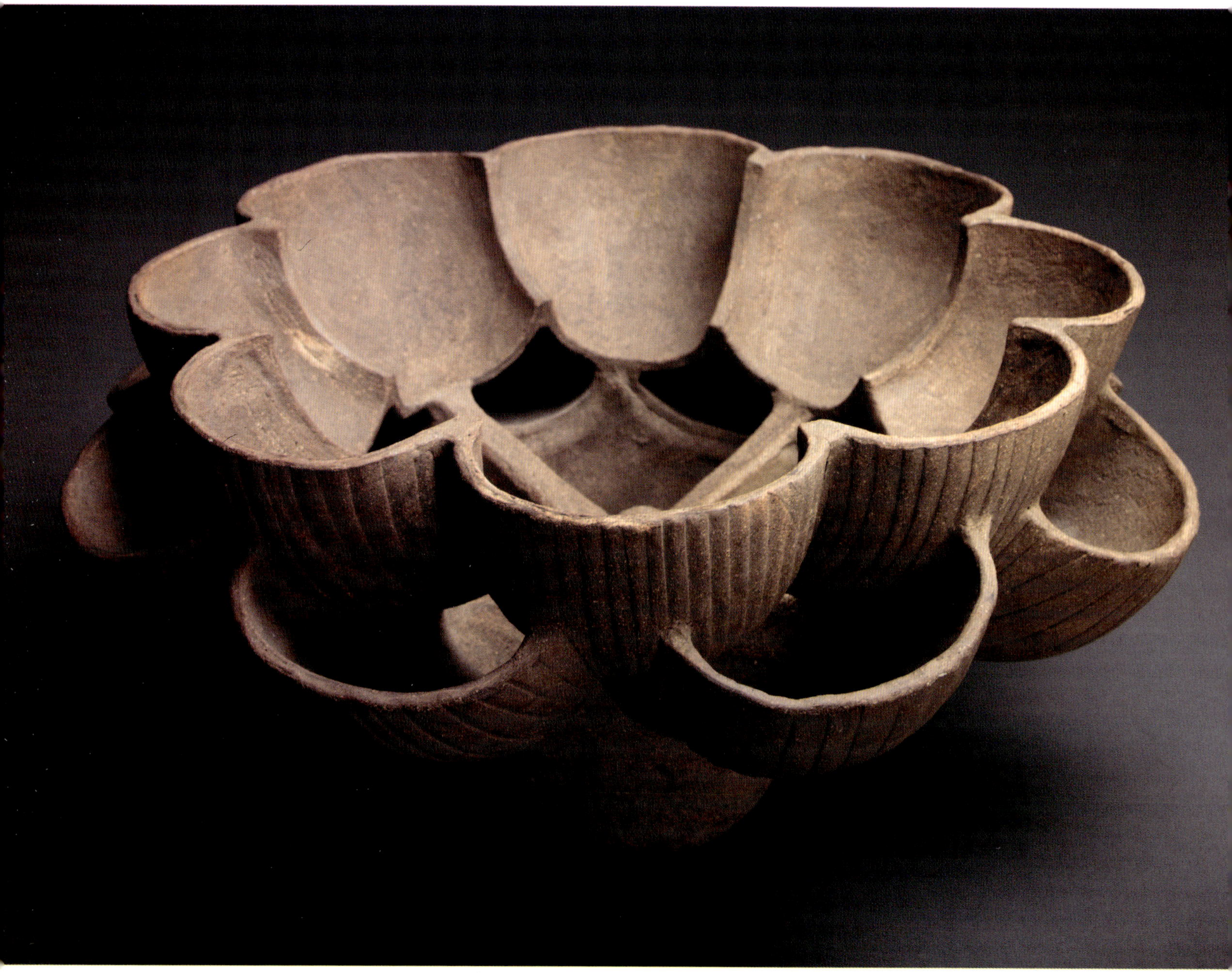

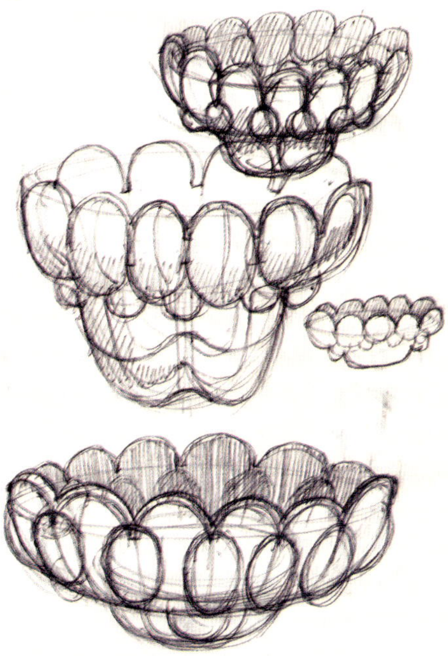

Fig. 126. *Pod* hanging planter, 1977
Unglazed Jersey brick clay
10 x 23 inch diameter
Collection of William and Catherine Daley

Fig. 127. Pod planter concept sketch, 1977
Pen on tracing paper
10 x 8 inches

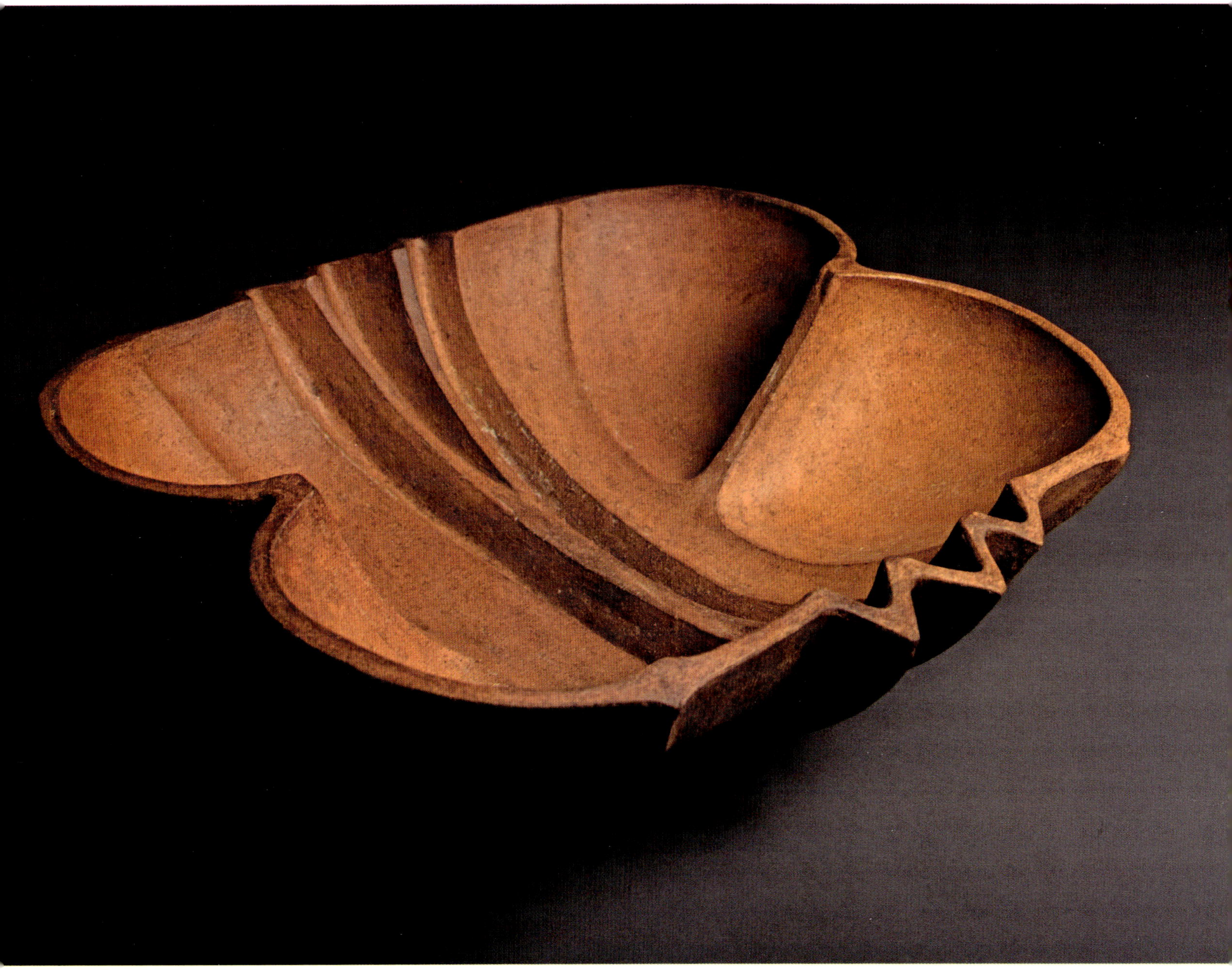

Photos: Amanda Schwab

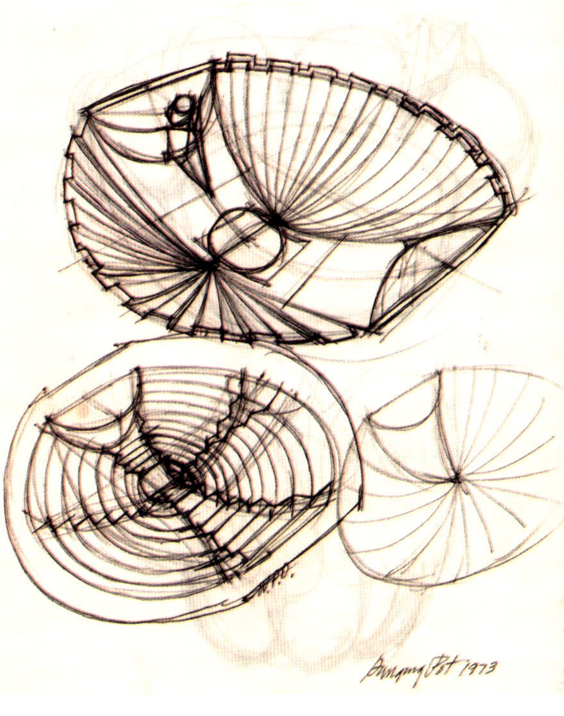

Fig. 128. *Hanging Pot*, 1975
Unglazed stoneware
5.75 x 20.5 x 20 inches
Collection of Françoise and Alain Schremmer

Fig. 129. *Hanging Pot* concept sketch, 1973
Pen on tracing paper
10 x 8 inches

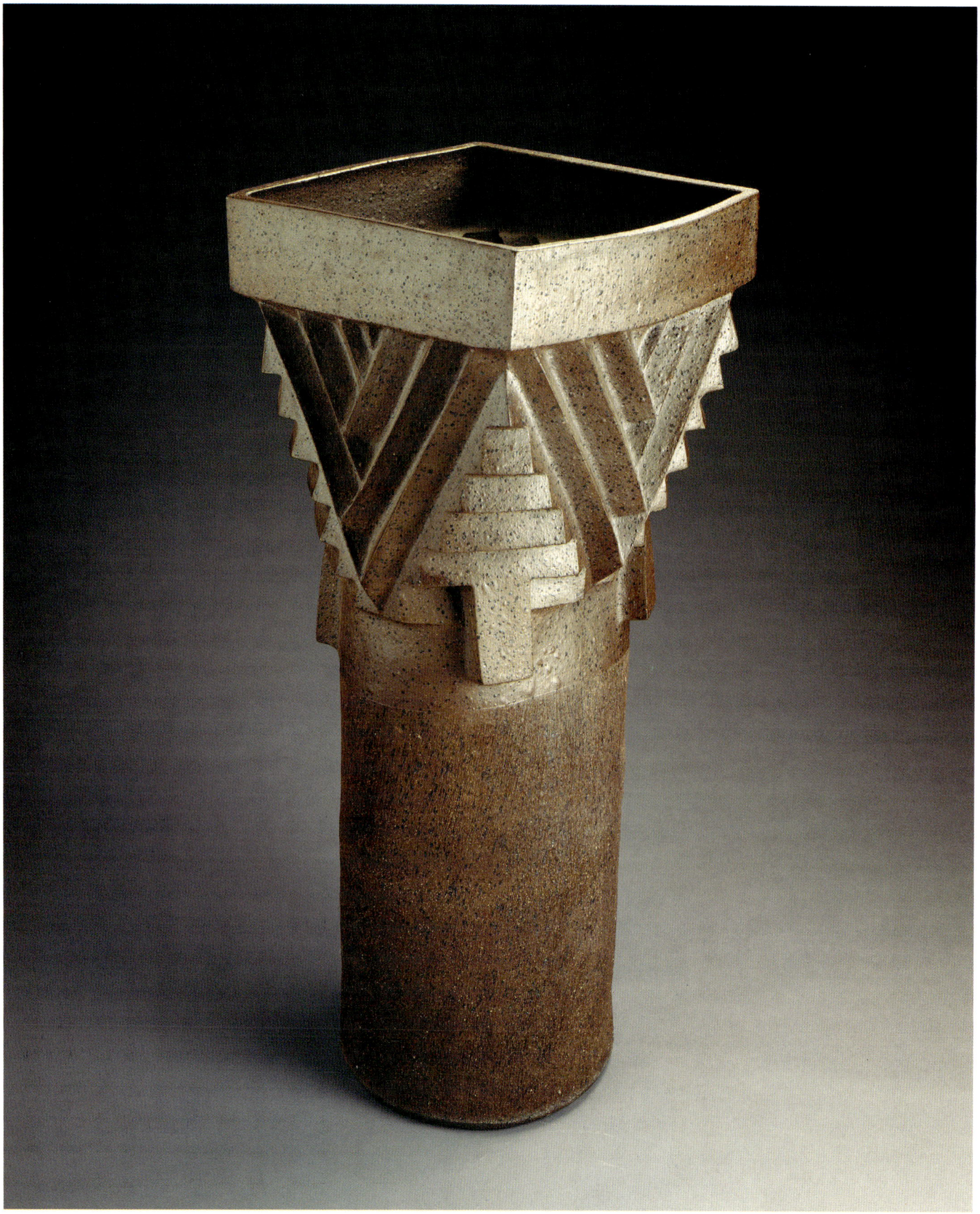
Photo: James Quaile

Fig. 130. *Vertical Floor Pot,* 1978
Unglazed stoneware
39 x 18 x 18 inches
Private collection

Fig. 131, opposite. *Shang Form,* 1977
Unglazed stoneware
33.5 x 21.5 inch diameter
Collection of Jerry and Dina Wind

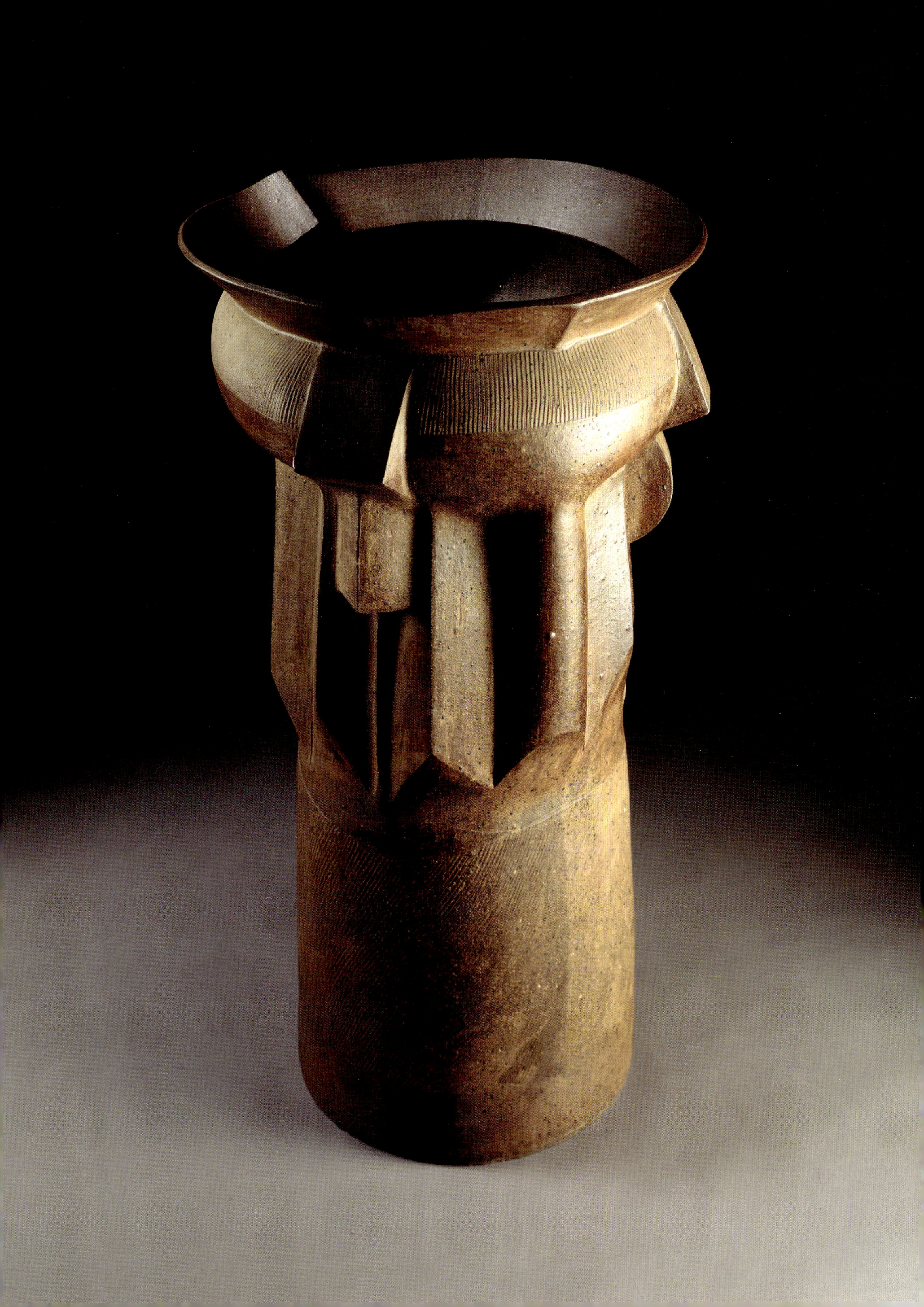

Photo: William Daley

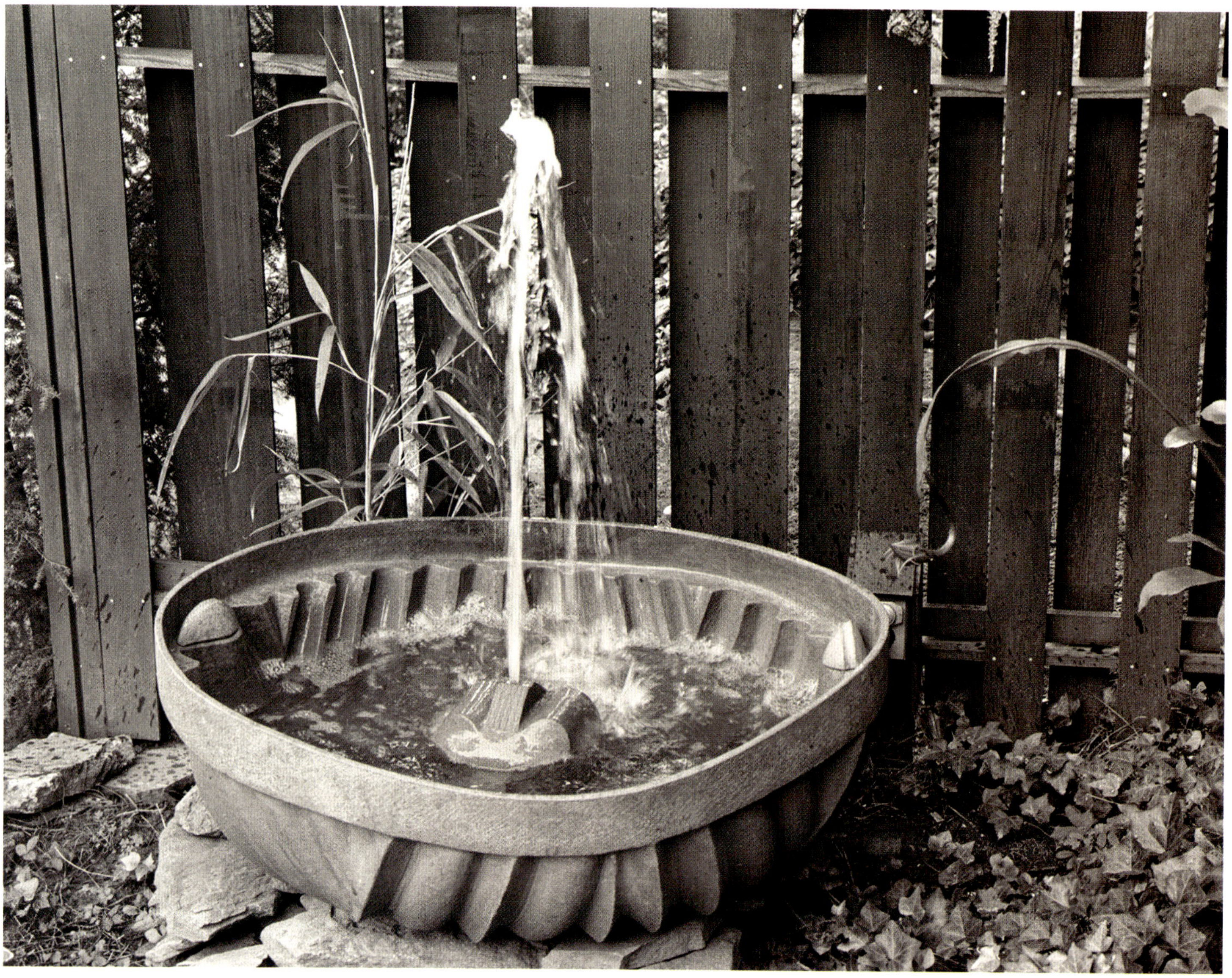

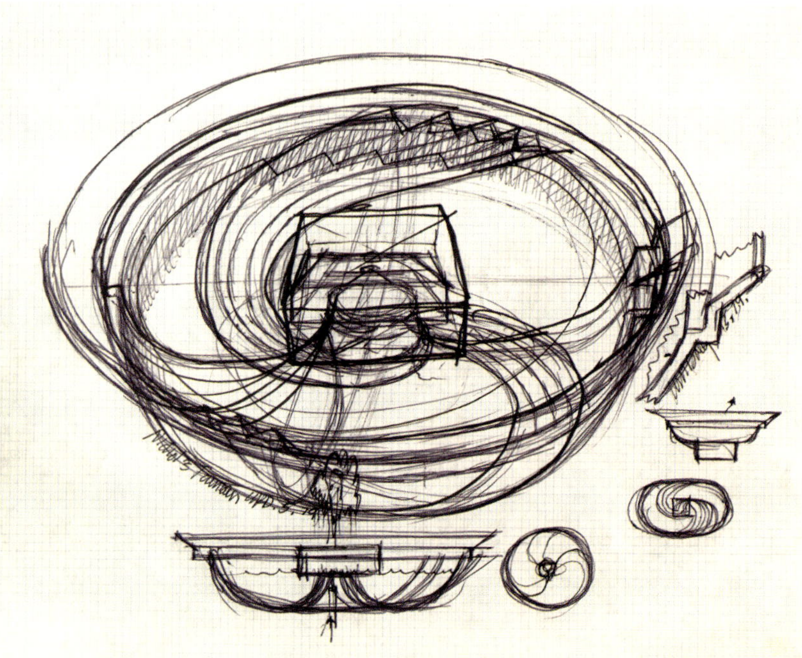

Fig. 132, above. Fountain, 1976
Unglazed stoneware
32 x 28 x 16 inches
No longer extant

Fig. 133, left. *Nichol's Fountain* drawing, 1979
Ink on paper
11 x 8.5 inches

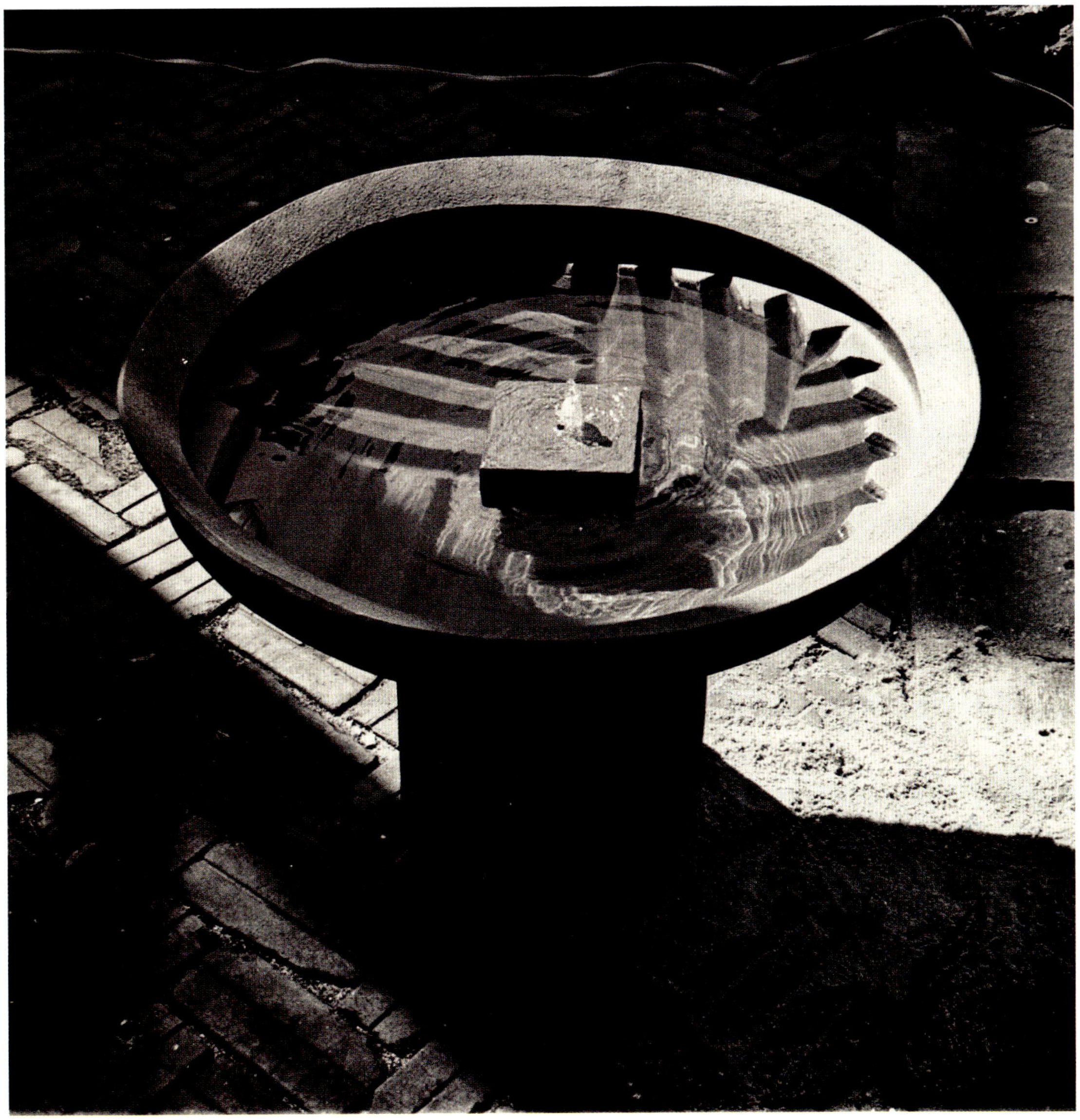

Photo: Charlotte Daley

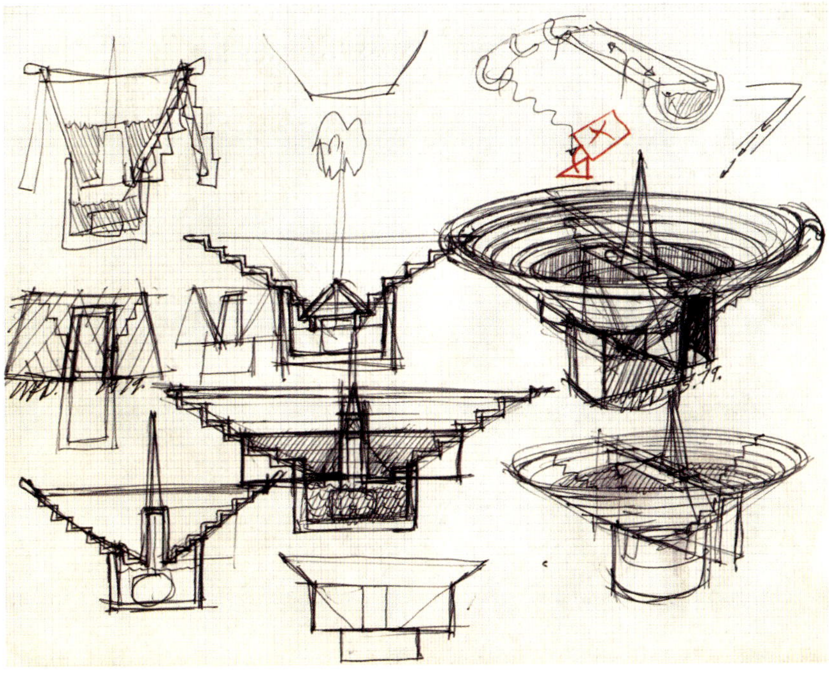

Fig. 134, top. *Fountain*, 1976
Unglazed stoneware
33 inches diameter
Collection of Thomas and Marjie Bershad

Fig. 135, left. Fountain sketches, 1979
Pen on paper
8.5 x 11 inches

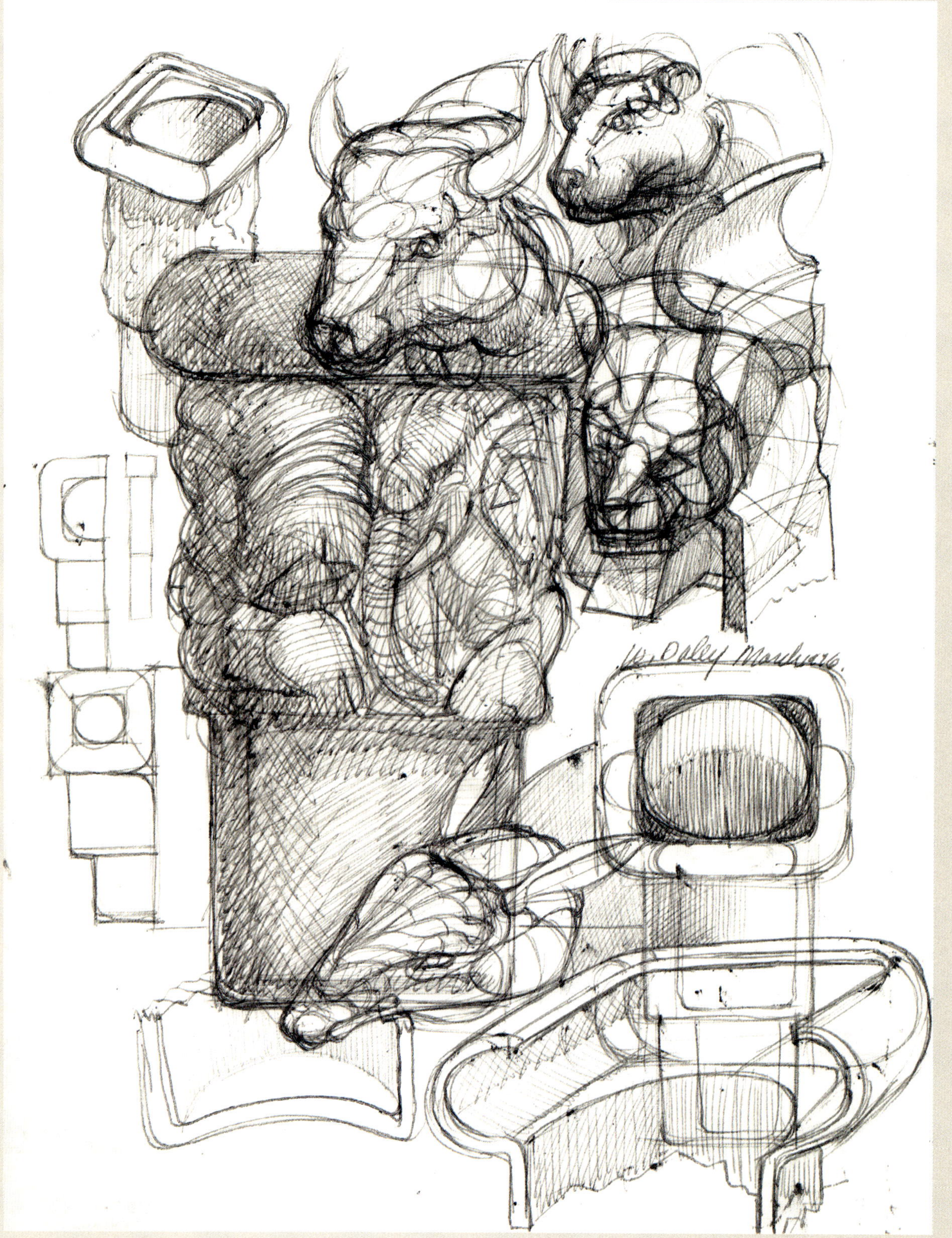

Fig. 136. Animal Pot sketch, 1976
Pen on paper
11 x 8.5 inches
Collection of Kenneth and Eleanor Heibert

Fig. 137, opposite. *Animal Pot,* 1976
Unglazed stoneware
28 inches height
Photographed in front room of
1625 Spruce Street, Helen Drutt Gallery
Collection of Helen Drutt English

Formerly at St. Charles Seminary
Philadelphia, PA

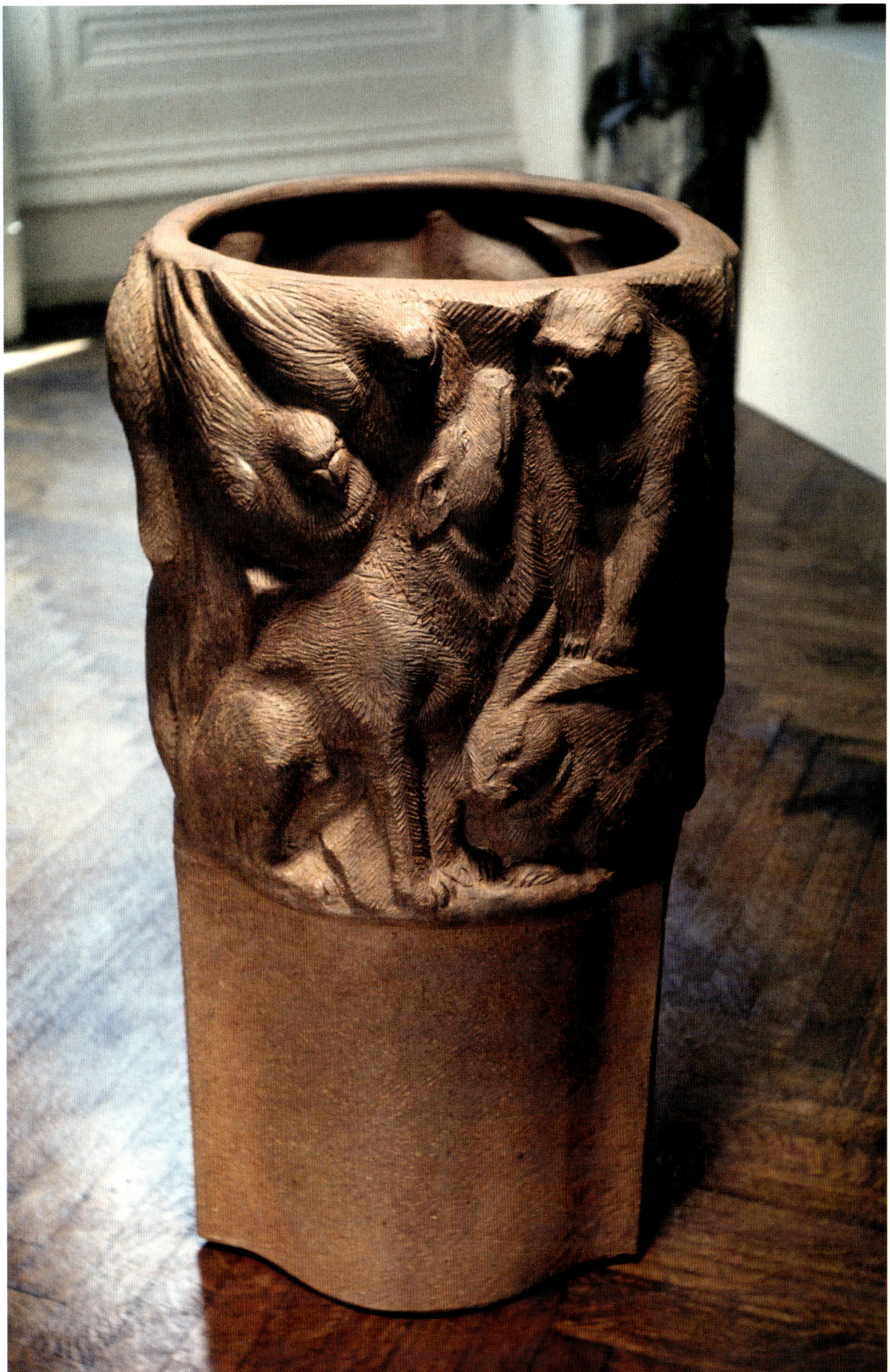

Photo: William Daley

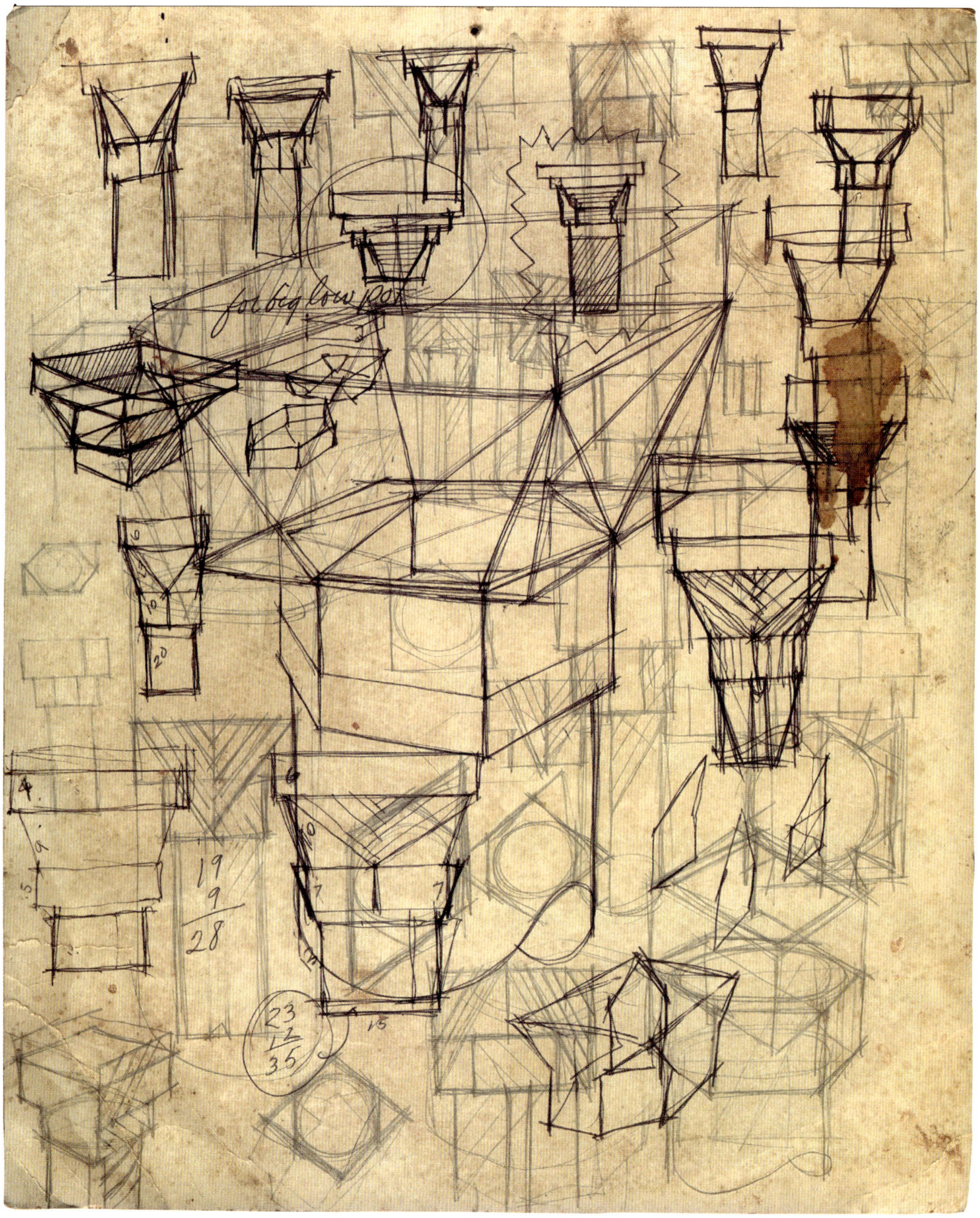

Fig. 138. Drawing
Pencil and pen on posterboard
12 x 10 inches
Collection of William and Catherine Daley

Fig. 139, opposite. *Hexed Square Up,* 1980
Unglazed stoneware
35 x 16 x 16 inches
Collection of Vincente Lim and Robert Tooey

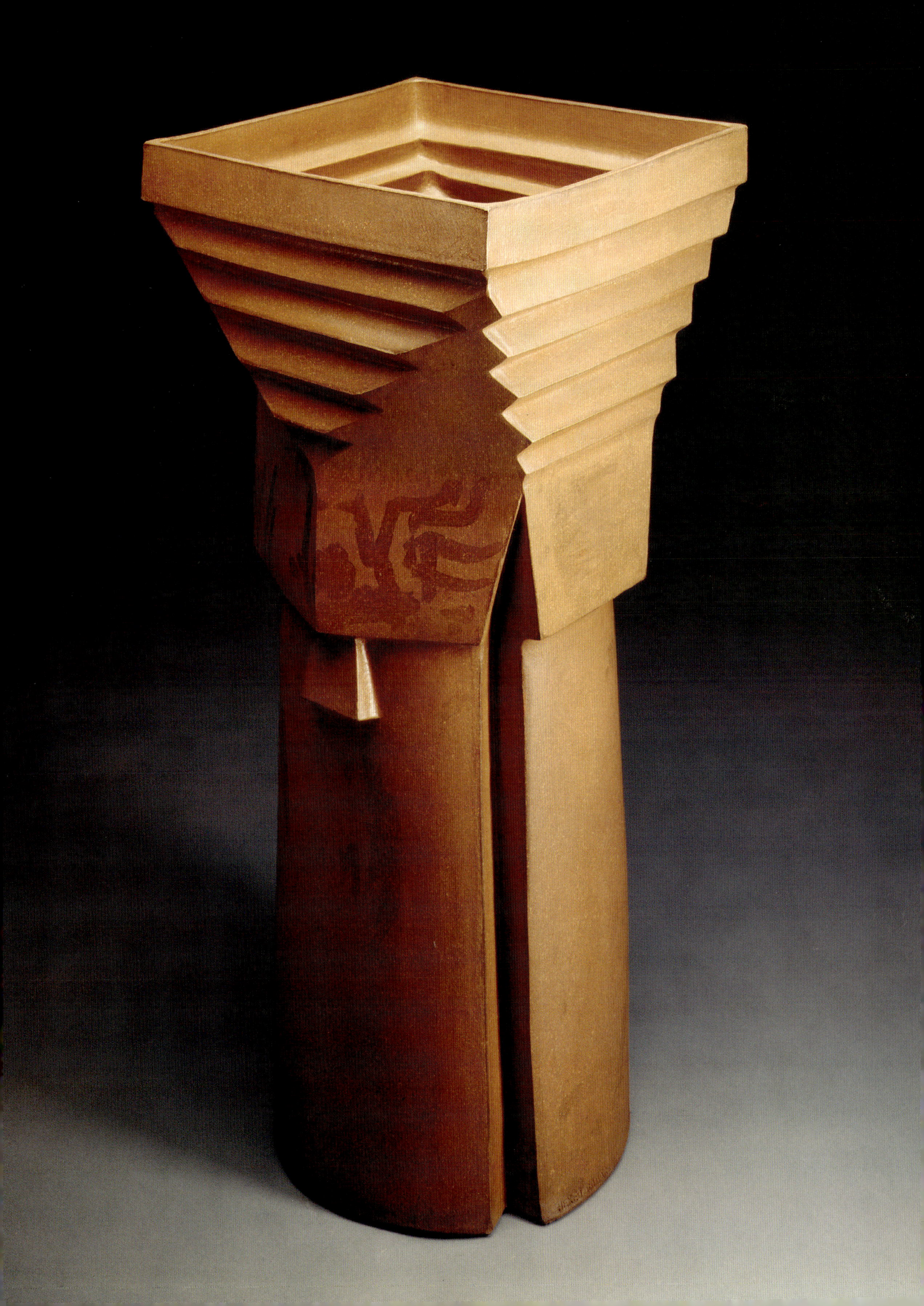

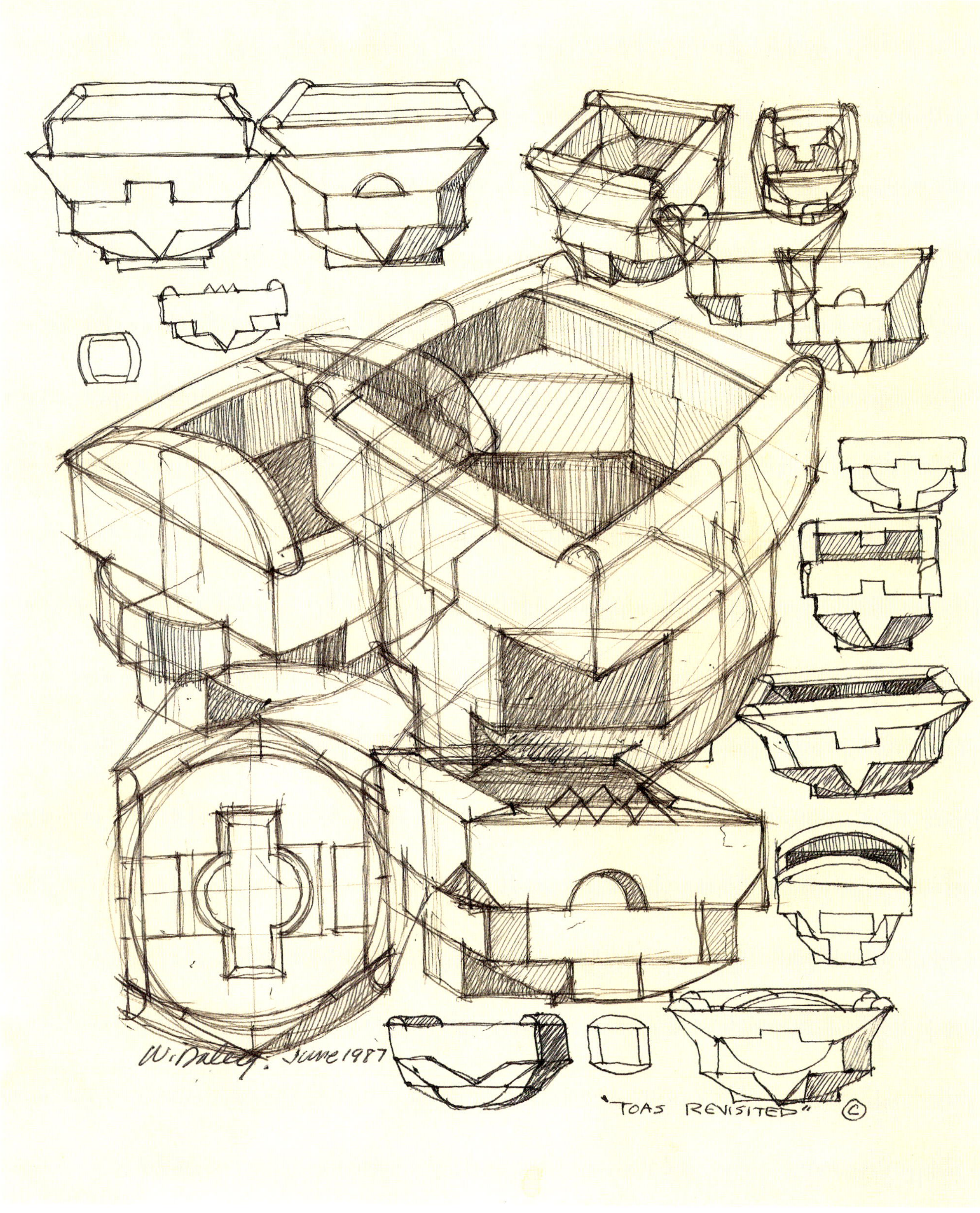

Fig. 140. *Toas Revisited* drawing, 1987
Ink on paper
14 x 11 inches
Collection of William and Catherine Daley

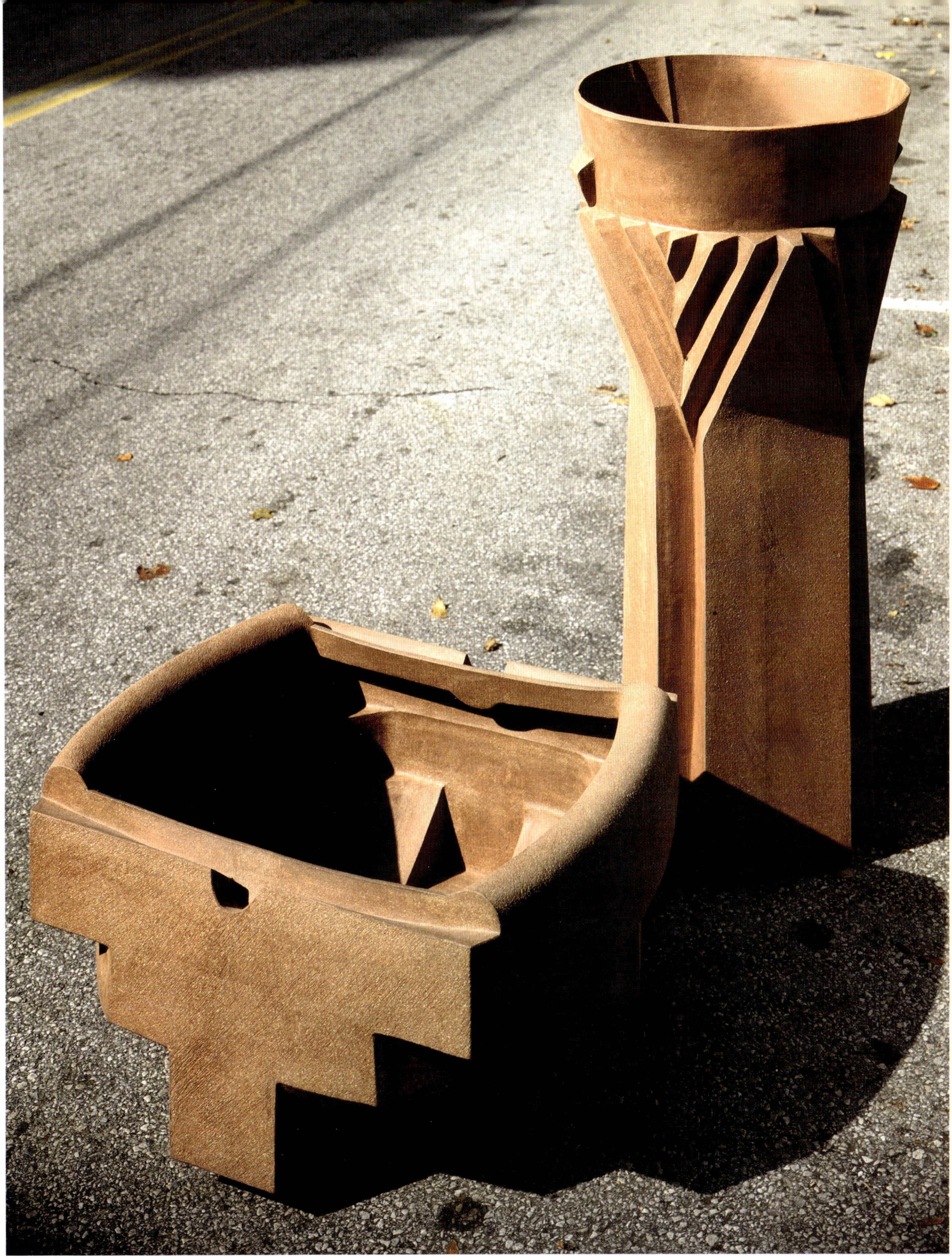

Photo: John Carlano

Fig. 141. *Taos Procession* – Low pot, 1980
Unglazed stoneware
14.5 x 24 x 28.5 inches
Signed William and Catherine Daley

Palmed Shift – Vertical pot, 1980
41.5 x 19 x 19 inches
Collection of Gene and Sueyun Locks

Photo: John Carlano

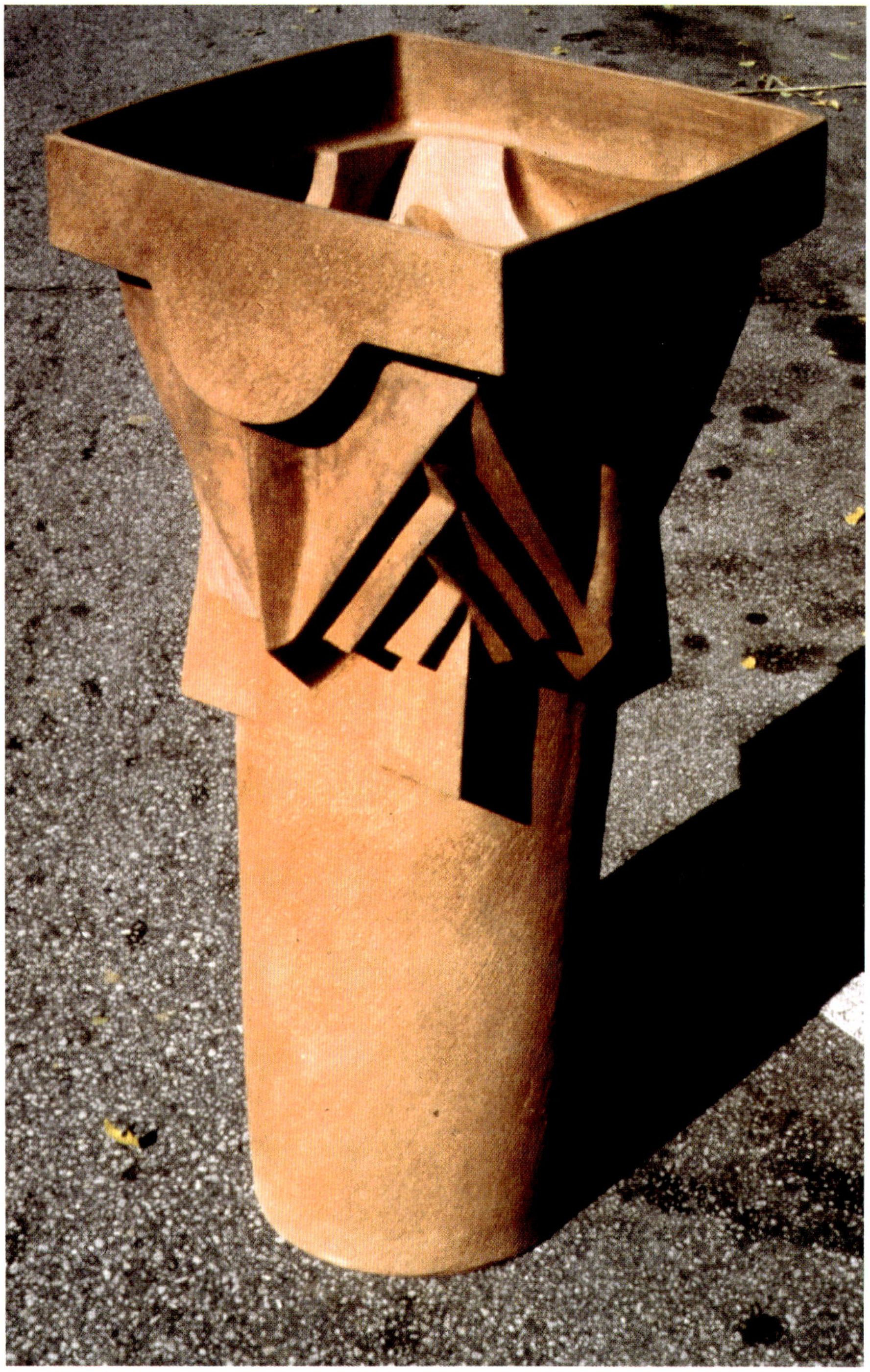

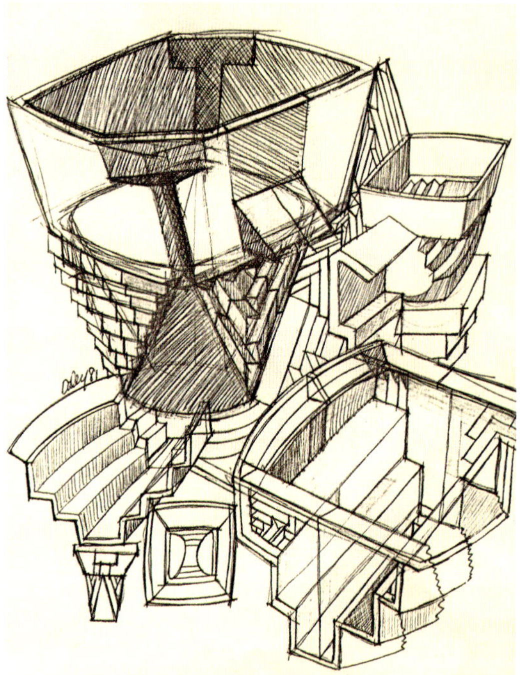

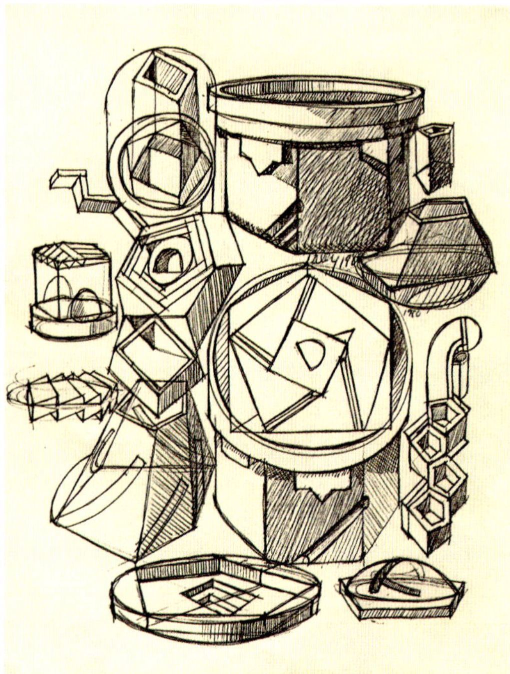

Fig. 142. *Columnar Form,* 1978
Unglazed stoneware
39.5 x 18 x 18 inches
Collection of Stedelijk Museum
Hertogenbosch, The Netherlands

Fig. 143, top right. Drawing, pot forms
14 x 11 inches

Fig. 144, bottom right. Drawing, *Lazarus I*
Tank pot
14 x 11 inches

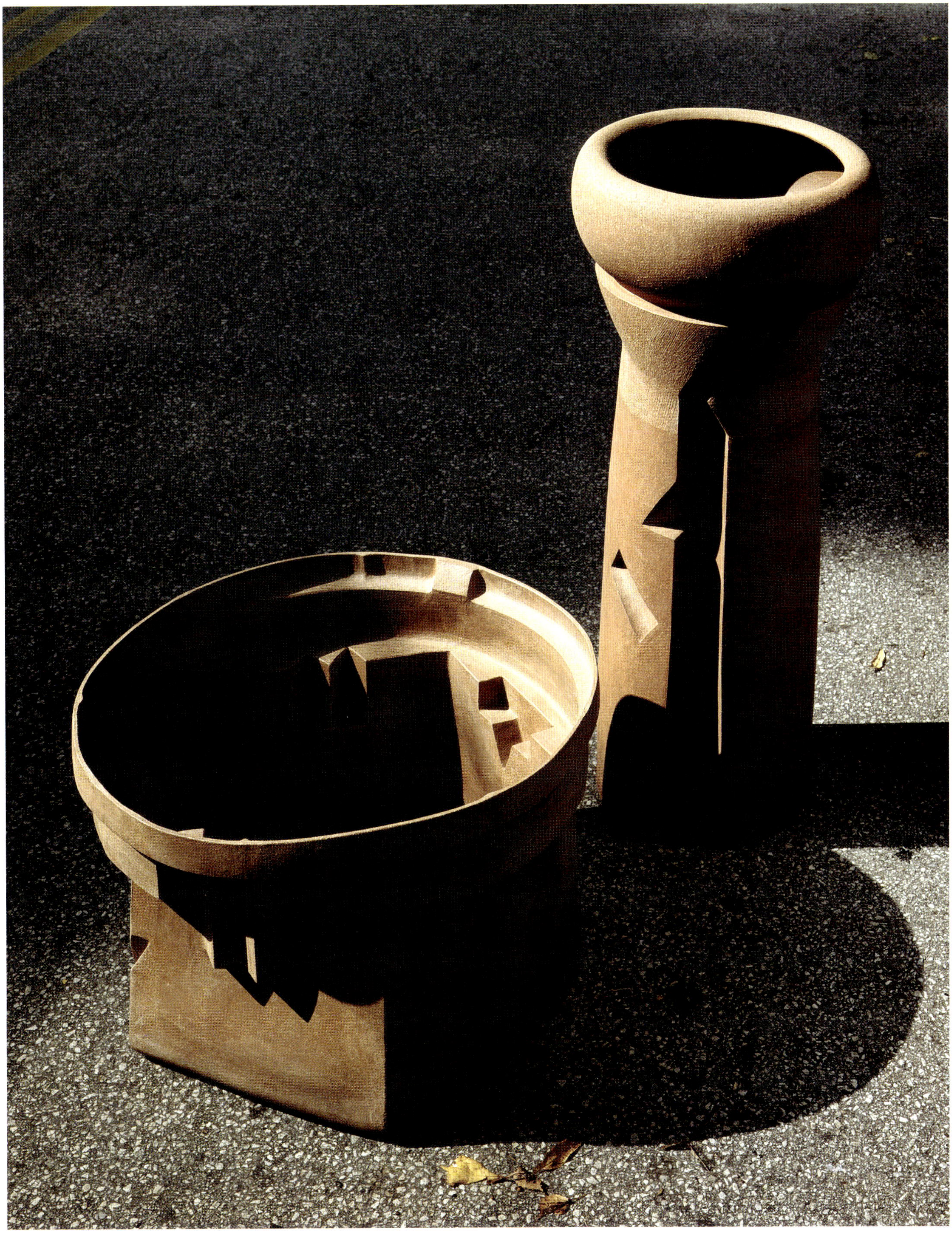

Photo: John Carlano

Fig. 145, left. *Lazarus 1* – Tank pot, 1980
Unglazed stoneware
14.5 x 25.5 x 25.5 inches
Collection of Peter Sterne

Onagas Passage – Vertical pot, 1980
Unglazed stoneware
38.5 x 18.5 x 18.5 inches
Collection of Edna Sloan Beron

Fig. 146. *Pentagonal Return*
Tank pot, 1980
Unglazed Stoneware
15 x 24 x 24 inches
Collection of Charlotte Daley

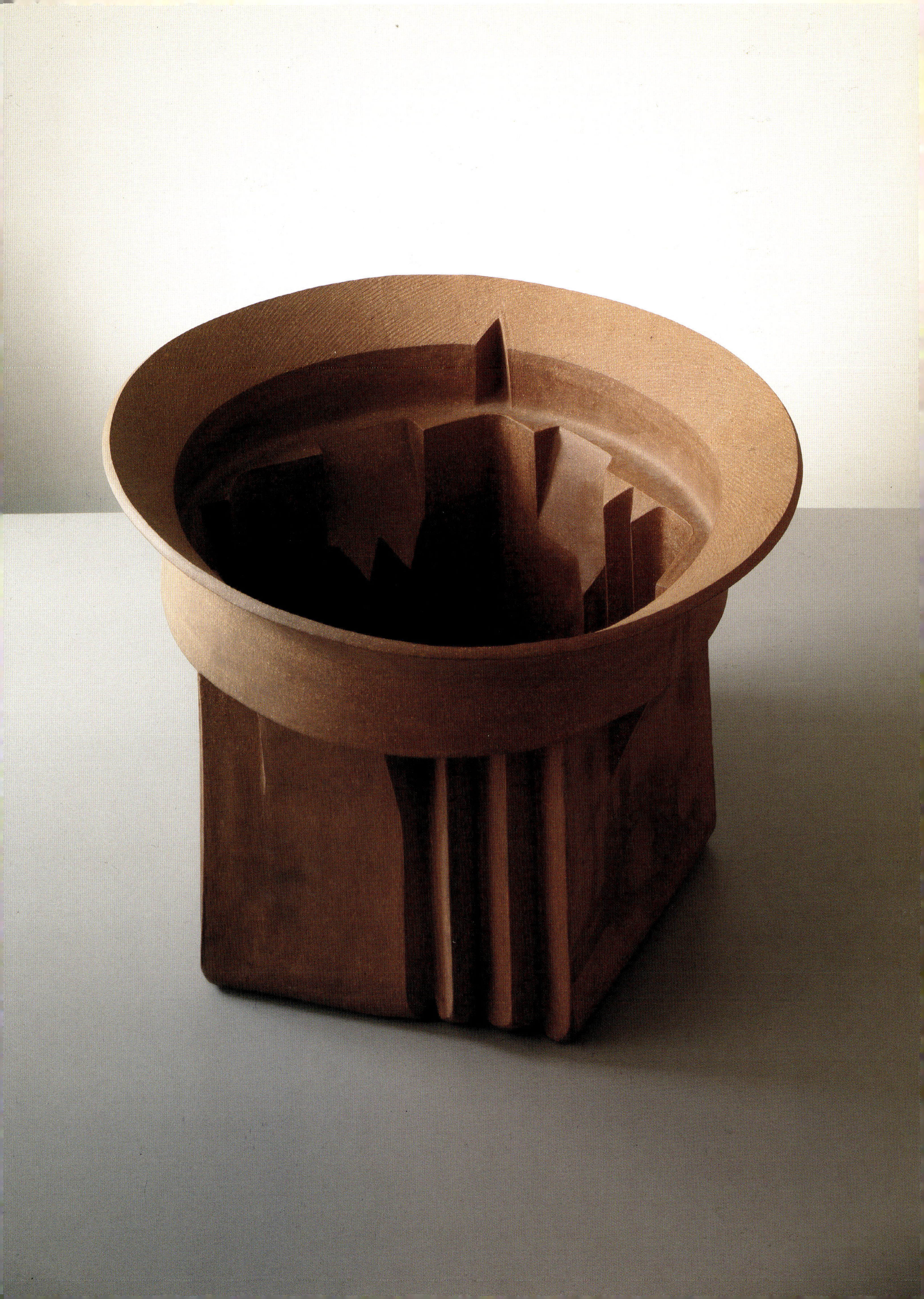

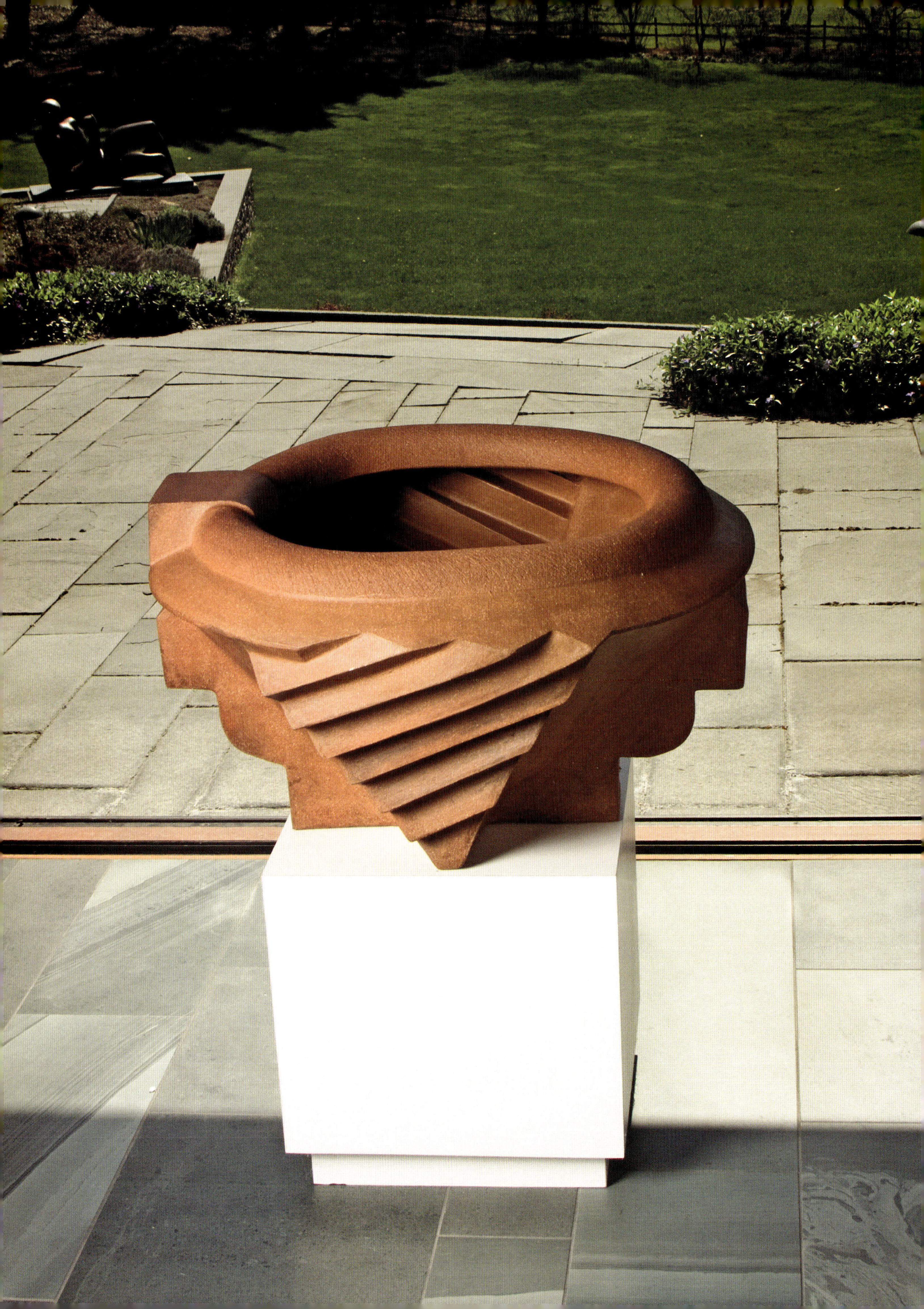

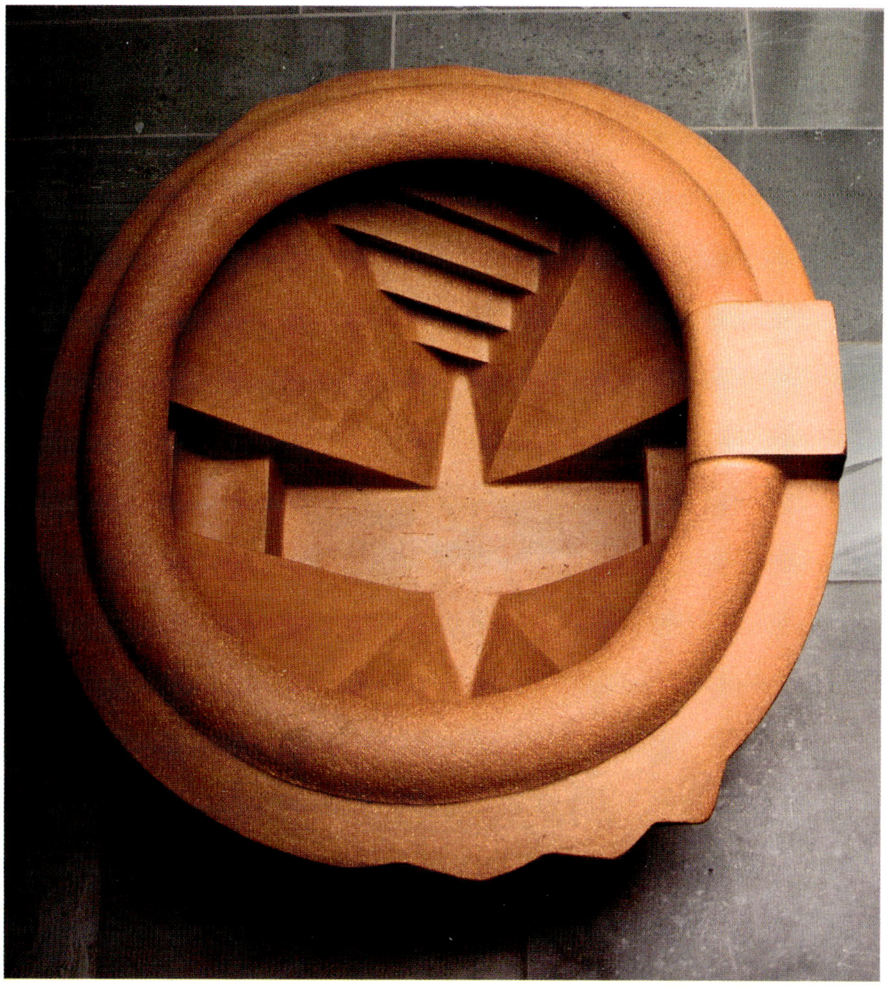

Fig. 147, opposite. *Cryptic Oval,* 1980, side view
Fig. 148, above. View from above
Unglazed stoneware
14 x 25 x 25 inches
Collection of Marlin and Ginger Miller

Photos: James Quaile

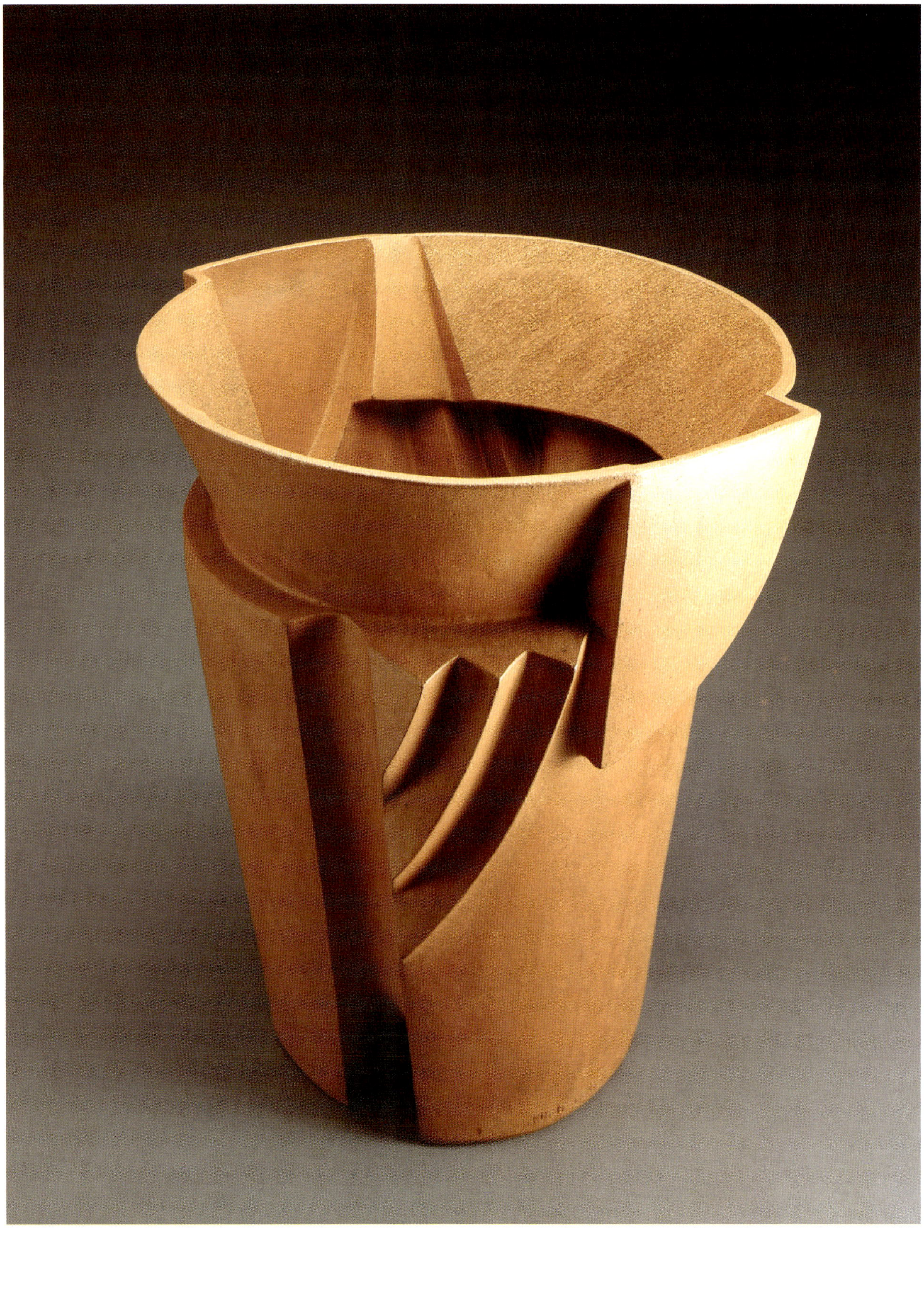

Fig. 149. *Kufic Center,* 1981
Unglazed stoneware
21 x 23 x 21 inches
Collection of Thomas and Donna Weaverling Daley

Fig. 150, opposite. *Passed Inn,* 1981
Unglazed stoneware
33 x 20 x 16.5 inches
Collection of William and Catherine Daley

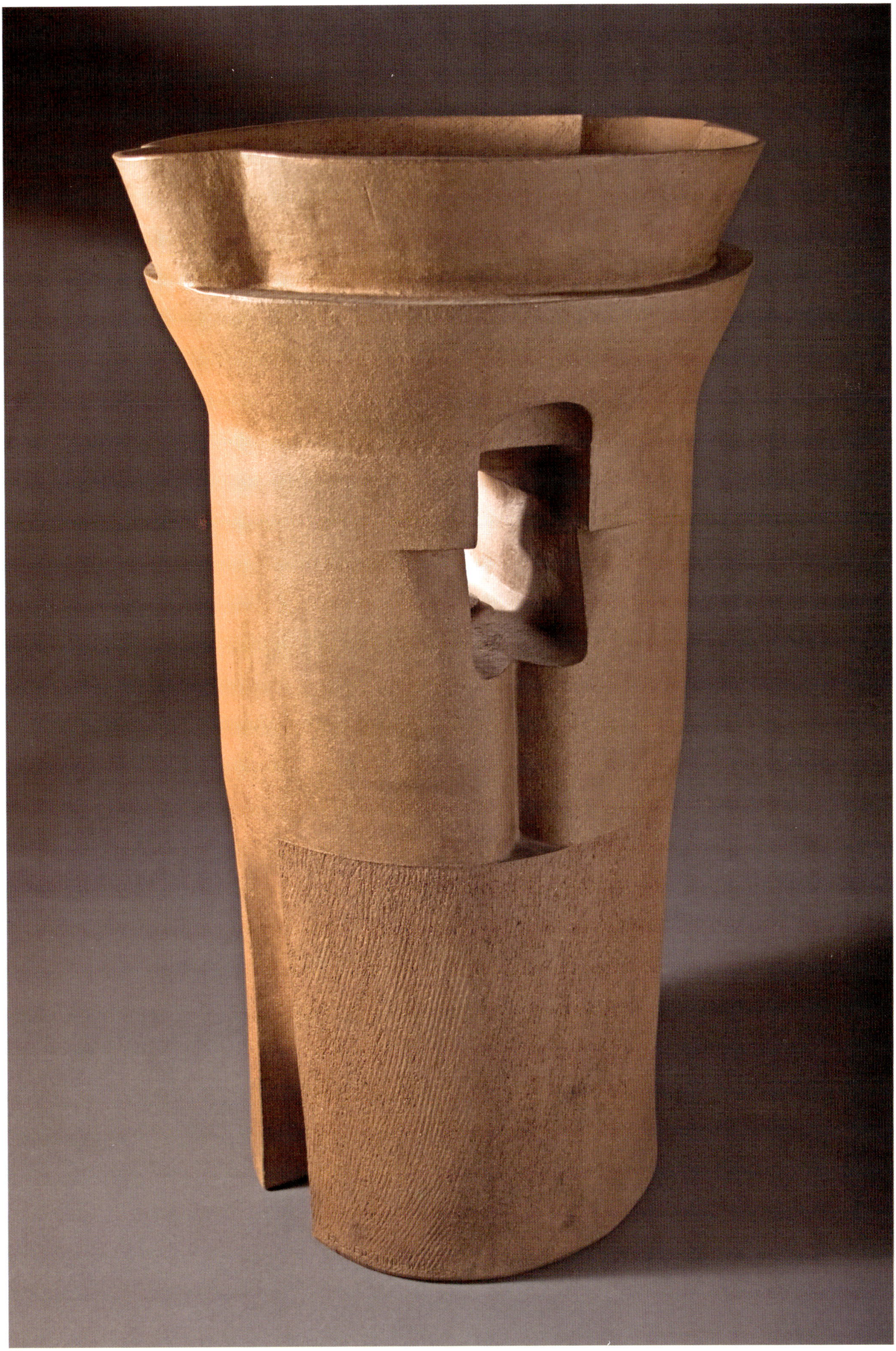
Photo: Amanda Williams

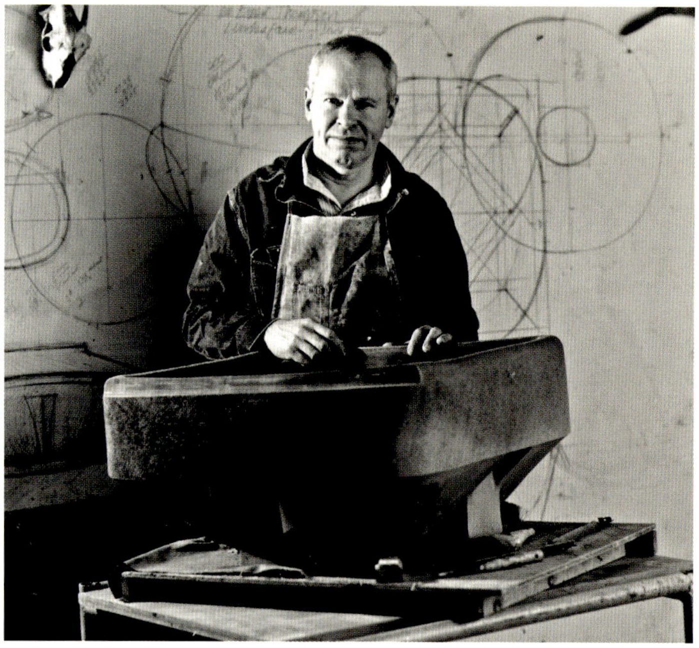
Photo: Charlotte Daley

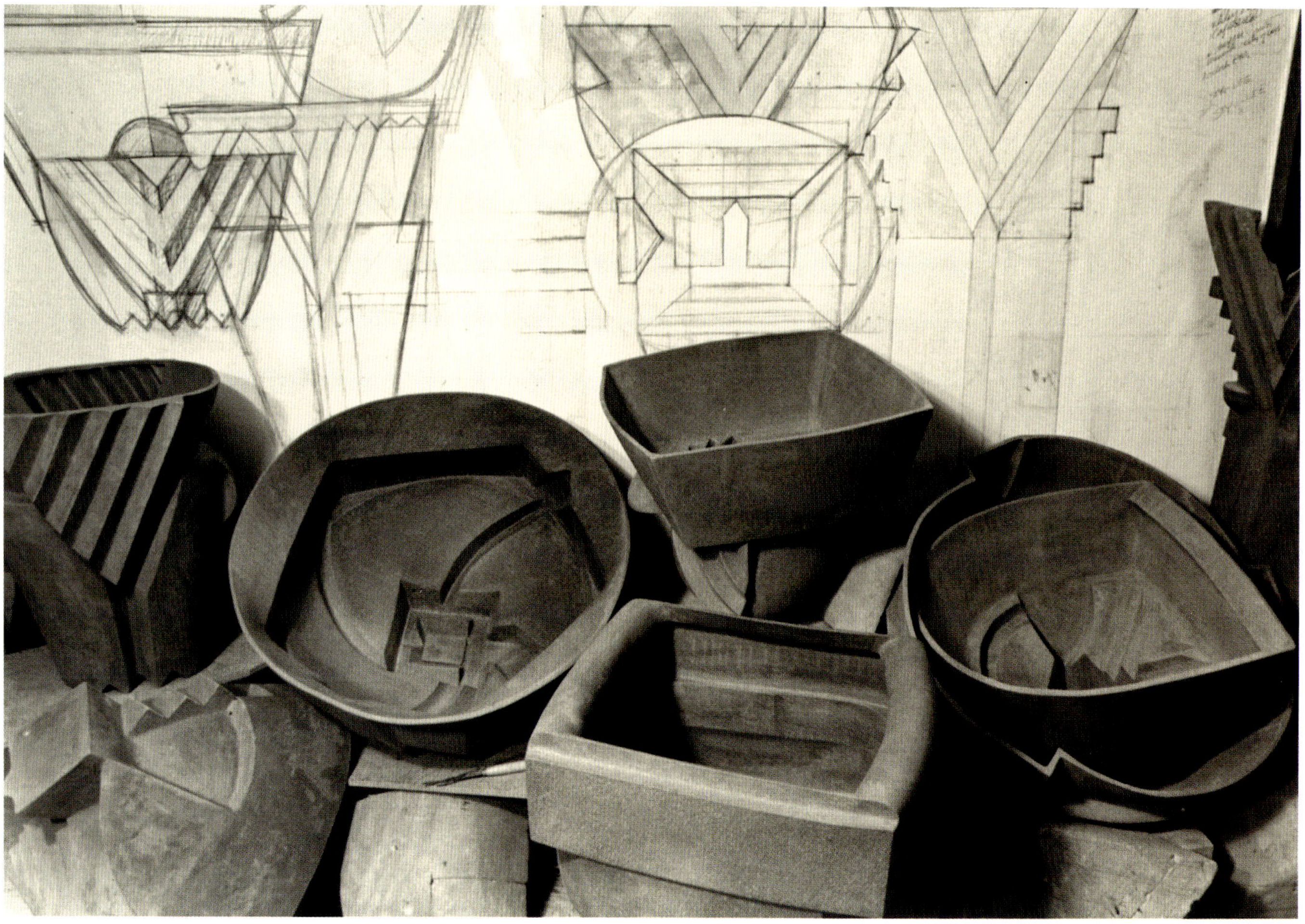
Photo: Sharon Carn

Fig. 151, top. William Daley in his studio
building *Danced Place,* 1982
12.75 x 27.5 x 27.5 inches
Collection of Stedelijk Museum
Hergtogenbosch, The Netherlands

Fig. 152, above. Ceramic works, 1983
Pots arranged by artist in his studio in front of the
Drawing Wall in preparation for an exhibition at
the Braunstein/Quay Gallery, San Francisco, CA

Photo: James Quaile

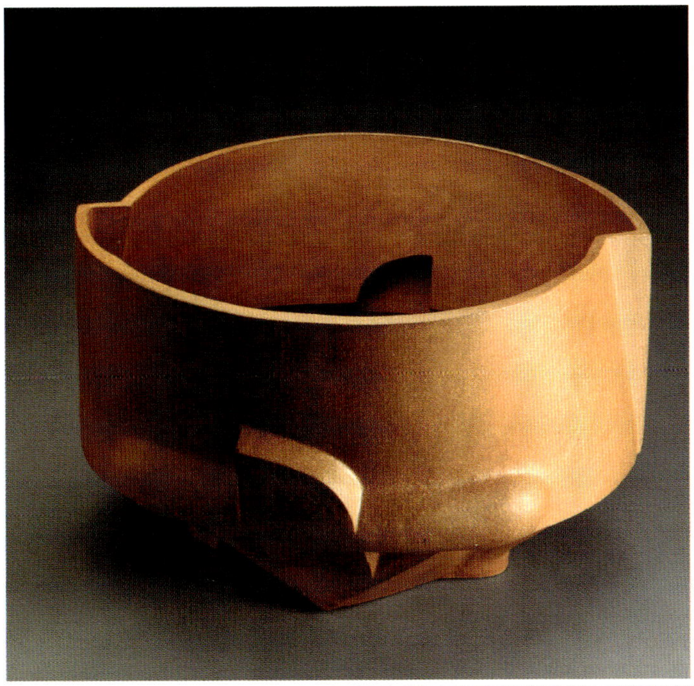

Photo: James Quaile

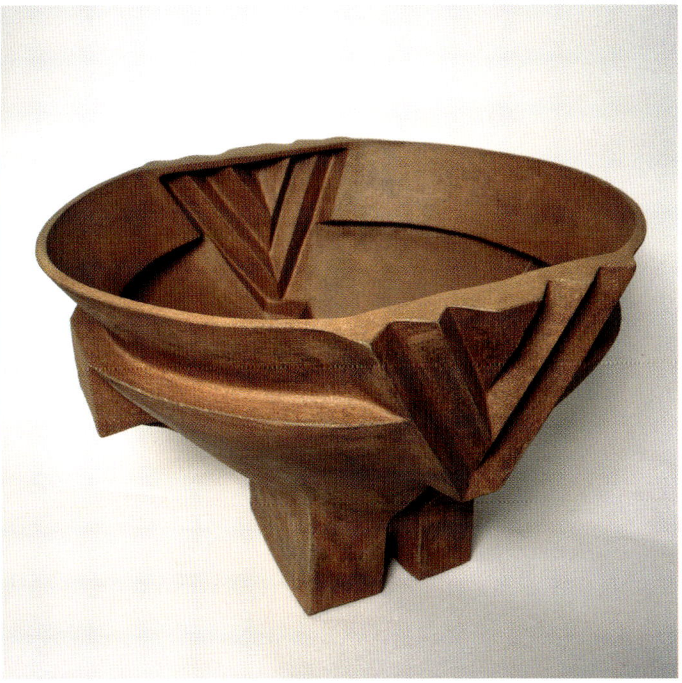

Photo: Amanda Schwab

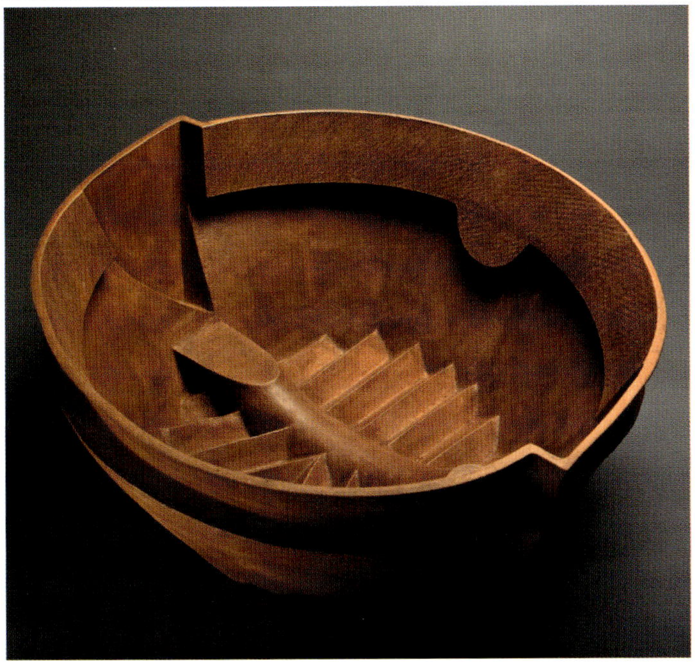

Photo: James Quaile

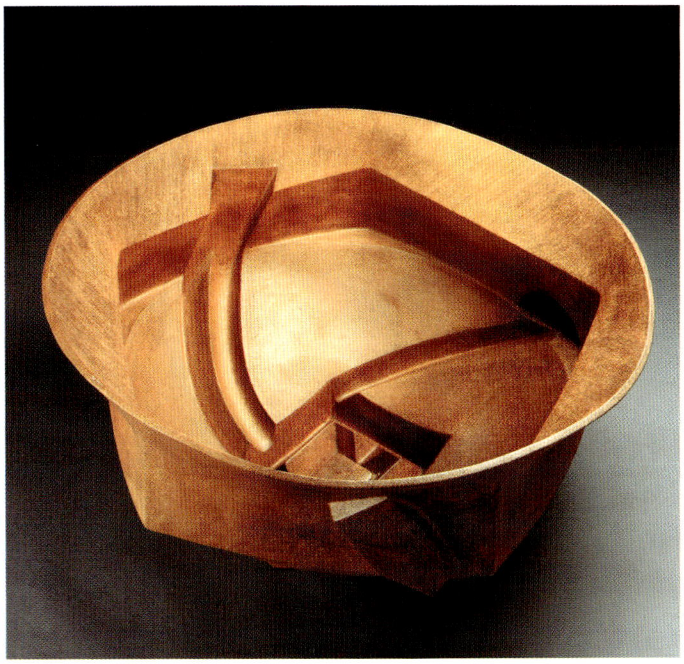

Fig. 153, top left. *Oval Ark,* 1983
Unglazed stoneware
16.5 x 25 x 24 inches
Collection of Drs. Harry and Betty Gotlieb

Fig. 154, bottom left. *Turners Court,* 1983
Unglazed stoneware
12.5 x 23 x 27.5 inches
Exhibited Crafts USA world tour
Named for Robert Turner
Collection of Thomas and Donna Weaverling Daley

Fig. 155, top right. *Shang Play,* 1983
Unglazed stoneware
17 x 25 x 26 inches
Collection of US News and World Report

Fig. 156, bottom right. *Pentagonal Destination,* 1984
Unglazed stoneware
15.5 x 31 x 31 inches
Signed: William and Catherine Daley
Private Collection

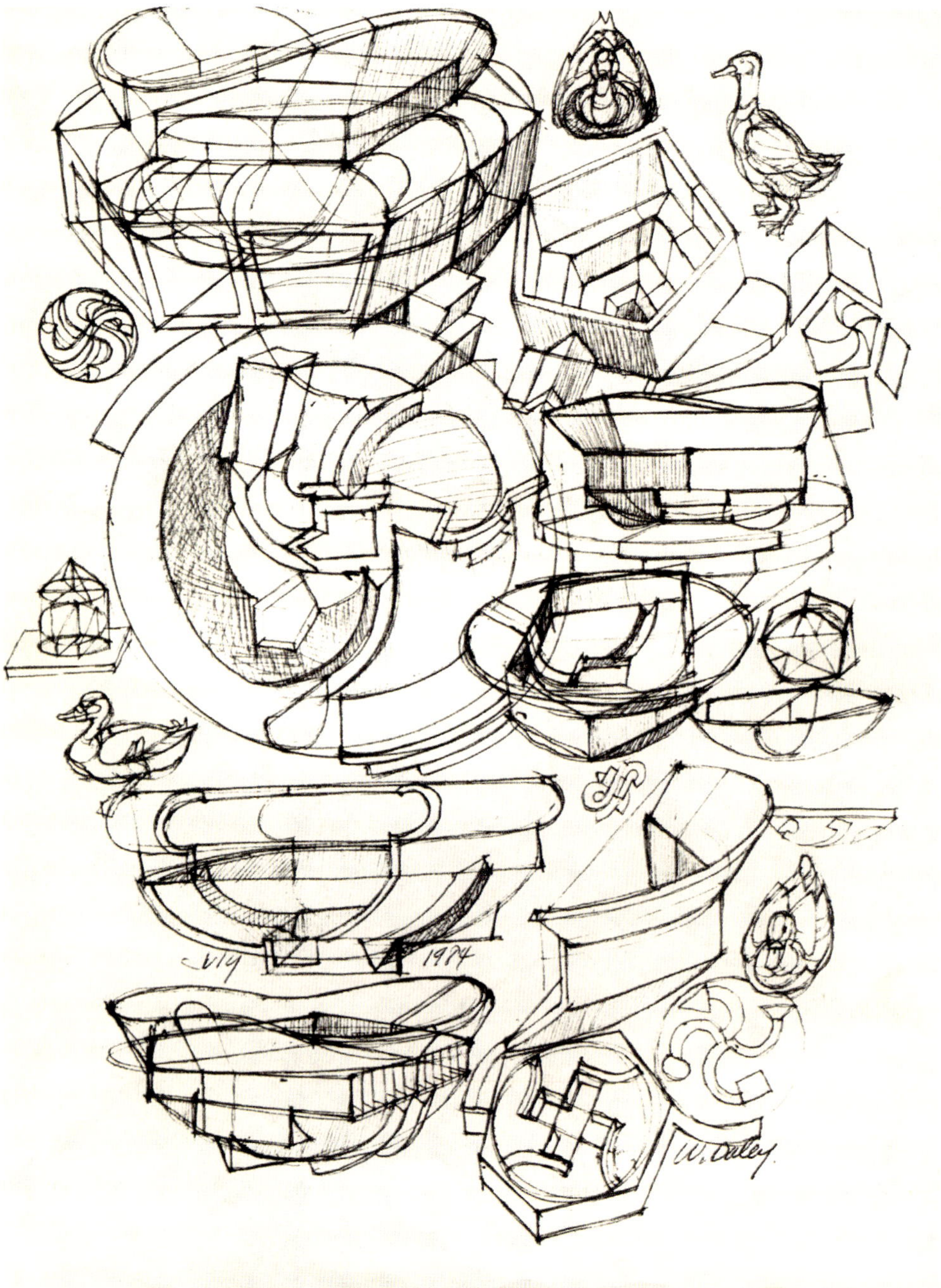

Fig. 157. Pot sketch, 1984
Pen on paper
14 x 11 inches
Collection of William and Catherine Daley

opposite:

Fig. 158, top left. *Friendly Location,* 1981
Unglazed stoneware
18 x 24 x 20 inches
Collection of Helen W. Drutt English

Fig. 159, top right. *Triaxial Procession,* 1982
Unglazed stoneware
21 x 25 x 21 inches
Collection of Helen and Jack Bershad

Fig. 160, middle left. *Tri-D Out,* 1983
Unglazed stoneware
20 x 18 x 21 inches
Collection of Vincente Lim and Robert Tooey

Fig. 161, middle right. *Brenden's Arc,* 1983
Unglazed stoneware
14 x 20 x 22 inches
Collection of George and Amanda Schwab

Fig. 162, bottom left. *B's Place,* 1985
Unglazed stoneware
12 x 33.5 x 33.5 inches
Collection of Clayarch Gimhae Museum,
Gimhae, Korea

Fig. 163, bottom right. *Apsed Entry,* 1984
Unglazed stoneware
16 x 22 x 22.5 inches
Signed William & Catherine Daley
Private collection

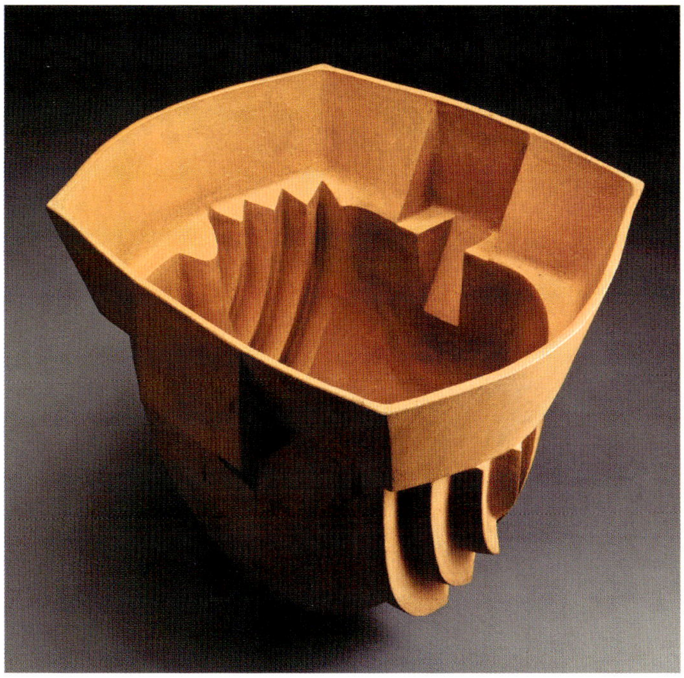
Photo: John Carlano

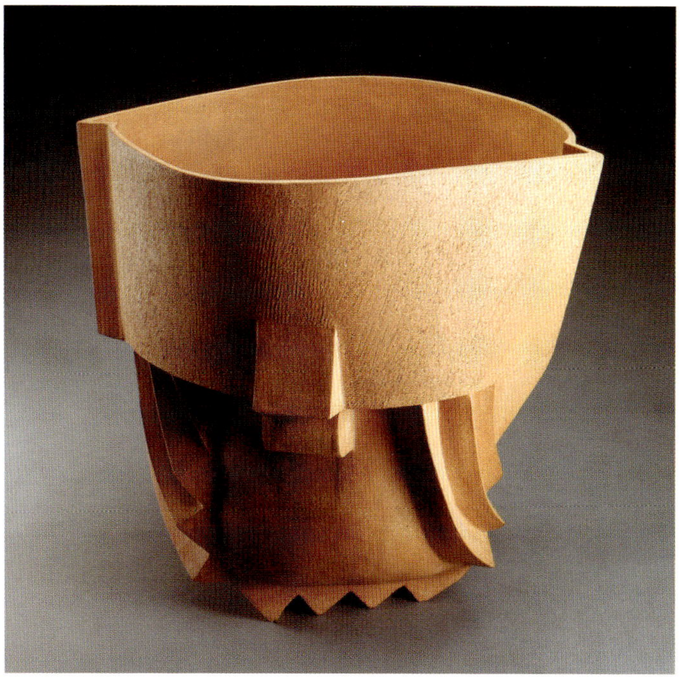
Photo: James Quaile

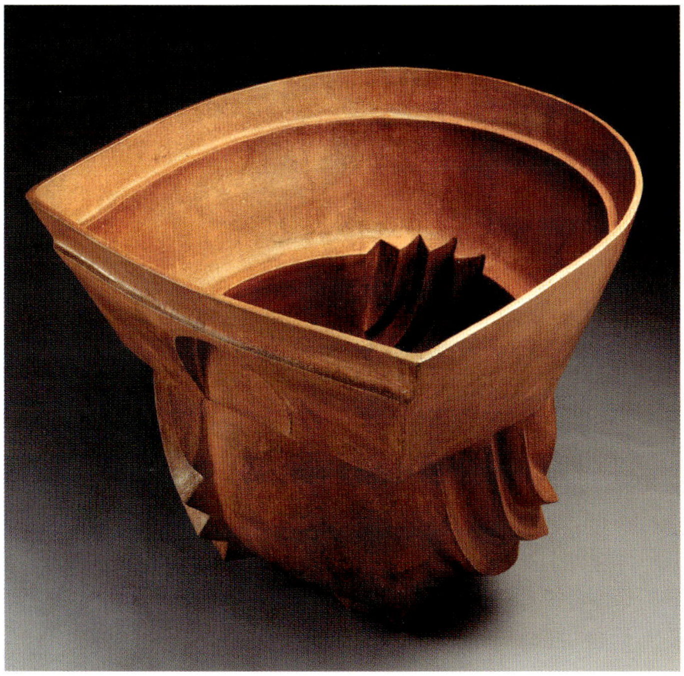
Photo: James Quaile

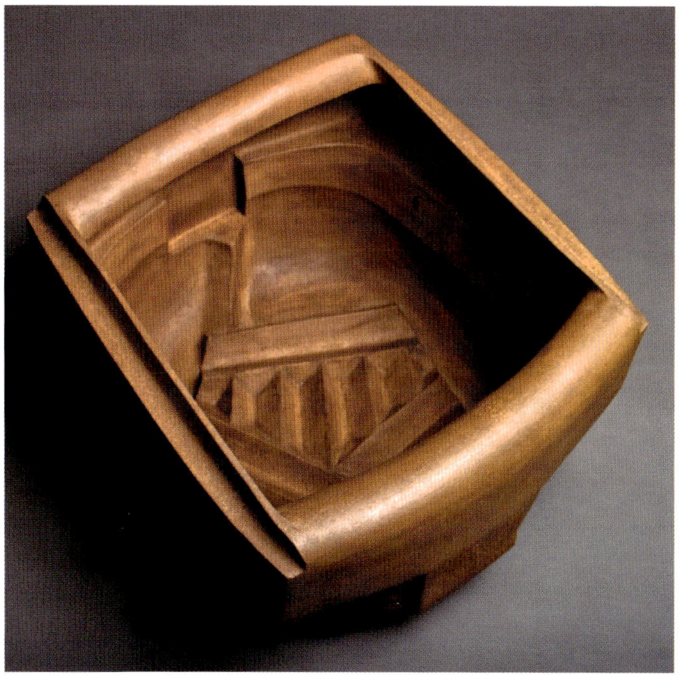
Photo: Amanda Schwab

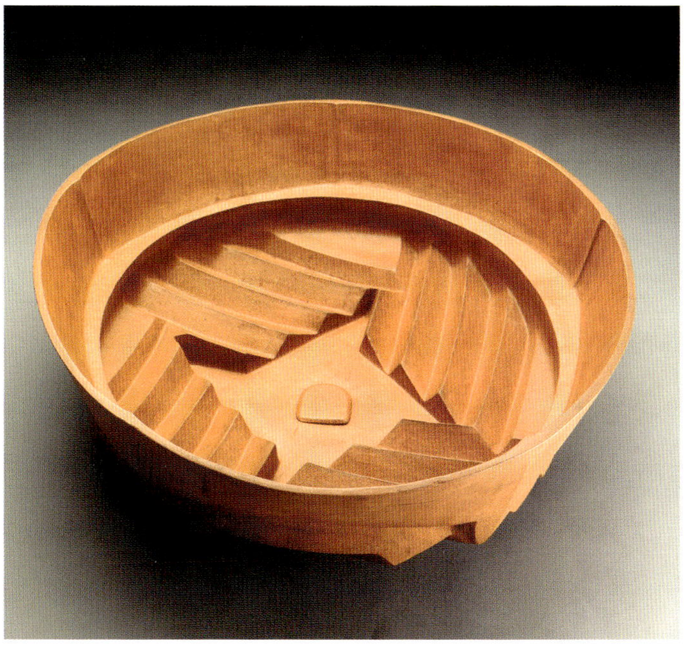
Photo: James Quaile

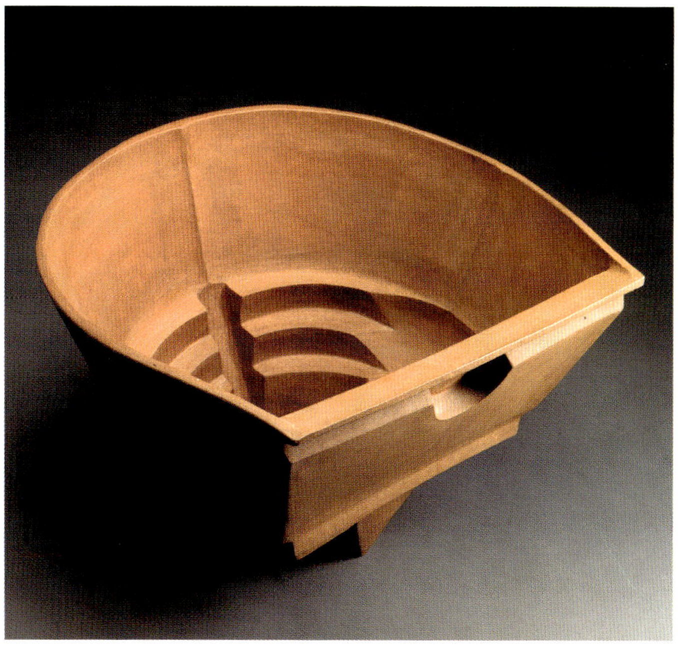
Photo: James Quaile

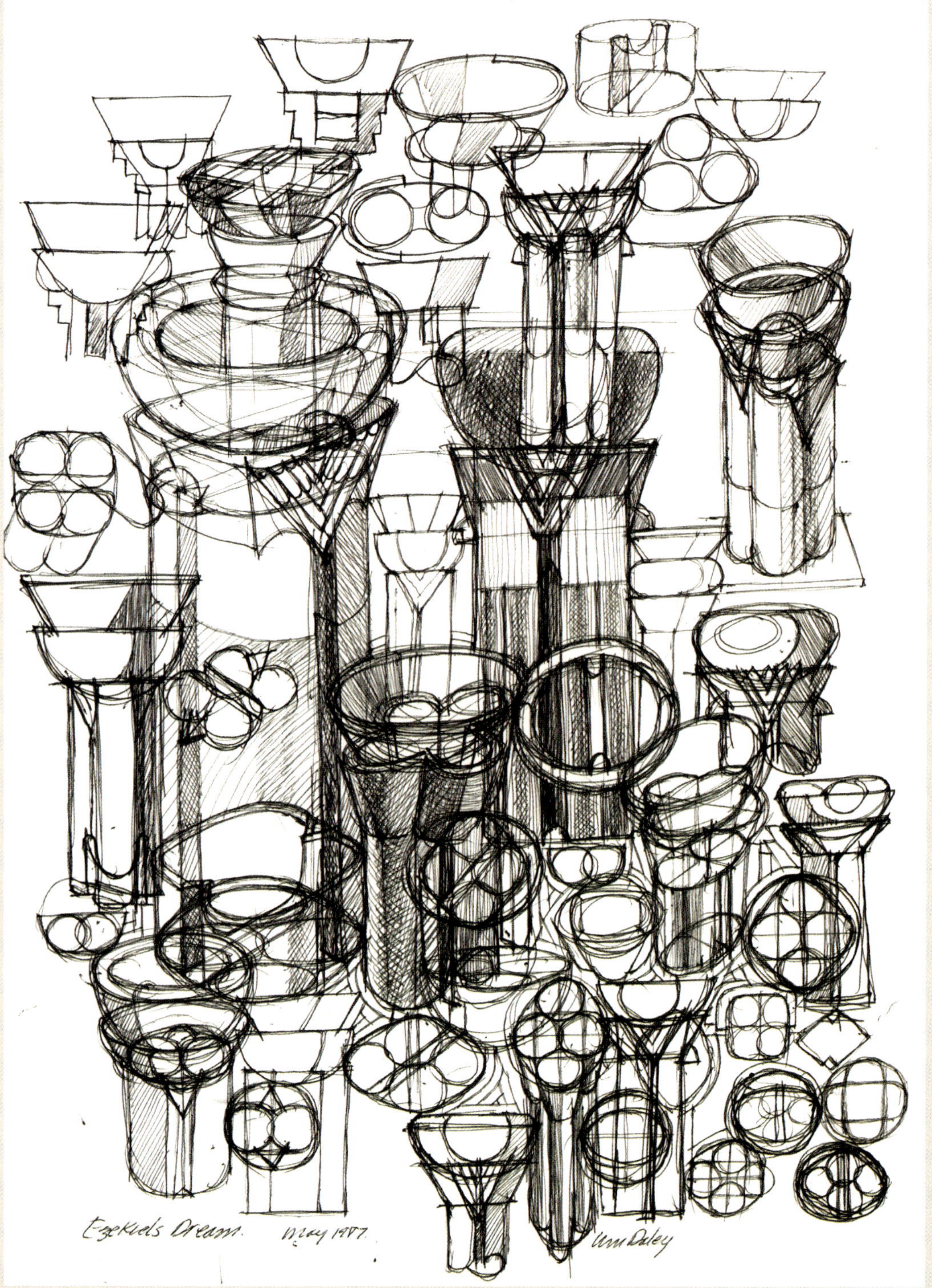

Fig. 164. Drawing, *Ezekiel's Dream*, 1987
Pen on bristol paper
12 x 9 inches

Fig. 165. *Coupled Place*, 1988
Unglazed stoneware
40 inches tall x 35 inches wide oval top
Collection of Jane and Leonard Korman

Photo: James Quaile

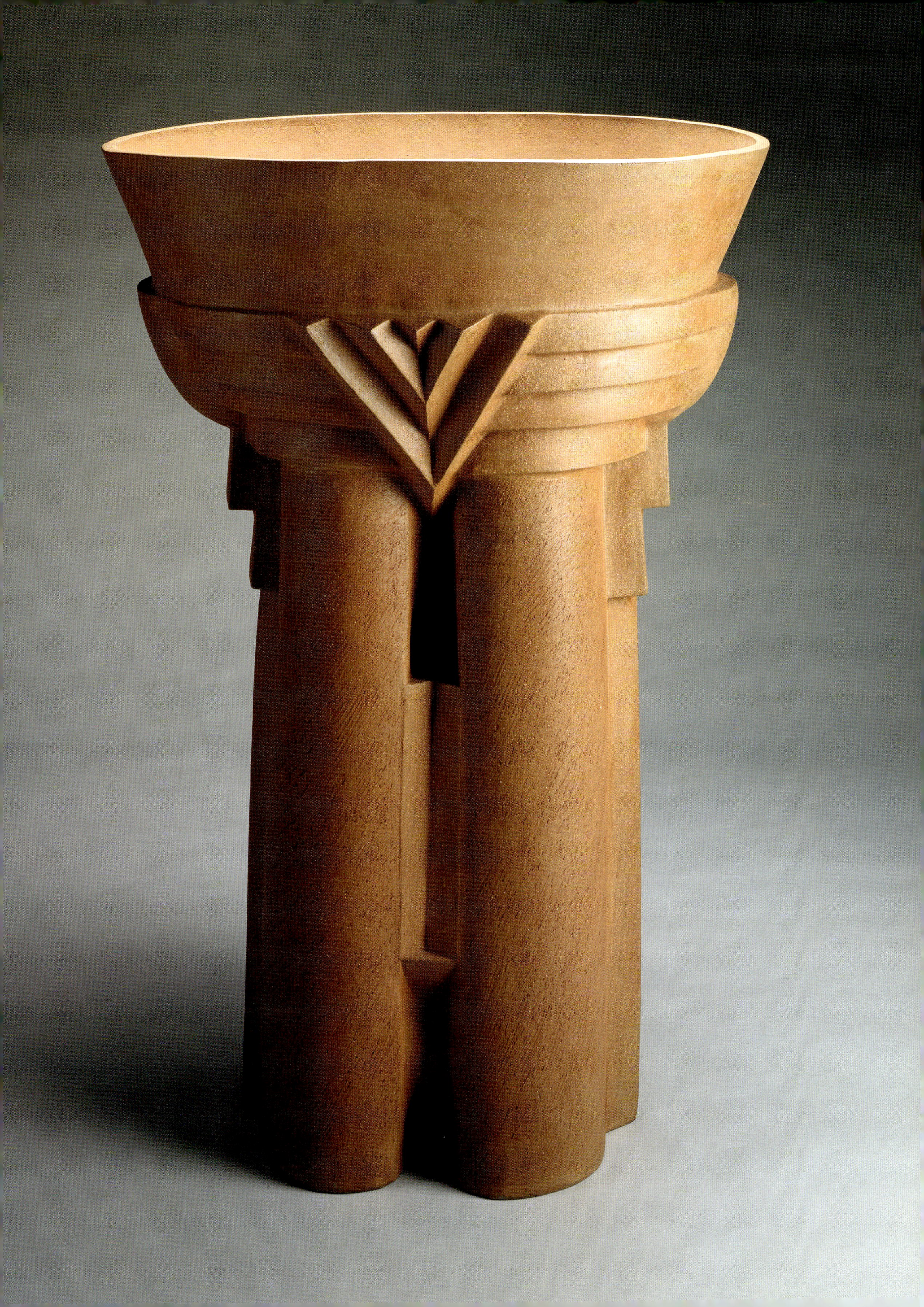

Photo: Amanda Schwab

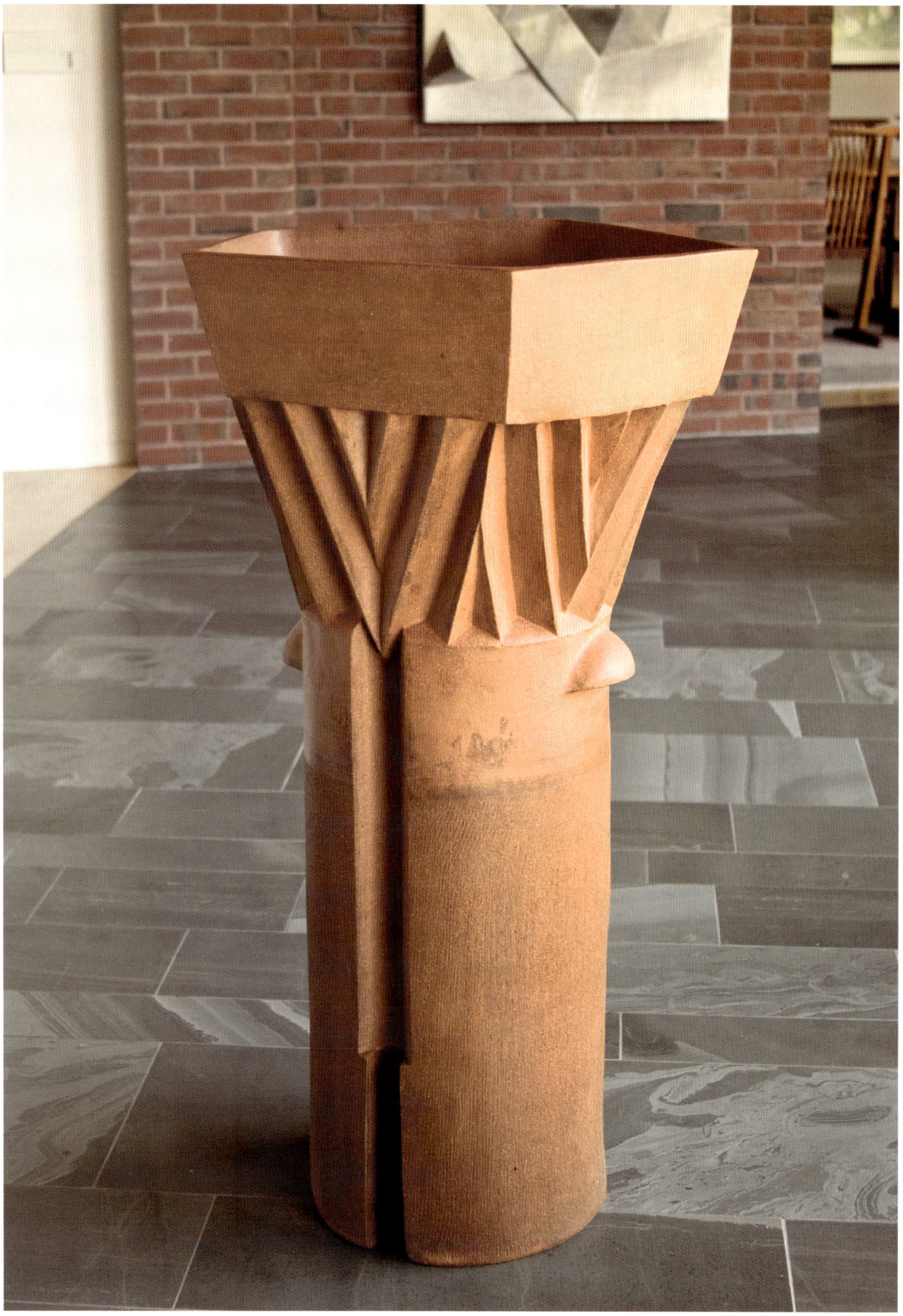

Fig. 166. *Pentagonal Column,* 1986
Unglazed stoneware
40 x 21 x 21 inches
Collection of Marlin and Ginger Miller

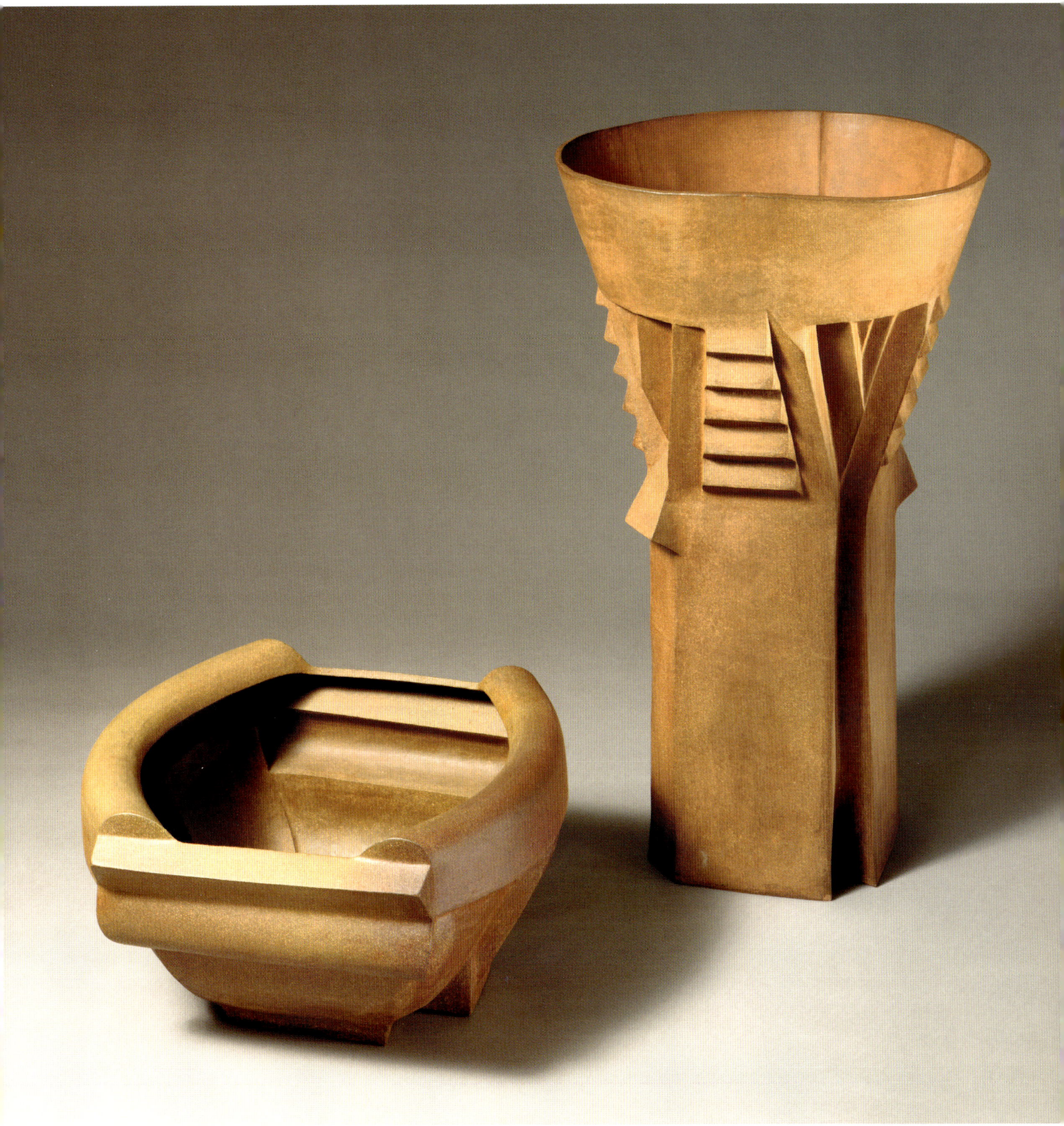

Fig. 167. *Prides Crossing,* 1985 – low pot
Unglazed stoneware
15 x 22.5 x 14 inches
Collection of the Metropolitan Museum of Art,
New York, NY

Squared Up, 1985 – vertical pot
Unglazed stoneware
39 x 23.25 inches diameter top
Collection of the Metropolitan Museum of Art,
New York, NY

Photo used with permission of the Metropolitan Museum of Art

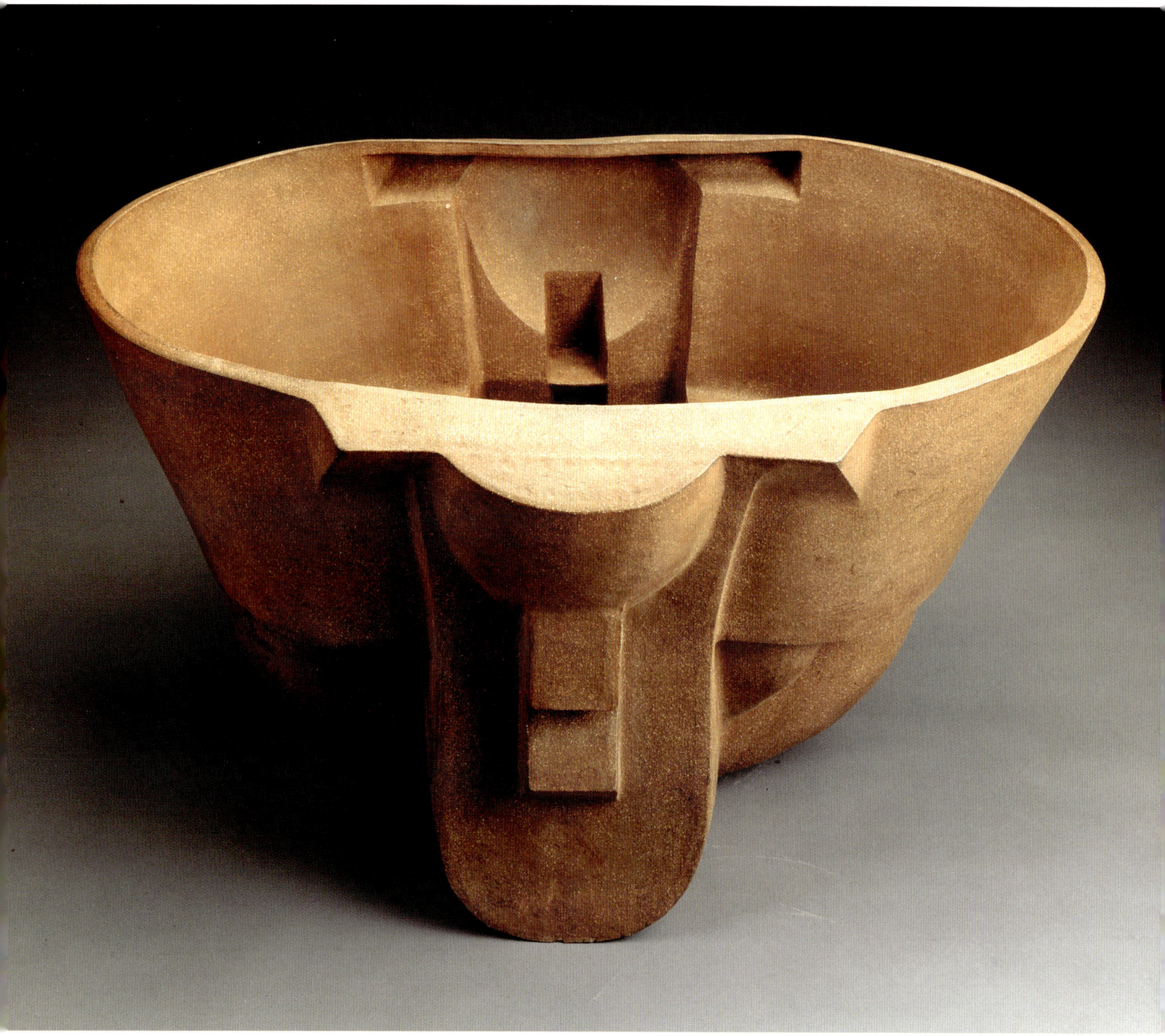

Fig. 168, above. *Cross Up,* 1986
Unglazed stoneware
16 x 22.5 x 33 inches
Collection of Alfred and Rica Fenaughty

Fig. 169, opposite.
Oval Step Tank, 1987, view from above
Fig. 170. Side view
Unglazed stoneware
16.5 x 23.75 x 30.25 inches
Collection of William and Catherine Daley

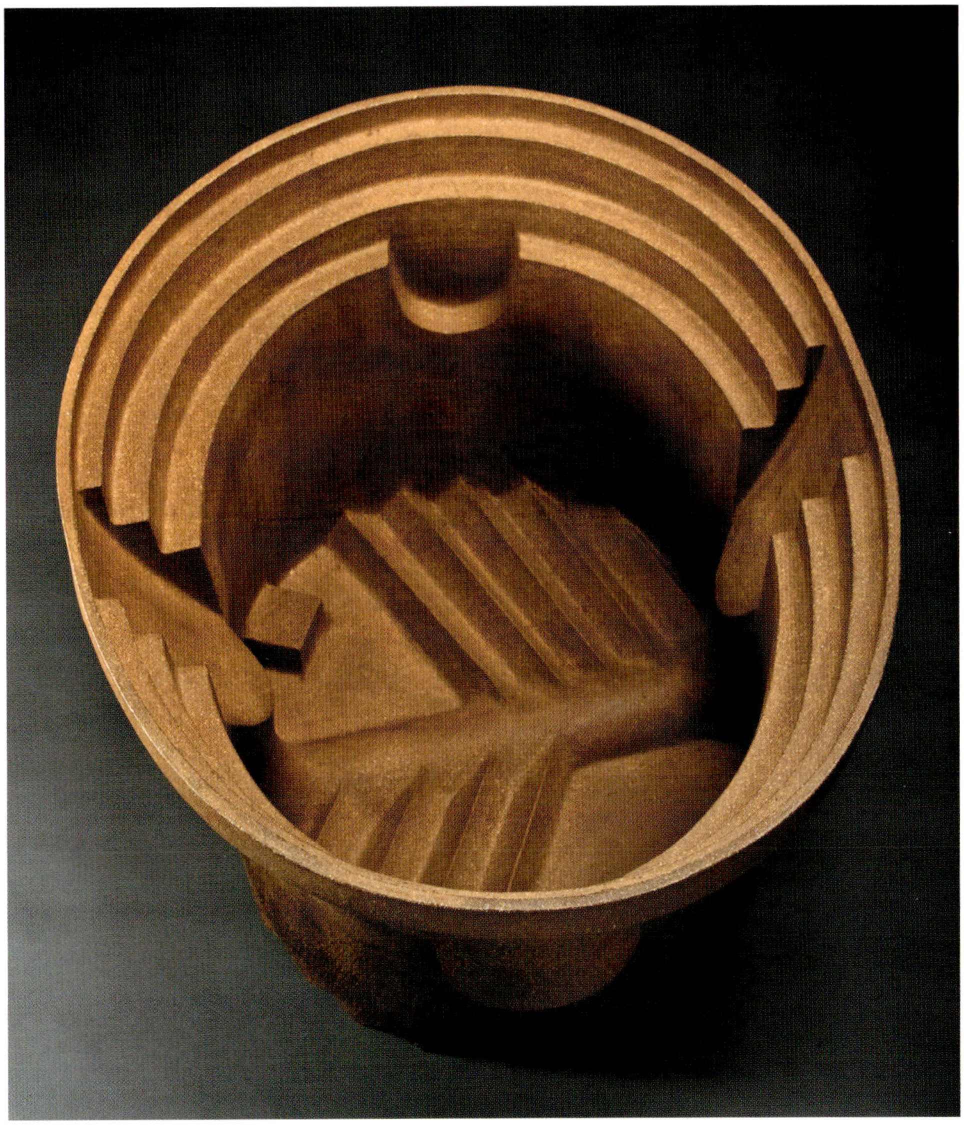

Photos: Amanda Schwab

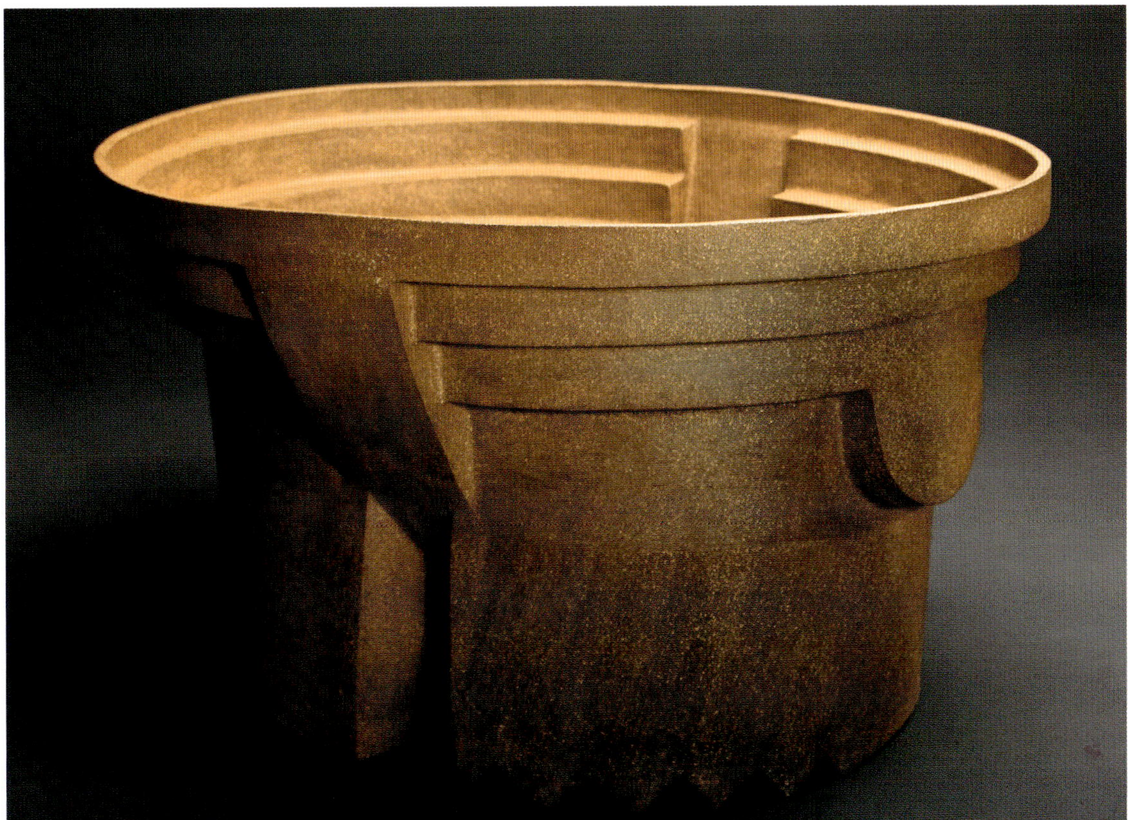

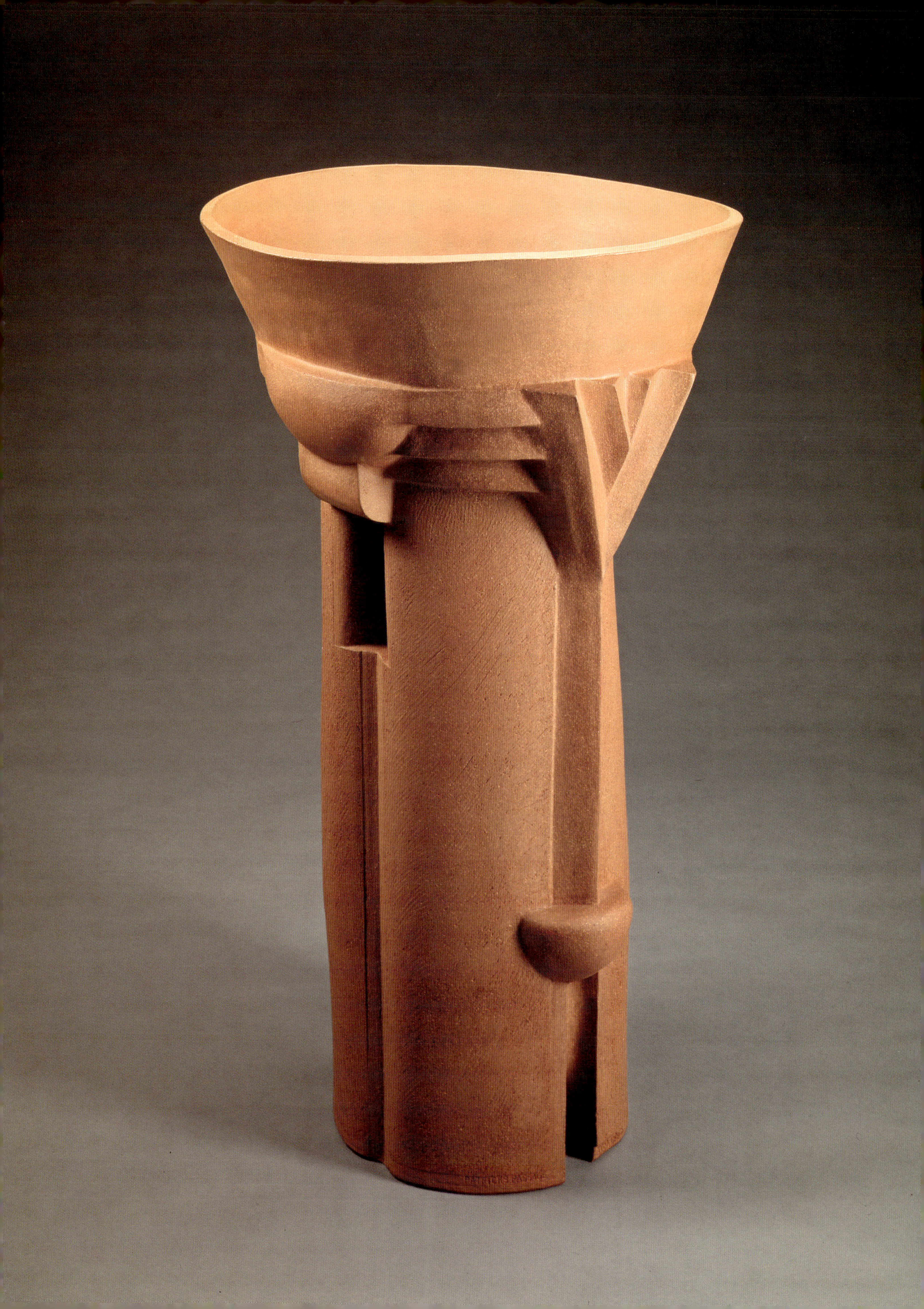

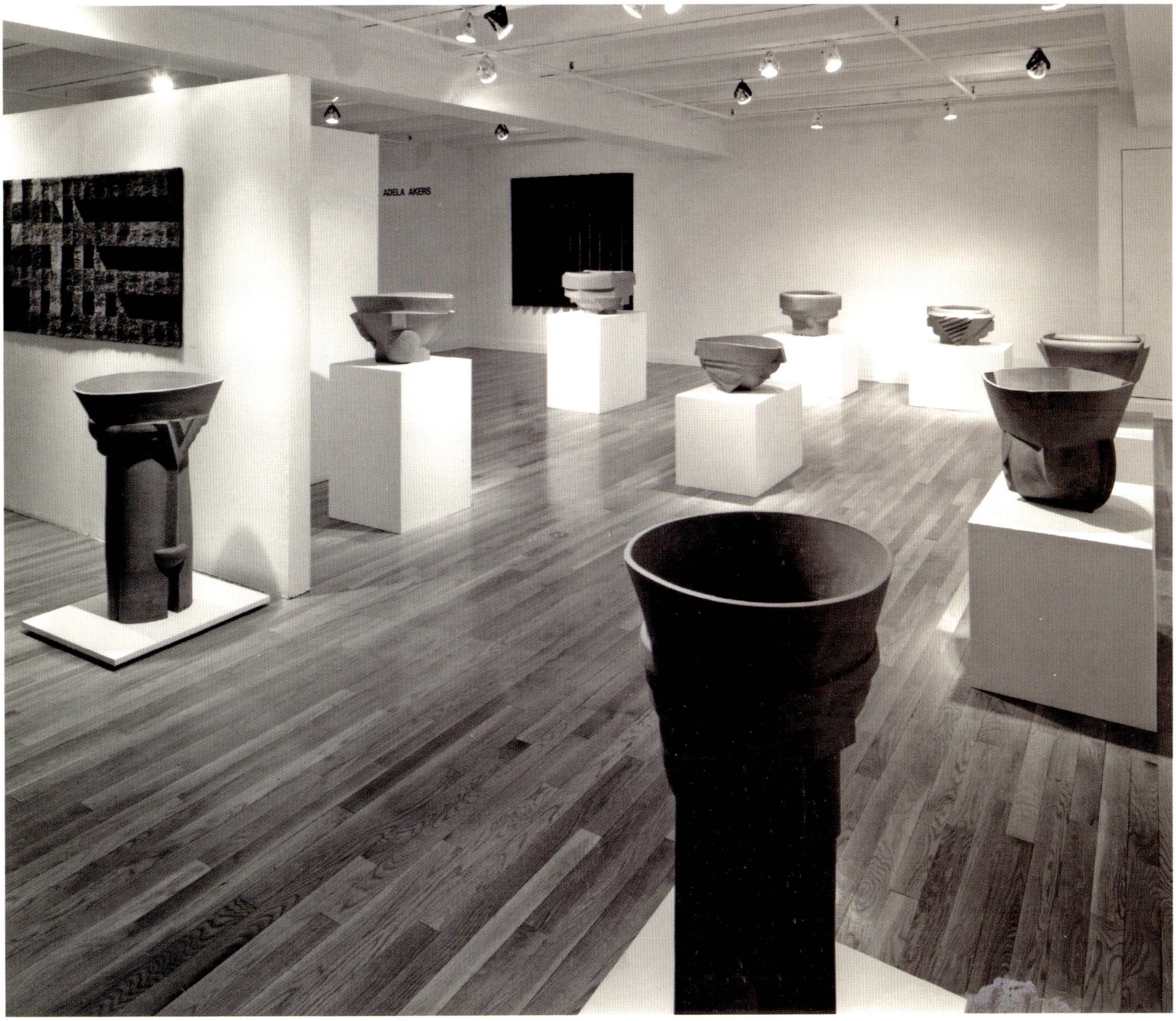

Photo: William Daley

Fig. 171, opposite.
Patrick's Passage, 1989
Unglazed stoneware
39 x 23 x 25 inches
Private collection

Fig. 172. Various pots on exhibit at the Helen Drutt Gallery, Fifth Avenue, New York, NY, 1990

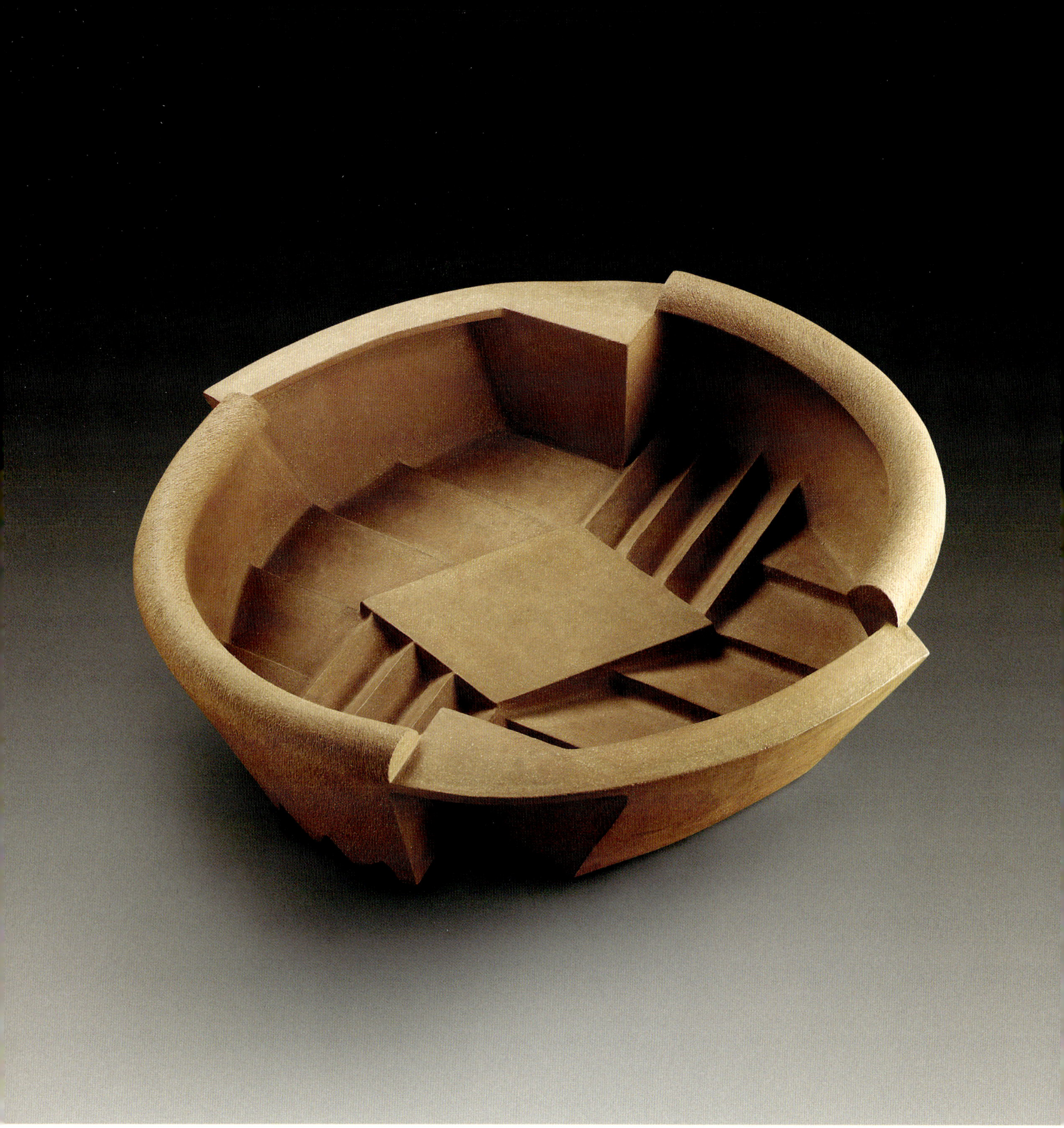

Fig. 173. *Mandarin Garden,* 1990
Unglazed stoneware
7 x 24 x 26 inches
Collection of William and Catherine Daley

Fig. 174, opposite. *Apsed Oval,* 1990
Fig. 175. Side view
Unglazed stoneware
13 x 22 x 24.5 inches
Collection of Vincente Lim and Robert Tooey

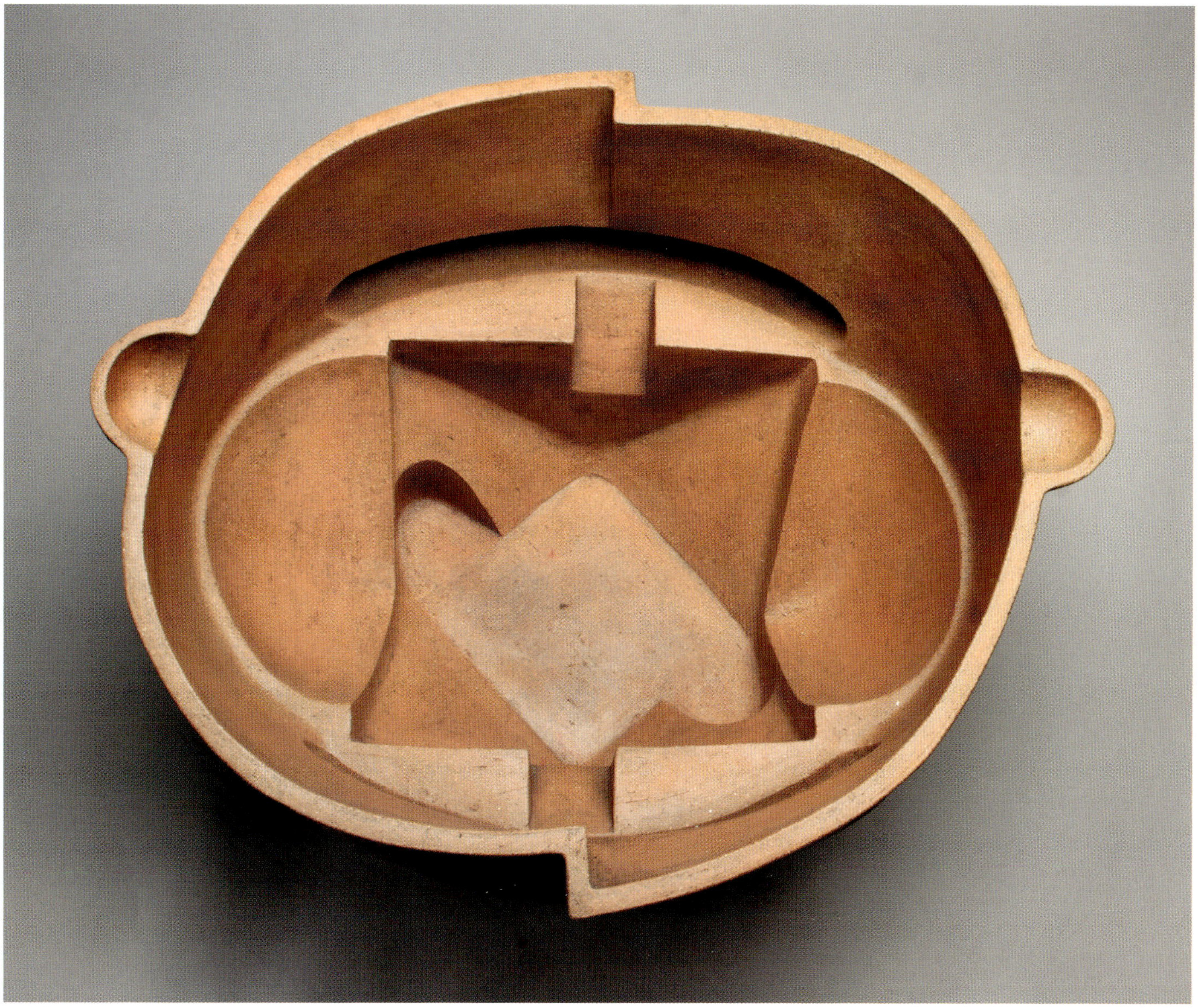

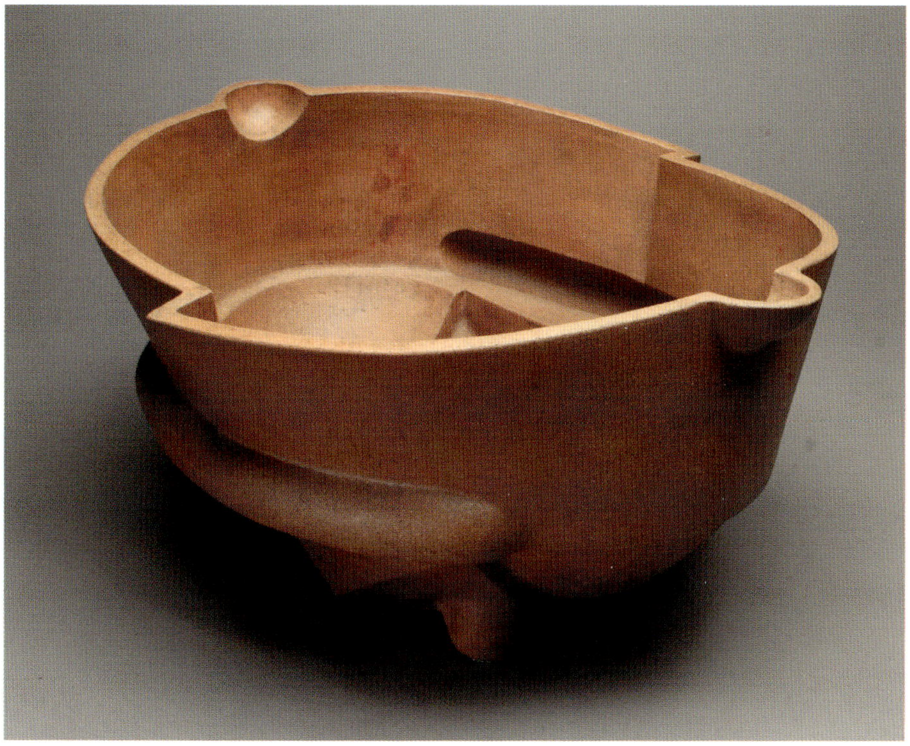

Photos: James Quaile

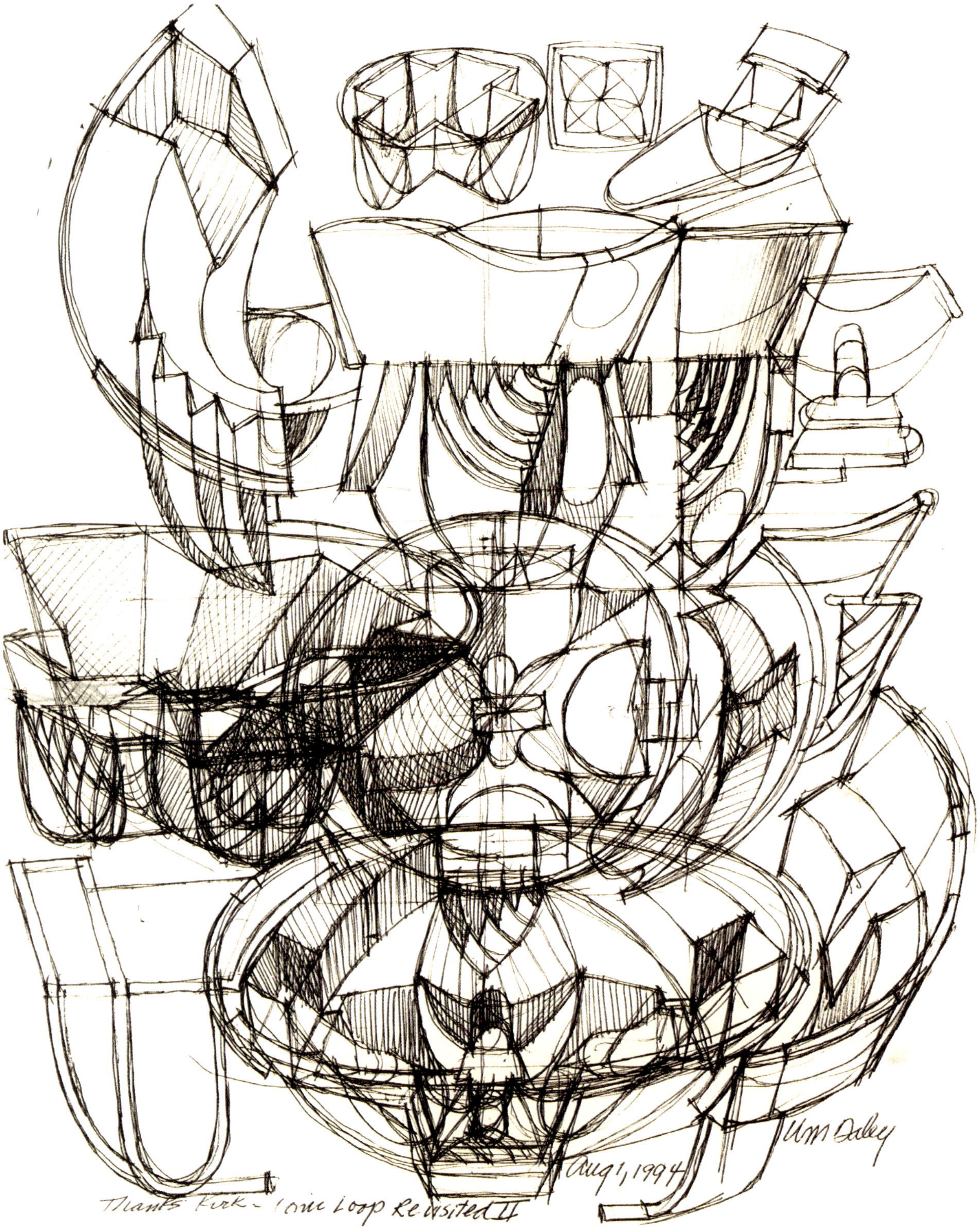

Fig. 176. Drawing, *Conic Loop Revisited II,* 1994
Pen on paper
14 x 11 inches

Fig. 177, opposite.
Vesica One, 1991
Fig. 178. Side view
Unglazed stoneware
7 x 24 x 26 inches
Collection of Warren Seelig and Sherri Gibson

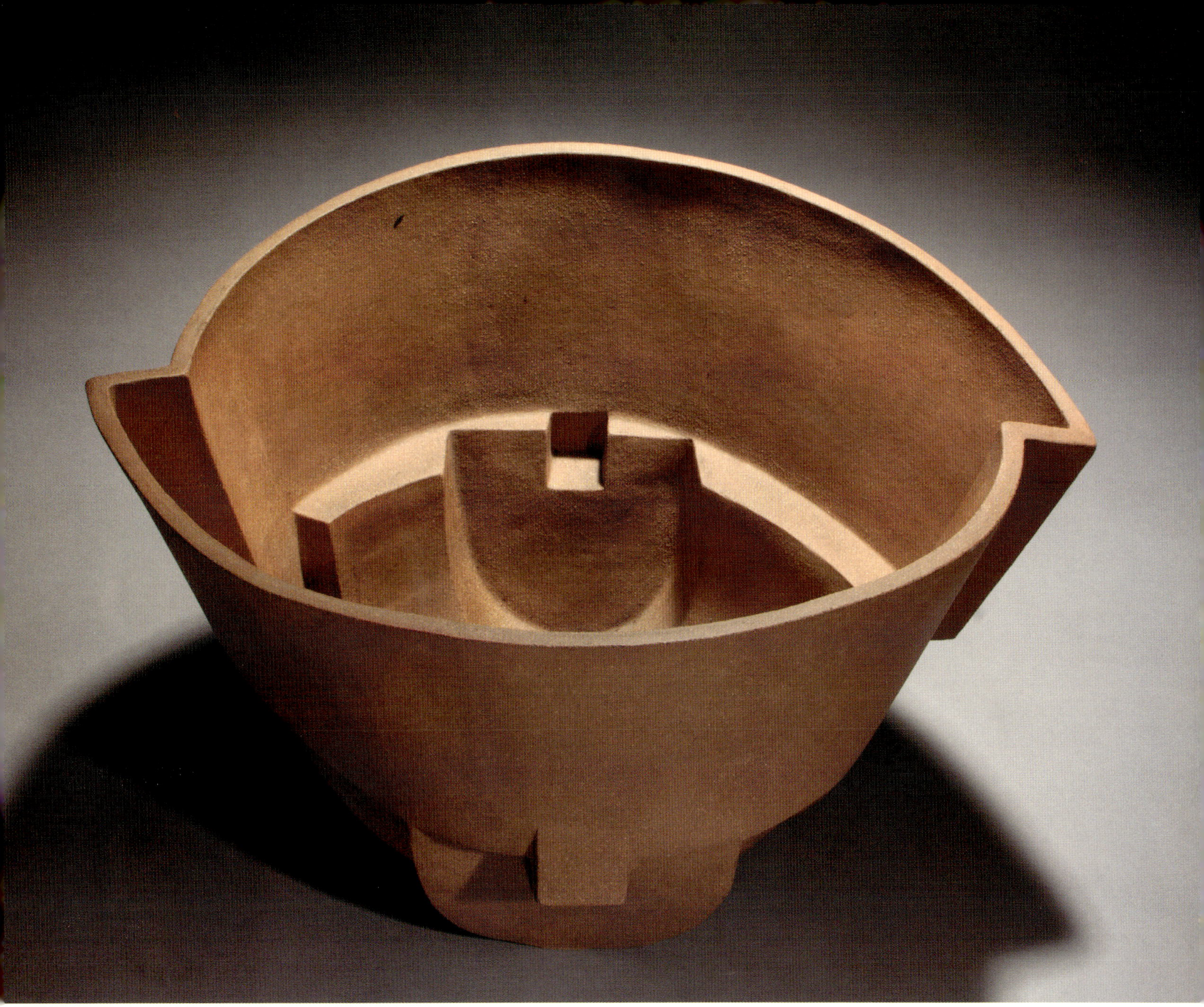

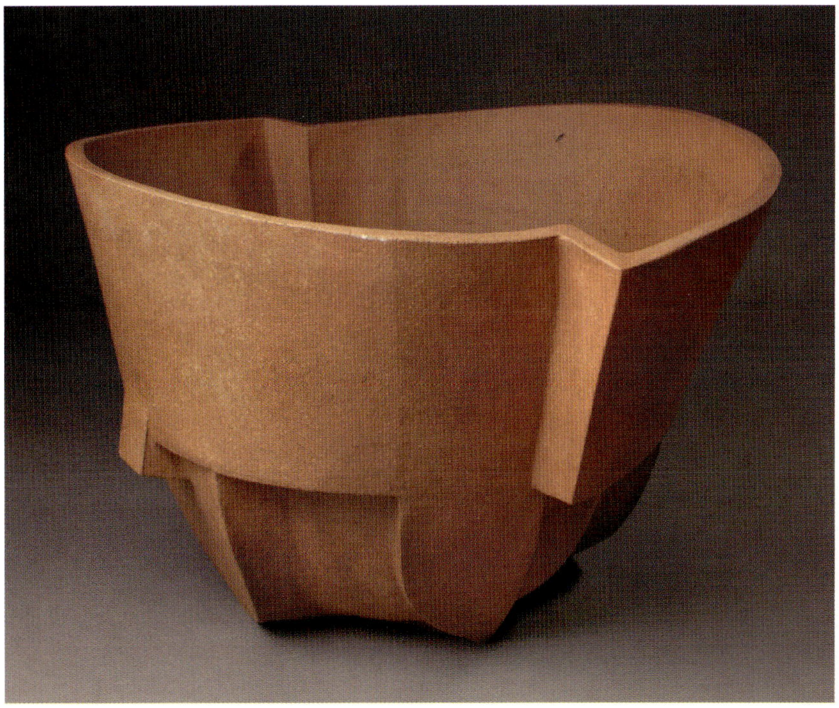

Photos: Amanda Schwab

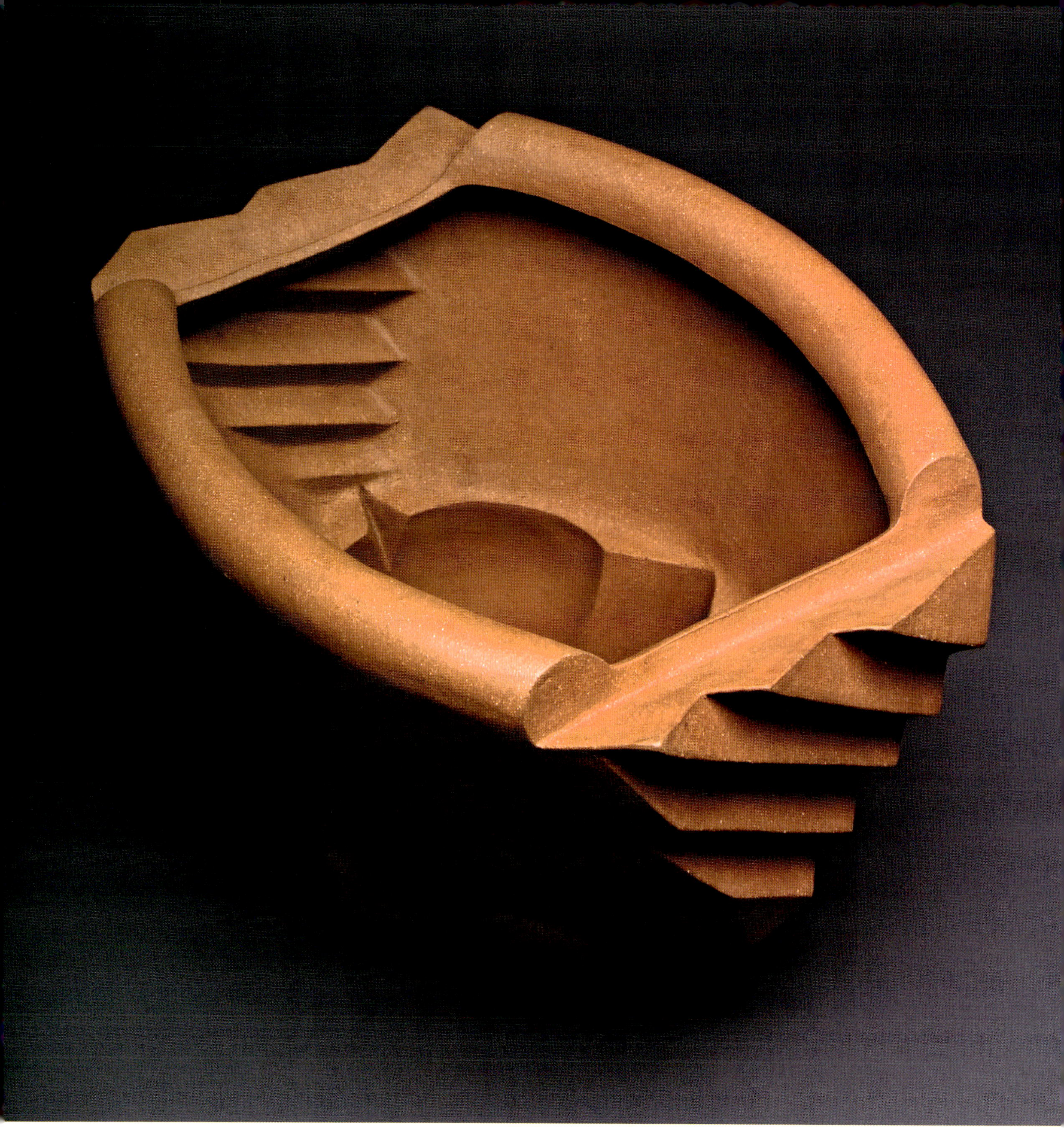

Fig. 179. *Guardian Chamber,* 1992
Unglazed stoneware
18 x 22 x 18 inches
Collection of George and Barbara Daley Schwab

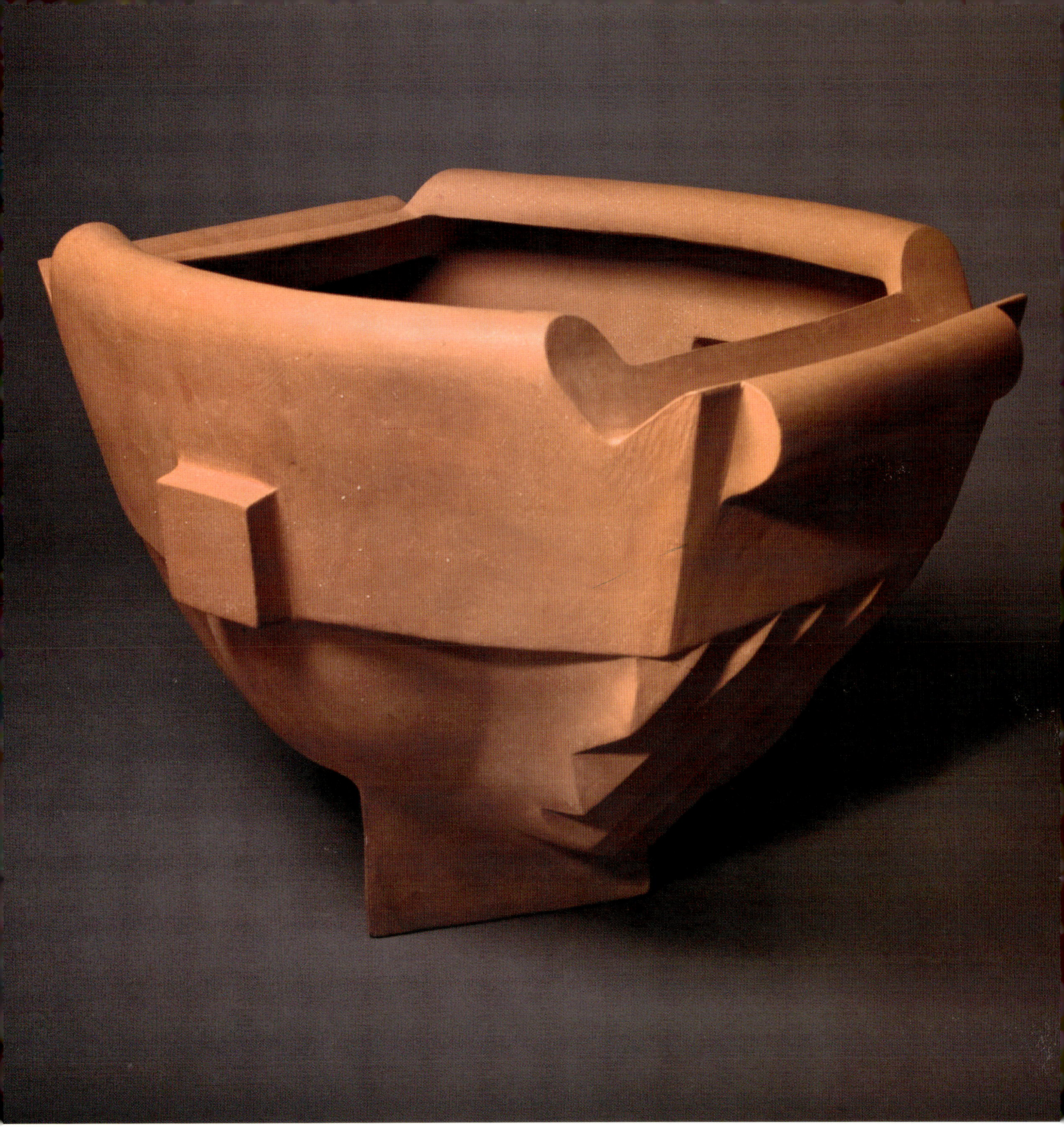

Photos: Amanda Schwab

Fig. 180. *Magister Inn,* 1993
Unglazed stoneware
17.5 x 19.5 x 26.25 inches
Lettered along bottom outer edge:
Magister Inn, Wm Daley, Feb 1993
Collection of Liam Thomas Daley

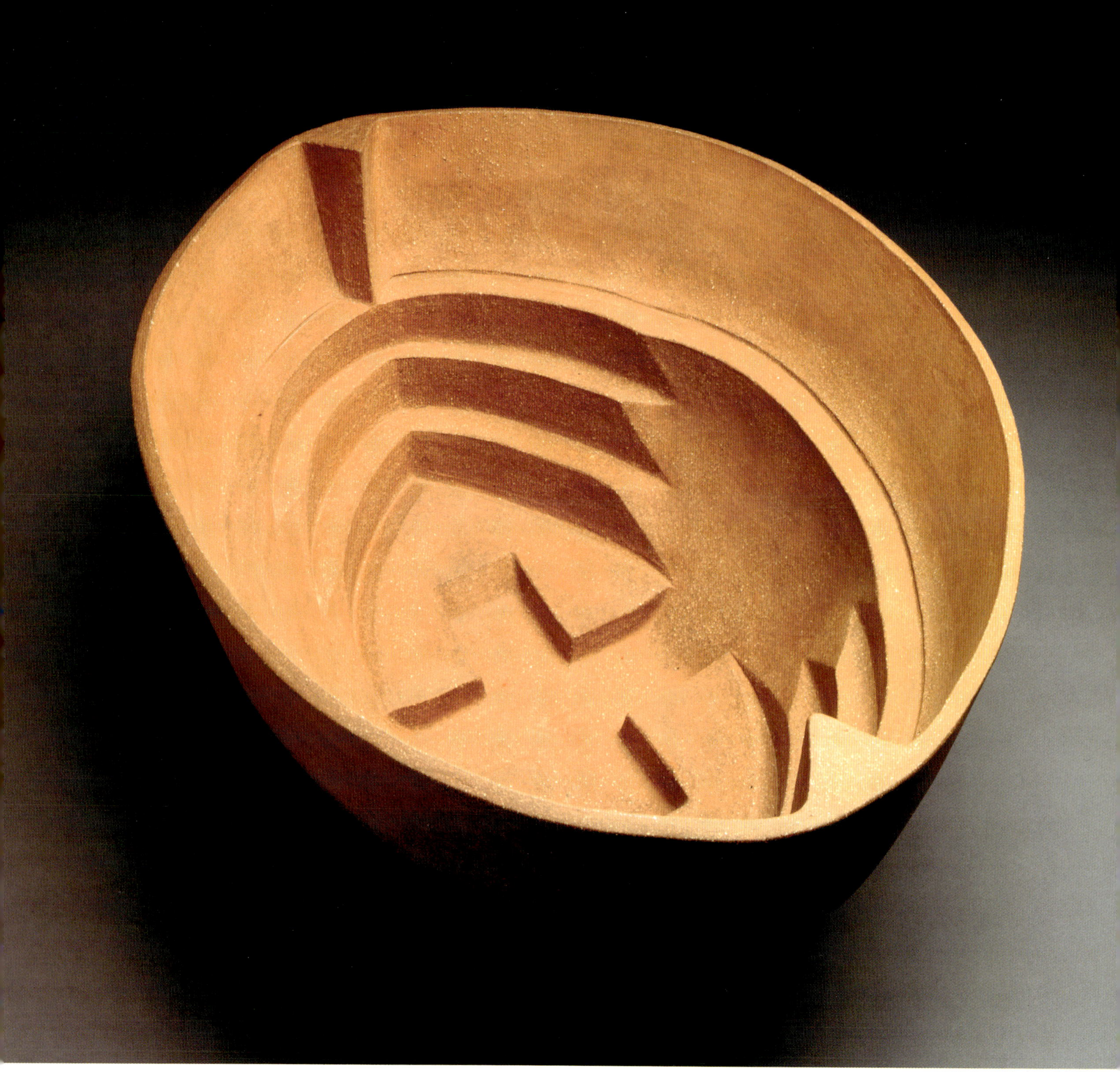

Photo: James Quaile

Fig. 181. *Vesica Pot,* 1992
Unglazed stoneware
14 x 17 x 21 inches
In memory of Alexandra Apperson
Collection of Brigadier General (Ret.) Jack A. Apperson

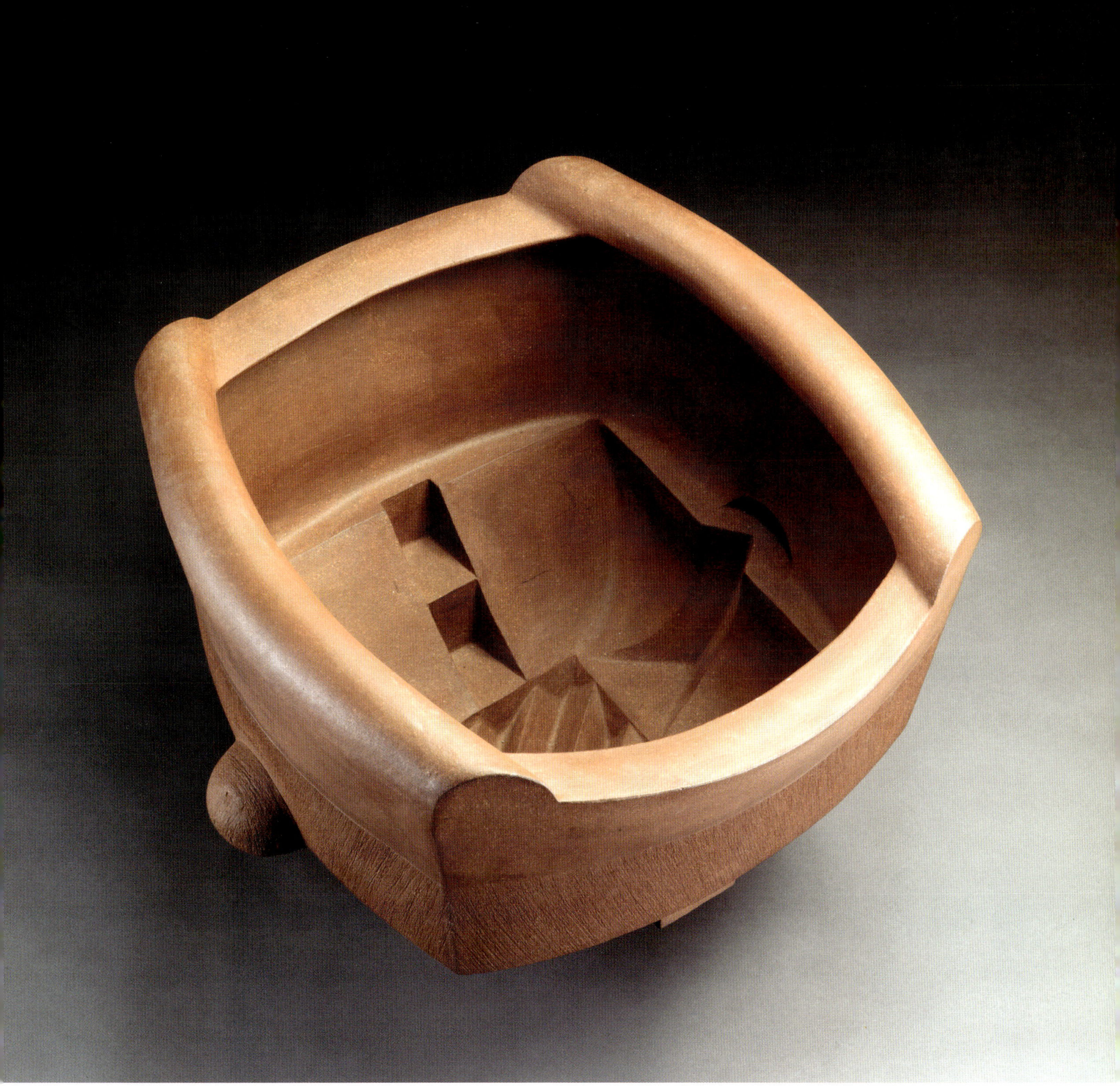

Photos: James Quaile

Fig. 182. *Passing Place,* 1998
Unglazed stoneware
14.5 x 21.5 x 20 inches
Collection of Rhode Island School of Design
Museum of Art

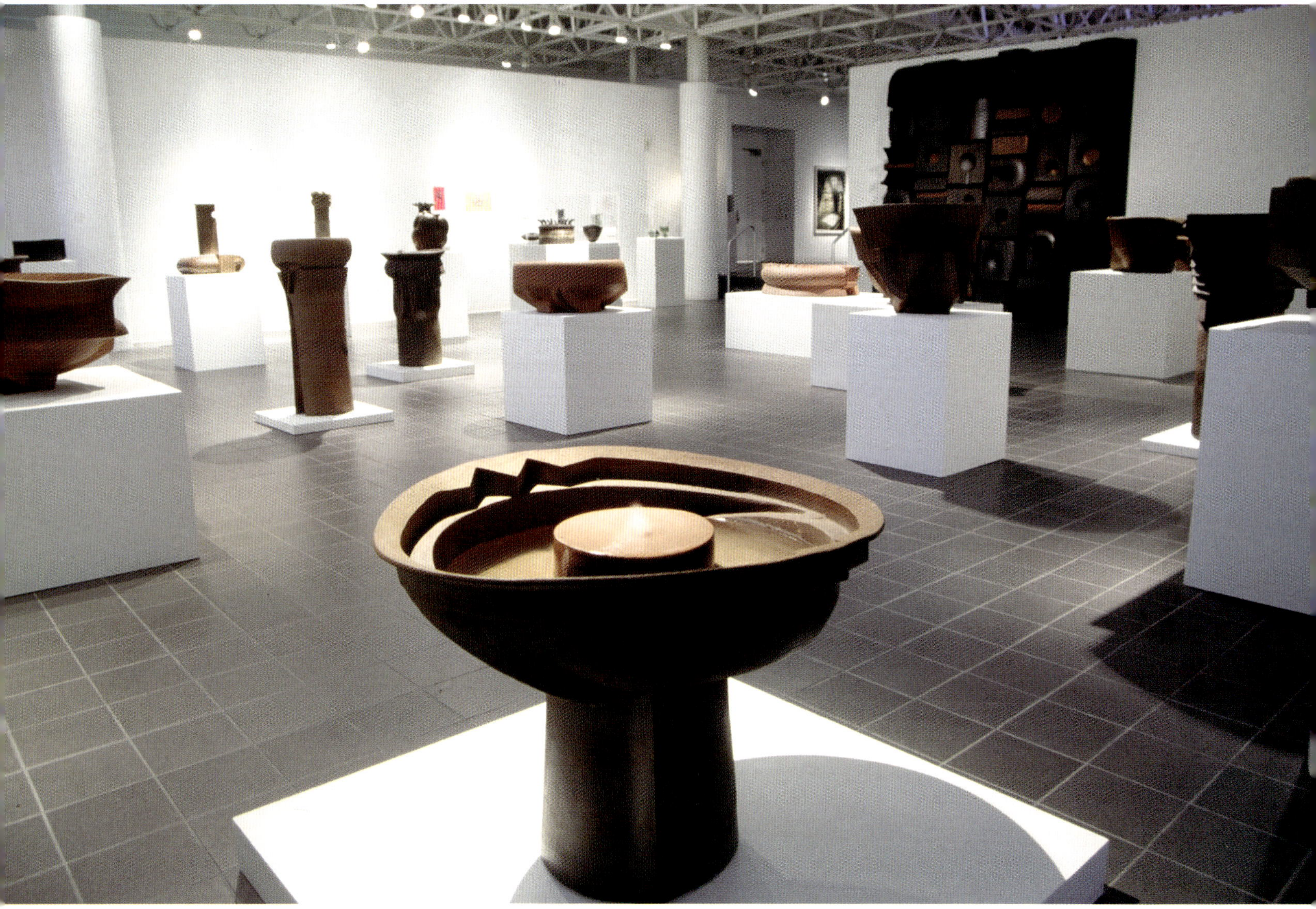

Fig. 183. *William Daley: Ceramic Works and Drawings* show, 1993 at the Levy Gallery for the Arts, Moore College of Art and Design Philadelphia, PA

Traveled to: The Renwick Gallery, Washington D.C., 1994; The Arkansas Arts Center, Little Rock, 1994; and The American Craft Museum, New York, 1996

Above: Fountain. *Celtic Remembrance*, 1979
Unglazed stoneware
22 x 29 x 26 inches
Constructed in three pieces
Private collection

Long-Overdue Retrospective

Edward J. Sozanski, Art Critic

The Daley show of 35 ceramic vessels, 20 drawings, and a large wall sculpture was curated by Mathew Drutt, son of Helen Drutt English and a curator at the Solomon R. Guggenheim Museum in New York. After Moore, it will travel to the Renwick Gallery of the National Museum of American Art in Washington (Feb 18 to April 17), to the Arkansas Arts Center in Little Rock (May 15 to July 4) and to Seoul, South Korea.

While the show features the work that followers of contemporary ceramics know best, it isn't a comprehensive retrospective. Daley has also executed a number of large sculptural pieces. Because of space and other logistical limitations, only one of these, a wall commission for South African Airways, is included.

Daley has filled his niche in contemporary ceramics with large, architectonic vessels. They're usually unglazed stoneware, their colors ranging from an orange-red to red-brown.

They're very basic vessels in the way their character derives from the manipulation of volume and line. (Daley calls them "pots," but that seems too plebeian for such regal forms.) Their interiors are just as important as their exteriors but they should be, because the two are essentially mirror images.

One can find a few symbols, such as the cross, in Daley's vessels, but they don't play a major role. Perturbations of symmetry and the integrity of the form are his primary concerns, and he has expressed them in a variety of ways. With some vessels, he has elongated the foot into a columnar pedestal, which transforms a shelf piece into an imposing floor piece. Many of his vessels terminate in a broad, flared rim; in others, the rim turns inward. Some have just a terminal edge.

The constant in Daley's mature work is the dialogue between inside and outside, usually articulated with patterns of serrated ridges, arcs, and protrusions. His designs carry the eye up the outside and down the inside, then around the walls and finally to the rim.

The large vessels can suggest either man-made structures, perhaps in a ruined state, or natural landscape features. But regardless of what might have inspired the artist, they do not evoke anything directly. They are pure expressions of traditional vessel language achieved through meticulous hand building rather than by raising a form on the potter's wheel.

For Daley fans, the most intriguing part of the exhibition should be the work he made before the large architectonic vessels emerged in the mid-late 1960s. The earliest works in the show are three small glazed pieces from the mid 1950s, derived from the vessel but suggesting a symmetry related to the body. Subsequently, the pieces grow progressively larger, glaze is abandoned, and the vessel becomes a kind of arena for developing spatial dynamics.

Daley's vessels are in one sense primitive in their tactility and natural finish and in another as refined and thoughtfully conceived as Song bowls. His show should look as handsome in Seoul, where the National Museum of Contemporary Art owns two pieces, as it does here.

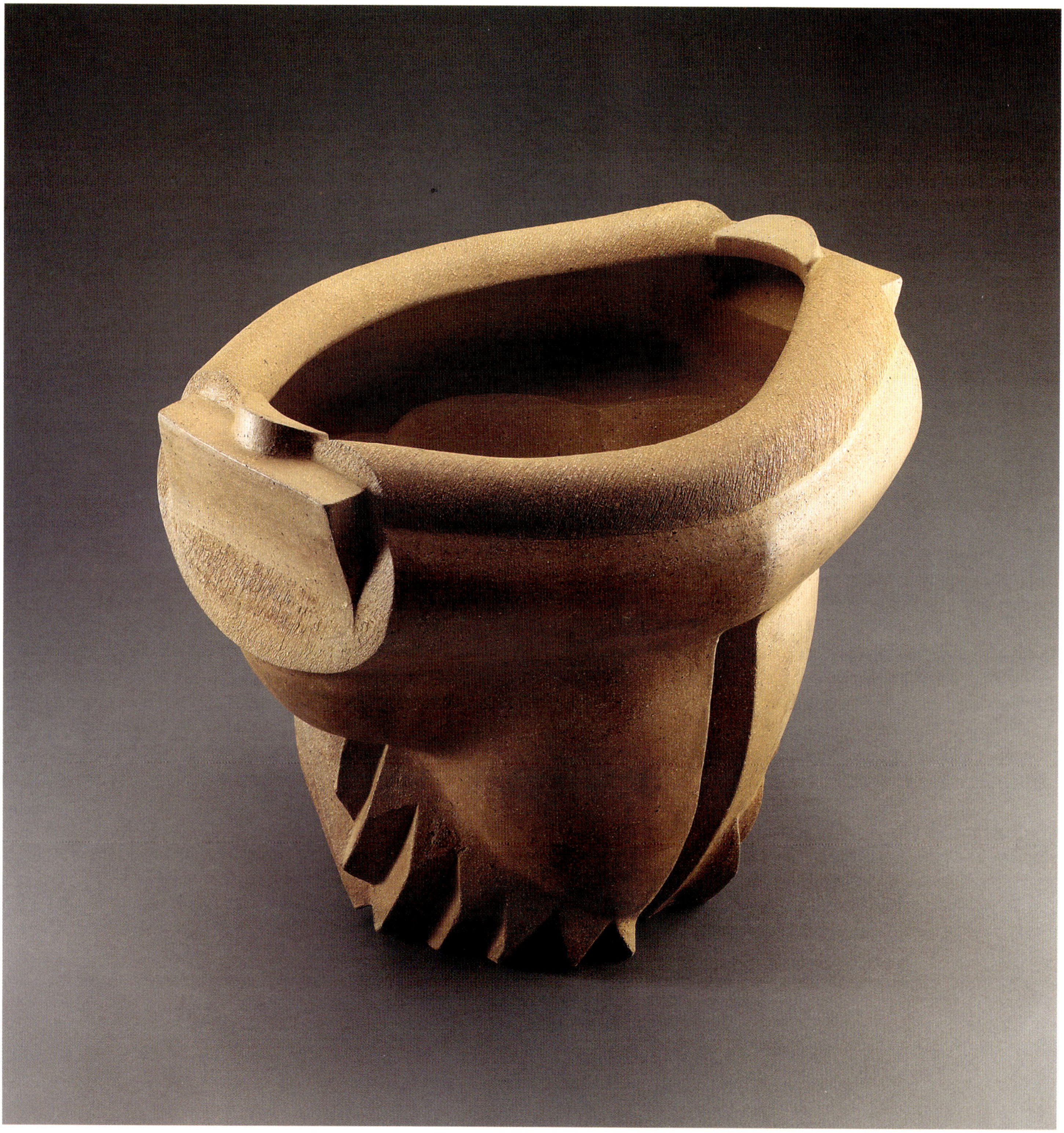

Photo: James Quaile

Fig. 184. *Petras Passing Vesica Pot,* June 2001
Jersey Buff stoneware and
Medicine Hat red clay
20 x 23 x 17.5 inches
Collection of Stanley and Mary Ann Snider

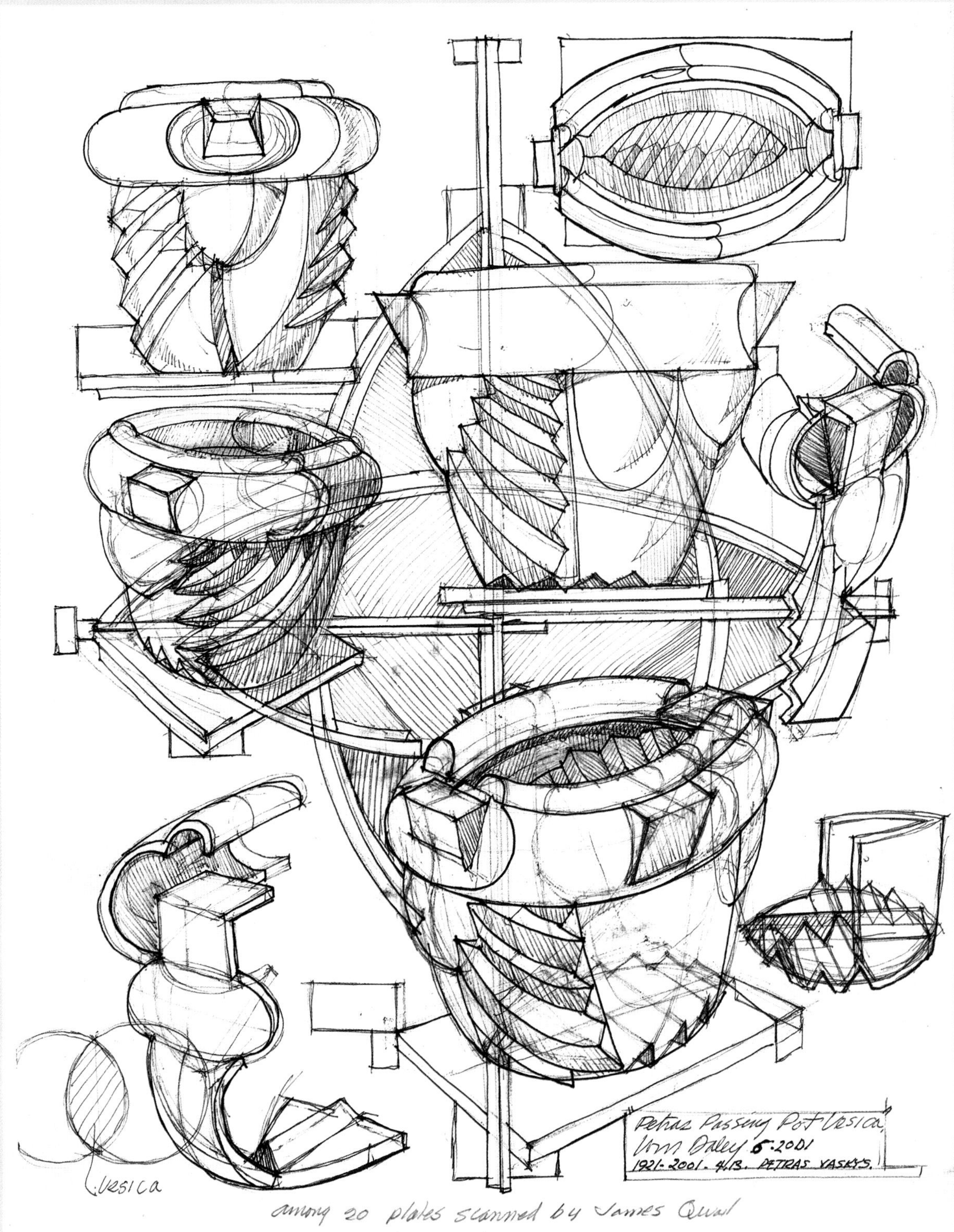

Fig. 185. Drawing, *Petras Passing Vesica,* 2001
1921–2001 Petras Vaskys
Ink on bristol paper
14 x 11 inches

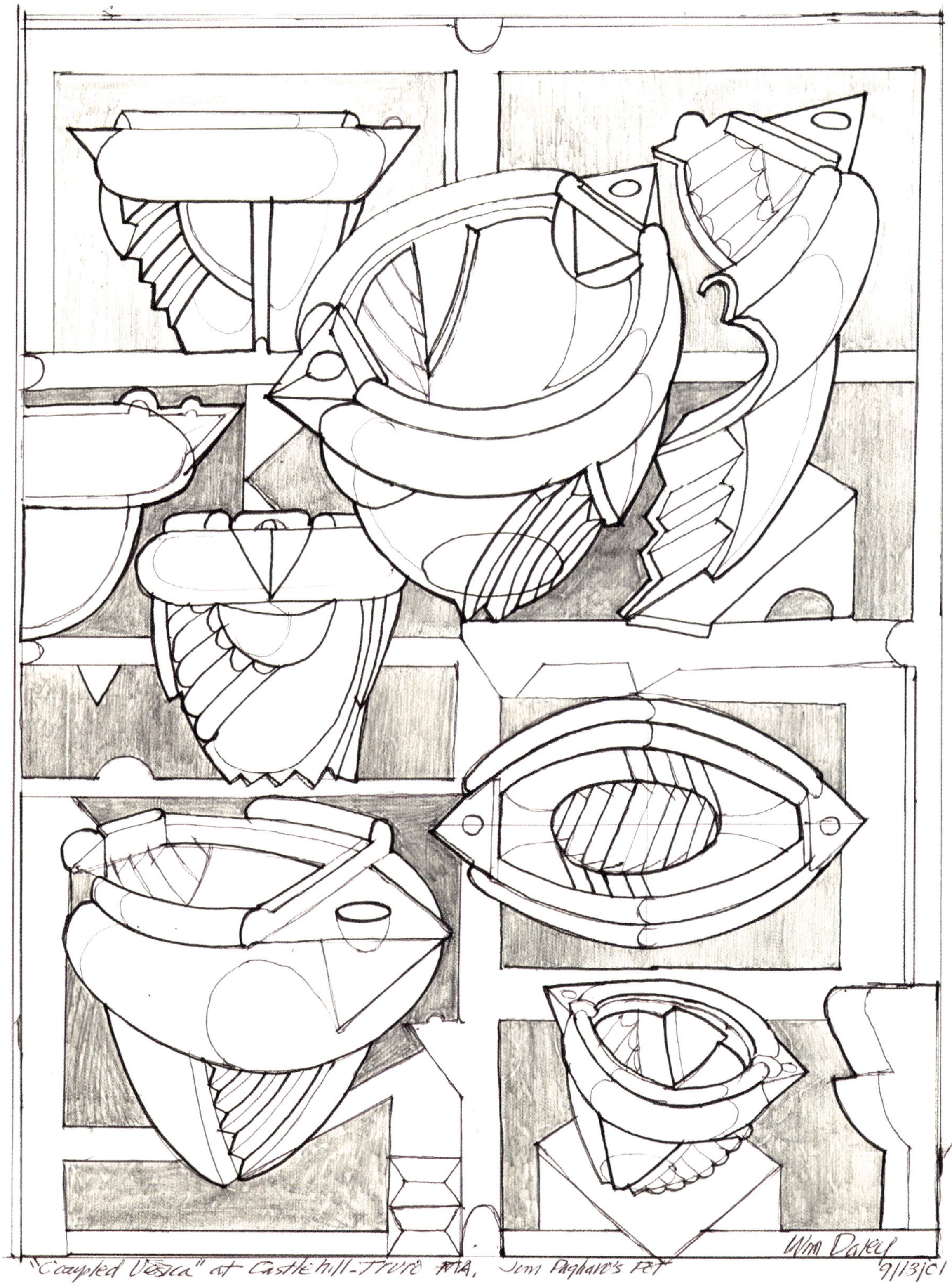

Fig. 186. Drawing, *Coupled Vesica,* 2001
Ink and pencil on bristol paper
14 x 11 inches

Fig. 187, opposite. *Coupled Vesica,* 2001
Unglazed stoneware
22 x 21 x 27 inches
Collection of James and Susan Pagliaro

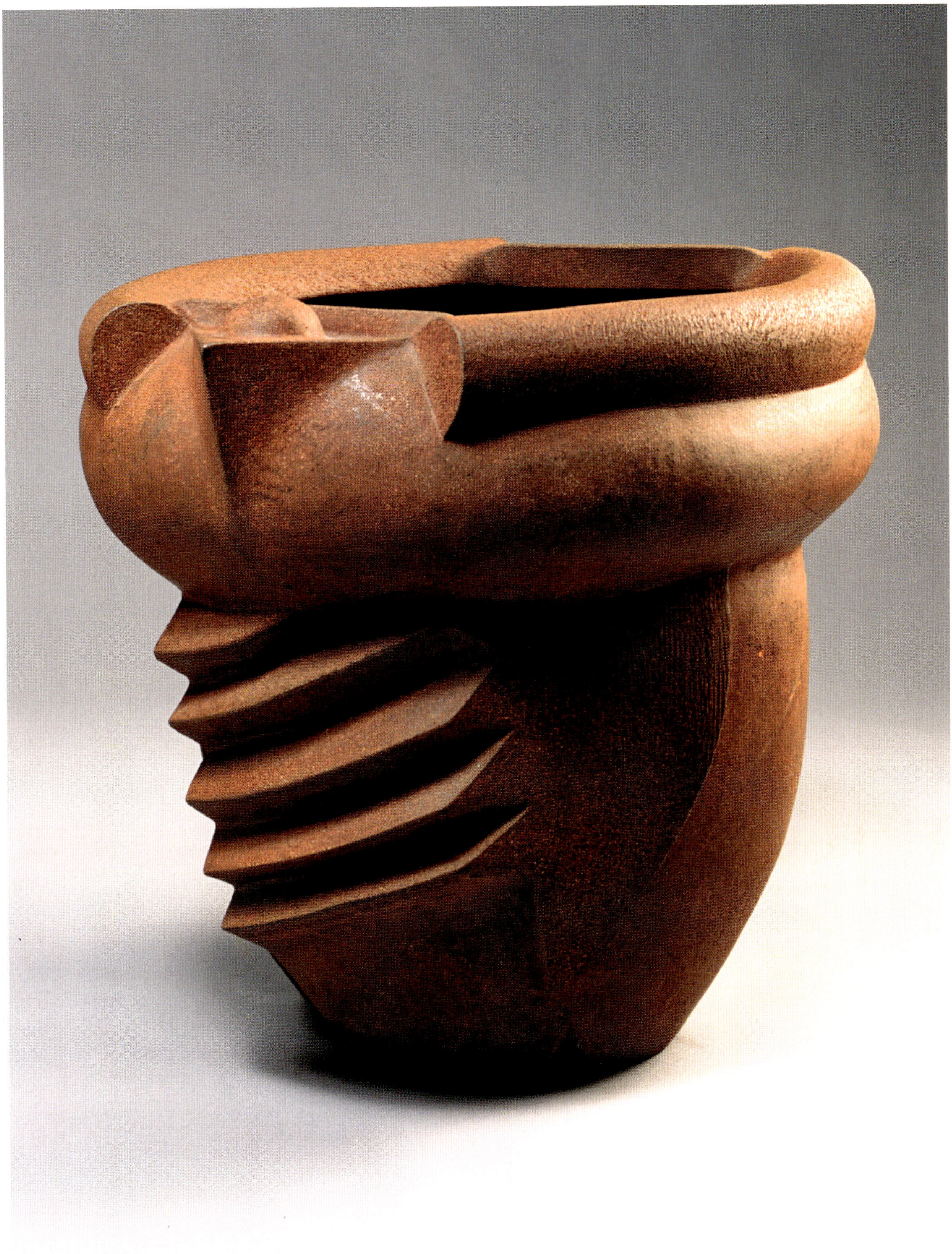

Photo: John Carlano

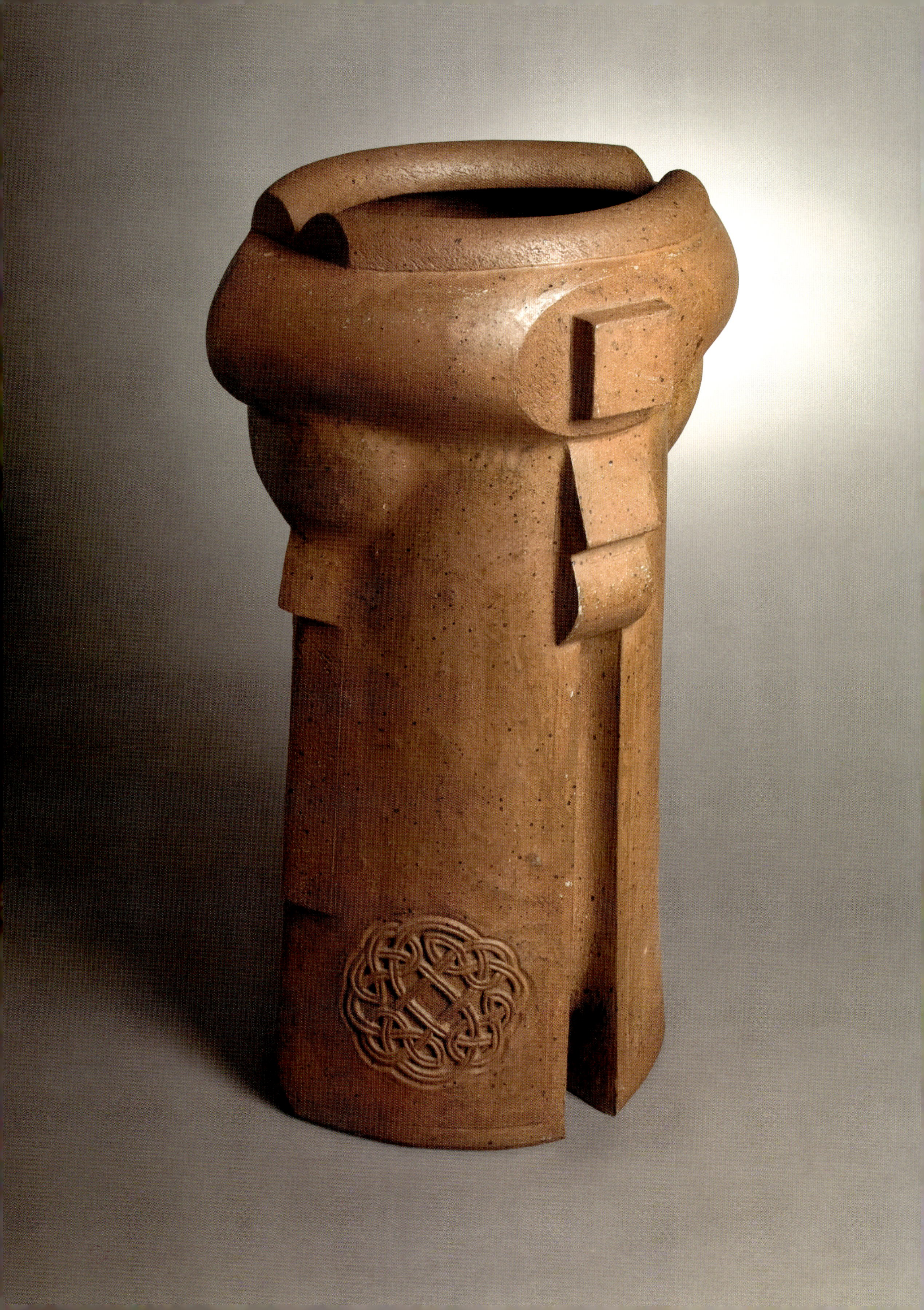

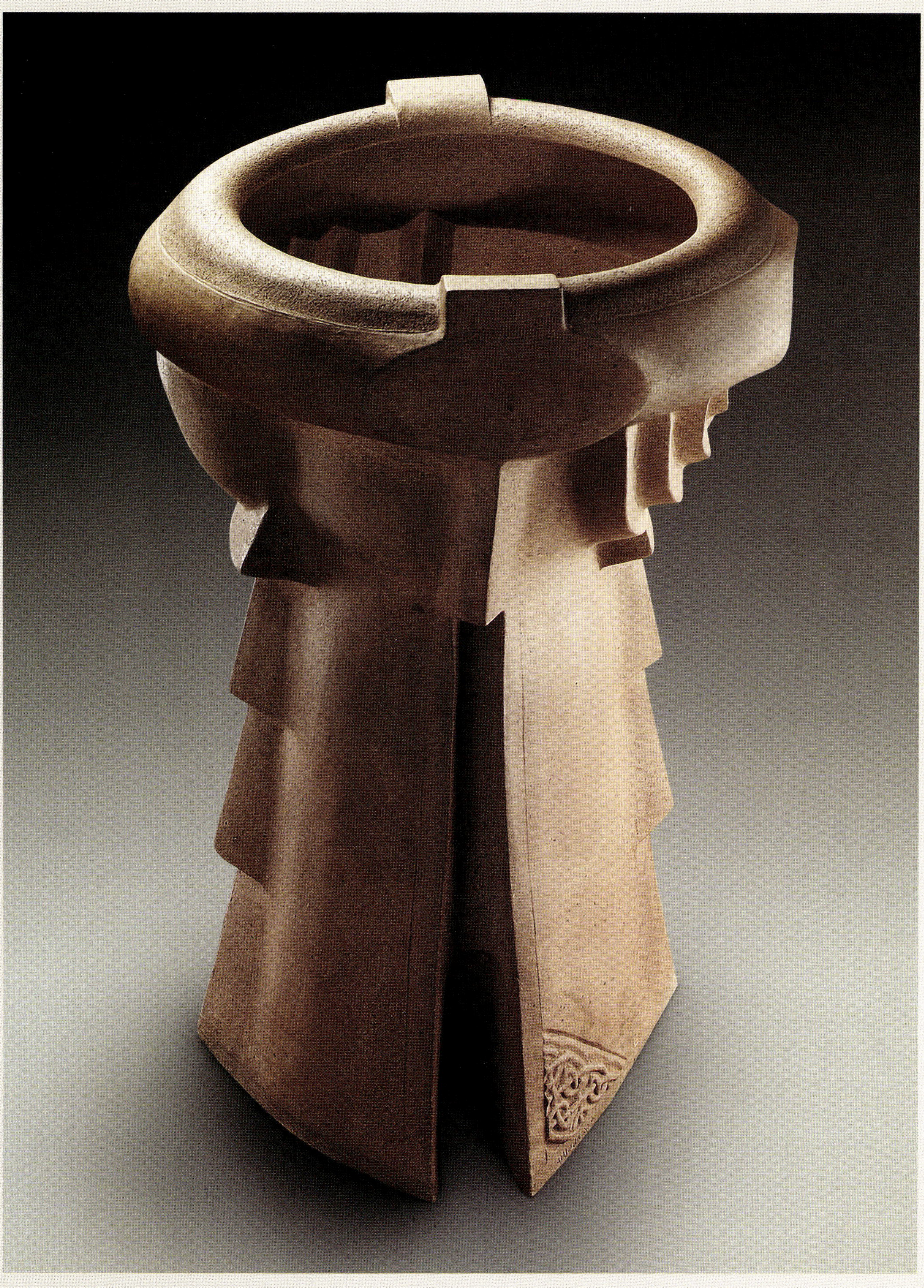

Photo: James Quaile

Fig. 188, opposite. *Celtic Vesica,* 2002
Unglazed stoneware
34 x 22 x 22 inches
Collection of Charlotte and Stephen Martin

Fig. 189. *Haystack Vesica,* 2003
Unglazed stoneware
37 x 21 x 22 inches
Collection of Candice B. Groot

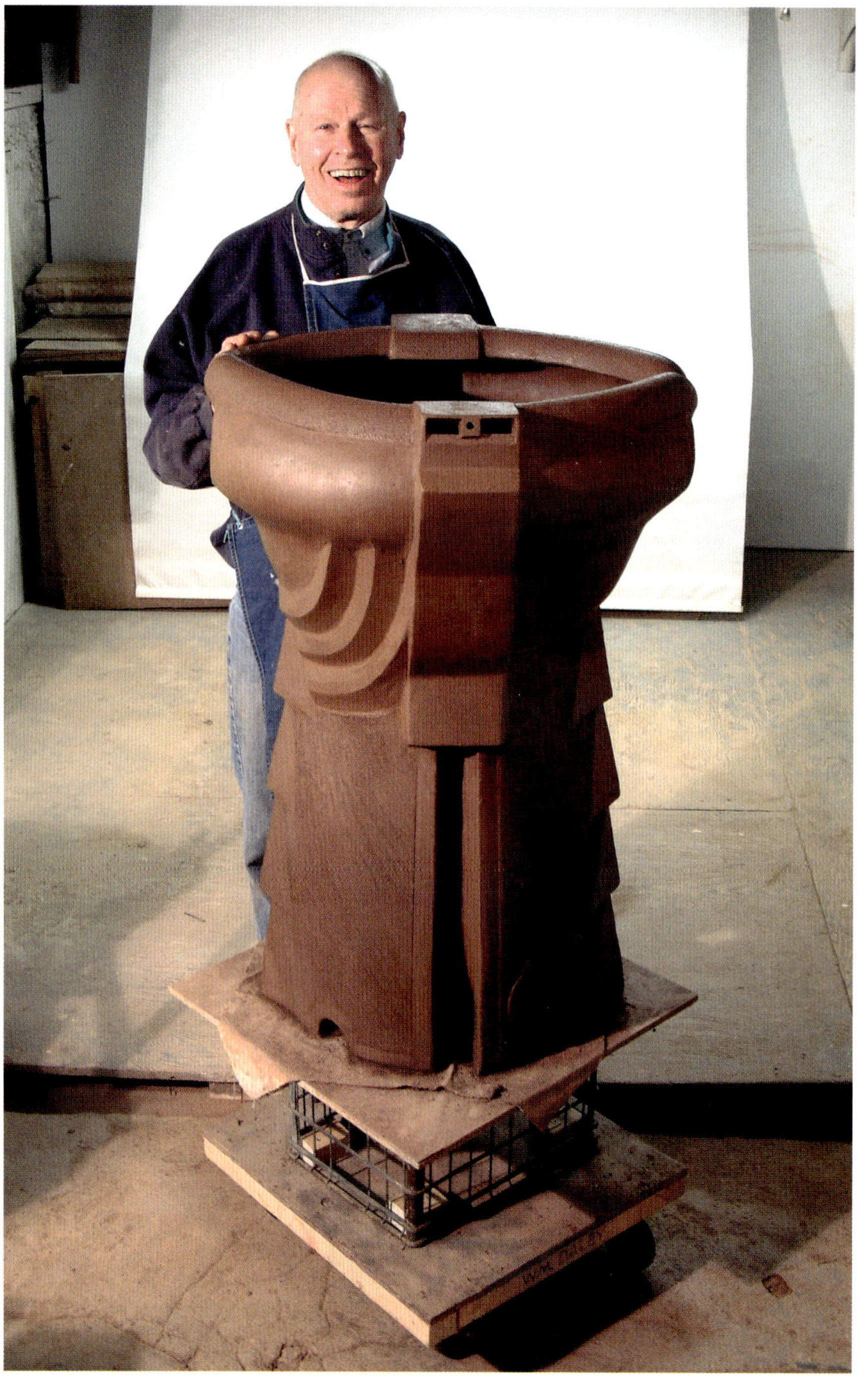

Photo: Wayne Bates

Fig. 190. *Runic Vesica,* 2003
William Daley finishing unfired stoneware pot in his basement studio, Elkins Park, PA
Pot was lost in firing

Fig. 191, opposite. *Orphic Vesica,* 2003
Unglazed stoneware
36.75 x 27 x 21.5 inches
Collection of the Arizona State University Art Museum

Exhibited American Crafts Council Awards Exhibition SOFA, Chicago

Photo provided by Arizona State University Art Museum

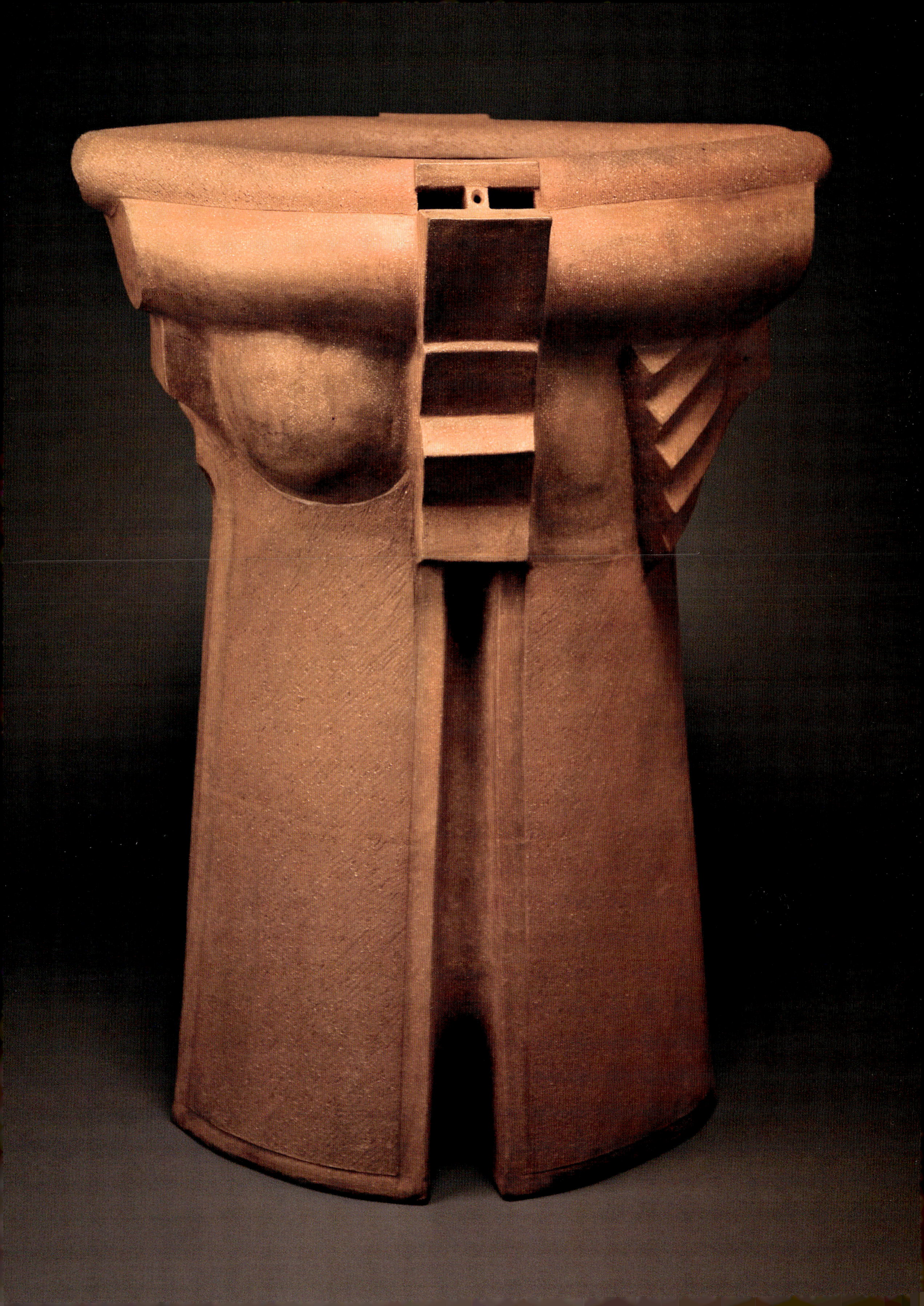

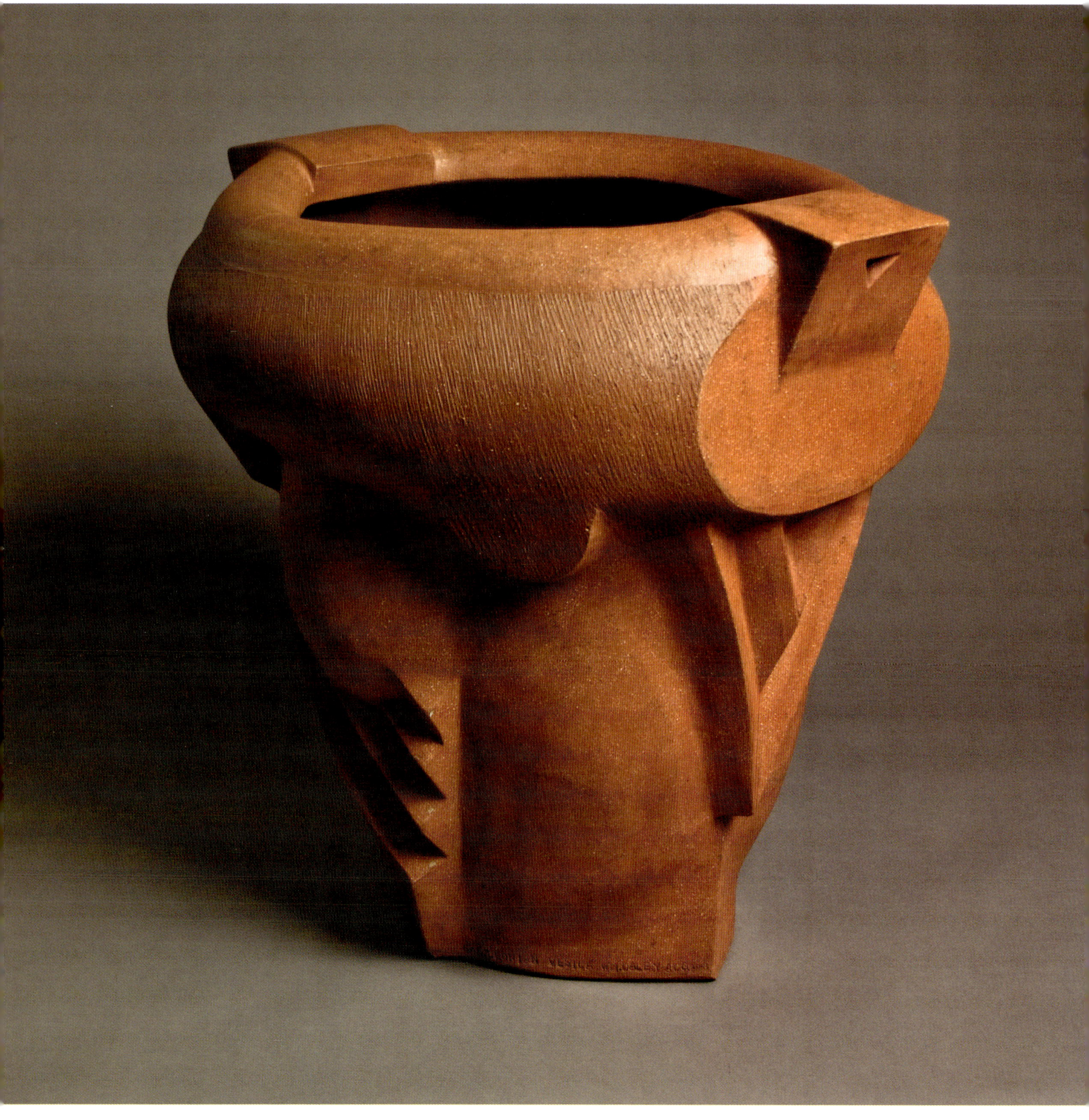

Fig. 192. *Apollonian Vesica,* 2003
Unglazed stoneware
23 x 23 x 29 inches
Collection of Daniel Jalboot and Jennifer Peterson

Exhibited American Crafts Council
Awards Exhibition SOFA, Chicago

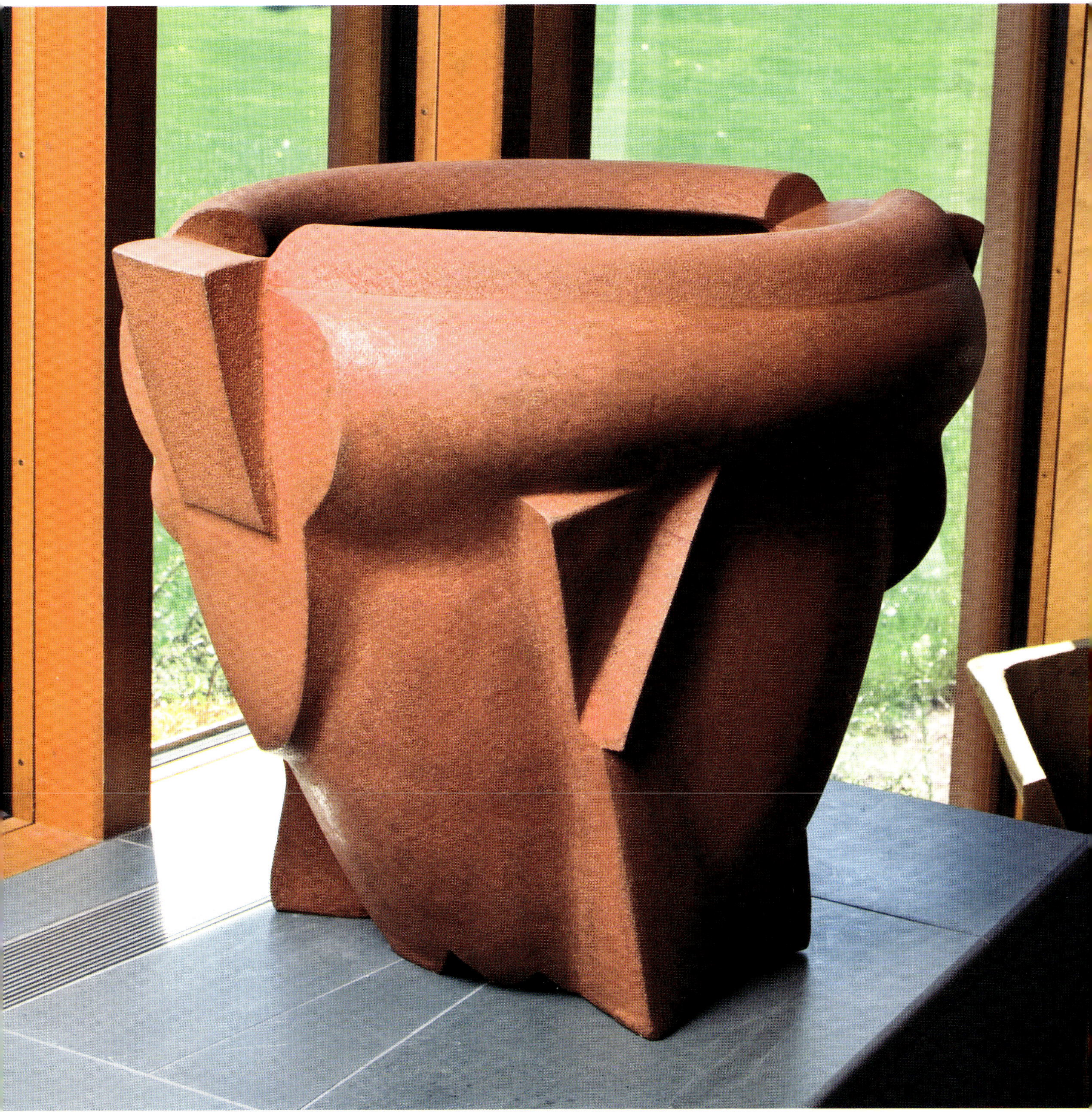

Fig. 193. *Vesica Mundi,* 2005
Unglazed ground stoneware
22.5 x 21.5 x 29 inches
Collection of Marlin and Ginger Miller

Photos: Amanda Schwab

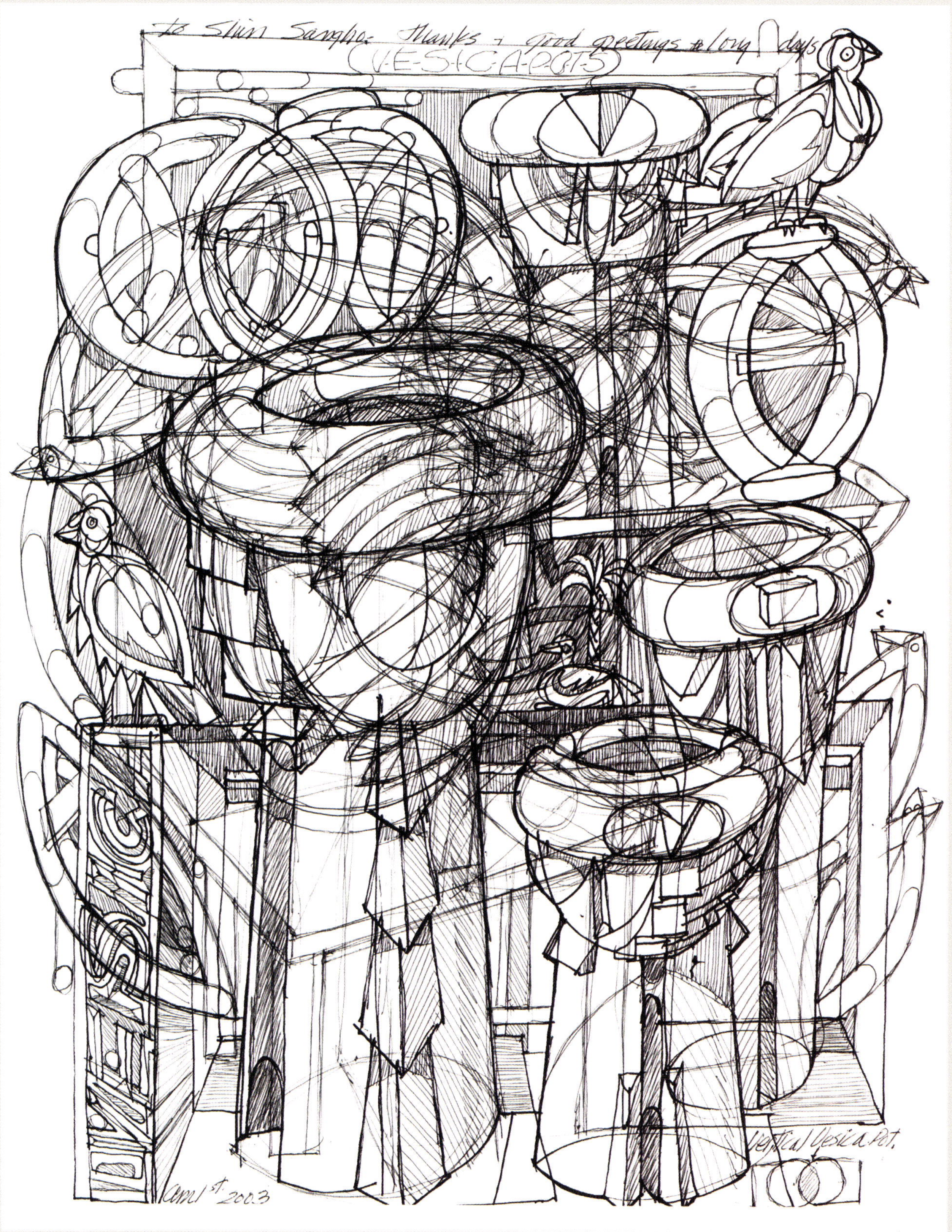

Fig. 194. Drawing, *Vertical Vesica*, 2003
Ink on bristol paper
14 x 11 inches

Fig. 195, opposite. *Phoenix Vesica*, 2004
Unglazed stoneware
30 x 24 x 19 inches
Collection of Alfred and Rica Fenaughty

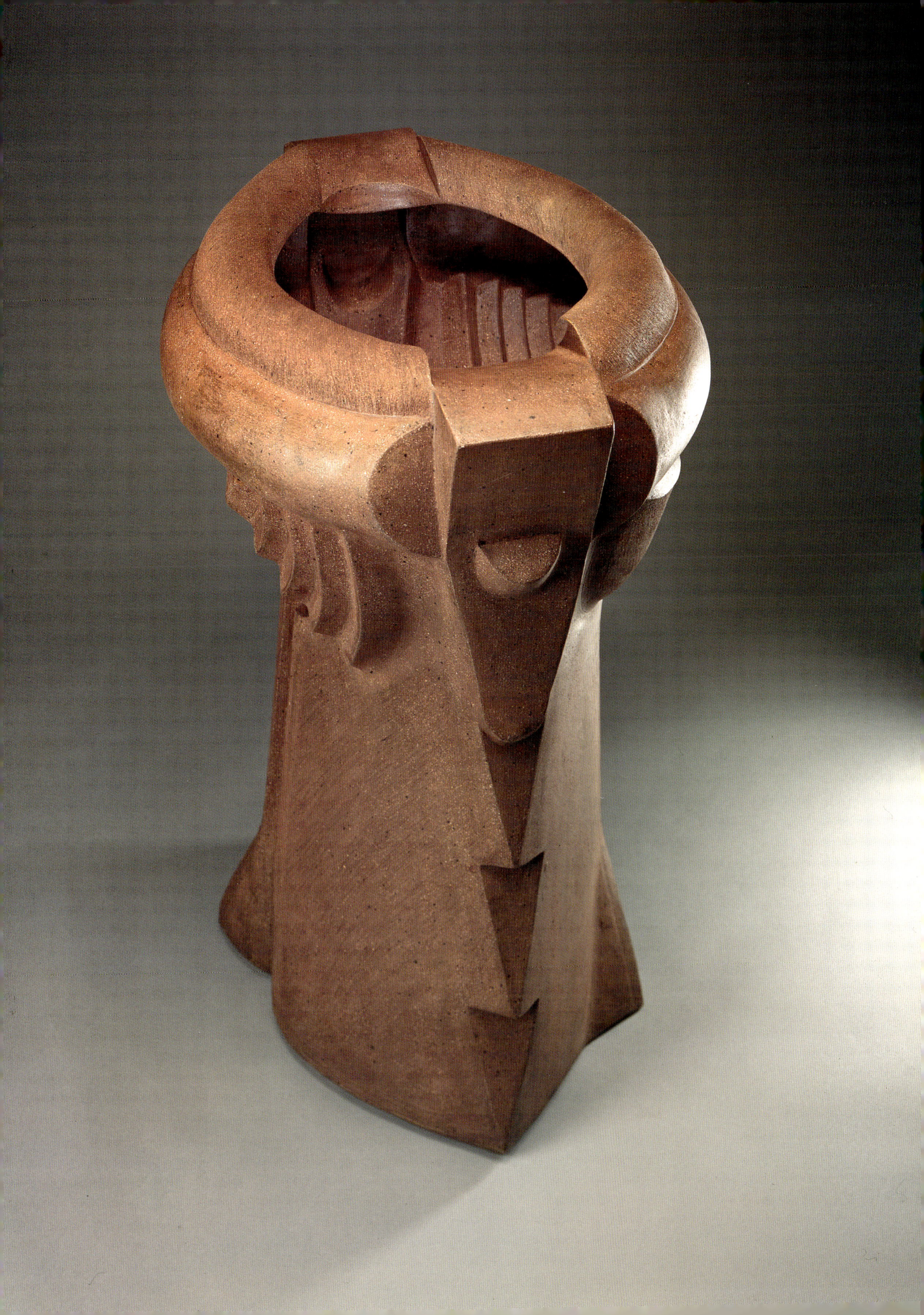

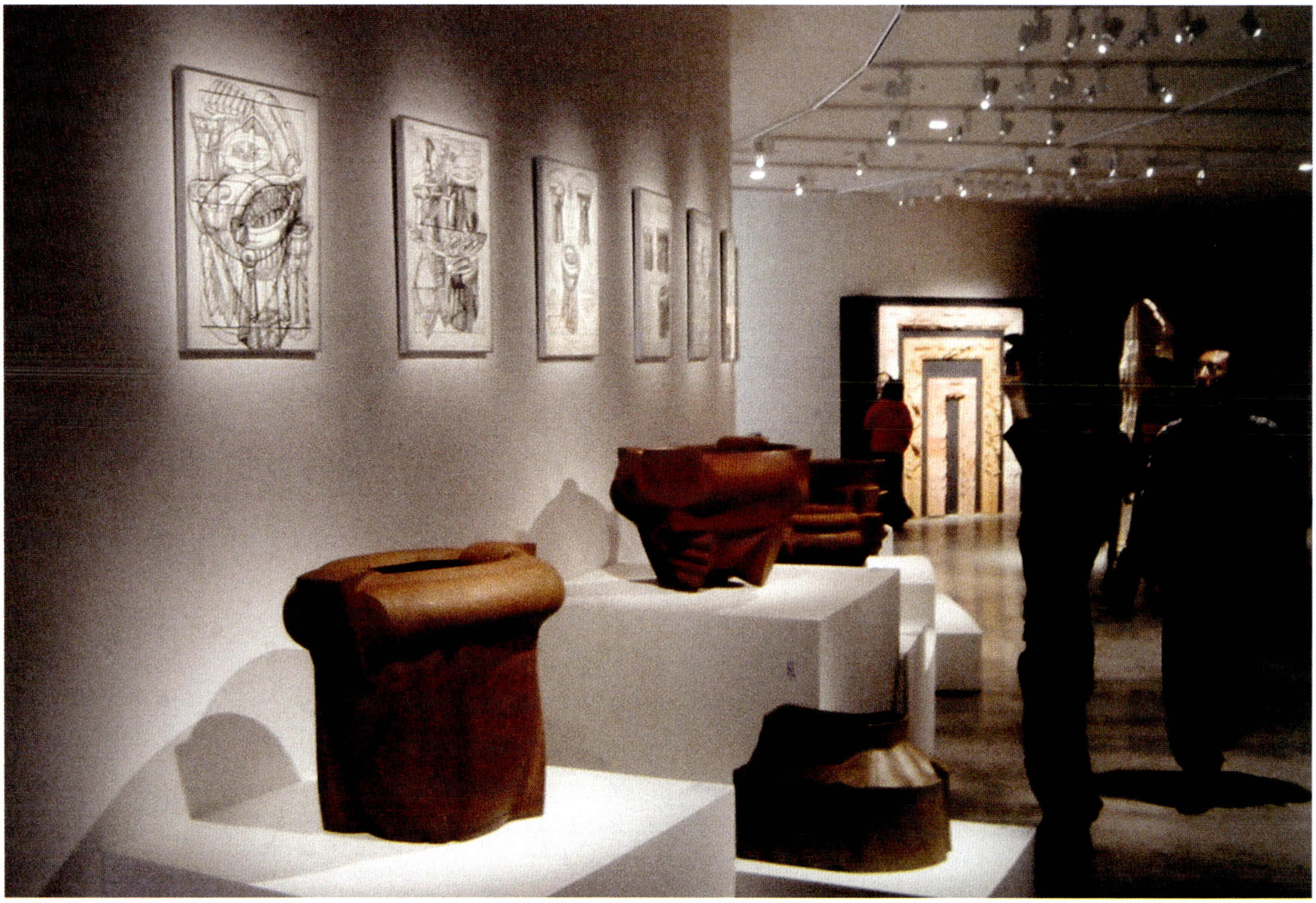

Fig. 196. Display of William Daley ceramic vessels and drawings as part of the inaugural exhibition 2006
Clayarch Gimhae Museum, Gimhae, Korea

Fig. 197, opposite. Pictured, four of ten drawings in a presentation suite
24 x 18 inches
Collection of the Clayarch Gimhae Museum, Gimhae, Korea

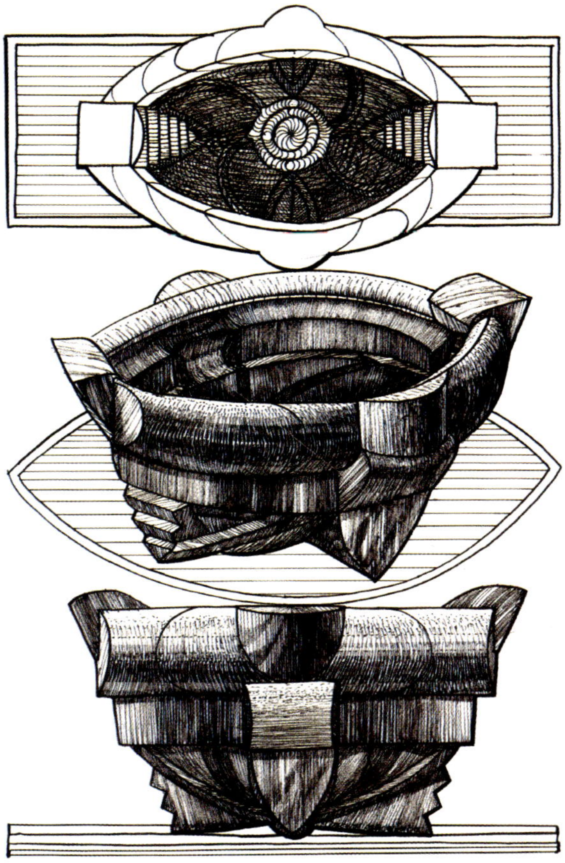

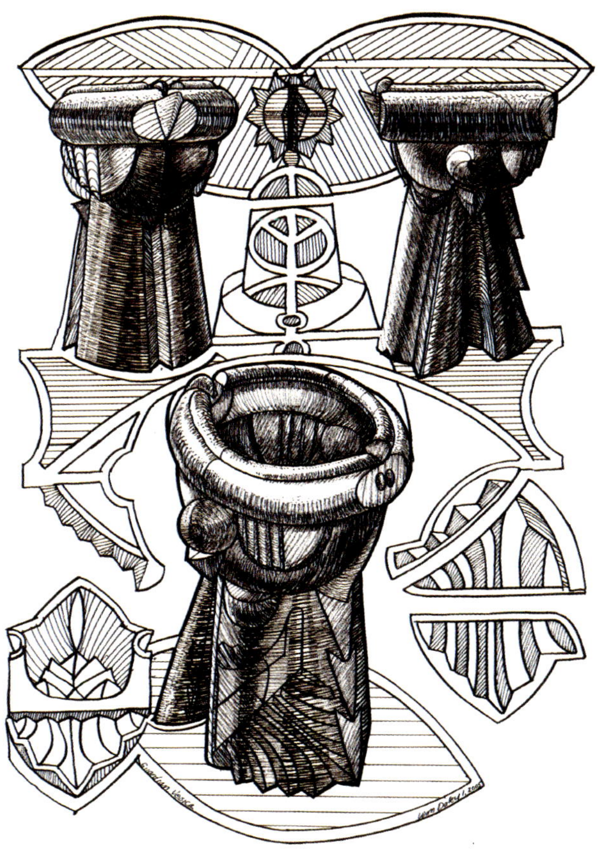

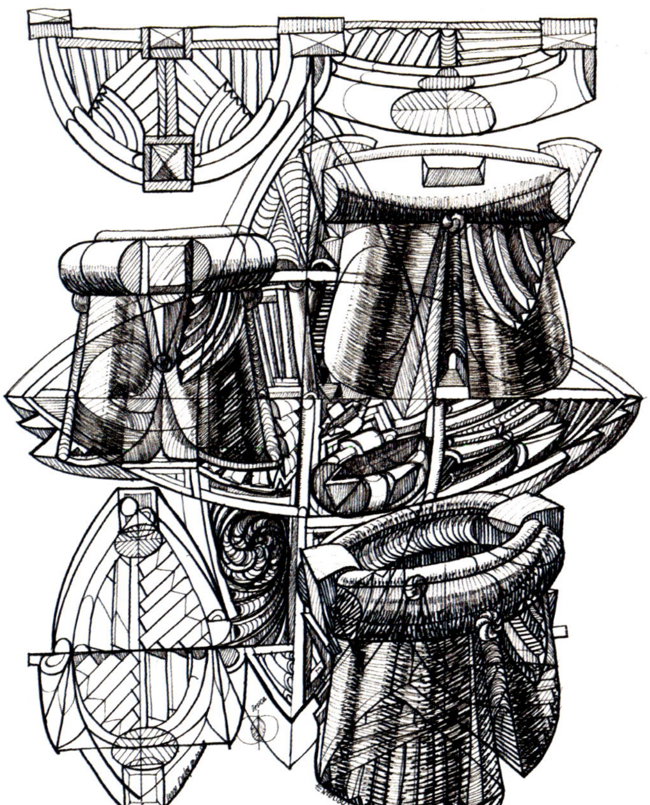

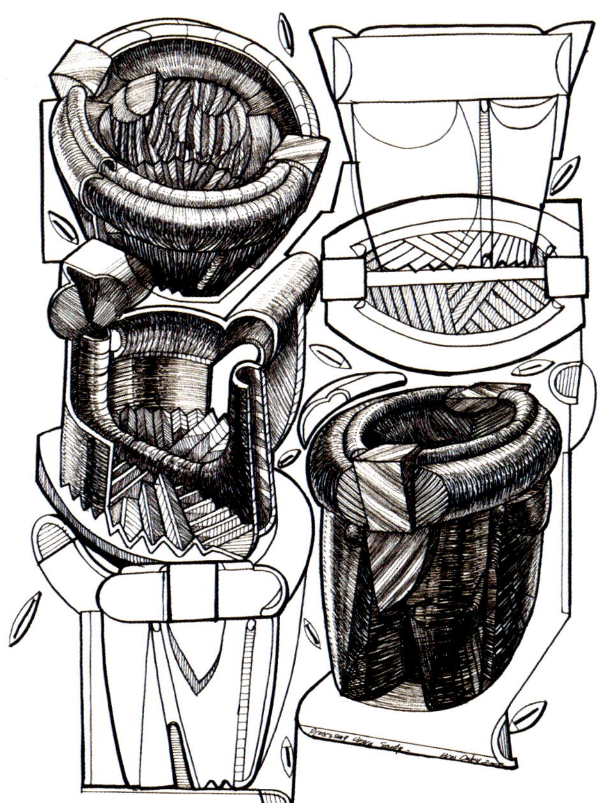

Photos: John Carlano

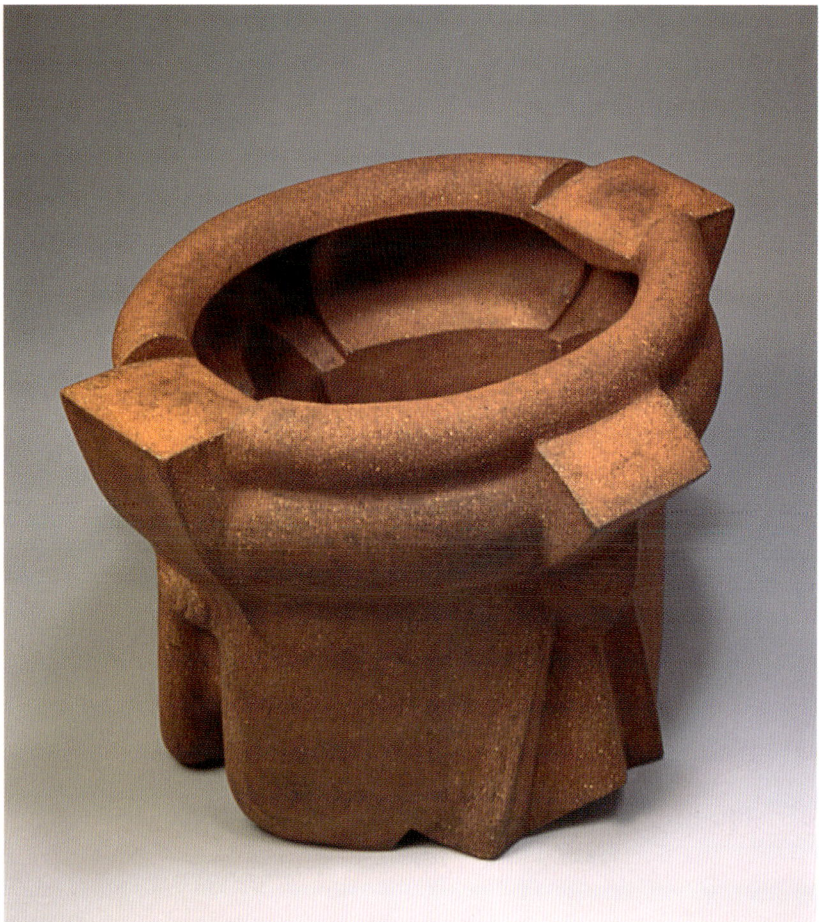

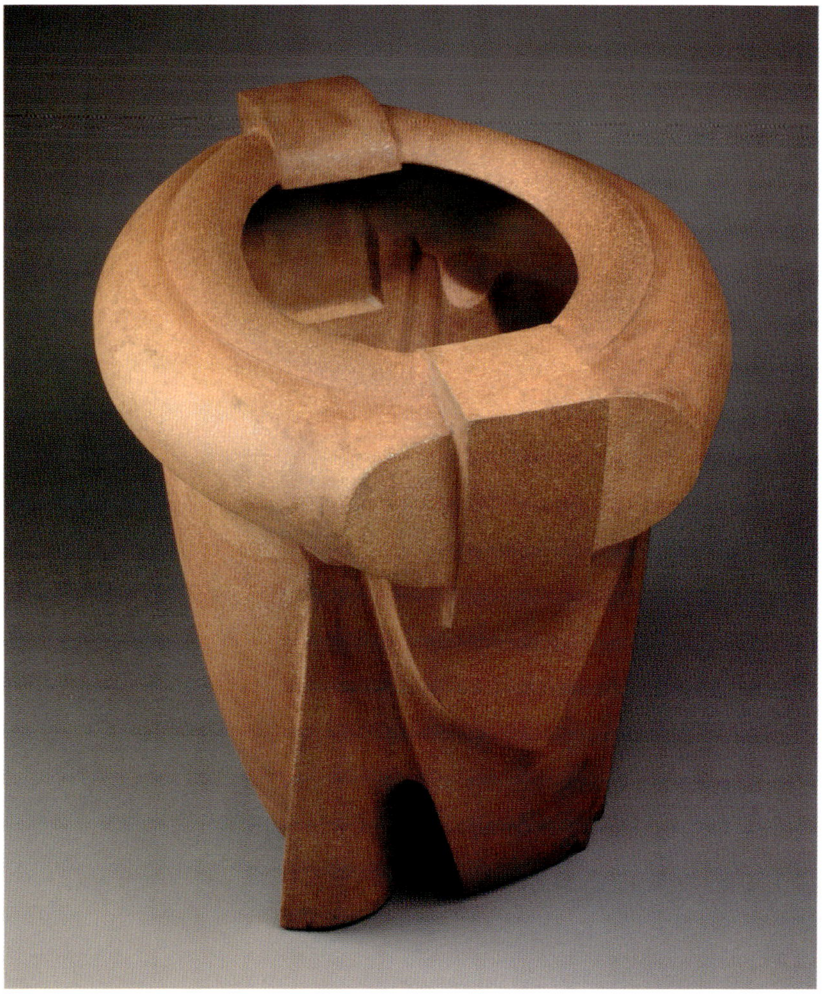

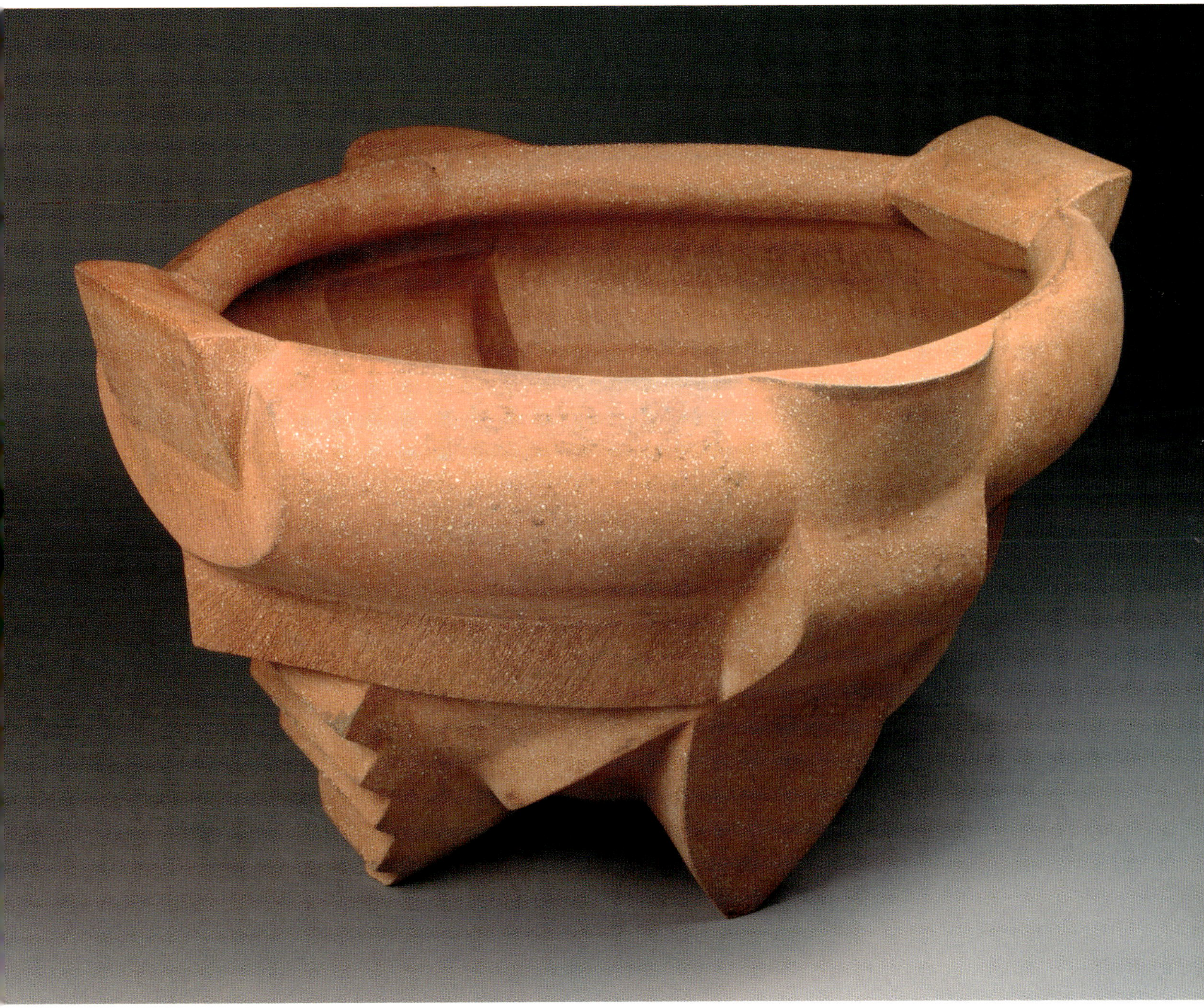

Photo: Jchn Carlano

Fig. 198, opposite top. *Exterious Vesica,* 2004
Unglazed ground stoneware
28 x 18.5 x 25 inches
Collection of Clayarch Gimhae Museum
Gimhae, Korea

Fig. 199, opposite bottom. *Processing Vesica,* 2005
Unglazed stoneware
18 x 22 x 28 inches
Collection of Clayarch Gimhae Museum
Gimhae, Korea

Fig. 200. *Oracular Vesica,* 2005
Unglazed stoneware
15.5 x 25 x 29.5 inches
Collection of Clayarch Gimhae Museum
Gimhae, Korea
Gift from the Artist

Photo: John Carlano

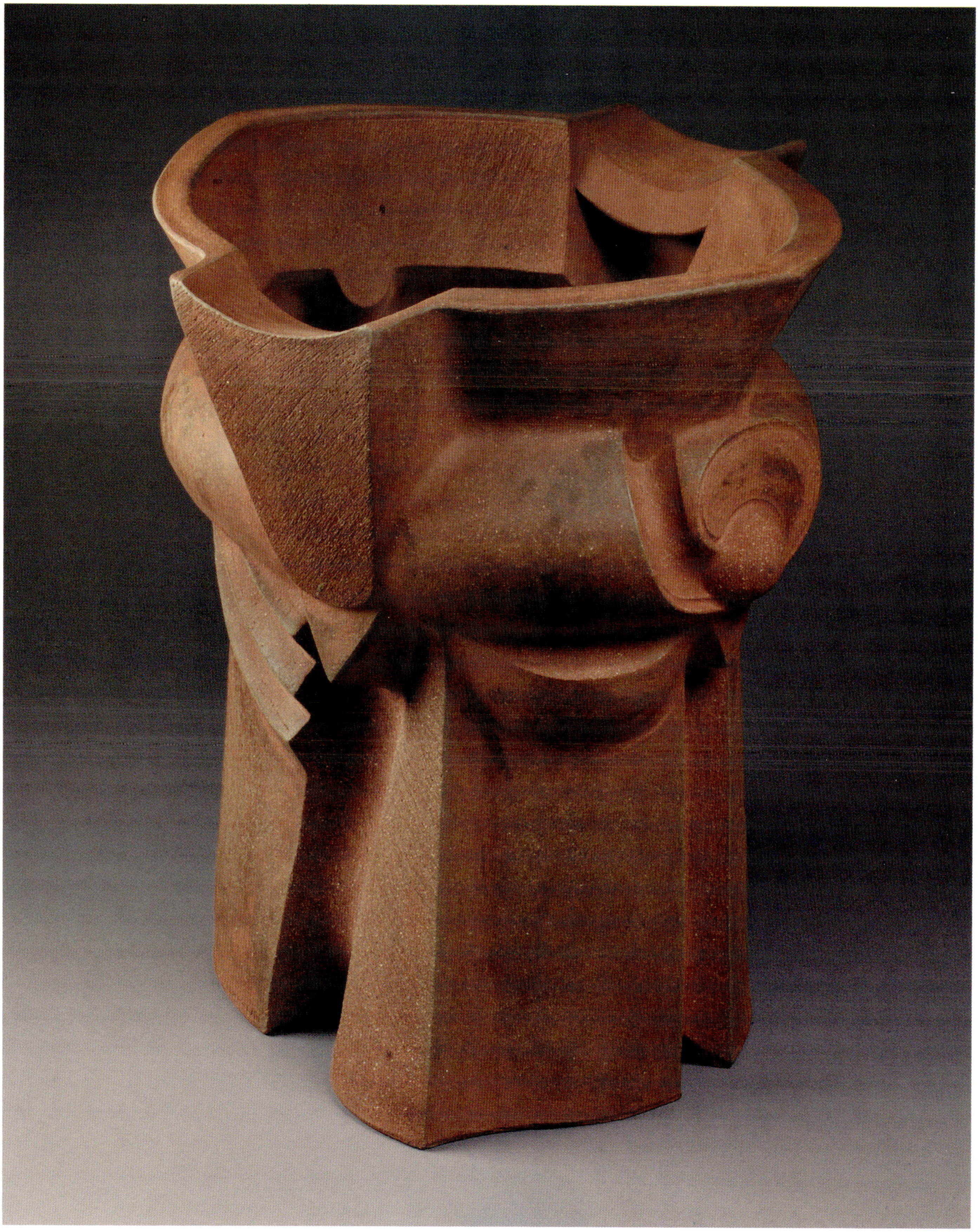

Fig. 201. *Nuptual Vessel,* 2006
Unglazed stoneware
26.5 x 24 x 25.75 inches
Collection of Clayarch Gimhae Museum,
Gimhae, Korea

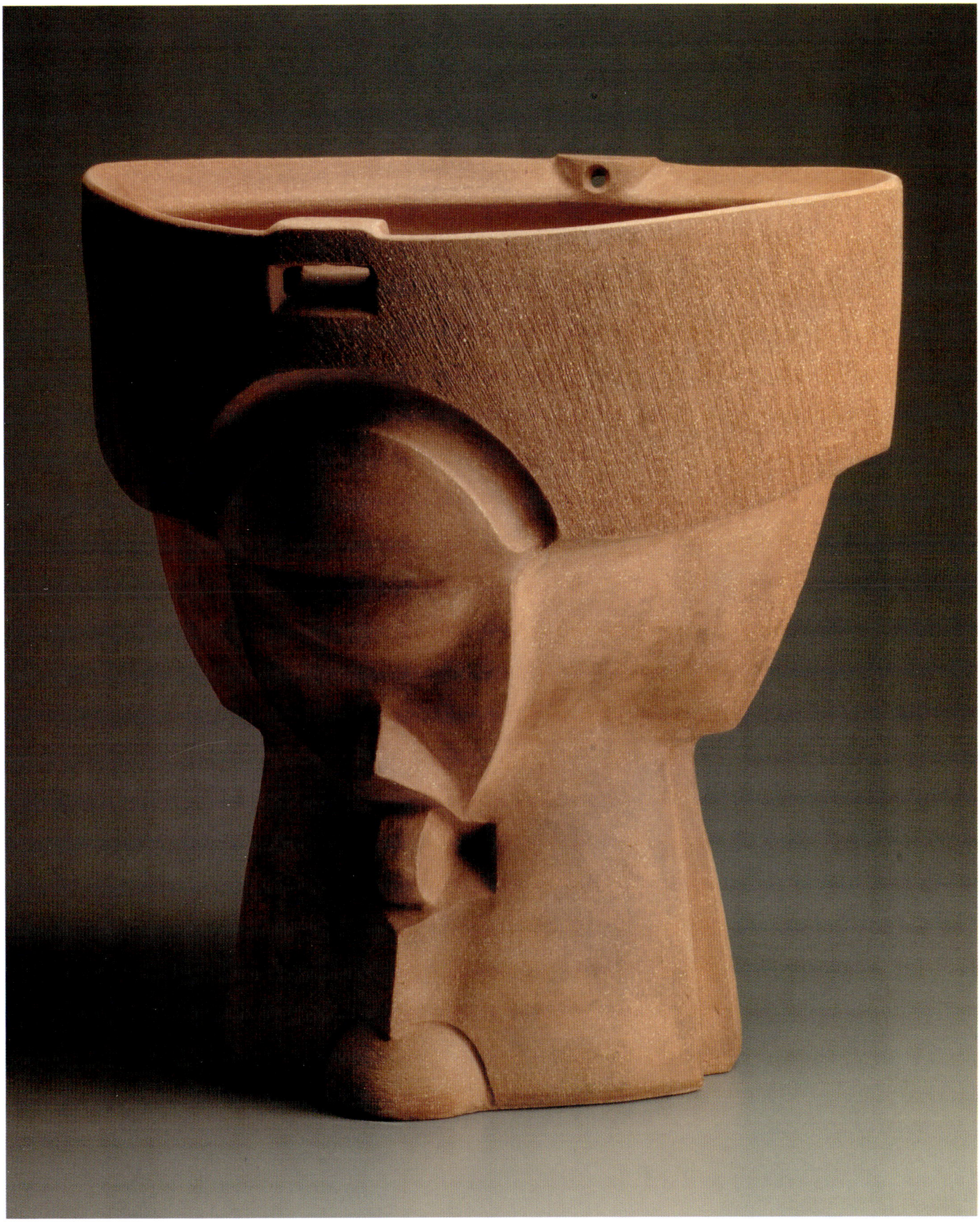

Photo: John Carlano

Fig. 202. *Cyclops Eyecup Revisited,* 2007
Unglazed stoneware
24 x 20 x 23 inches
Collection of Marlin and Ginger Miller

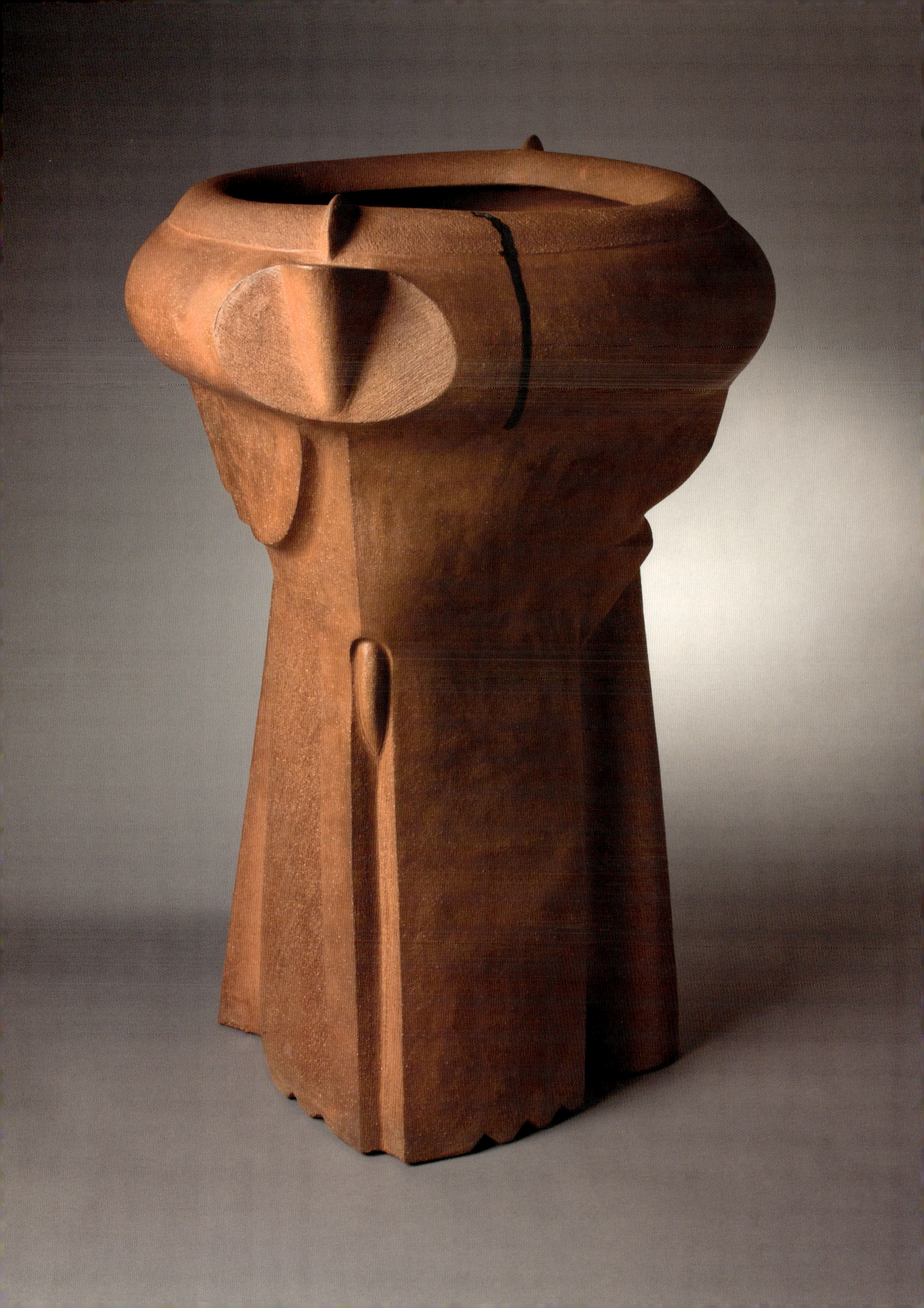

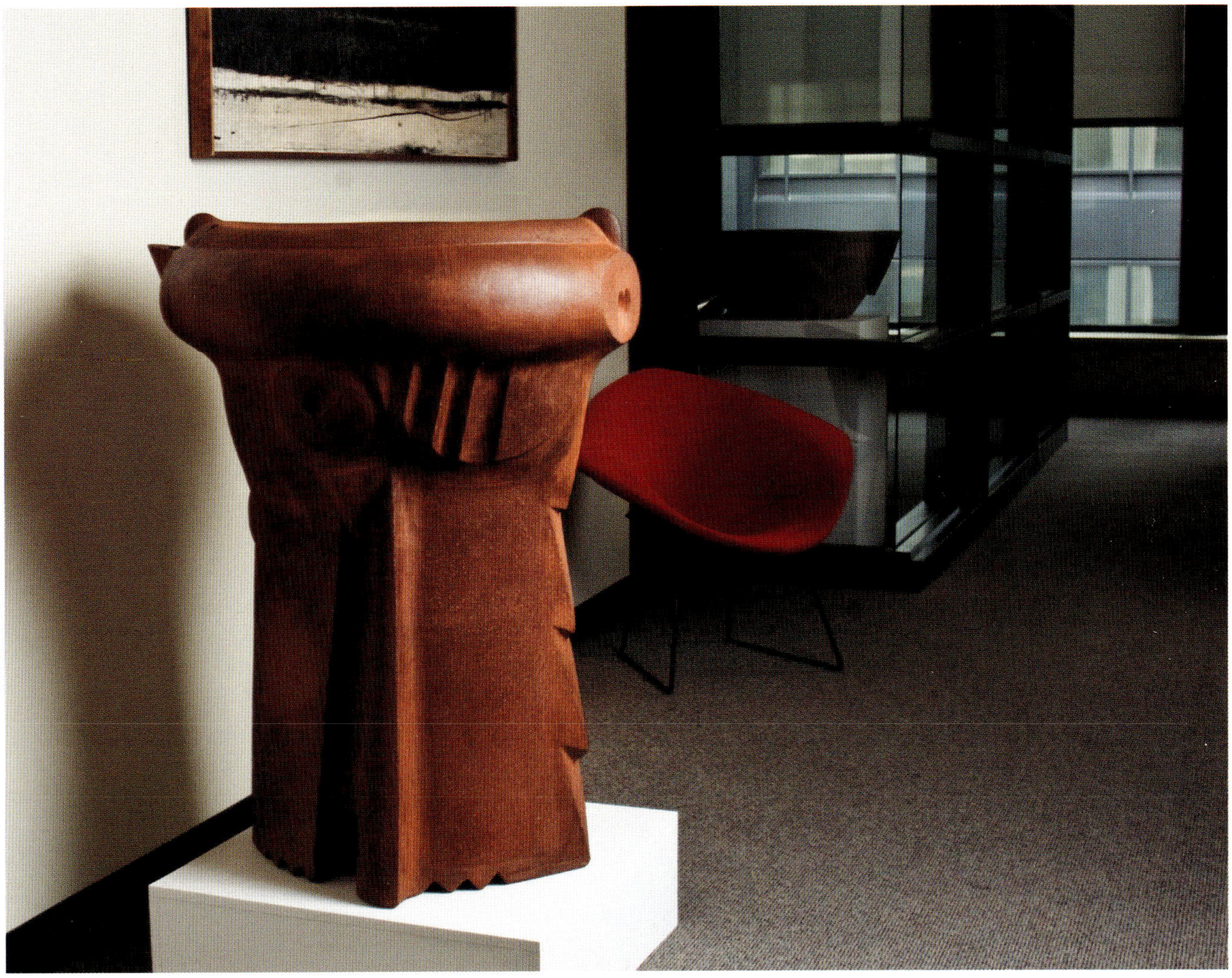

Photo: Amanda Schwab

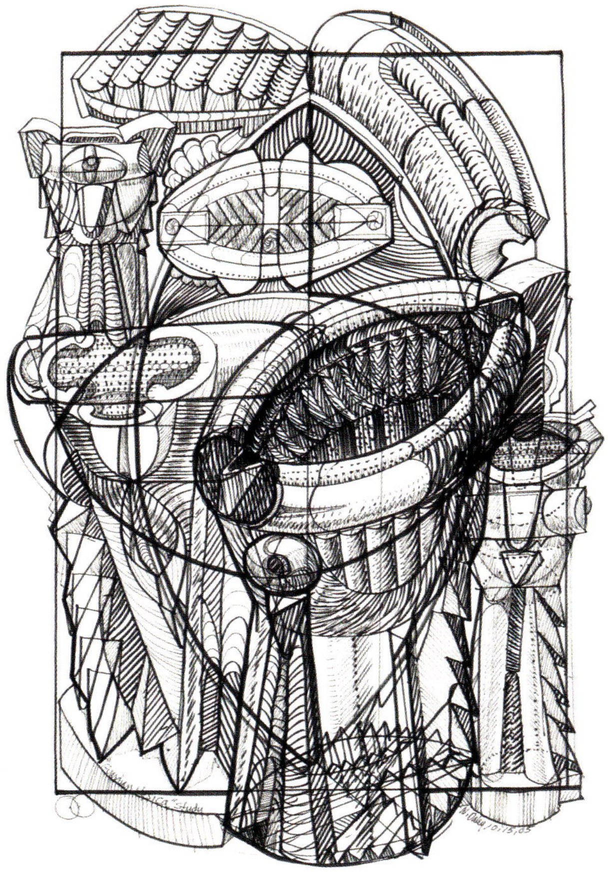

Fig. 203, opposite. *Guardian Vesica,* 2005
Unglazed stoneware
40 x 24 x 28 inches
Collection of Thomas Daley

Fig. 204, above. *Guardian Vesica,* 2005
In the office of Daley + Jalboot Architects, Philadelphia

Fig. 205, right. *Guardian Vesica* study, *2005*
Clayarch Gimhae Museum,
Gimhae, Korea

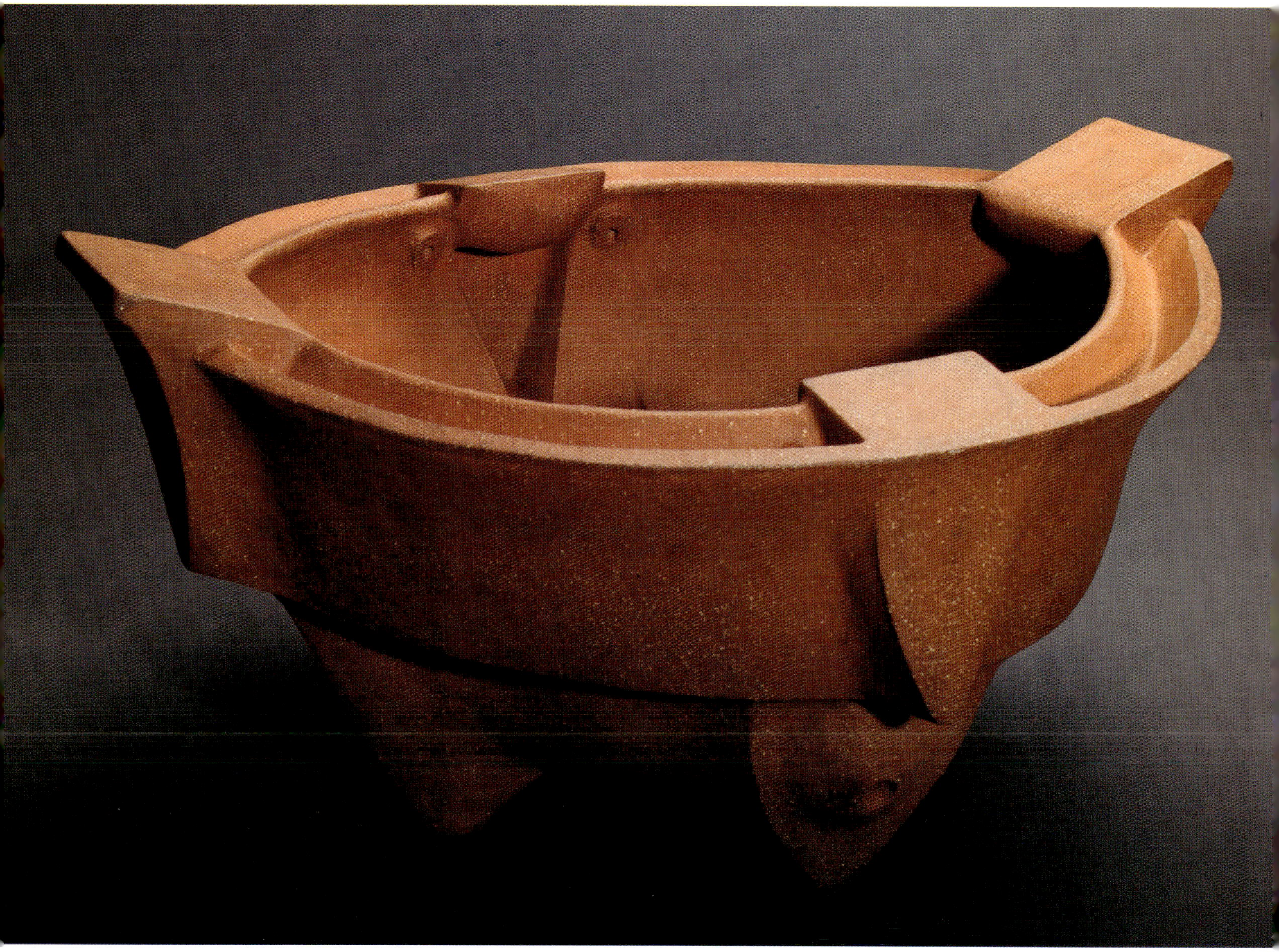

Fig. 206, above. *C-Venture Vesica,* 2008
Fig. 207, opposite. View from above
Unglazed ground stoneware
with white grog
14.5 x 28 x 23 inches

Exhibited American Crafts Council
Awards Exhibition SOFA 2010, Chicago
Courtesy of Perimeter Gallery

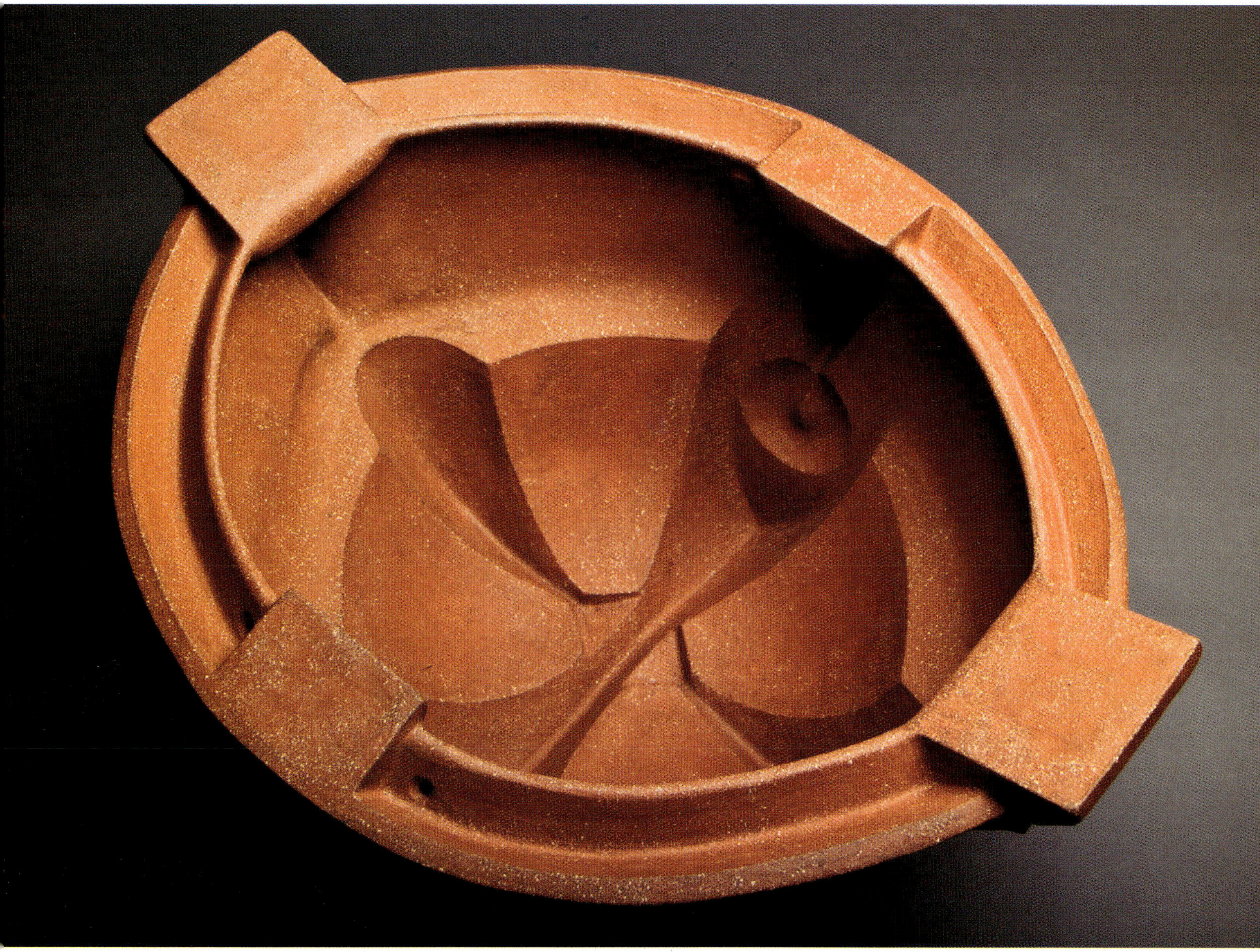

Photos: Amanda Schwab

Fig. 208, opposite. *Vesica Vesica II,* 2008
Fig. 209. View from above
Unglazed ground stoneware
12.5 x 20 x 24.5 inches
Collection of Alfred and Rica Fenaughty

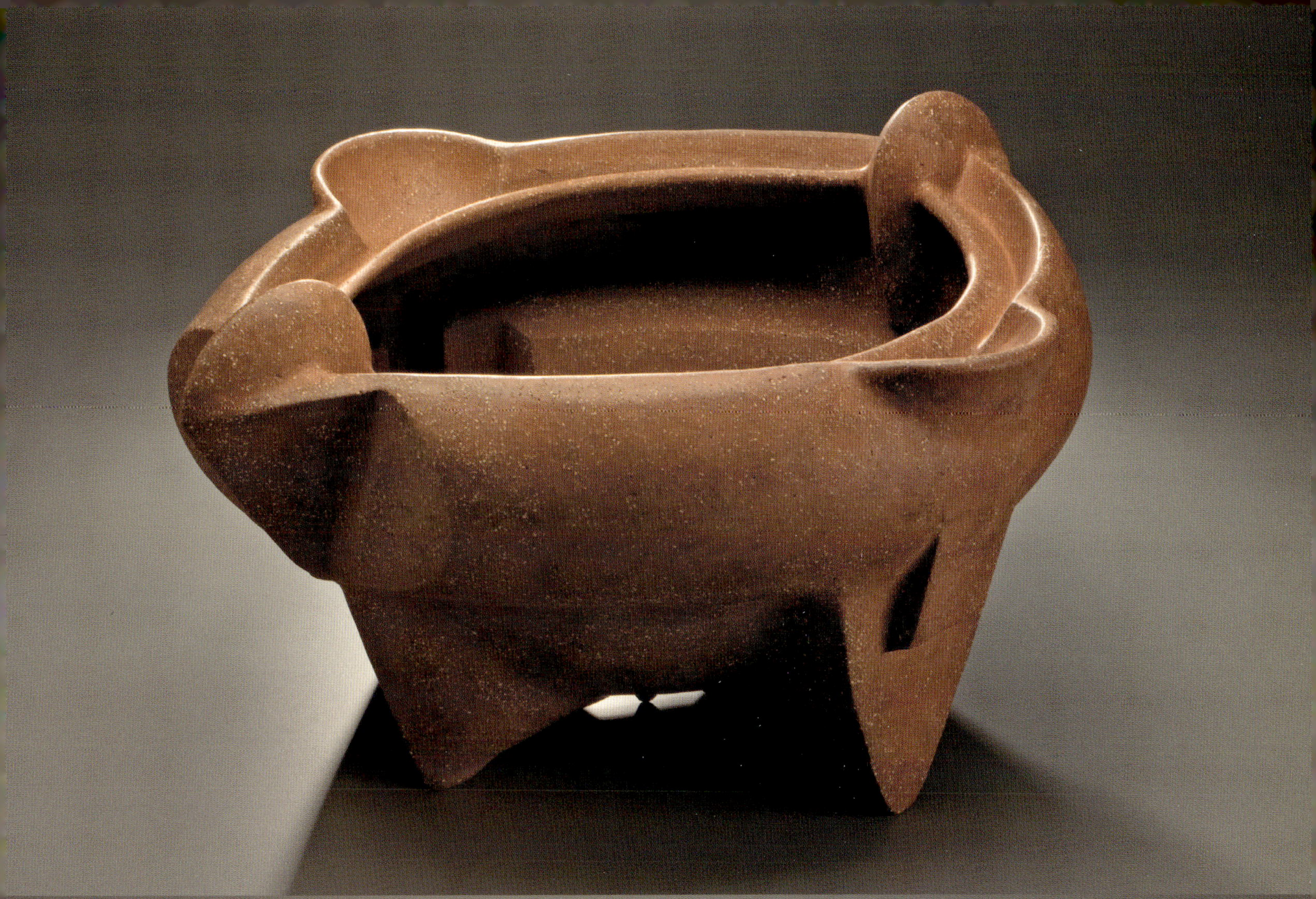

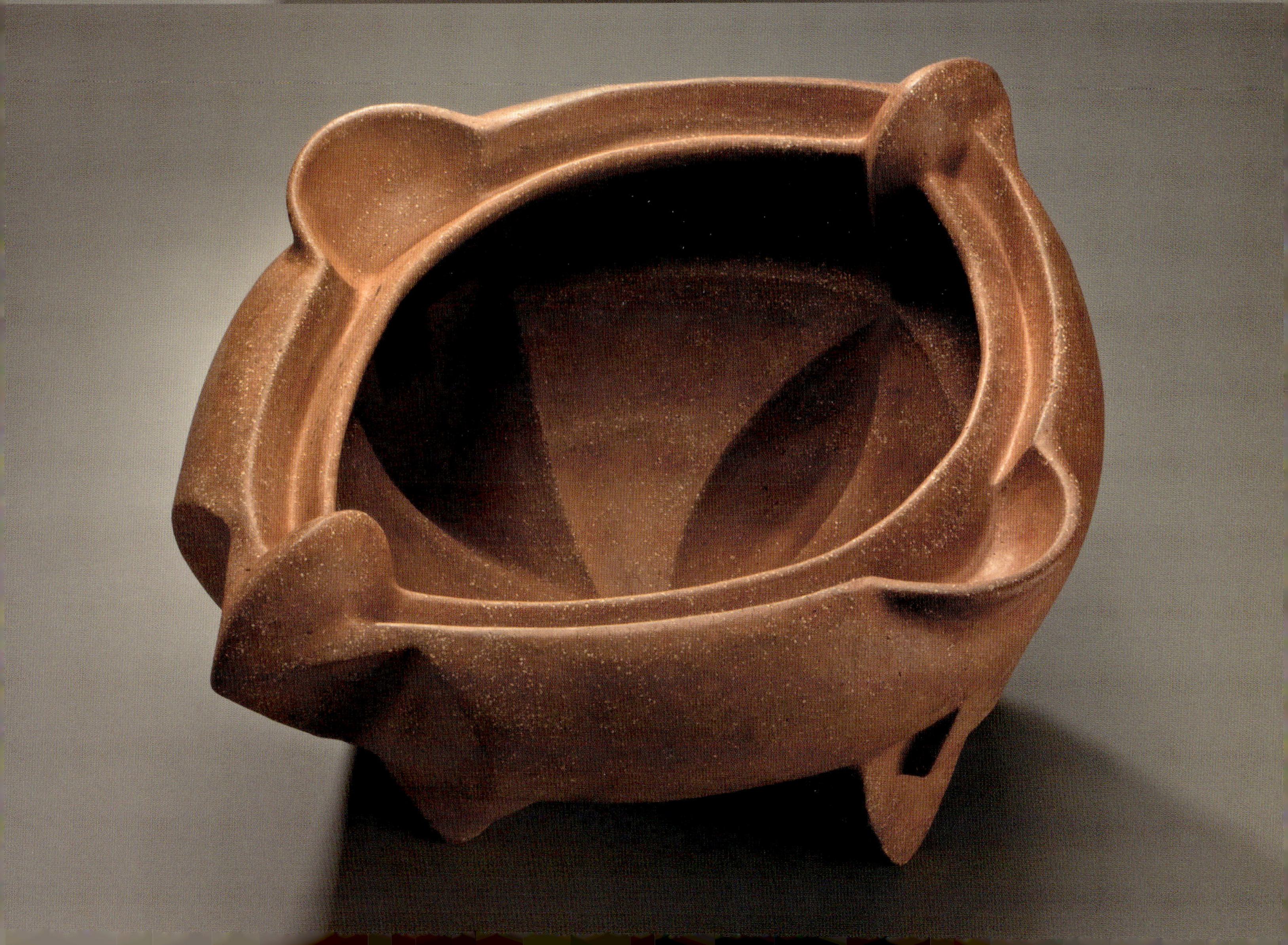

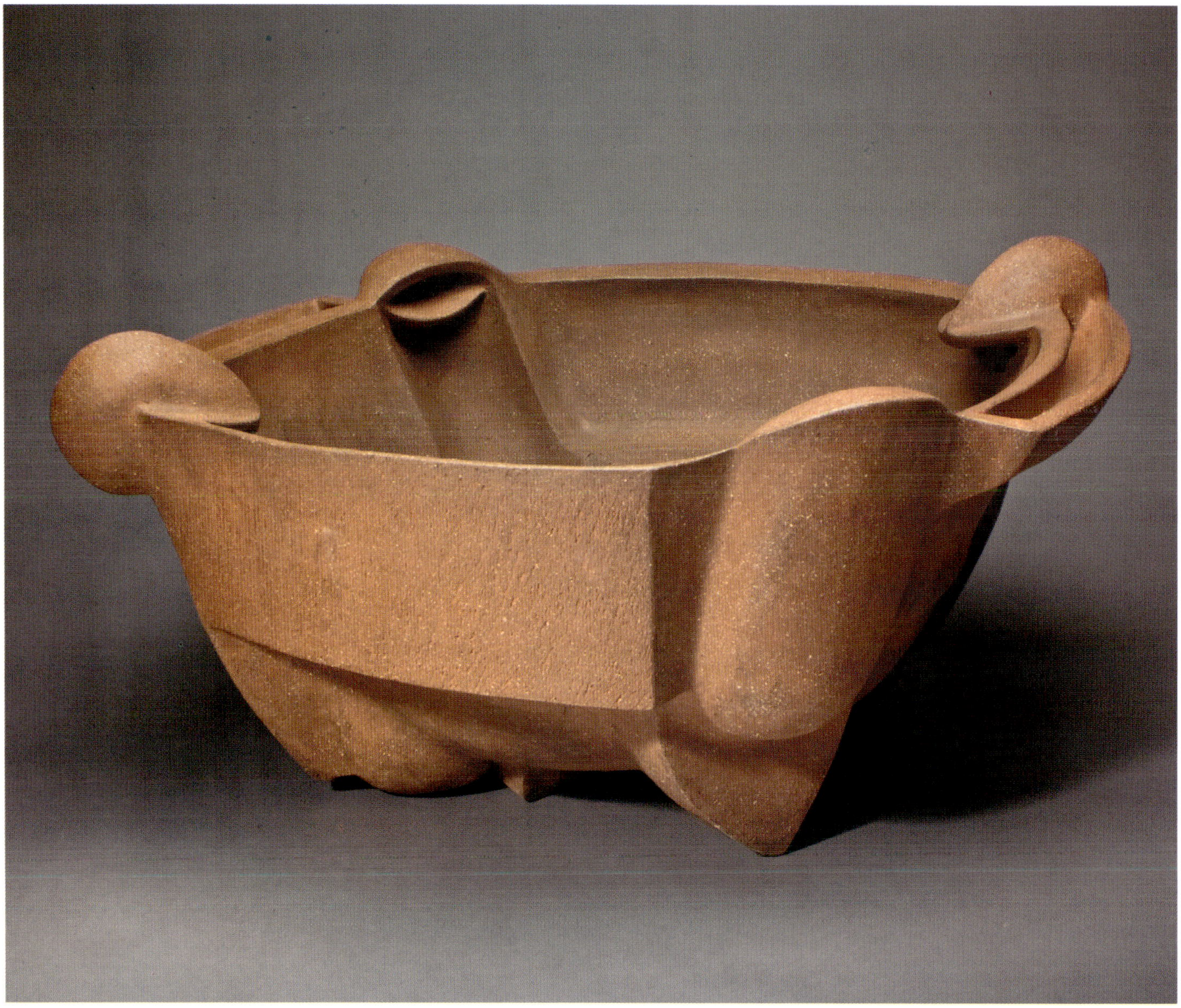

Photos: Amanda Schwab

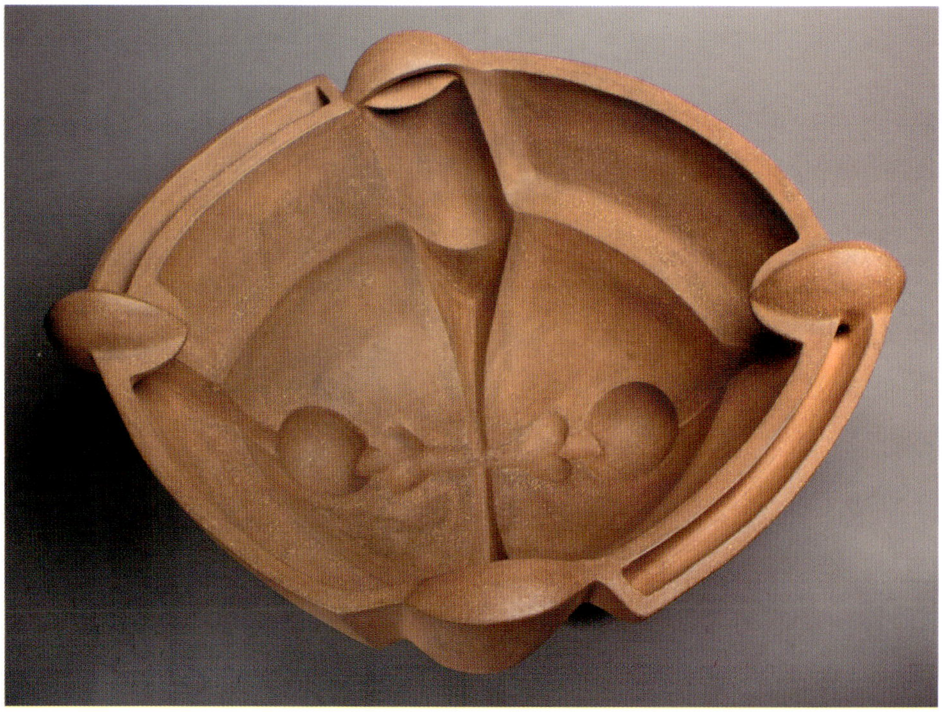

Fig. 210. *Azoic Vesica,* 2008
Fig. 211, right. View from above
Unglazed ground stoneware
14.5 x 27.5 x 32 inches
Collection of Alfred and Rica Fenaughty

Photos: John Carlano

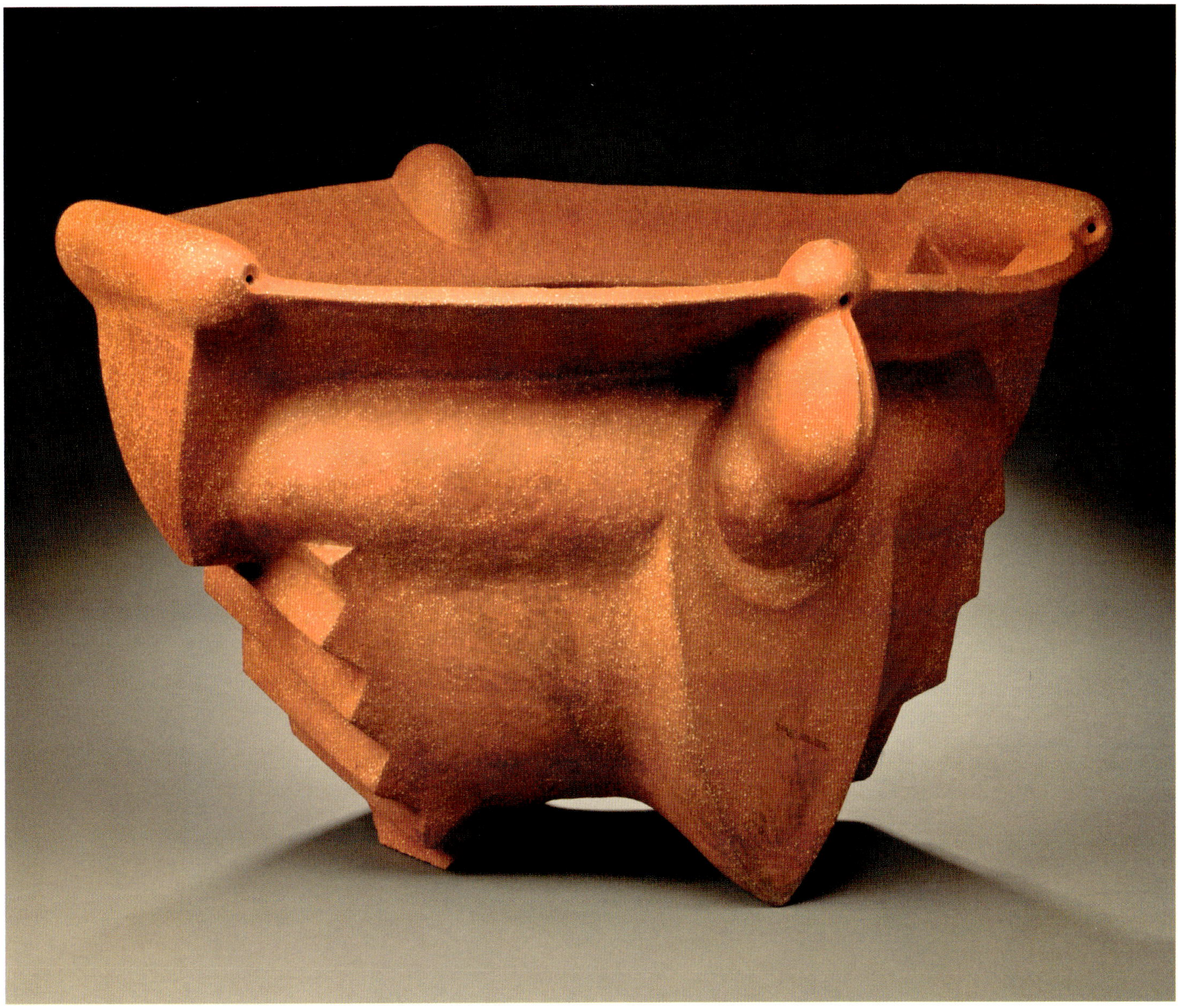

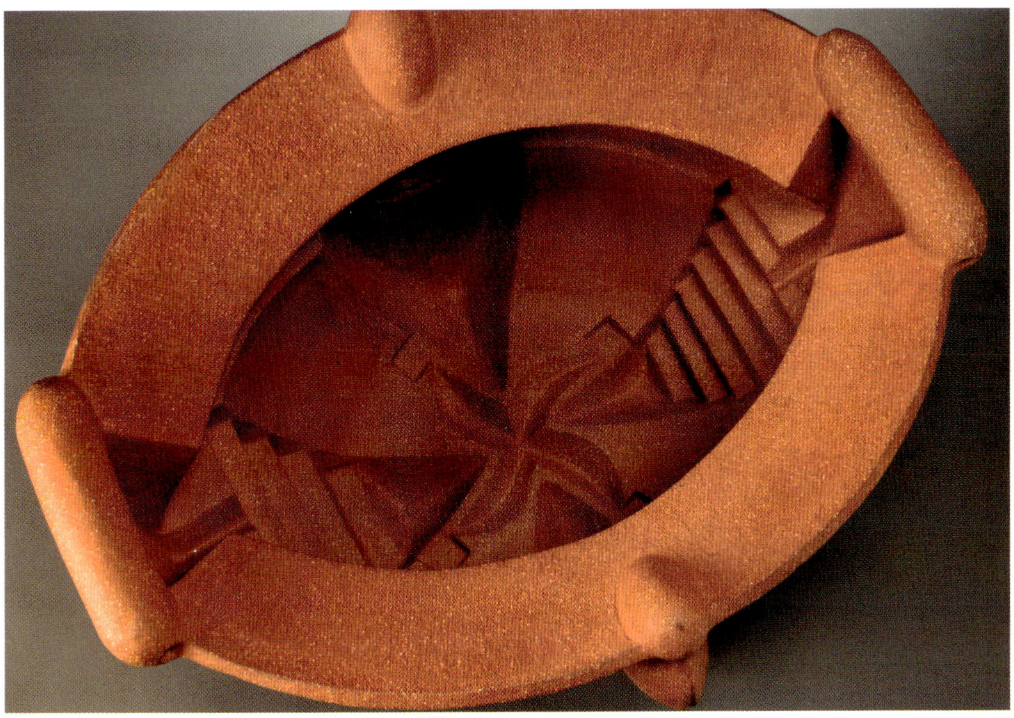

Fig. 212. *Crossed Vesica,* 2007
Fig. 213, left. View from above
Unglazed stoneware
20 x 28 x 33 inches
Collection of Alfred and Rica Fenaughty

Exhibited American Crafts Council
Awards Exhibition SOFA 2007, Chicago

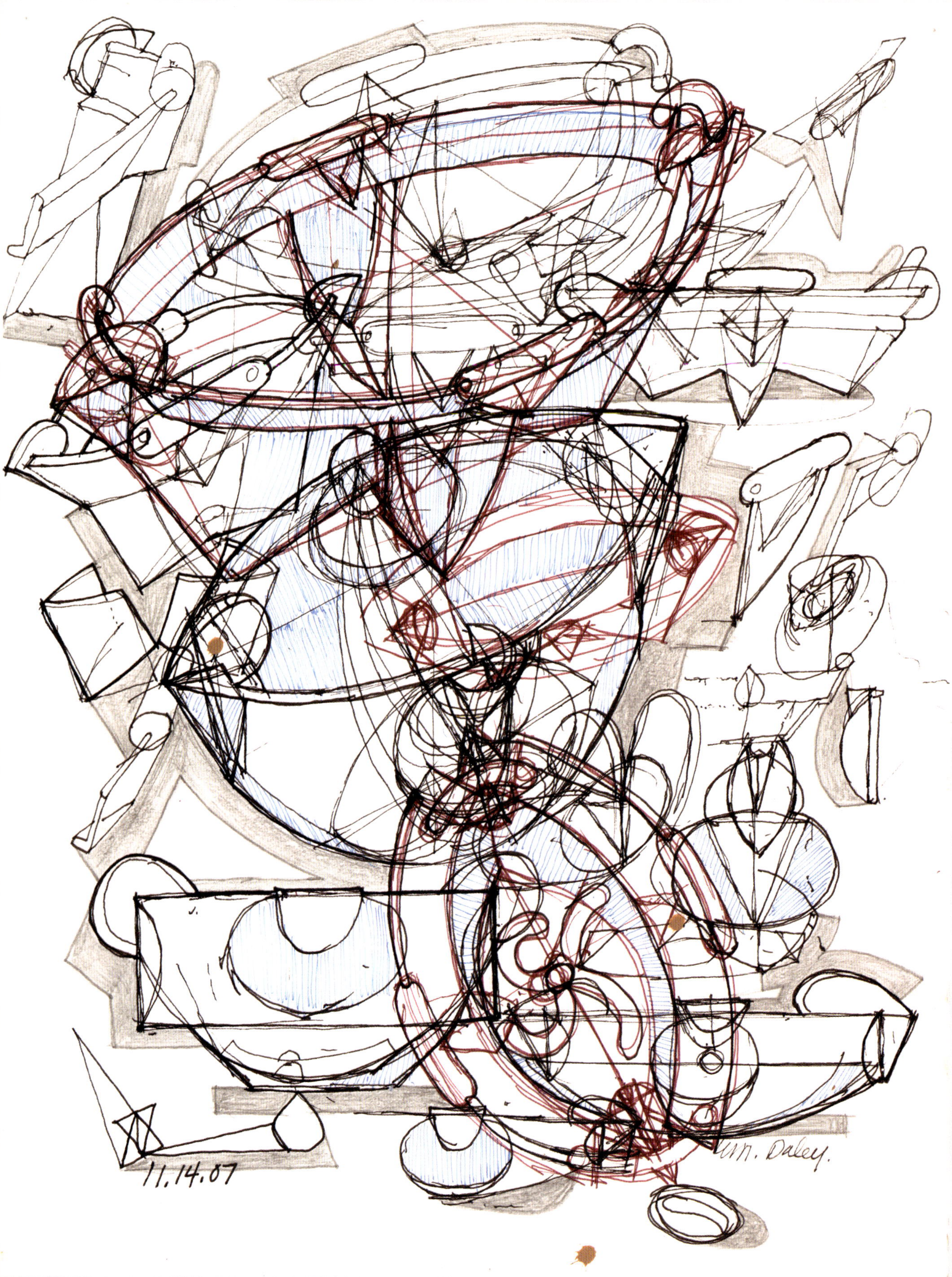

Fig. 214. Drawing, 2007
Pen and colored pencil on bristol paper
14 X 11 inches
Collection of Thomas Porett

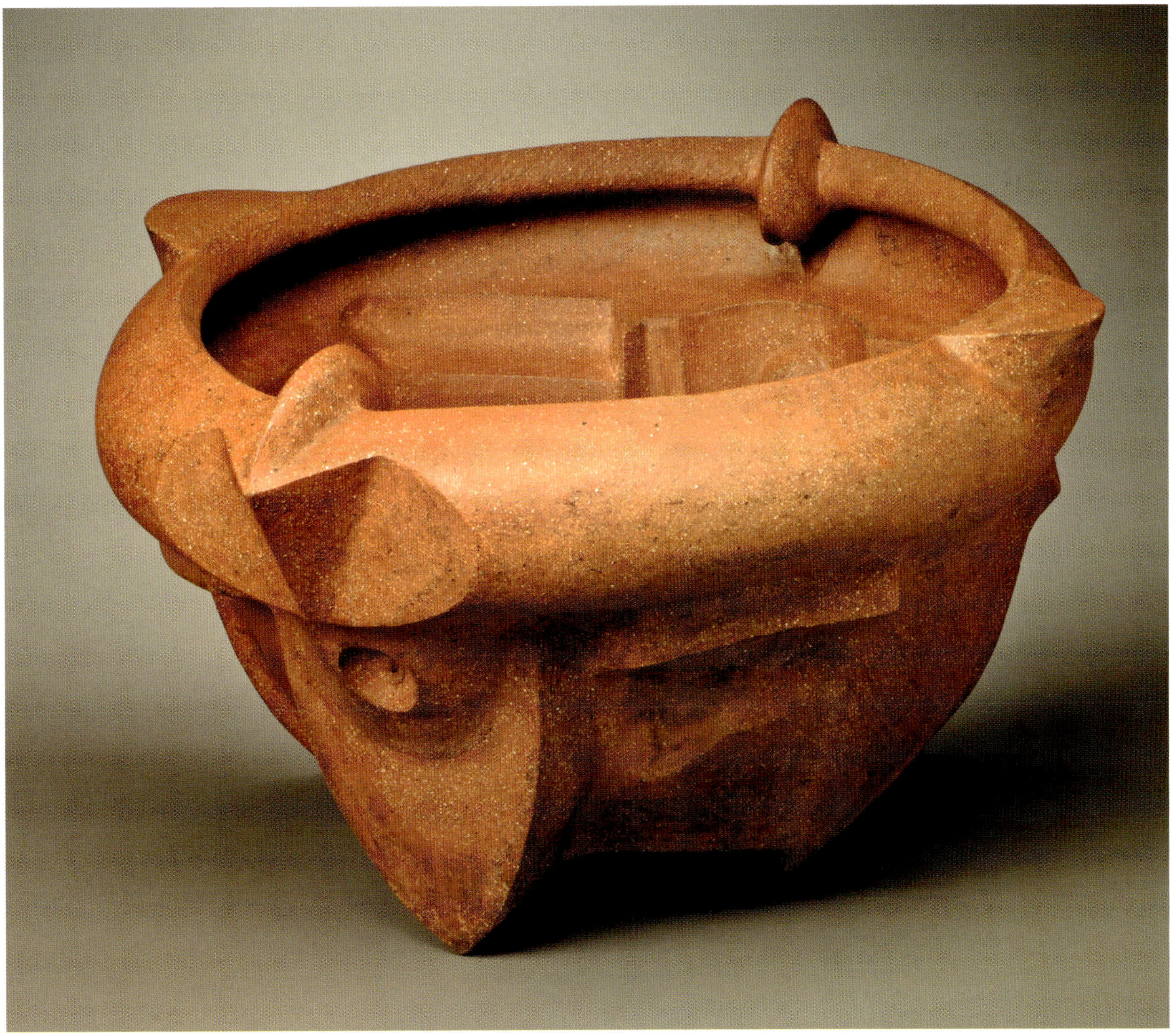

Photos: John Carlano

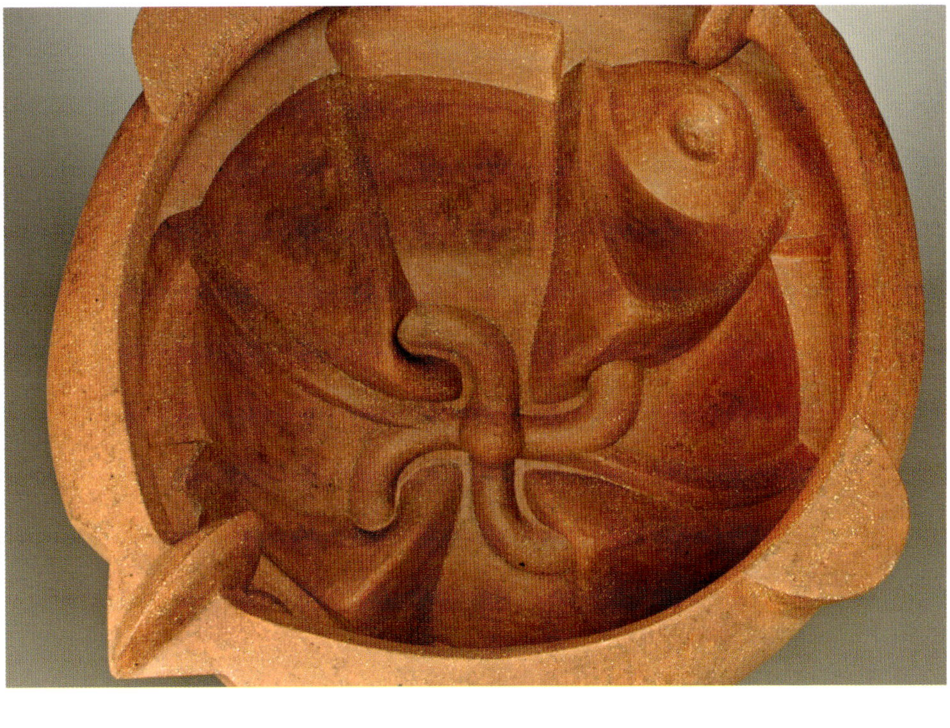

Fig. 215. *Venture Z Vesica,* 2007
Fig. 216, right. View from above
Unglazed stoneware
17 x 29 x 32 inches
Collection of William and Catherine Daley

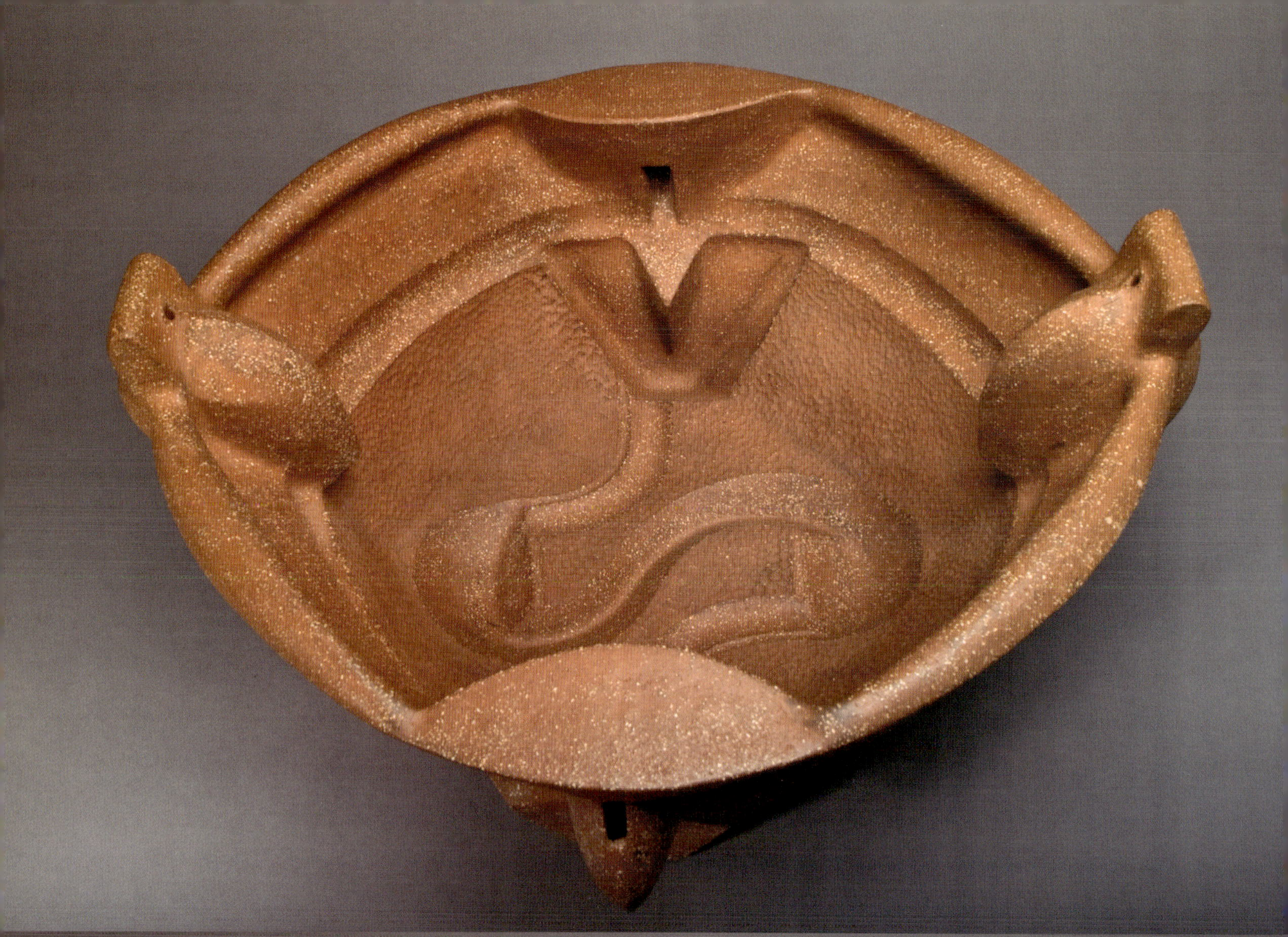

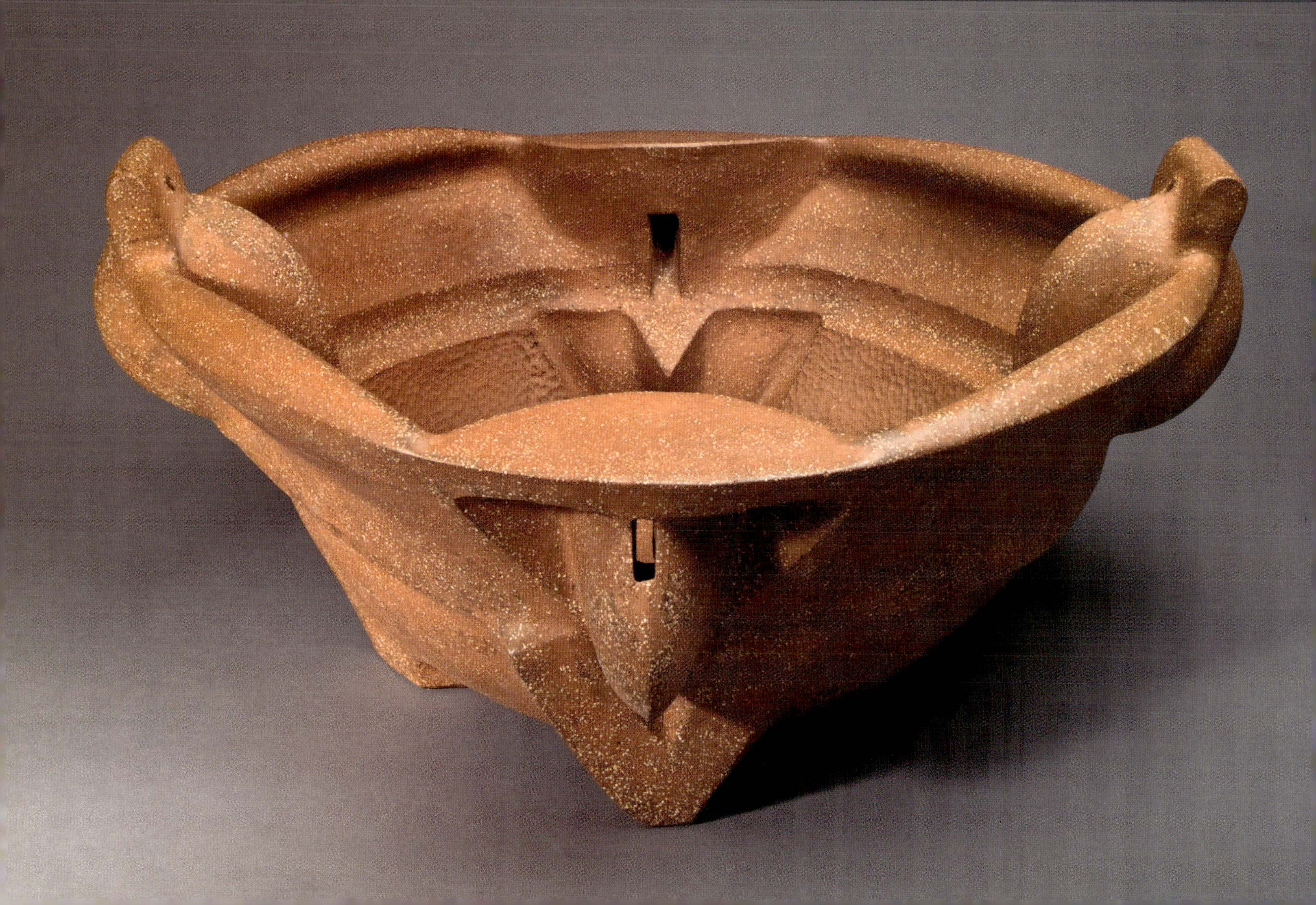

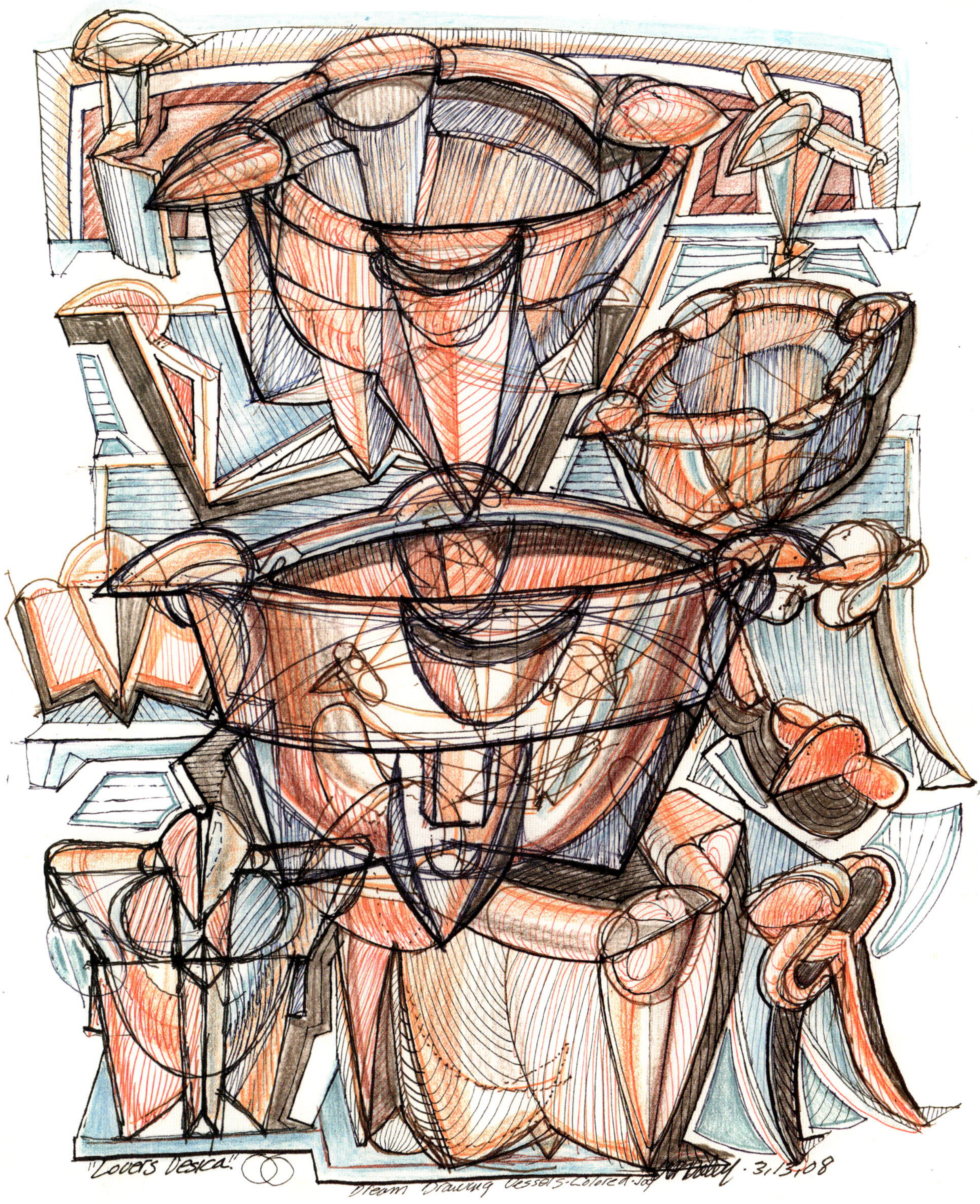

Fig. 217 opposite top. *Lovers Vesica*, 2008
Fig. 218, opposite. Side view
Unglazed ground stoneware
11.75 x 23.5 x 25.5 inches
Collection of Sara Jane Daley

Fig. 219. Drawings, *Lovers Vesica,* 2008
Colored pencil and markers
on bristol paper
14 x 11 inches
Collection of Charlotte Daley

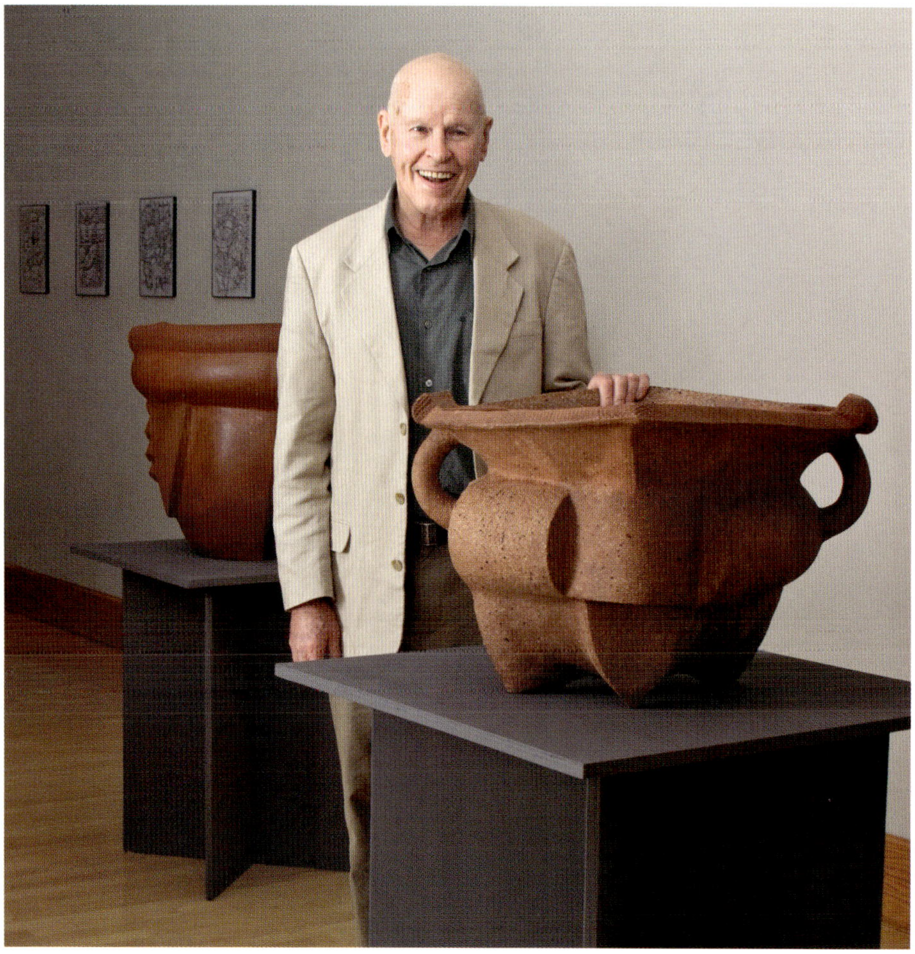

Fig. 220. William Daley, *Vesica Explorations*
List Gallery, Swarthmore College,
Swarthmore, PA
Pictured with *Sentinal Vesica,* 2009

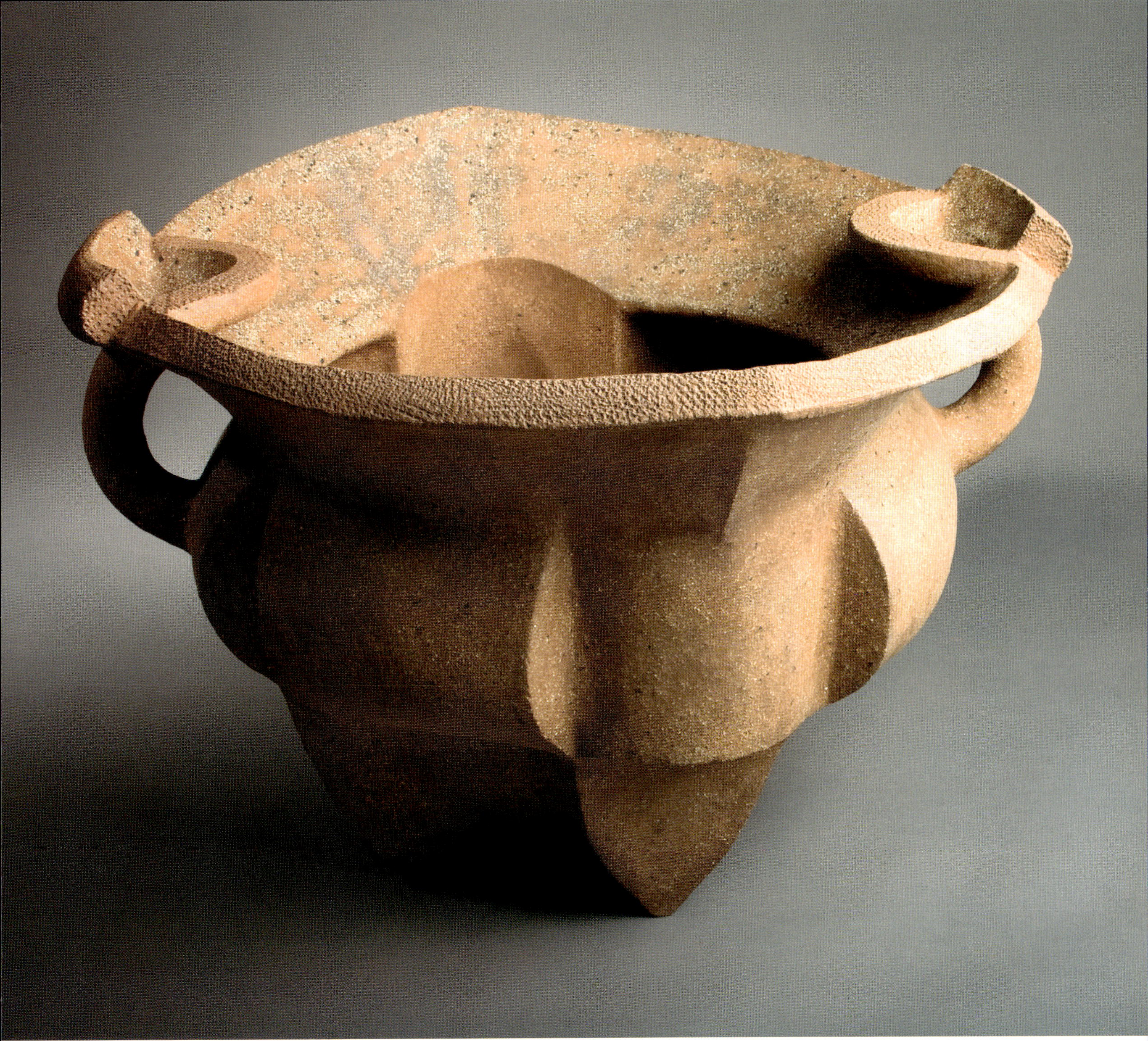

Fig. 221. *Sentinal Vesica,* 2009
Unglazed ground stoneware
16 x 26 x 29 inches
Collection of Jim and Susan Pagliaro

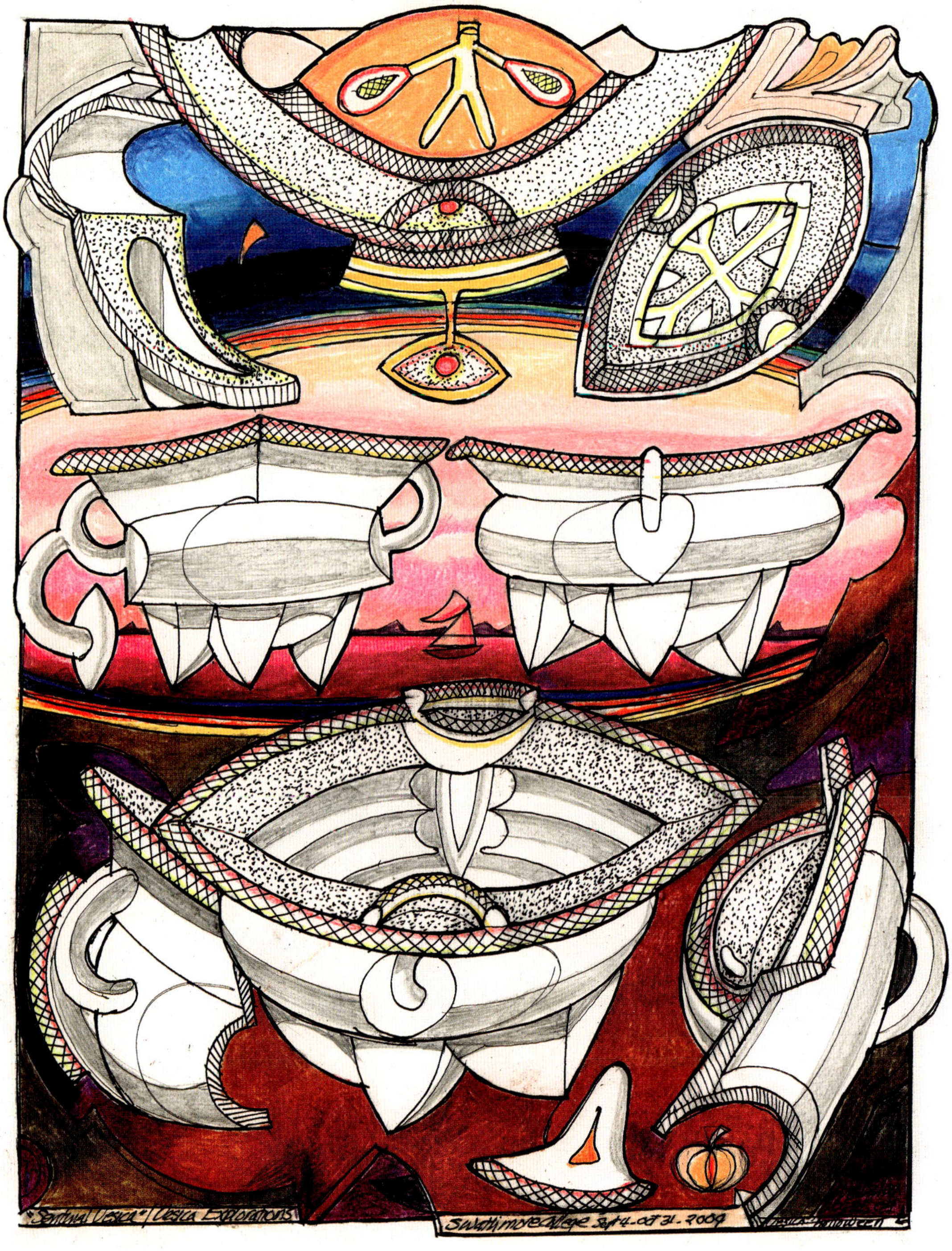

Fig. 222. *Sentinal Vesica/Vesica Explorations,* 2009
Drawing pen and colored pencils on paper
14 x 11 inches
Collection of George and Barbara Daley Schwab

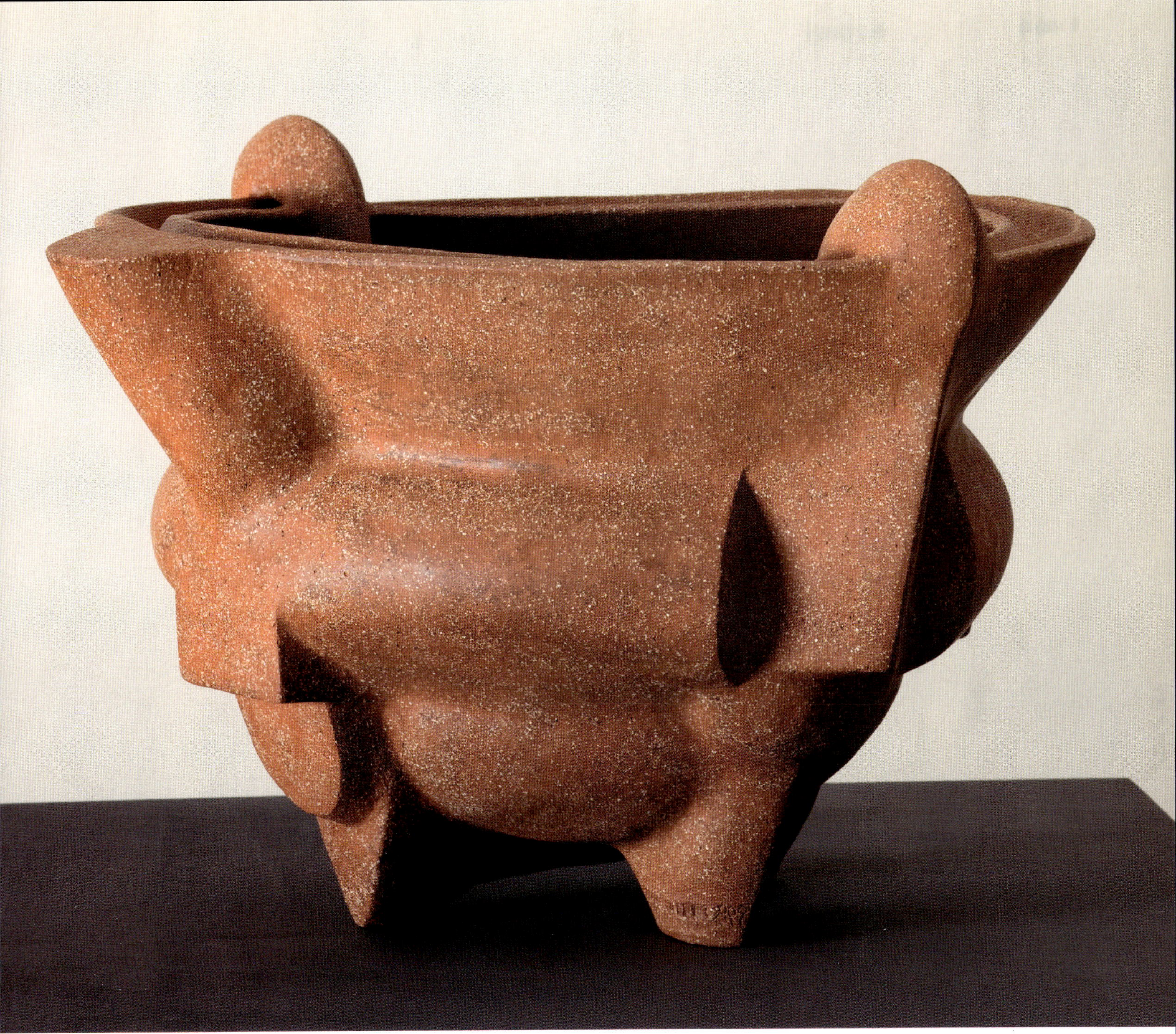

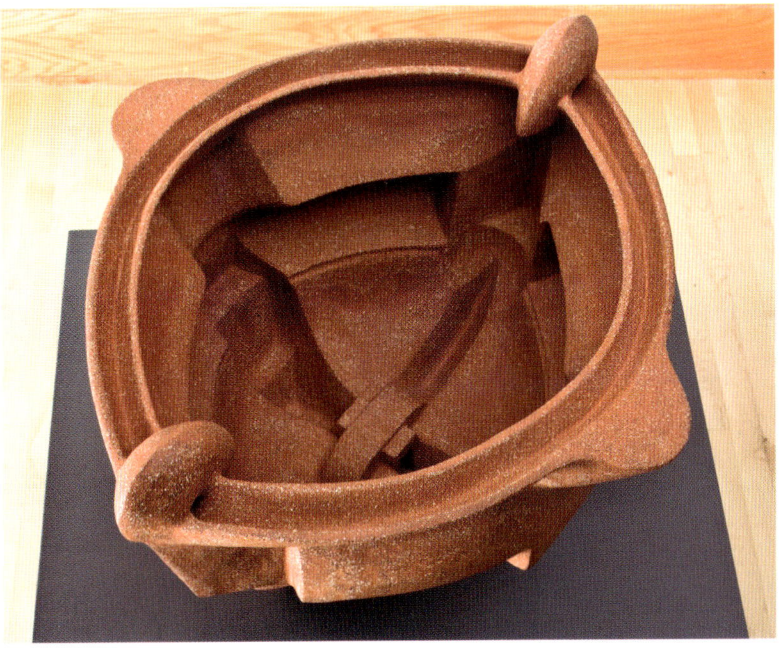

Fig. 223. *Cross Under*, 2009
Fig. 224, left. View from above
Unglazed ground stoneware
17 x 23 x 27 inches
Collection of Helen W. Drutt English

Building Pots and Making Buildings
A Visual Conversation

This presentation is a collaboration for a building between a Father and Son, William Daley, Artist, and Thomas Daley, Architect.

It started from two nearly simultaneous events. The first, a comment passed on to us from a Northern Clay Center concept meeting,"why don't we make our dream building like a Bill Daley Pot." The second spark was an invitation from Clayarch Gimhae Museum in Gimhae, Korea. The occasion was to celebrate the opening of a new center for the advancement of Ceramics and Architecture by hosting an international symposium and exhibition.

The building's design premise is based on the NCC ceramic education program, and a theoretical urban site located at the juncture of the two small streets and bound on its north side by the party wall of an adjacent building. The plan employs brick masonry construction, in the tradition of "The Philadelphia School," as an anchor to enclose the forms evolved from decades of Daley pots, forms based on the morphology of the clay wall's invitation to "Inside-Outside" habitation, and architecture.

This sketch model, like the pots and drawings that describe its hypothetical becoming, is an outcome of The Northern Clay Center, The Clayarch Gimhae Museum, where it was presented as Thomas Daley's offering to the symposium.

In a deeper sense, it is a special Vesica of shared birthdays and a passion for creating telling enclosures that embody the core of Architecture and Ceramics.

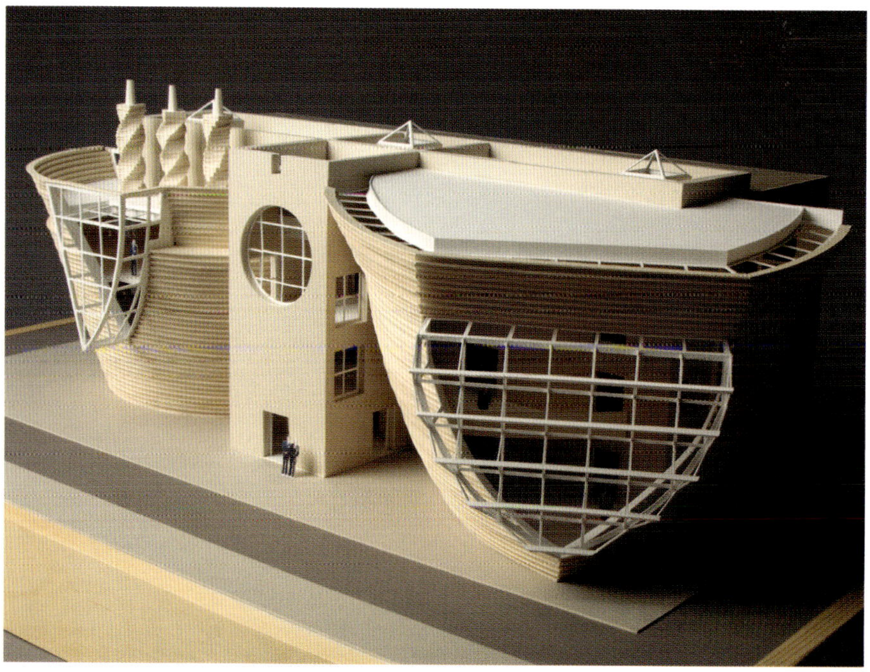

Photo: Amanda Schwab

Fig. 225. Sketch model
Thomas W. Daley AIA
Daley + Jalboot Architects
as presented at the Clayarch Gimhae Museum's Ceramics and Architecture symposium in Gimhae Korea

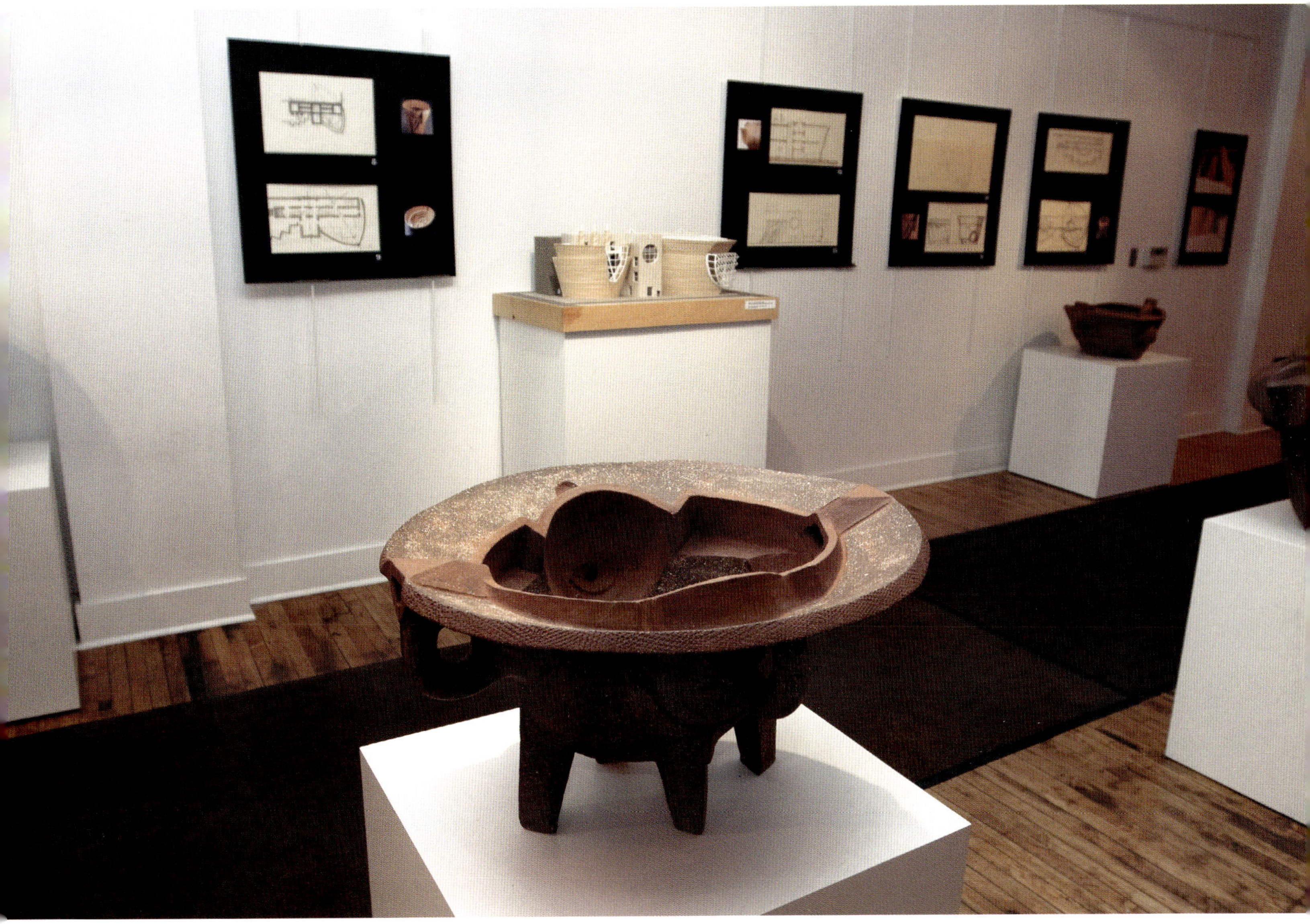

Fig. 226. *Joshua Cistern Vesica,* 2009
Unglazed ground stoneware
18.5 x 25.75 x 29 inches

Daley and Daley – A collaboration
Exhibit – March 29, 2010
Philadelphia Center for Architecture
In conjunction with the 25th Anniversary
NCECA Conference, Philadelphia, PA

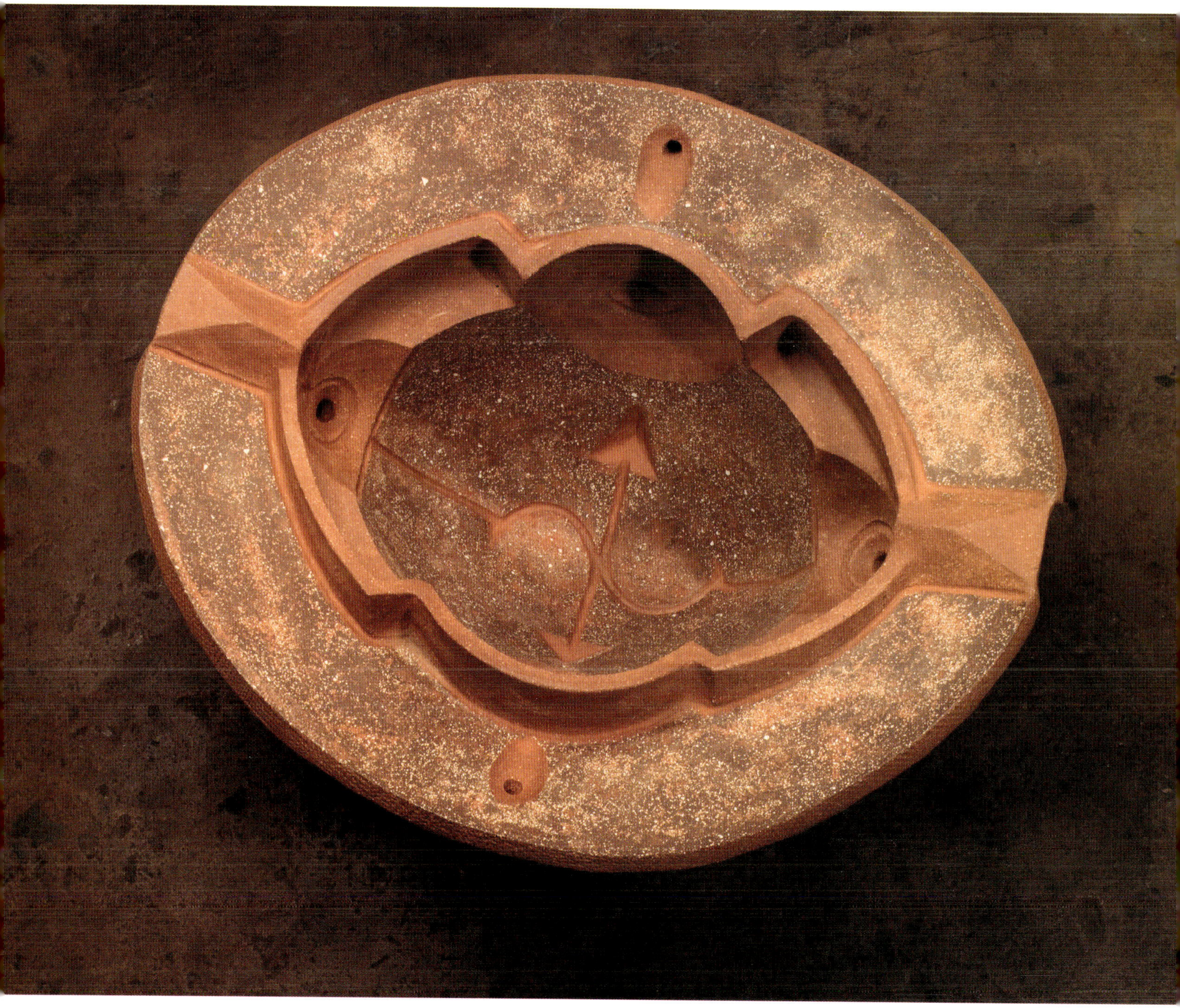

Fig. 227. *Joshua's Cistern,* 2010
Fig. 228, opposite. Side view
10 x 26 x 30 inches
Unglazed ground stoneware
Collection of Karyn Johnson Boyd

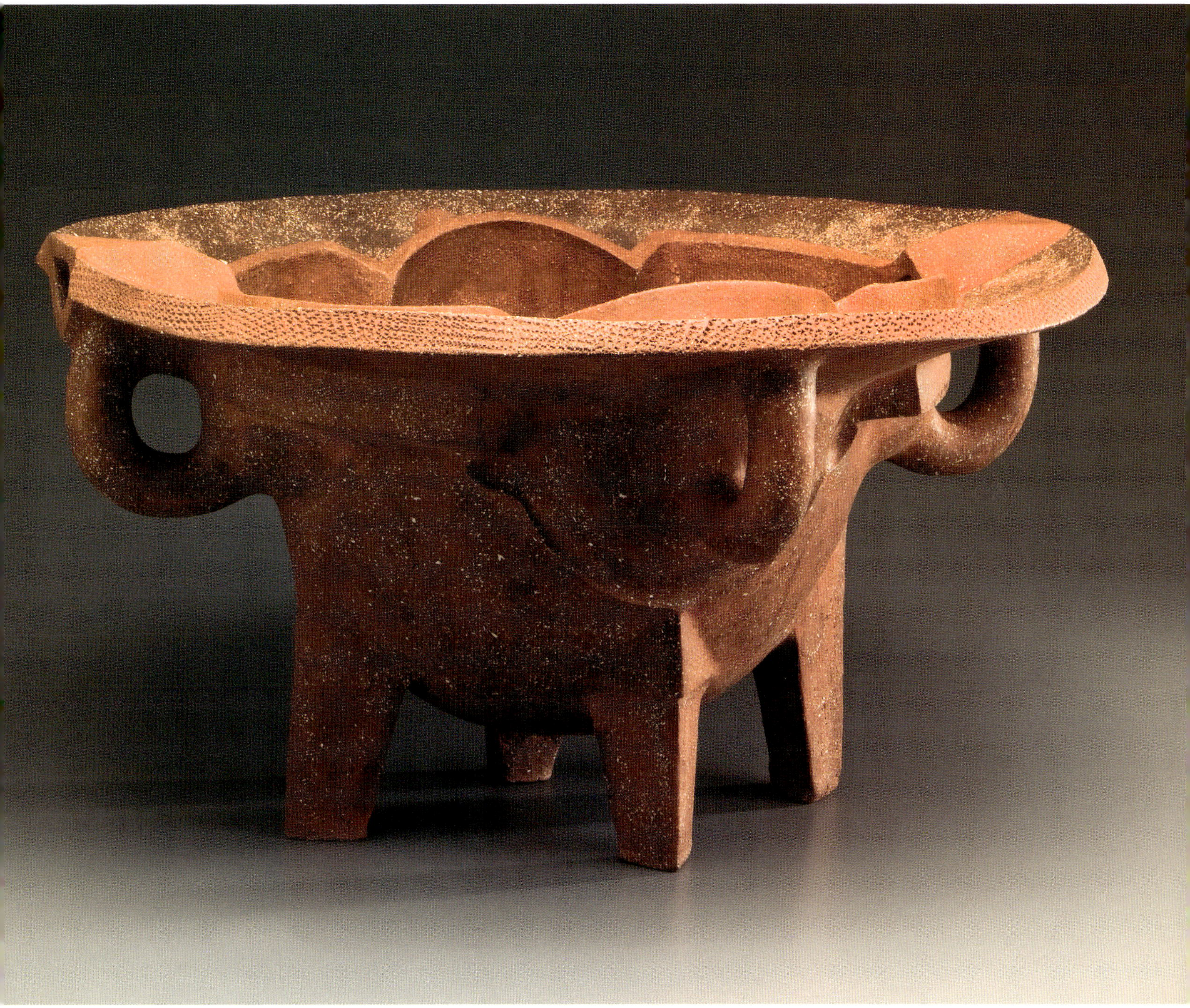

Photos: John Carlano

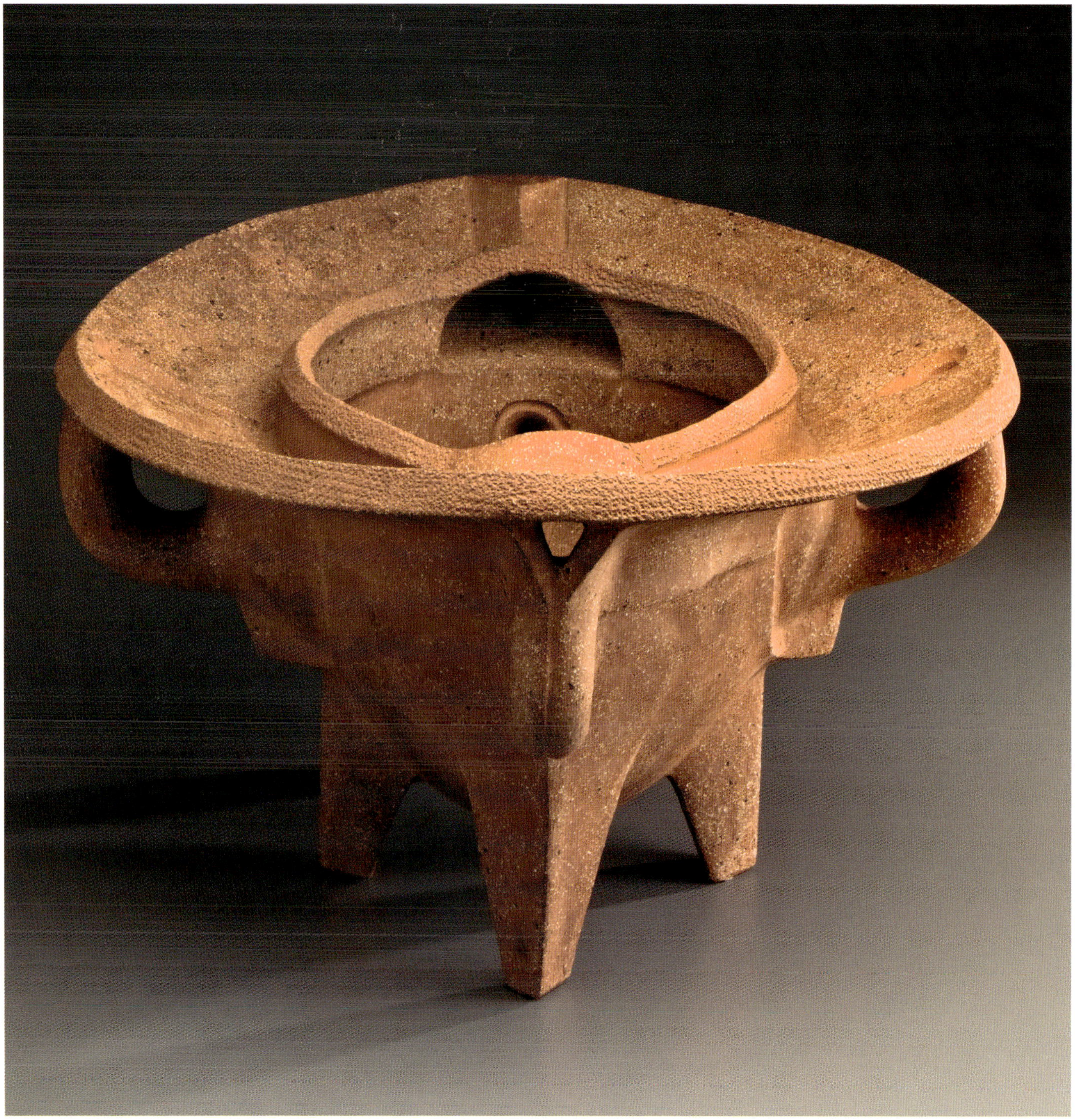

Photos: John Carlano

Fig. 229. *Joy-full Libation Vesica,* 2010
Unglazed ground stoneware
Collection of William and Catherine Daley
In tribute: William Parry, Richard Reinhardt, Petras Vaskys

Fig. 230, opposite.
Libation Cistern Vesica, 2010
Fig. 231. Side view
Unglazed ground stoneware
11 x 23.5 x 28.5 inches
Collection of Charlotte and Stephen Martin

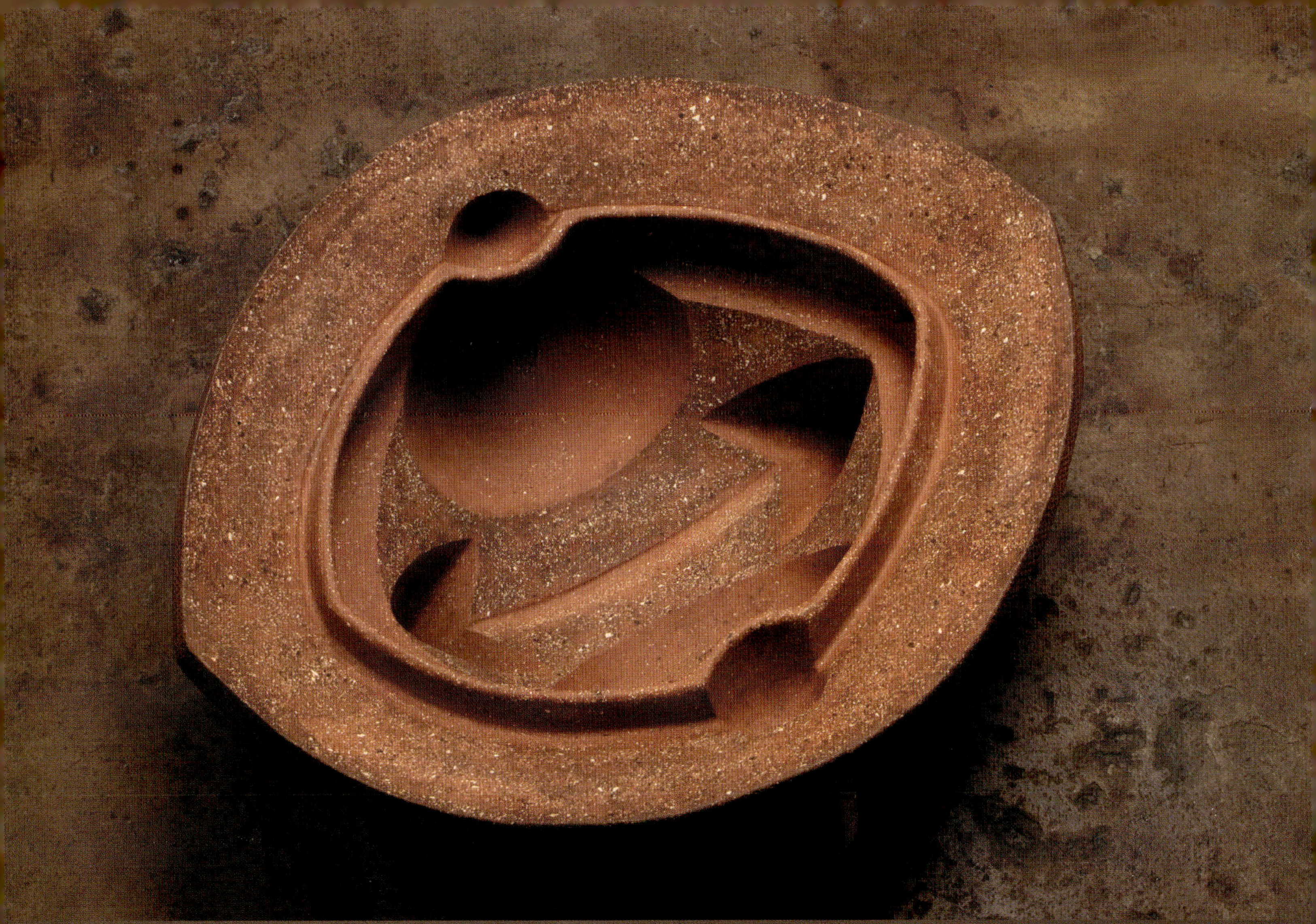
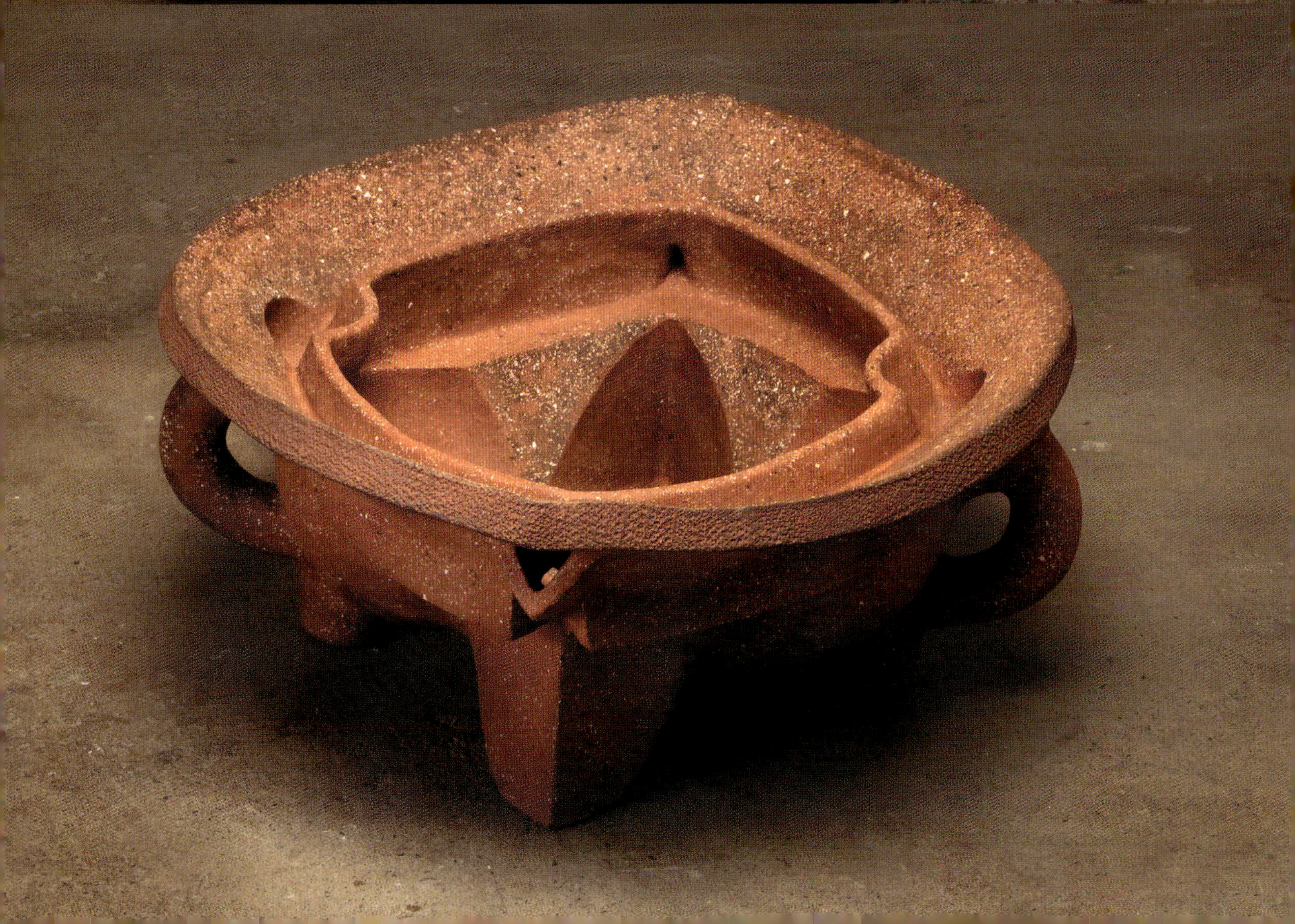

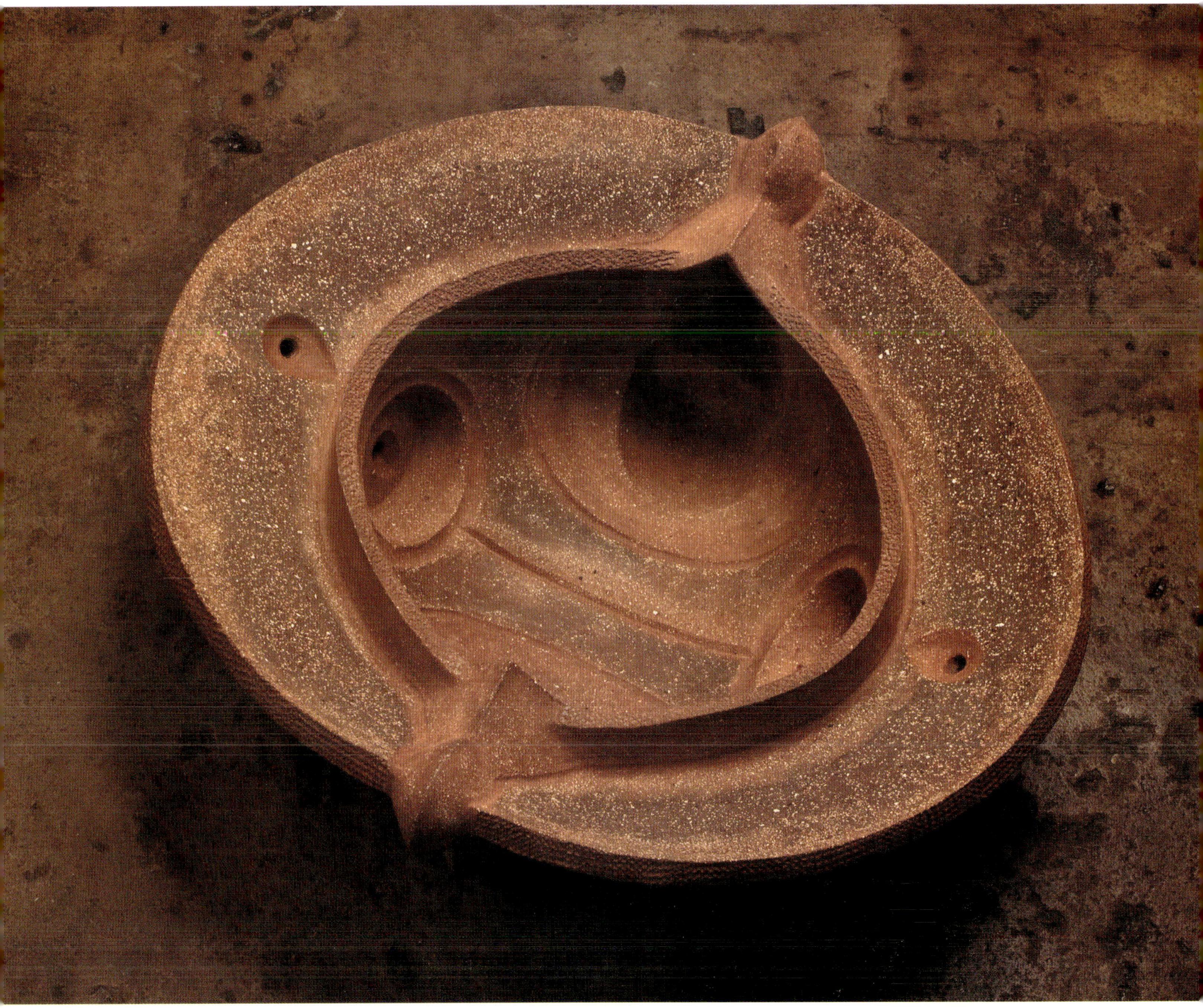

Fig. 232. *Joy's Cistern,* 2010
Fig. 233, opposite. Side view
Unglazed ground stoneware
10 x 26 x 28 inches
Collection of William and Catherine Daley

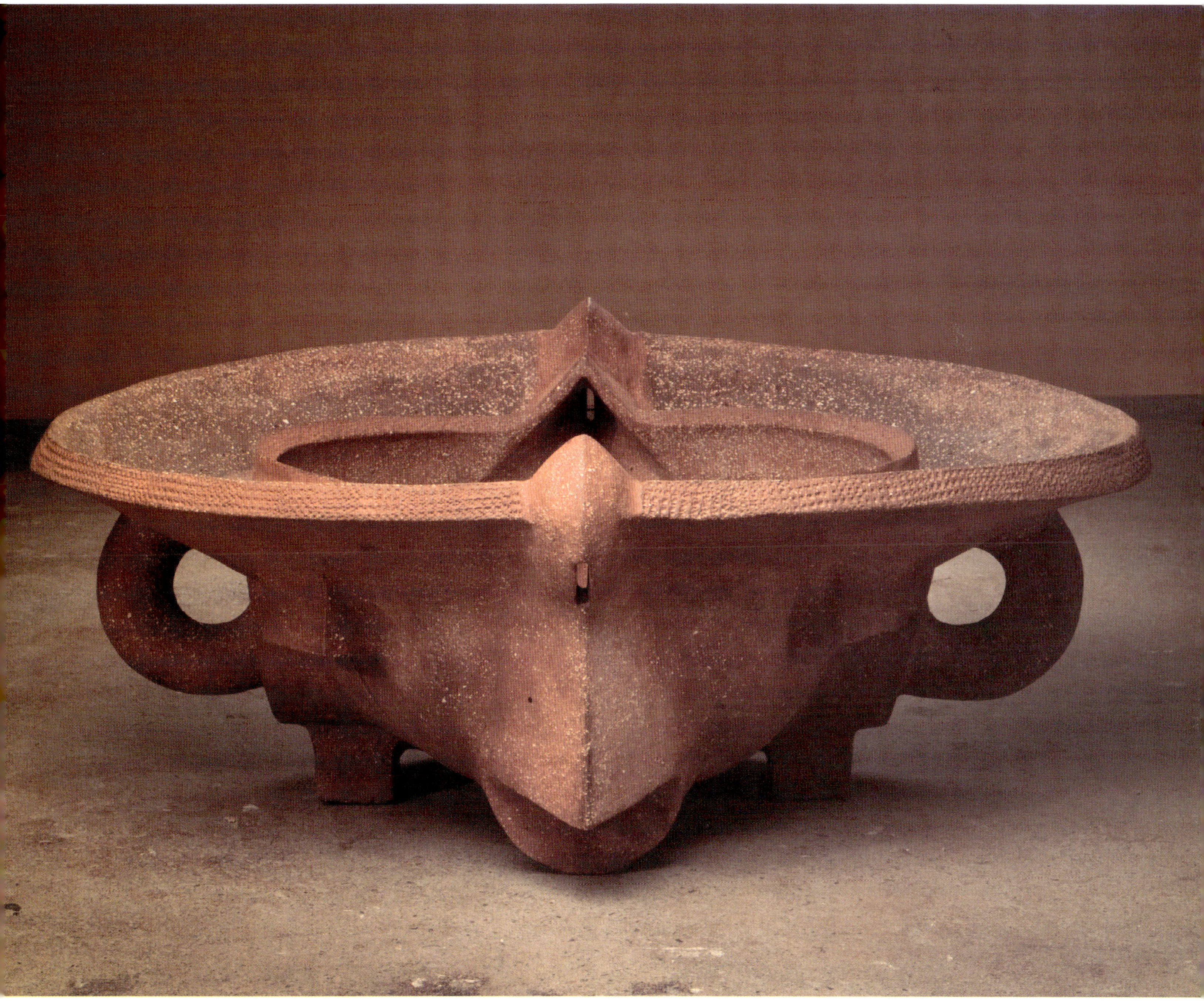

Photos: John Carlano

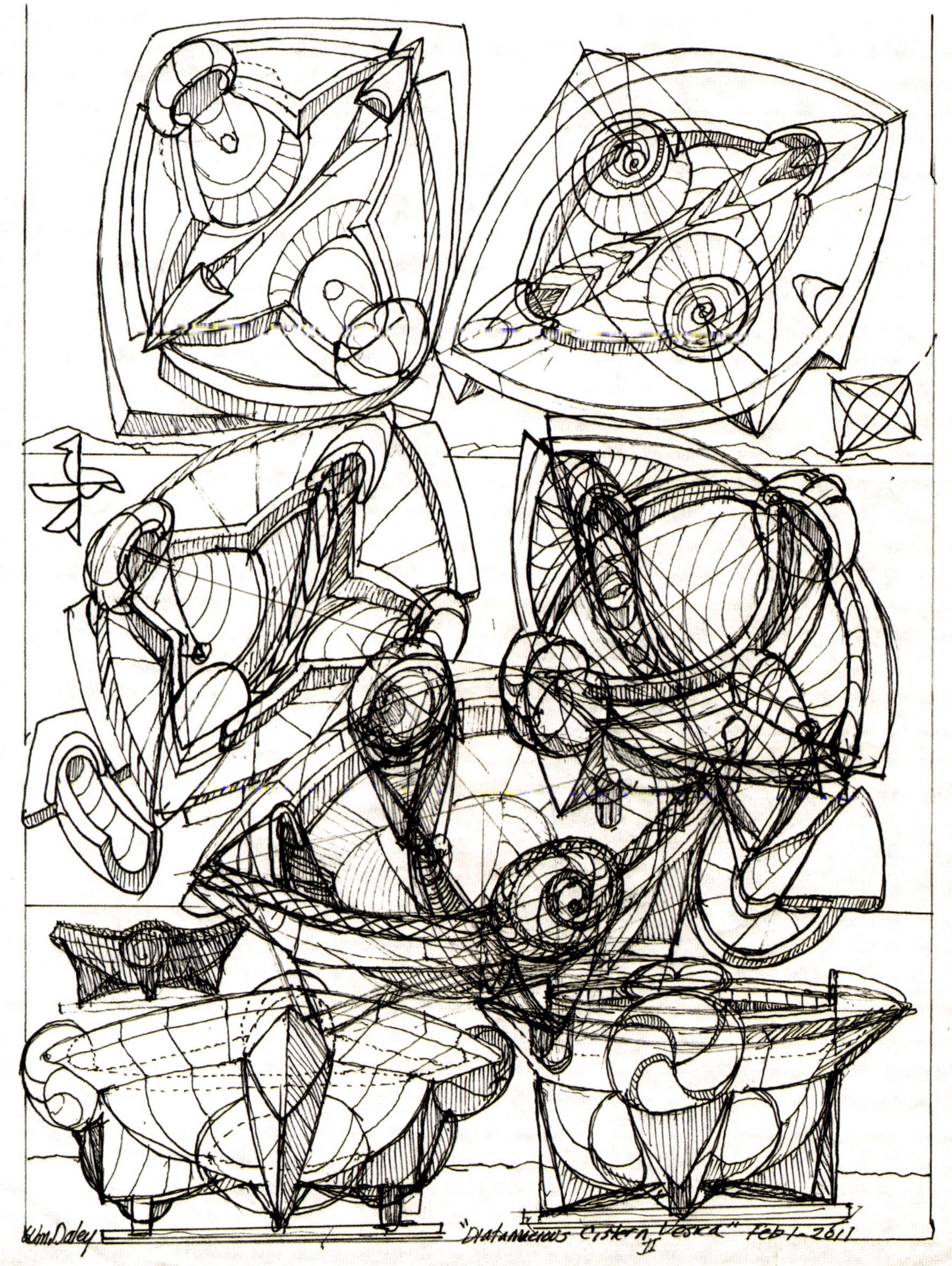

Fig. 234. Drawing, *Cistern Vesica,* 2011
Pen on paper
14 x 11 inches
Collection of William and Catherine Daley

Photos: John Carlano

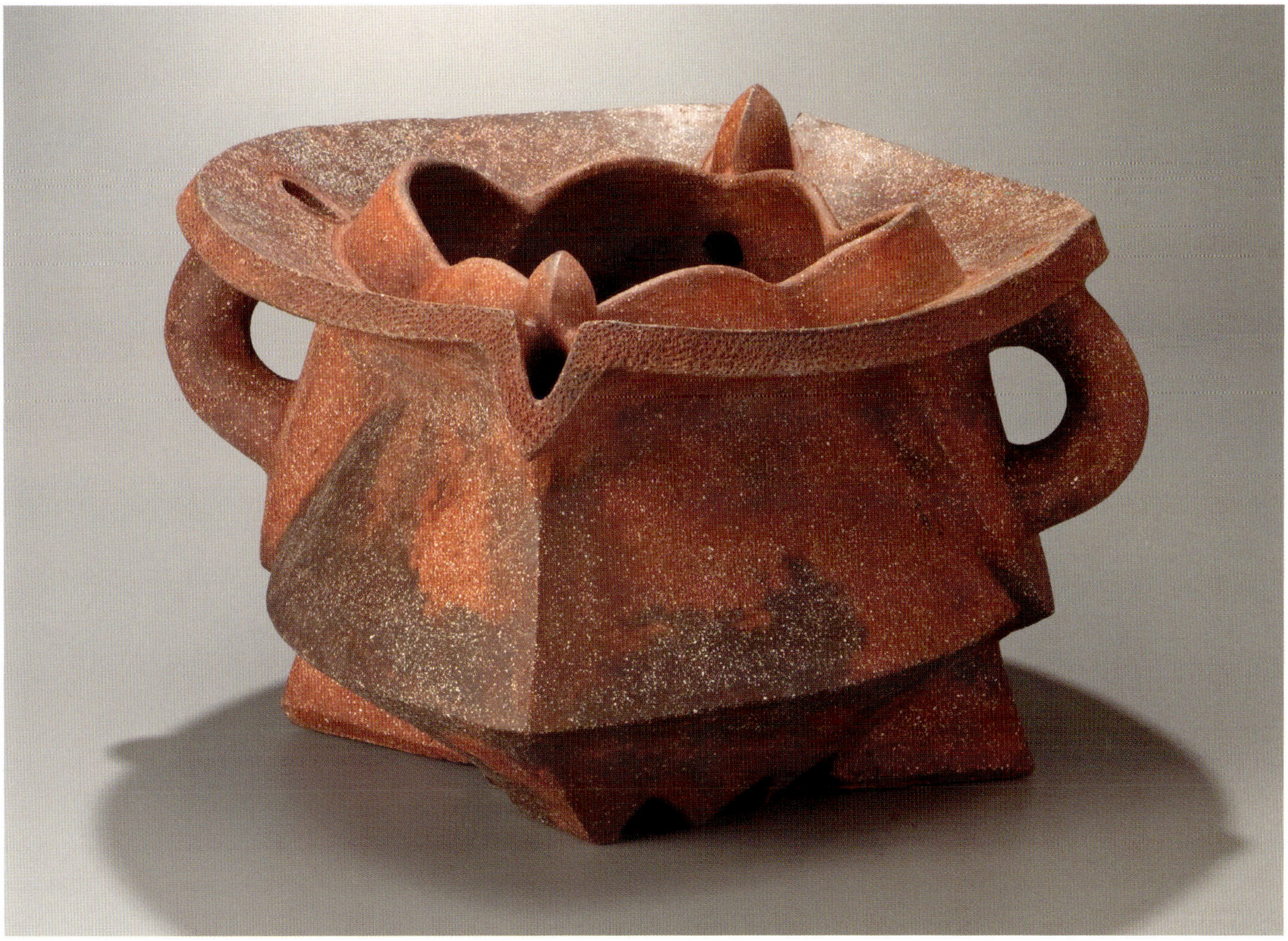

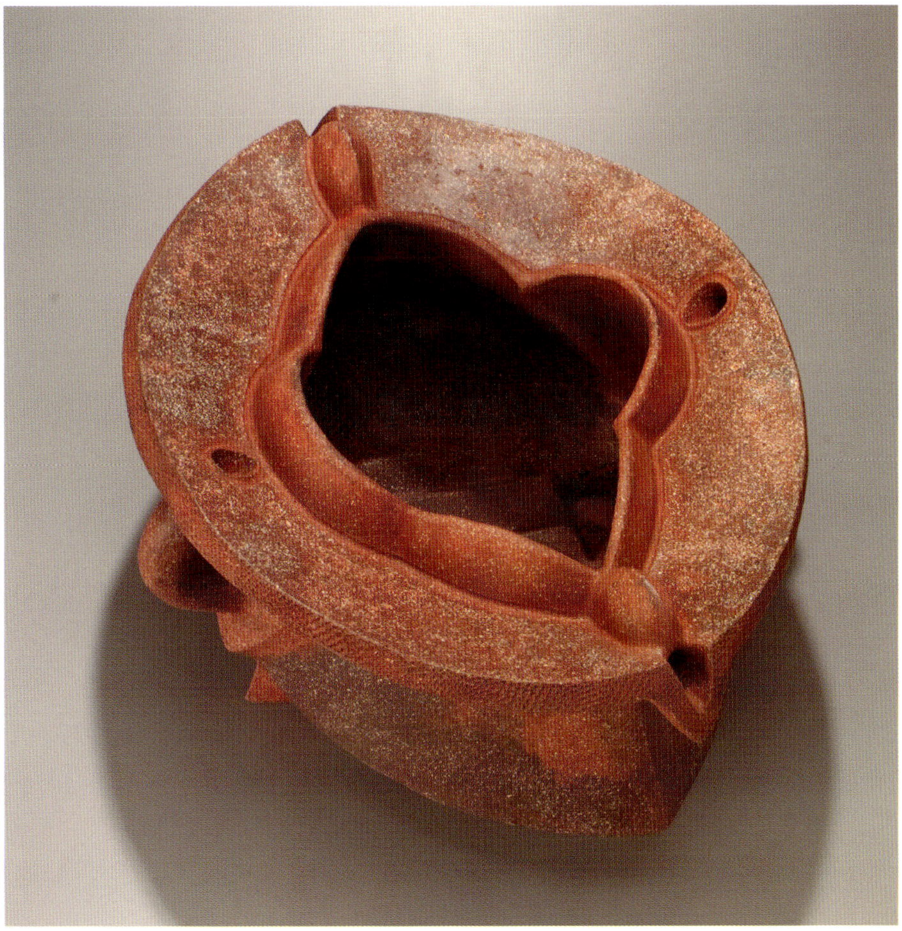

Fig. 235. *Terrain Cistern,* 2011
Fig. 236. View from above
Unglazed ground stoneware
16 x 26 x 28 inches
Collection of William and Catherine Daley

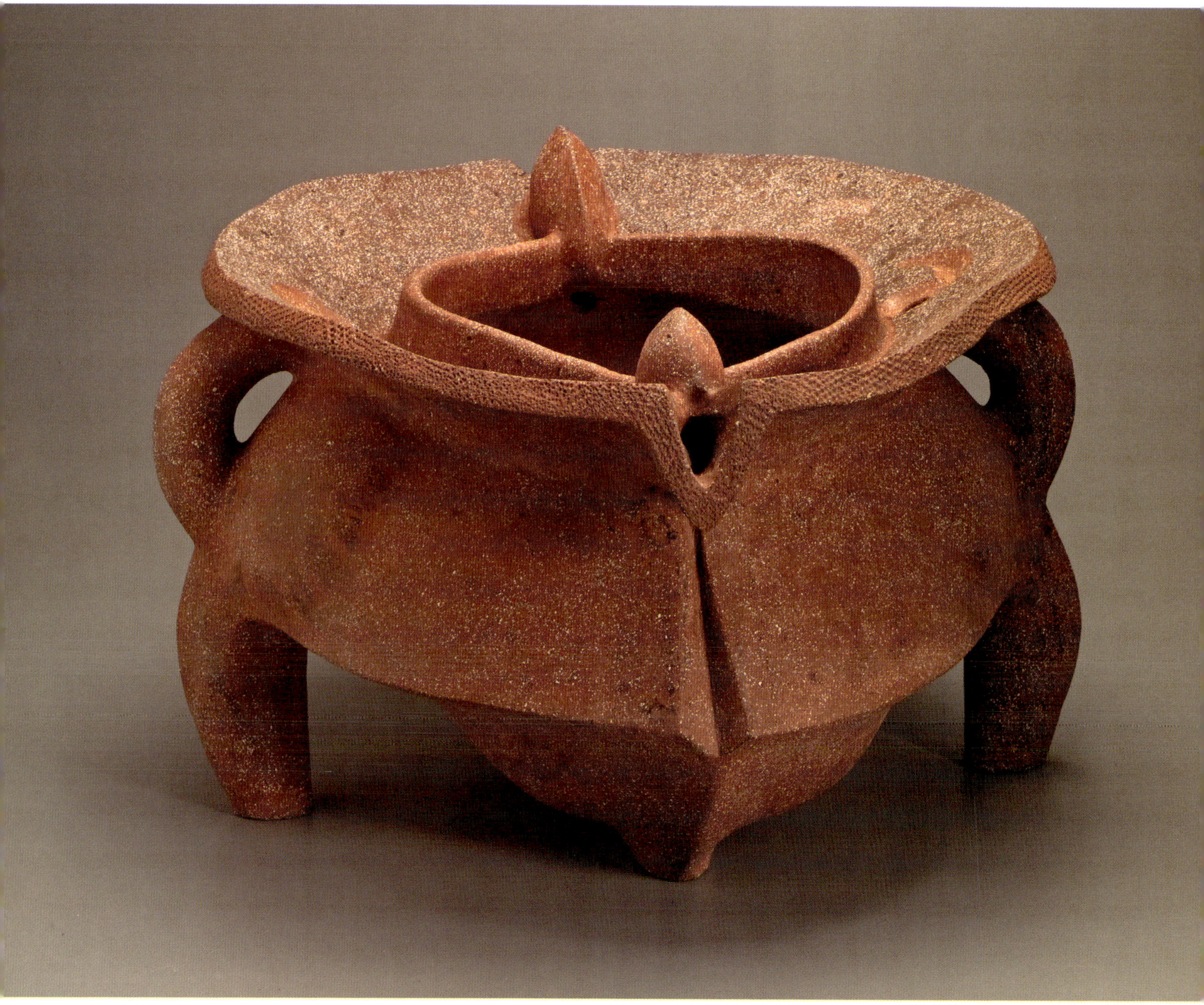

Fig. 237, above. *Animate Cistern,* 2011
Fig. 238, opposite. View from above
Unglazed ground stoneware
15.25 x 25 x 25 inches
Collection of William and Catherine Daley

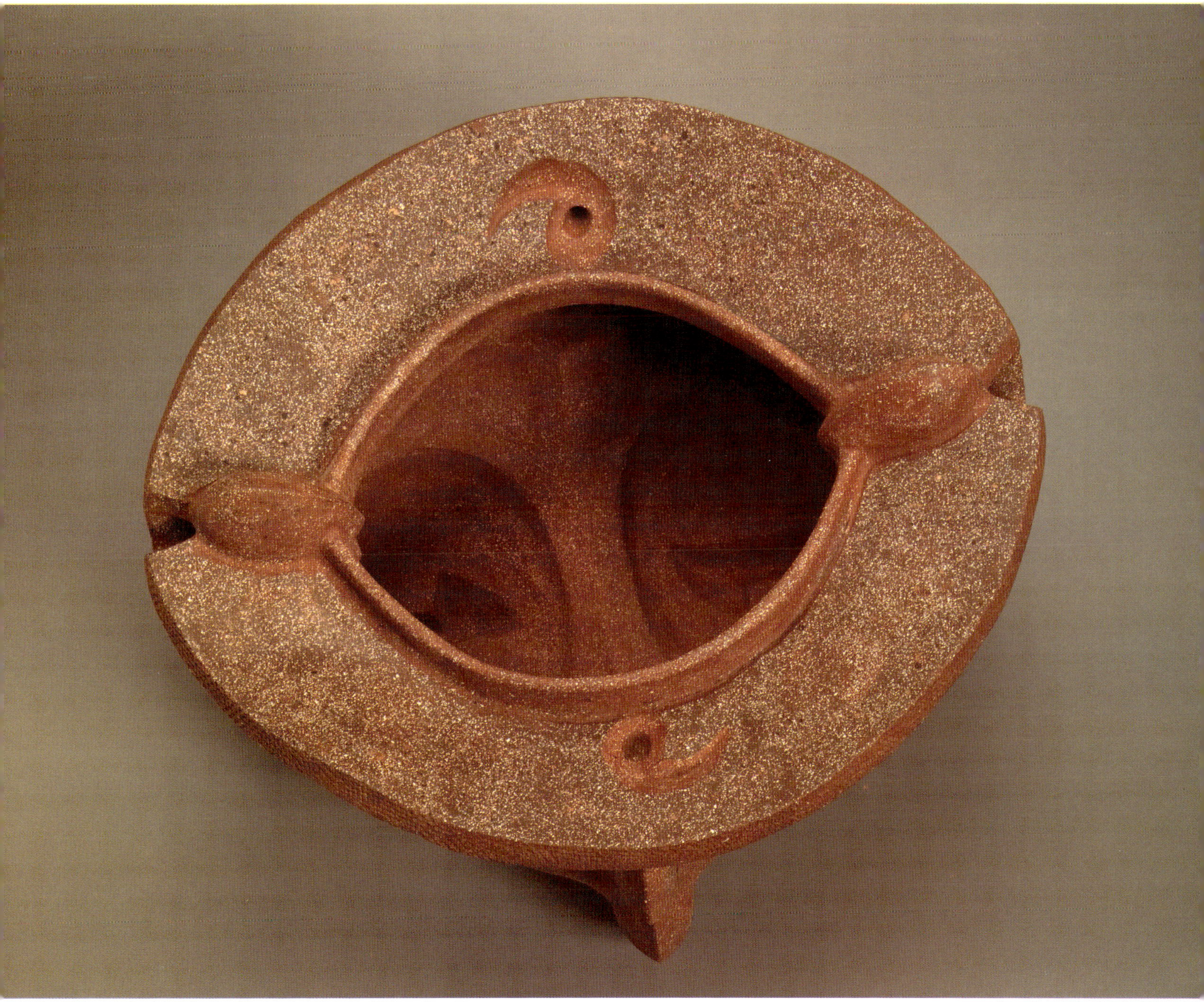

Photos: John Carlano

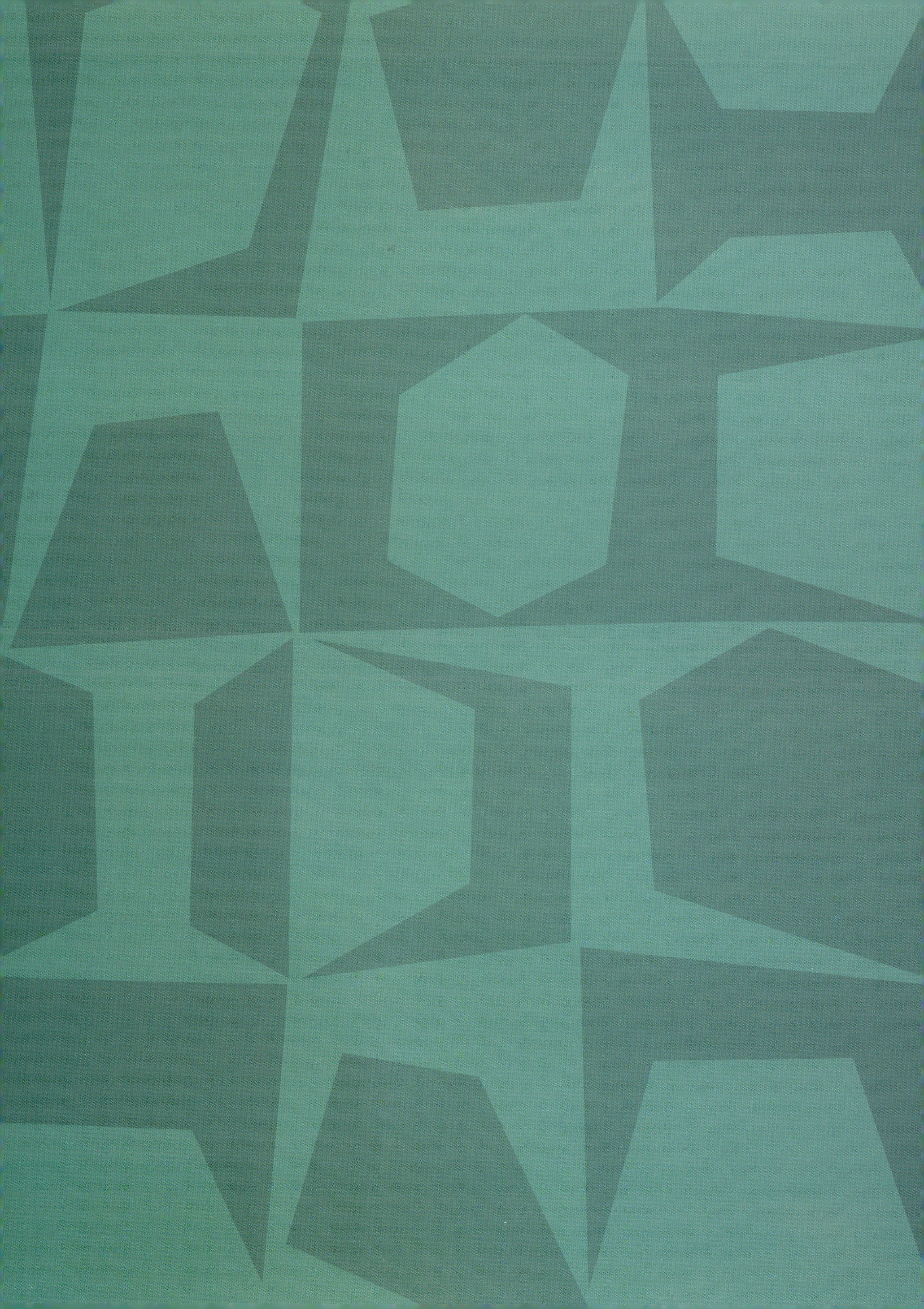

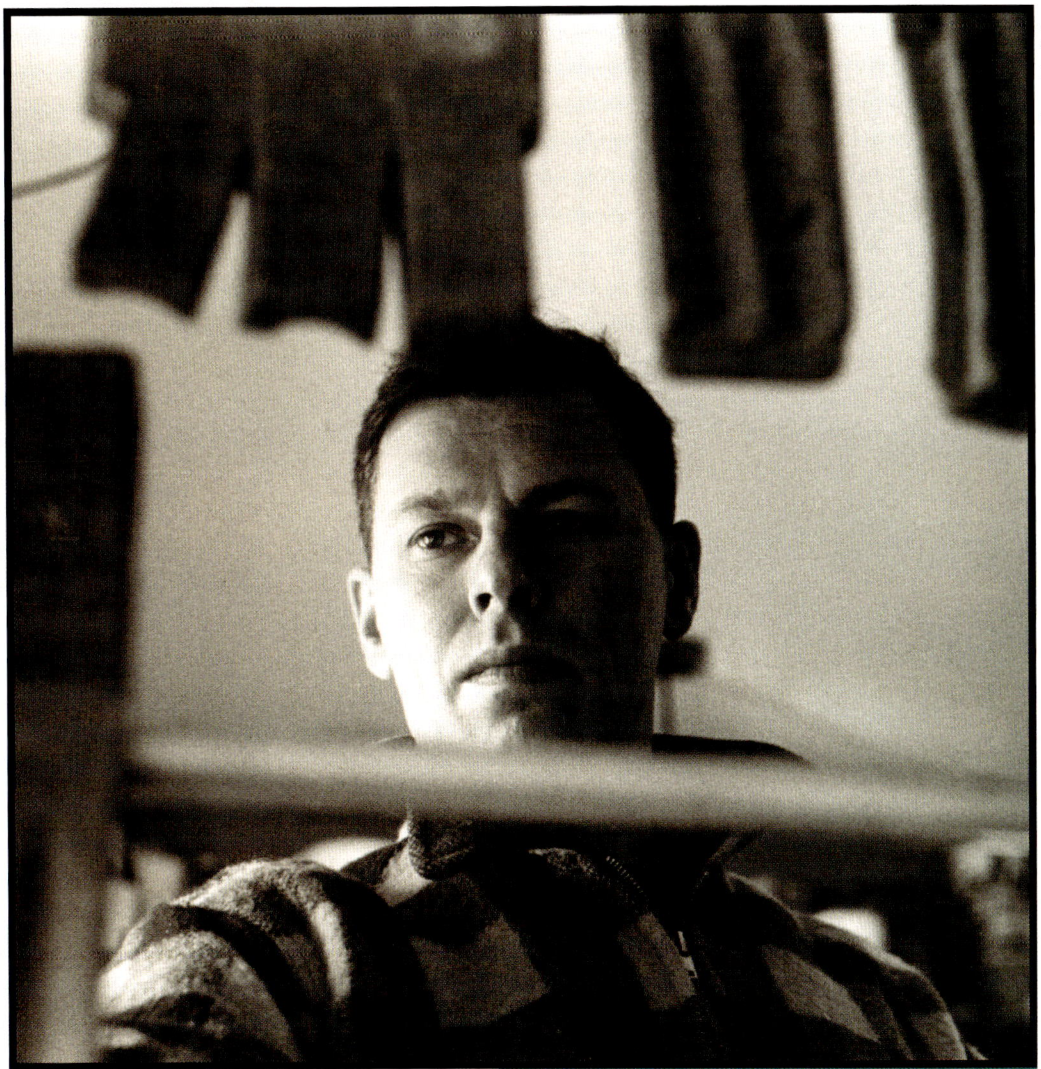

Photo: John Carlano

COMMISSIONS

Fig. 239. *Abacus*, 1962
Bill Daley working on ceramic, wood, and bronze screen commission for IBM Pavilion, Seattle World's Fair

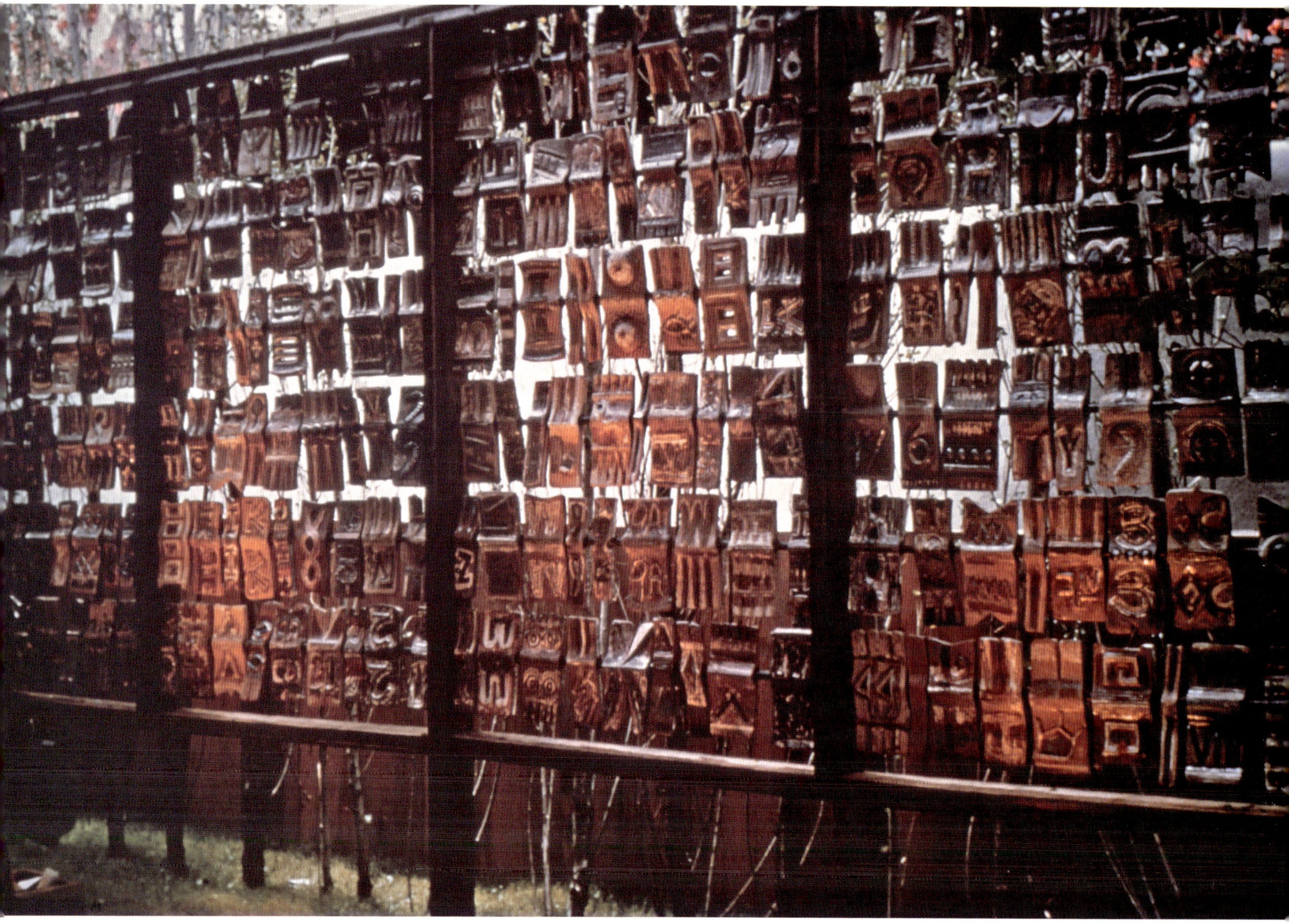

1962

IBM Screen – Seattle World's Fair

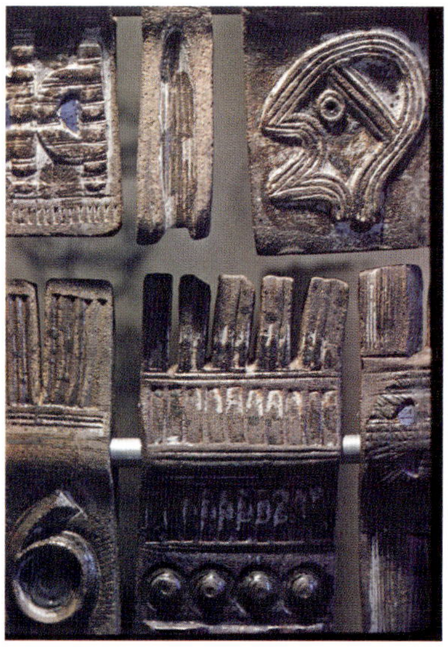

Fig. 240. *Abacus*, 1962
Ceramic, wood, and bronze screen commission for IBM Pavilion, Seattle World's Fair; Carreiro Design Associates, architects

Fig. 241, left. Detail of *Abacus* tiles

Fig. 242, opposite. Drawing for *Abacus* tiles

The IBM pavilion for the "Century 21" Seattle World's Fair in 1967 was designed by Charles Broudy of Carreiro Design Associates. It was conceived as a rectangular open space for children and their parents to explore the wonders of numeric computation.

Covered by a Mansard roof and IBM Logo and supported by rugged, redwood post and linteled timbers set above a gray slate floor, it was open on all sides. The fourth short side was bordered by a close row of slim tall aspen trees, planted just outside the roof's generous overhang. The shaded, high, inside space, with its timbered ceiling, was spot lit by a series of white and brightly colored post and panel Game Mazes. They were friendly labyrinths for the exploration of counting in tens to compliment the computer's binary numerics. The aspen-lined side, beyond the slate floor, was given over to moss and was the location of a ceramic screen. Homage to the abacus, the screen was comprised of two hundred and twelve tile entablatures set in a redwood frame six rows high. The tiles were mounted to slide on bronze rods. Together they celebrated the genius of our world's ancient and modern numeric computation signs and symbols.

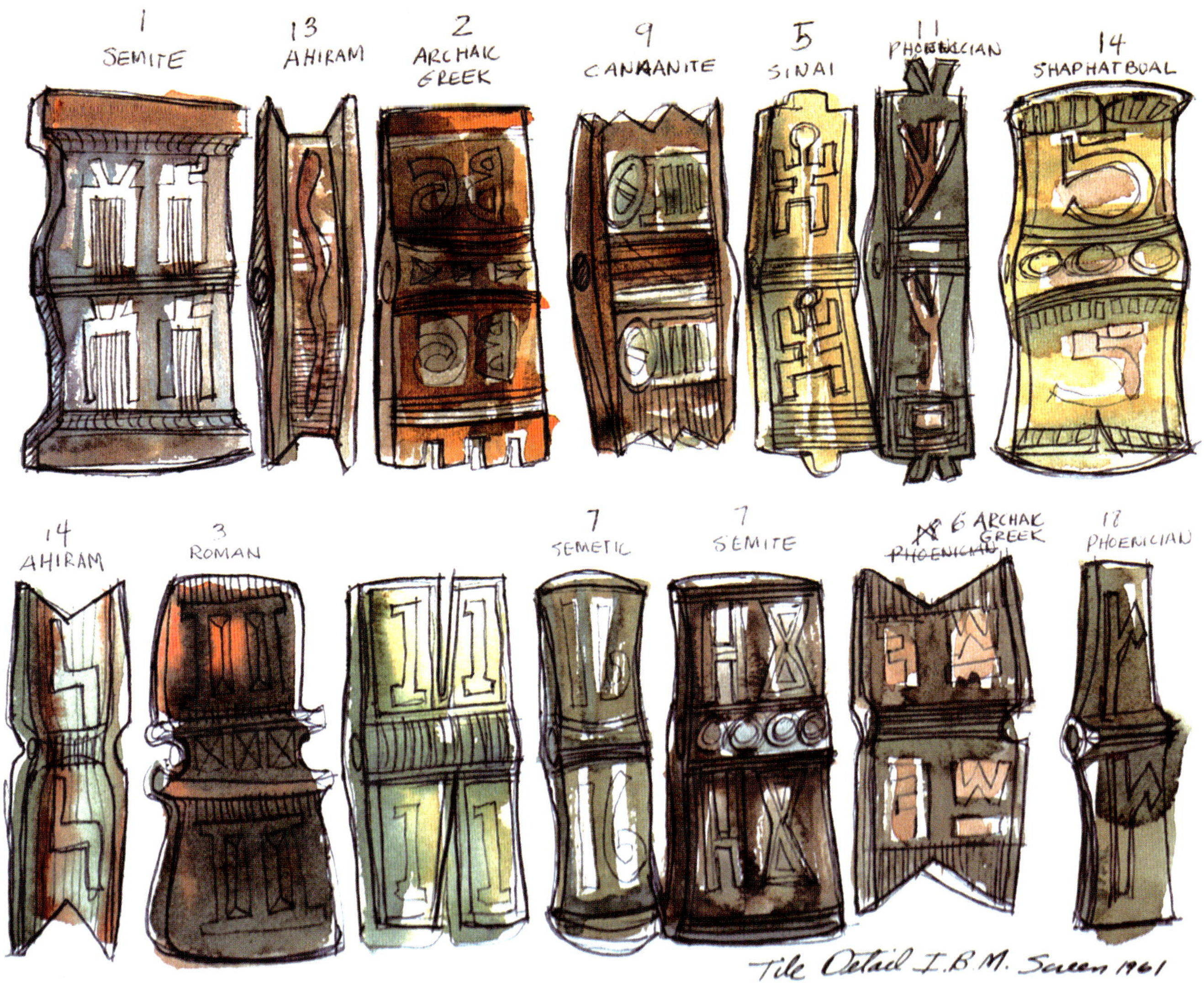

When I was awarded the commission, I began by drawing my hands, noting the archetypal symbols for tracking gain or loss. I withdrew to draw in museums and libraries, trying to find a place somewhere between knotted strings and Roman Numerals. There, I filled pages with telling glyphs and icons from Egypt to Babylon and Islam, from Judea to Greece and with Alexander to India. Catherine's Viking ancestors' runes led me to Hibernia, Scotia, Brittany, and Russia. The Conquistadors took me to Mayan and Aztec Mezzo-America. And so it went. As I scribed for signs to offer "my take" on the most telling numeric symbols as primal place holders, I felt enabled to venture and risk quantifying wonder. I strived to employ mindful meanings in clay as an old pursuit. Cuneiform and the silicone chip are two ends of a media processing continuum made from clay to computerized computation.

To form a tile abacus icon as clay, George Greenamyer, my talented student sculptor assistant, and I evolved a way to laminate two rectangular sheets of soft clay with a length of broom stick between them. When the icon modeled tile stiffened enough by drying, the sandwiched broom stick was removed creating a tunnel. It allowed the tile, after firing, to slide on a rod simulating the abacal computation mechanics. The IBM screen's icon tiles are near twelve inches high and vary in width depending the icon's needs. They are made of iron-bearing, stoneware clay fired in varying reduction atmospheres; modeled, carved, textured, and partially brushed with oxides. Clay slips and clear glaze amplify their relief and incised images.

We rearranged and relisted them often in studio trials before they traveled with me to be placed with their Key list. A world icon montage for the fairgoer's pleasure, set before the aspens in Charles Broudy's beautiful IBM playground/pavilion. It pleased me "to the power of ten" when I later learned that after the fair was over, IBM gifted its "Abacus Homage" to a new City of Seattle Public Library.

— William Daley

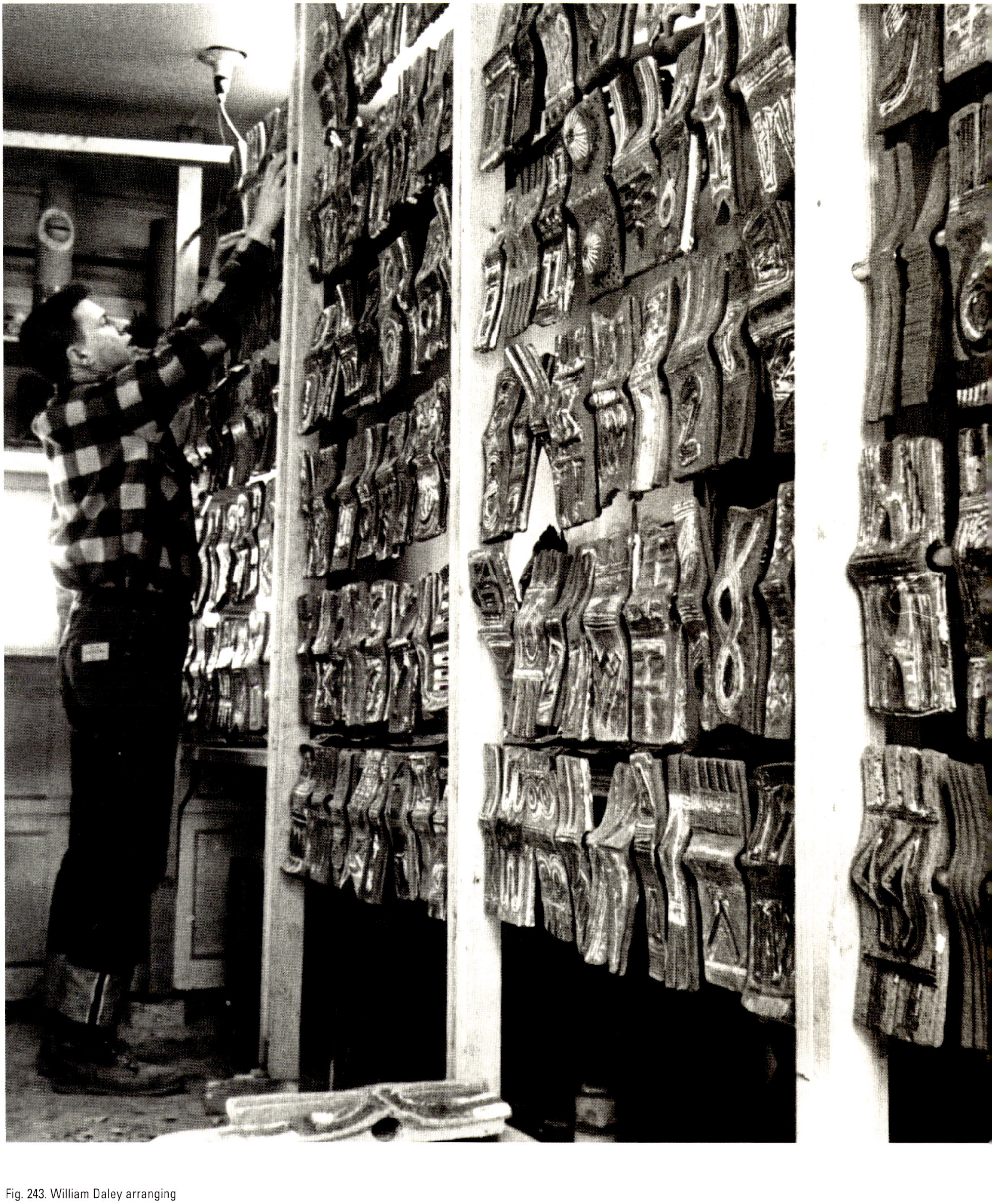

Fig. 243. William Daley arranging *Abacus* clay tiles in trial installation of the IBM Wall in his Feasterville, PA studio, 1962

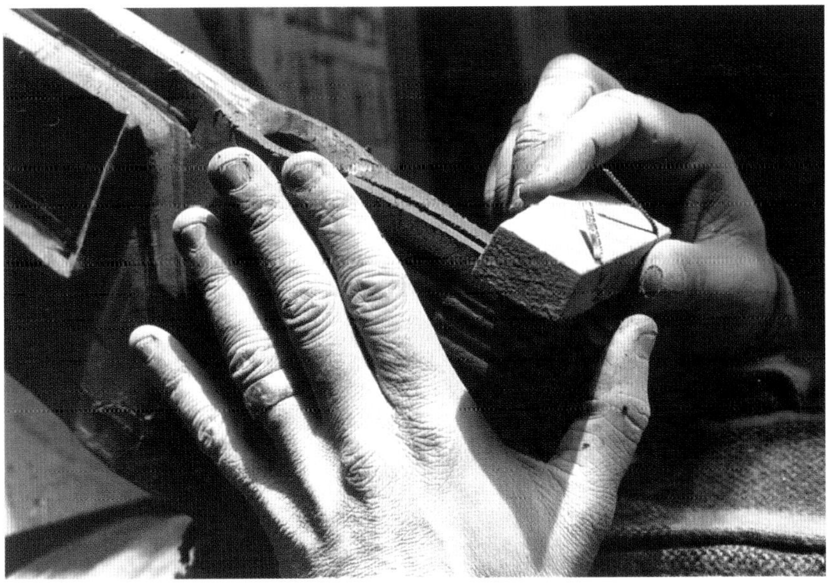

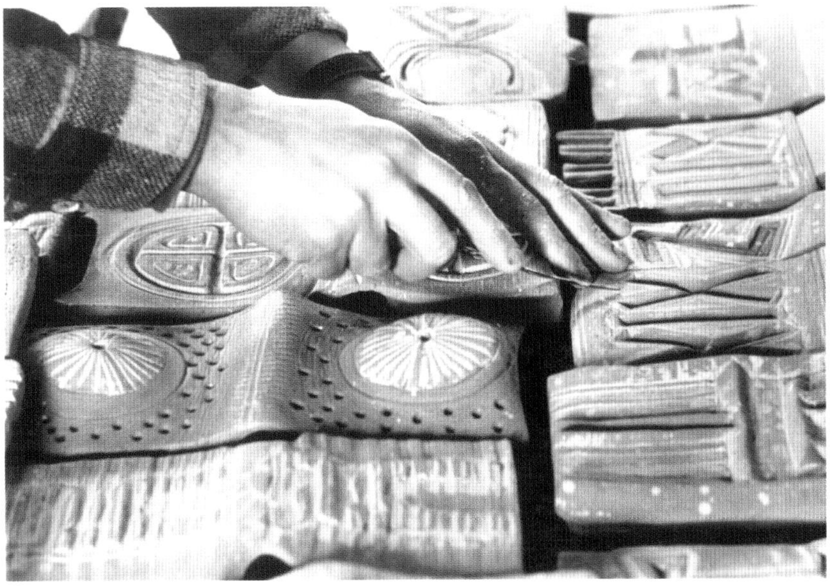

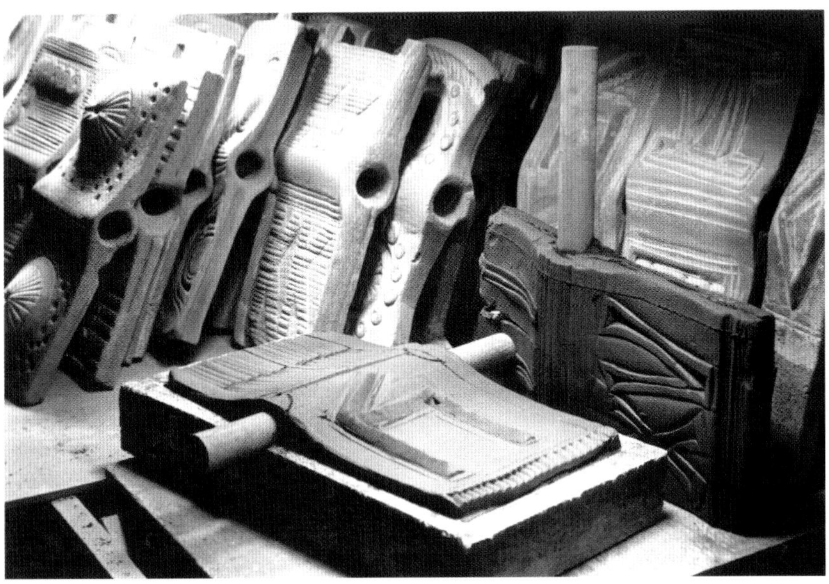

Fig. 244, top. Bill articulating the edge of the tile with custom designed tool

Fig. 245, middle. Bill making a final touch on the tiles before firing

Fig. 246, bottom. Tile sitting in a plaster mold

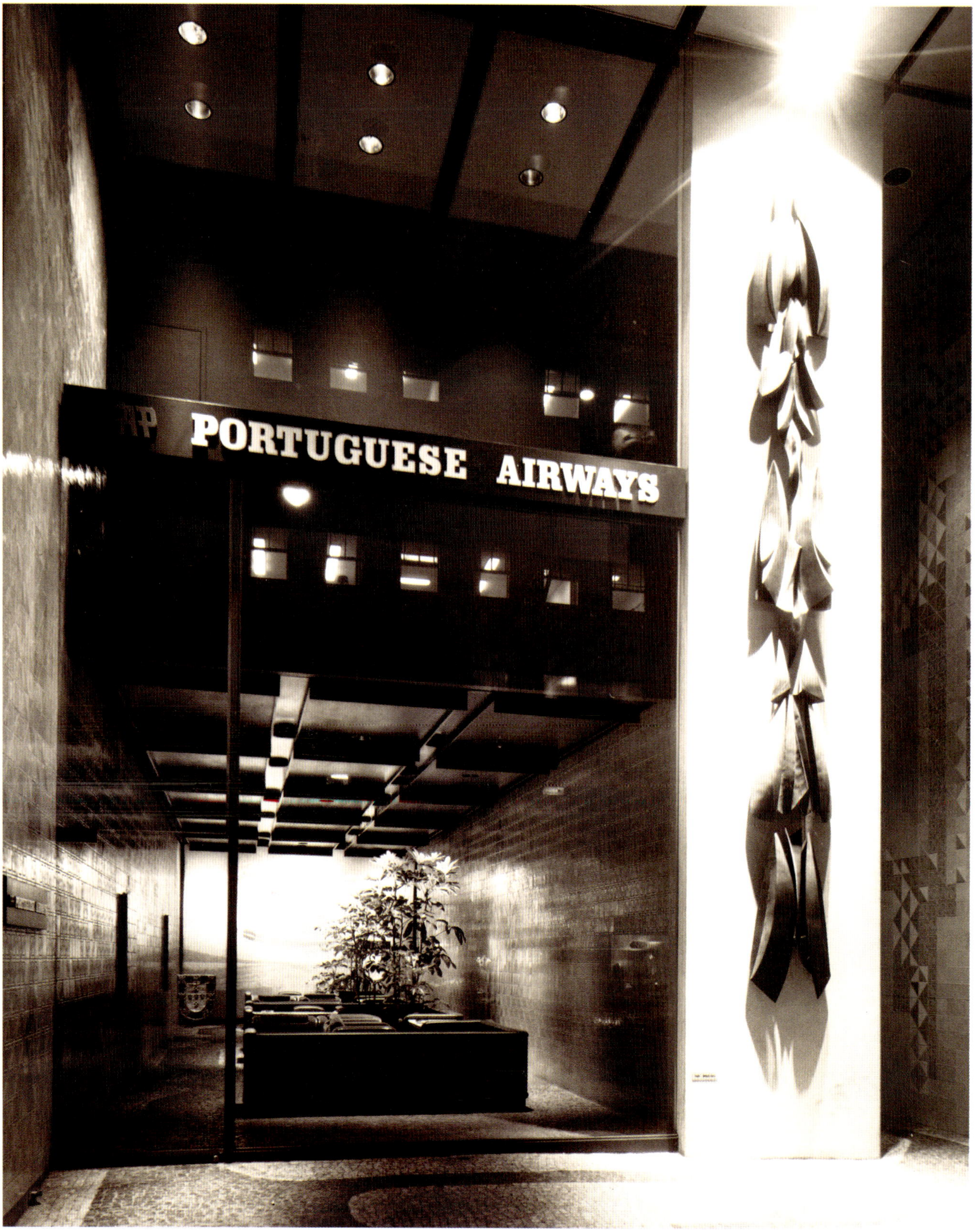

Photo: William Daley

1967

Portuguese National Airlines

Fig. 247. *Flight,* 1967, sheet bronze sculpture
Commission for Portuguese National Airlines,
5th Avenue, New York
Wilke Davis Mitchell, architects

Photo: Jack Simon

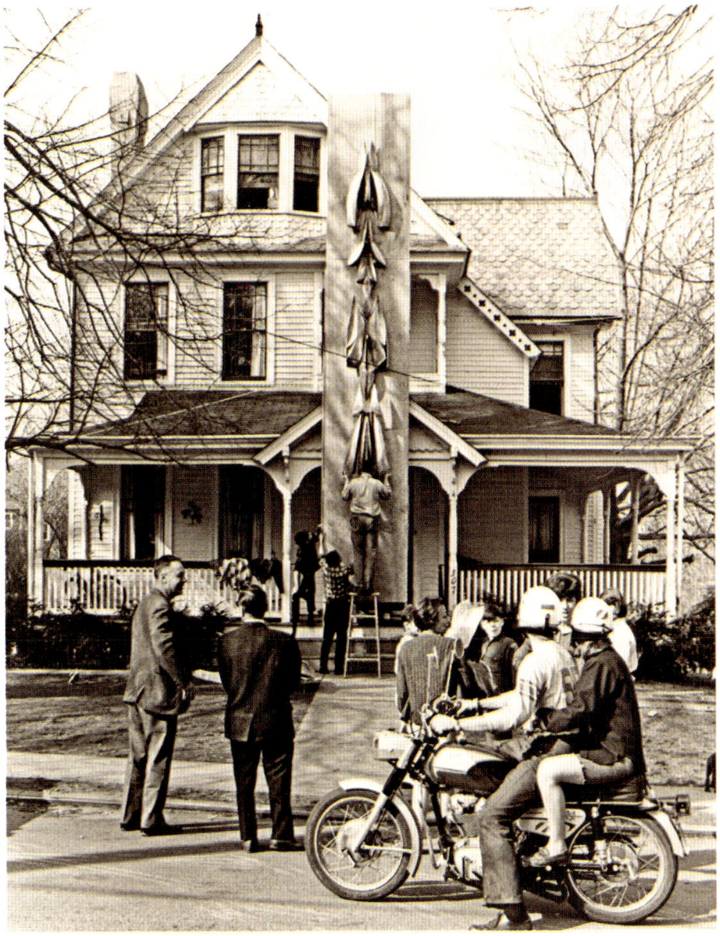

Photo: William Daley

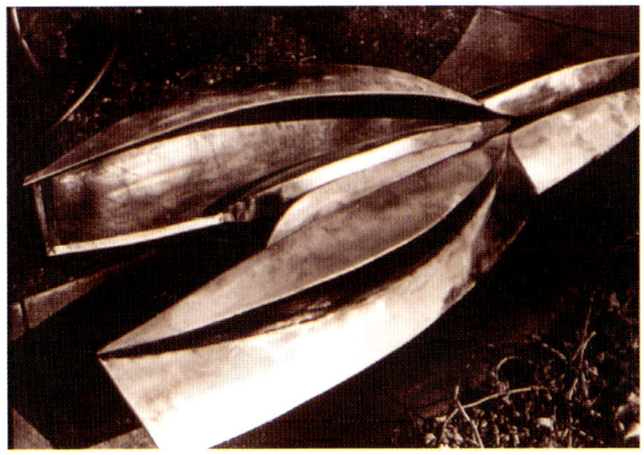

Fig. 248, above. Trial installation on the front of the Daley house in Elkins Park

Fig. 249, above right. Detail of section of the sculpture

In 1967, the design firm of Wilke Davis Mitchell invited me to collaborate with them designing the façade of Portuguese Airways' Manhattan, New York, sales office on Fifth Avenue. My assignment was to evolve a sculpture entitled *"Flight"* for the facade's vertical marble panel. The panel, thirty feet high, defined the right side of the glass and marble facade.

Flight's avian ideation began with scaled sketches and small maquettes of cut and assembled thin sheets of soft foundry wax. By forming planar crescent, wing, and blade shapes in left/right pairs, I could coax the soft wax forms to become shallow dishes. When I joined their long edges, they became shallow airfoils suggesting parts of sleek aerial flight forms. When I arranged them in bilateral pairs, they evoked winged bodies, seabirds, sails underway, racing hulls, or supersonic fuselages. Reviewing the findings we three picked five wax models from twelve for me to develop. Unlike Icarus, my goal was to have the wax became polished bronze flight forms soaring on marble air.

I began preparing tests by carving long shallow convex oval bowls in scrap oak lumber. I successfully learned to hammer annealed sheets into concave contours and created sleek convex forms. I sought their sweetest cross sectional curvature and draft, and I checked the forgings against a full-scale wall drawing. Reassured, I then glued up two five-quarter stair treads to carve a mighty oak concave drafting mold. Cutting bronze parts and pieces with my Beverly shears, then brazing them as bilateral pairs was a joy. Given my modest metal forging skills, to have telling flight forms emerge without re-annealing was on the cusp of the miraculous. The most challenging yet satisfying task was forming and brazing the triangular reinforcing edge moldings on the five elements.

Grinding and polishing was a hypnotic and tedious interlude before the grand Saturday morning trial ascension of *Flight's* pentagonal mission. Set to take place on a thirty-foot high plywood mock up attached to the front of our gabled house, I took the elements of the sculpture from my cellar studio and arranged them in order on the front lawn. Petras Vaskys, my close colleague, and I raised his thirty-foot ladder to mount each element in turn from top to bottom. We used cut paper patterns to position, mark, and drill the elements' attachment rods. Gingerly holding on, I raised each element upward along the ladder's rails and fitted them in the rod holes, and Petras bolted them to the plywood back from another ladder.

By evening, all five avian wonders were in full-flight facing west and glowing in the late June Saturday's setting sun. Across the street sitting on a wall, gazing east, smiling, and smoking, Petras and I were ecstatically exhausted. Jack Simon, Petras's friend and our colleague, was still taking photographs. Neighbors came out, cars slowed down, and a couple stopped on a motor cycle. Our children and their friends kept their profound, all-day play watch. Early Sunday morning, Cate and I crossed the street to see the flying five in shadow with the sun rising behind our house, and we knew it all was good.

I am deeply indebted to the caring council of silversmith Richard Reinhardt, Joseph Richardson's acolyte and my colleague and metal mentor. The assuring gift giver, that meta-morphing primal materials from "mud" to metal is a wondrous continuum. All the rest is social commerce.

— William Daley

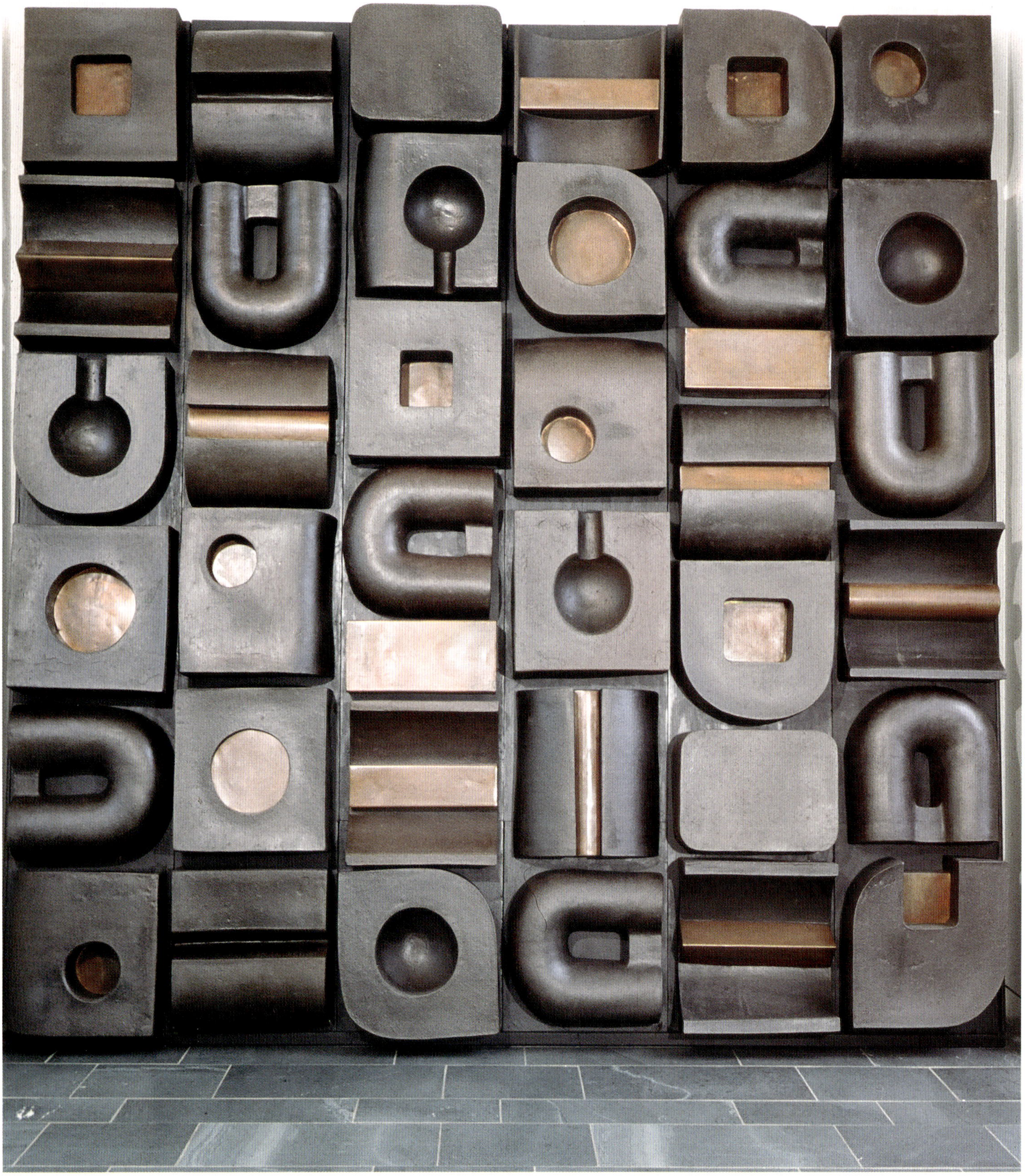

1970

South African Airways/Arrow Corporation, 1970

Fig. 250. Stoneware and copper interior wall was originally created for the South African Airways building Fifth Avenue, New York.
Wilke Davis Mitchell, architects

Wall was reconfigured for traveling exhibition and is currently installed in the lobby of Arrow Corporation, Reading, PA

Fig. 251, opposite. Bill standing in front of the wall during trial installation

Remembrance of making the wall

I began this letter four months ago and stopped working on it for some reason. Now it's the 6th of October. I am painting the bottom half of our great old house and coming around to completing the last side. I should just get finished before it's too cold. While I paint, I enjoy letting my head take journeys as my hands do my father's work. On the ladder today, I started thinking about the Haystack board meeting last week and how satisfying it was to be with everyone, seeing the library going up, and hearing the reports on our sound progress. Haystack is a magical place. It's grand to have it going into the next century with such vitality. Haystack thoughts led to thinking of you, your new house, the wall, *American Craft* doing an article on the Arrow collection, and my wall in its permanent home, when my unfinished letter to you about making it seeped back into my brain.

Saturday June 25th, wall installation day, was a very special day for me and I thank you for making it so. Thanks also to Carl and his super crew for making it go up so smoothly. Seeing the wall in its new space is a great joy. It fits so well, with such great long and short sight lines of coming down the inner hall or turning the corner of the outer hall and coming up the stairs, all offer wonderful ways to experience its presence – Thanks.

Coming home in the car, I started reminiscing with Cate about making it more than twenty-five years ago. Cate suggested that you might be interested in sharing some of the memories that are part of its history. When I modified the wall from 12x12 to 10x10 feet in October 1993 to make it suitable to travel for *"William Daley, Ceramic Works and Drawings,"* I had some misgivings about whether making it two feet smaller would change its psychic scale. My fears were overcome when I saw it installed in the Levy Gallery at Moore when the show opened. In November, I enjoyed seeing it at its new size and unchanged scale in each of the galleries where it was shown across the country. At the Levy Gallery in Philadelphia, it was in a white light-filled reflective space. At the Renwick in Washington, it was in a dark gray space dramatically lit from above very much like its original setting in the South African Air Lines Office on 5th Ave in Manhattan. In Little Rock at the Arkansas Art Center, it was shown in the Center's old mansion next to quartered oak doors and elaborate painted moldings. In the Museum of American Crafts in NYC, it was on the wall to be viewed coming up the grand staircase to the main gallery. On Saturday at 1:30, when the wall was completely installed at Arrow, I realized it was to have its future in a setting that surpassed all the others, including the one it was conceived for. Making it compress to 10x10 in six vertical sections for traveling was very auspicious considering that its permanent home was to be in a 10'8" wide space. For me, having it share the company of other master works by Nodell, Higby, Currier, Schira,

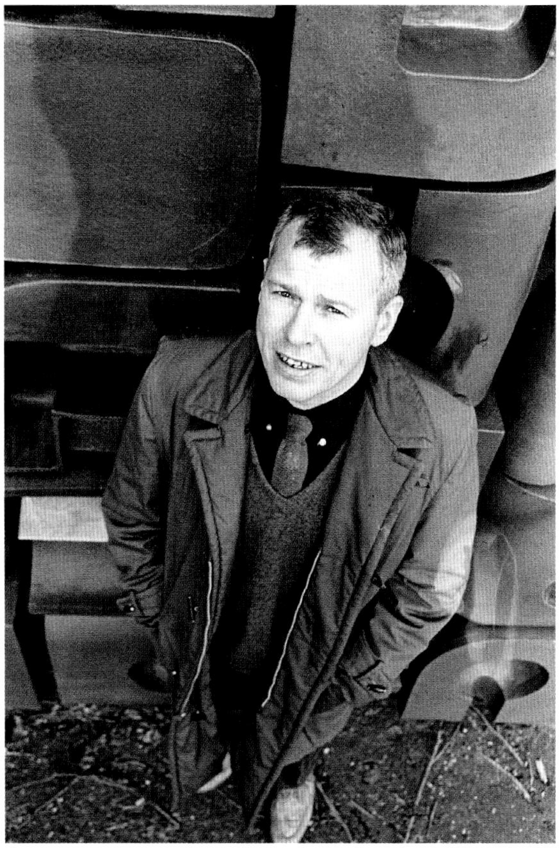

Photo: Dan Dailey

and Akers in an award winning building gives its site an added dimension, a satisfying plus to its future in such distinguished company.

The piece was commissioned by Wilke Davis Mitchell, a New York design firm, in 1969. Barry Wilke is a PCA (Philadelphia College of Art, currently University of the Arts) graduate and former student of mine. I began making sketches for him in October of 1969 and installed the wall in New York on May 9th 1970 with the help of Dan Dailey and John Meade, two more PCA graduates. We installed it on a weekend to avoid problems with the Construction Union. The wall was made as the focal point of the back wall of the sales office so that it could be seen walking down 5th Avenue. It was placed and sized to allow employees to walk around it on either side to go into the private offices behind the wall. The final design was approved after making a quarter-scale model of one line of modules in foundry wax. Right after getting an advance to begin, I vividly remember renting a truck, hiring students, and driving to the sheds at the Glenharry Brick works in Elm New Jersey. We went to break up and load dry clay. We then slaked it in mason's mortar pans, added grog to temper before pugging the clay in PCA's old vertical chamber Crosley.

I hired students to roll slabs using two, twenty-four inch cherry rolling pins that Dan Jackson turned for us. Students also worked slab building modules. The buff-colored stoneware jersey brick clay was great to work soft leather hard. It also had unbelievable green dry strength. Bruce Singer, John Meade, and I were the principal builders. There were six basic modules for which I made patterns of each module's pieces. I added clay bumps or holes and

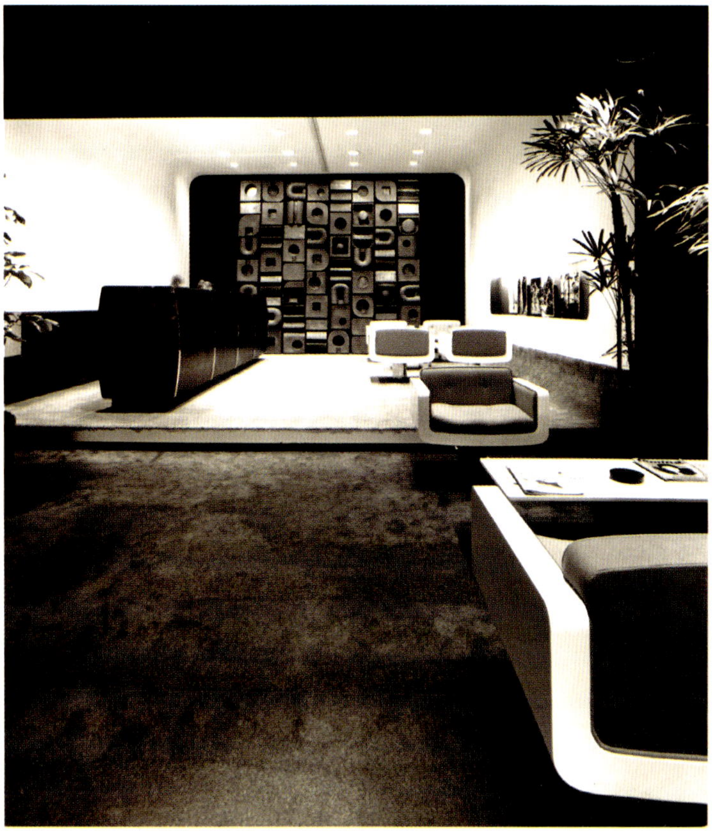

Fig. 252. Stoneware and copper interior wall in the South African Airways building, Fifth Avenue, New York. Wilke Davis Mitchell, architects

spaces for future copper inserts as I went along. We made many more modules than we needed to allow for mishaps in handling and firing. We built them on plywood boards which slid into a rack of thirty storage spaces that I made to hold the work in progress along the back wall of the studio. We hung plastic drop cloths over the rack's front to allow the pieces to dry slowly. Students were anxious to work for two dollars an hour and lunch. They worked on my day out from school when they could, but Saturday and Sunday were the mega production days. During the Christmas holiday from school, we worked a month, from eight to four every day. I usually worked a couple hours after dinner reviewing the production, planning on what to make next, touching up, and putting away the day's making. We each build a module a day, if someone else rolled out the slabs.

The most anxious part of the project was boxing the green ware in Elkins Park and driving it forty miles to Elm to fire on top of the bricks in the Glengarry tunnel kiln. I especially remember worrying about breaking modules on the bumpy, twisting work road from the highway to the kiln. We carried a small compressor and sprayed the air-dry modules beside the kiln with manganese dioxide mixed in a solution of Karo syrup and water to get shades of black in the oxidation fire. We did our work after the day shift left; placing the sprayed modules on top of the bricks was exacting. Spraying out hand marks after placing them discolored some of the green unfired bricks. The firing fee was established by the number of manganese sprayed, unsellable bricks that were created. It was exciting loading the modules on the linked carts which were to be hooked up in their turn and pulled on rails for four days through the tunnel kiln's heating and cooling cycle. We did four firings and they were all magical. Working as a team alone in the dim light of the covered open kiln shed, seeing the glow from the long rows of burner ports and feeling the exciting presence of combustions, the sound, and temperature of such a huge kiln that ran continuously day and night, weeks and months thrilled us.

As romantic outsiders (artists) we were witnessing what were for us the risks, mysteries and excitement of consigning mud to fire on a magnificent scale. Seeing the modules after the first firing was a great thrill. None of my fears about the forms slumping or warping happened; there were very few shrink cracks and the straight manganese as a once-fired glaze on green ware was perfect; it didn't run, it wasn't shiny, didn't crawl and the range of tones from deep brown to ebony to black celebrating myriad millions of South Africans. The workers unloading them with the cooled bricks were as pleased as we were. So was Jim Letty, a ceramic engineer and master of the works at Elm, who liked artists and often piggy backed their work through his fabulously controlled continuous fire.

When the four firings were back in the studio, the workers began to assemble them into a wall. I made many more modules than I needed thinking that the kiln would claim some. I also believed that I would have many unusable ones because of the usual glaze faults, especially from using my Daley one-ingredient glaze and applying it at the kiln site. After carefully inspecting them back in the studio, I was pleased to find that there were very few unusable modules. The next job was to make the copper additions and inserts for the modules that were designed to have them. I bought a second-hand Pecto brake for bending metal. The copper pieces were cut with a Beverly Shears, brazed together, and mechanically fastened with threaded rods and nuts to the clay where possible. Copper parts were glued in place with Epoxy PC-7 when mechanical means could not be used. Dan Dailey and I then constructed, against the side wall of our house, a twelve foot square, 3/4-inch marine plywood wall, on 2x4s with 18-inch centers and painted it matte black. The wall leaned against the house at the same incline as the wall to be installed in New York. We then carried all the usable modules out in the yard and arranged them on the ground near the wall in eight vertical rows, eight modules in a row. The rows were far enough apart so that we could walk up and down between them arranging and rearranging the sequences so that they read well up, down and side to side. I put a ten-foot step ladder below the center isle so we could climb up and look down to see what the wall would look like when it was vertical and mounted on the plywood.

We spent four days playing musical modules, taking one out for another, switching and swapping vertically, horizontally, and diagonally. We worked to achieve a balanced arrangement by worrying about the places of the six basic module shapes. We also worked to get the spotting of the copper-bearing units to work over the entire twelve-foot square. We spent hours reading the grid line by line, back and forth across, and up and down checking out each module's relationship to its neighbors to find the most engaging sequences. We also wanted the eye to skip around and find irregular passages to counter the regularity of the sixty-four place grid.

Playing head games on top of a ten-foot ladder, talking out the moves with Dan Dailey, a marvelously talented young artist, for a week in early spring was, for me, the highlight of the whole project. When I started making small scale wax modules to initially present the idea, I knew that the wall was to be about "Theme and Variation in Mud and Metal." On the ladder, I learned how rich the possibilities are for the variations of six times sixty-four (or doing the same thing over without doing the same thing twice.) As we saw it all come together, it was the first time we could see how wonderful what we worked on all winter really was.

Getting the modules mounted on the wall was time consuming, fussy but easy. We went line by line from top to bottom in each line starting on the left. There was some switching, but the time was taken up making and fitting the pine brackets to hang each module by bolting the bracket to the plywood. Each module has a clay lug for a bolt inside its bottom to prevent lifting the modules off the brackets. It took the best part of a week to mount them. Dan worked the backside putting on washers and nuts as I measured and drilled the bracket bolt holes. Holding the modules while he fastened them from behind was a numbing task, but one I preferred to screwing on quarter-inch burrs while standing on a ladder in the cramped space between the wall and the house. When we got them all up, I invited friends and neighbors to come and see it all together. I gave many of them one of the left over modules. It was also fun to take pictures of our children and their friends standing in front of it. Taking it down and packing it for New York was anticlimactic. Making brown paper strip patterns of all the bolt holes for each row's final installation was tedious. So was indexing all the cartons, modules, and brackets by row. Three of us loaded the U-Haul truck on Friday and drove it to New York on Saturday. We returned home safely Sunday evening, very tired and pleased that everything went as planned. The wall was installed. The plywood wall there was correctly built, the paper patterns were perfect, and our packing and truck driving skills were adequate. We even found a safe place to park a truck in Manhattan on Saturday night.

It seems hard for me to believe that these recollections are about events that occurred more than twenty-five years ago. I still savor the exhilaration of proceeding with assurance while working in doubt through each stage of the project, from design to final installation. It is this state of risk taking toward hoped and planned for but unknown resolution that makes working at being an artist so addictive. When a major work like the wall is completed and all the doubts are resolved, and the object is greater than the sum of its parts and their makers, then the satisfactions of retrospect in giving form to spirit are unsayable. Thanks for closing the connections between the stuff of creation and artists by being such a grand keeper of their dreams. Having the wall at Arrow completes the circle of creation for me. "Art is long and life is short."

I feel privileged to be part of both with you. Thanks.

William Daley
Oct. 23, 1996

Marlin Miller – graduate of Alfred University, Ceramic Engineering
C.E.O. Arrow Corporation, Reading PA

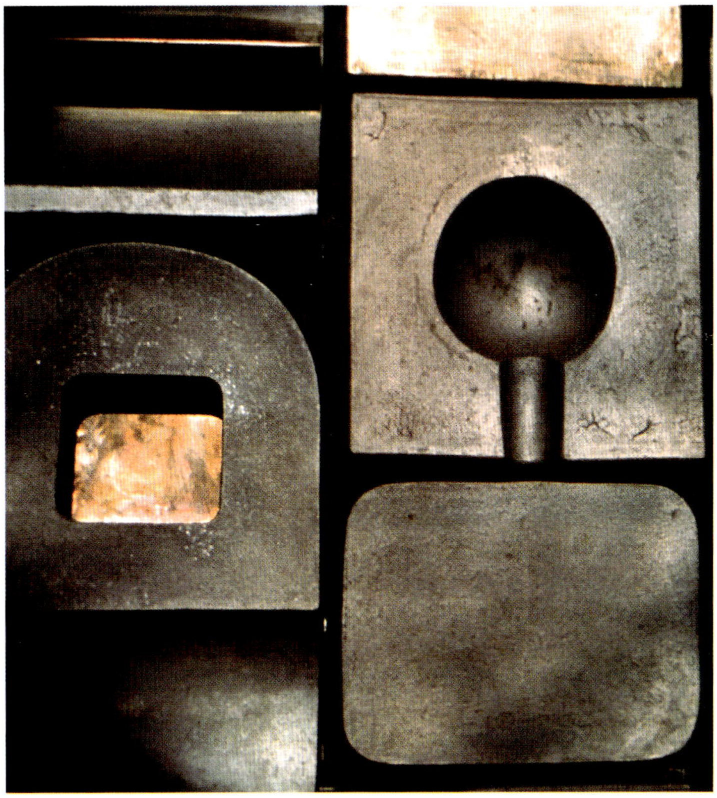

Fig. 253. Detail of South African Airlines/Arrow Corporation wall, 1970

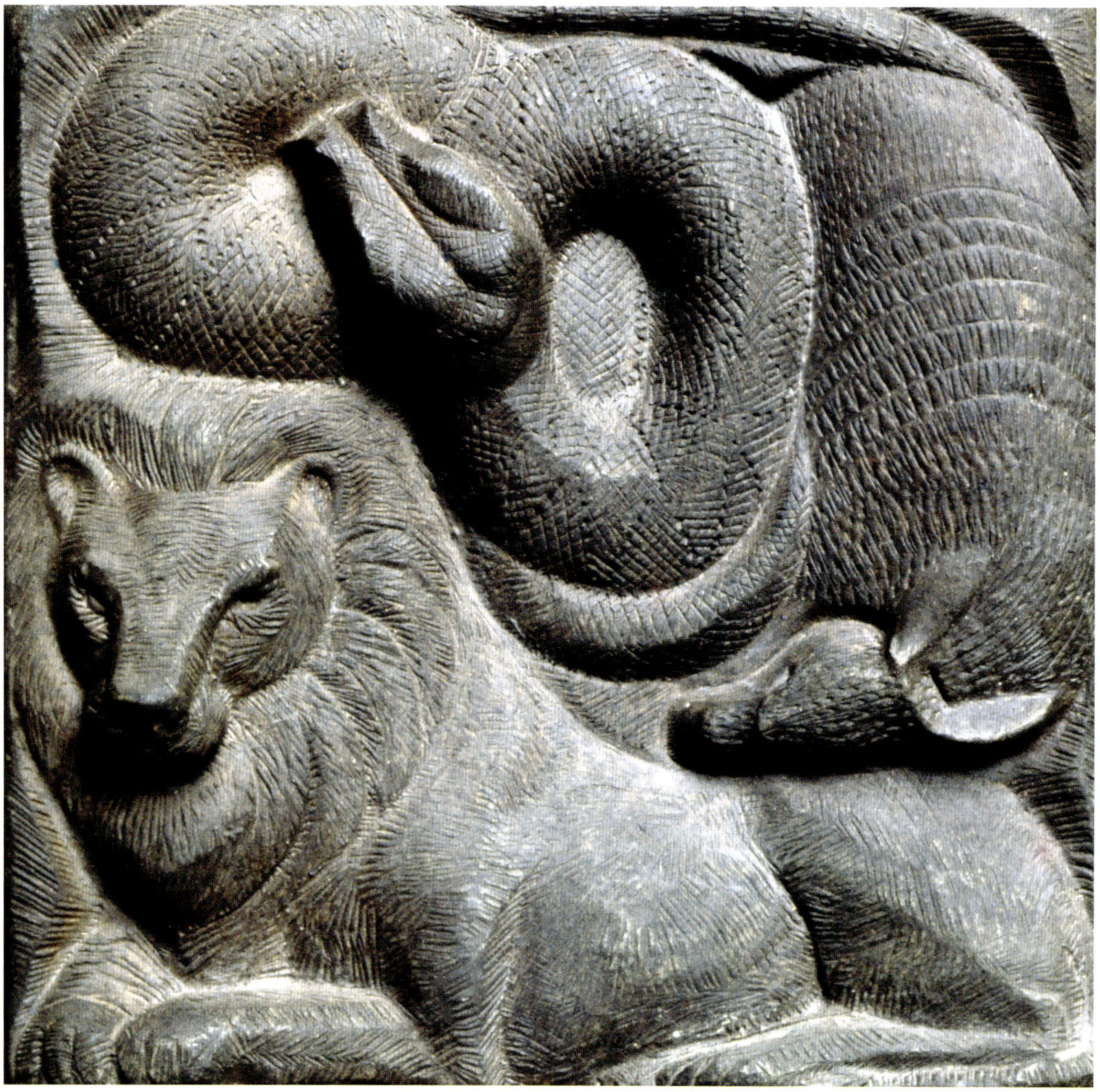

Photo: William Daley

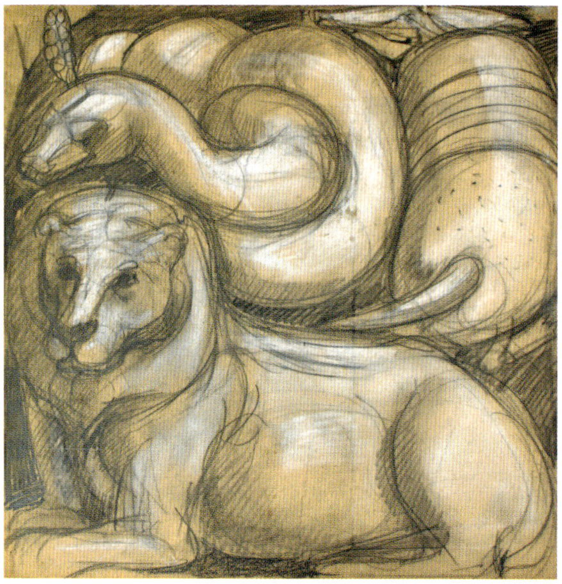

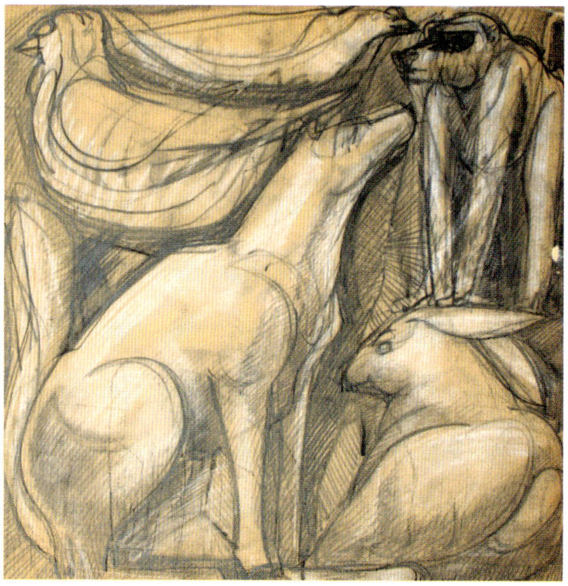

1971

Haddington Recreation Center

Fig. 254. One of nine *Animal Tiles,* 1971
Unglazed stoneware, 14 x 14 inches
Ceramic drinking fountain surround
Commission for Haddington Recreation Center in West Philadelphia
Sydney Jelenk, Architect

Fig. 255. Animal tile sketch with lion, snake, and armadillo.
Pencil and chalk on kraft paper

Fig. 256. Animal tile sketch with bird, dog, baboon, and rabbit.
Pencil and chalk on kraft paper

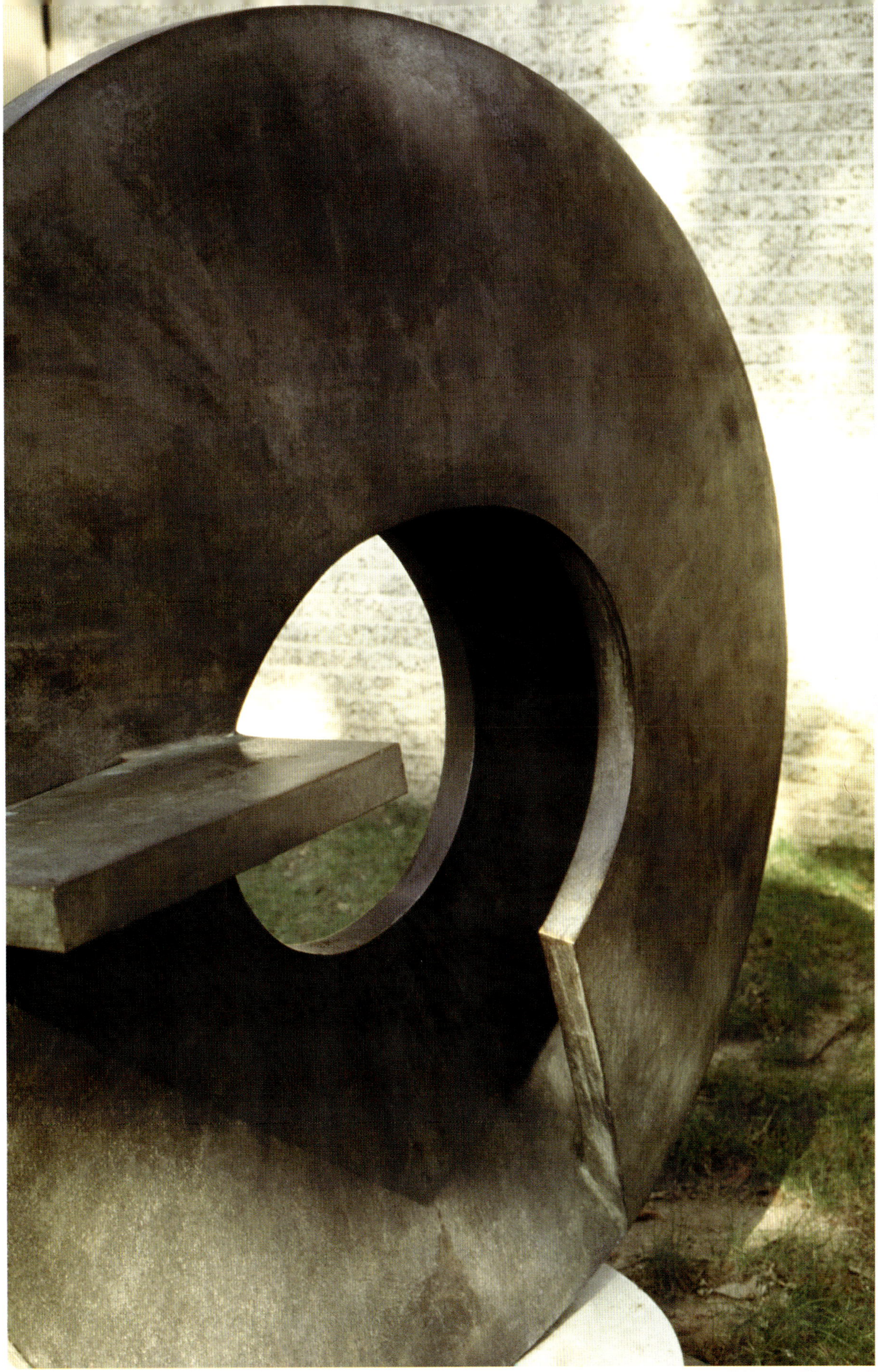

Photo: William Daley

1971

Germantown Friends School

Fig. 257. *Helix,* 1971
Bronze sculpture
Commission for Germantown Friends School,
Philadelphia, PA
Chappelle and Crothers Associates, Architects
(currently Daley + Jalboot Architects)

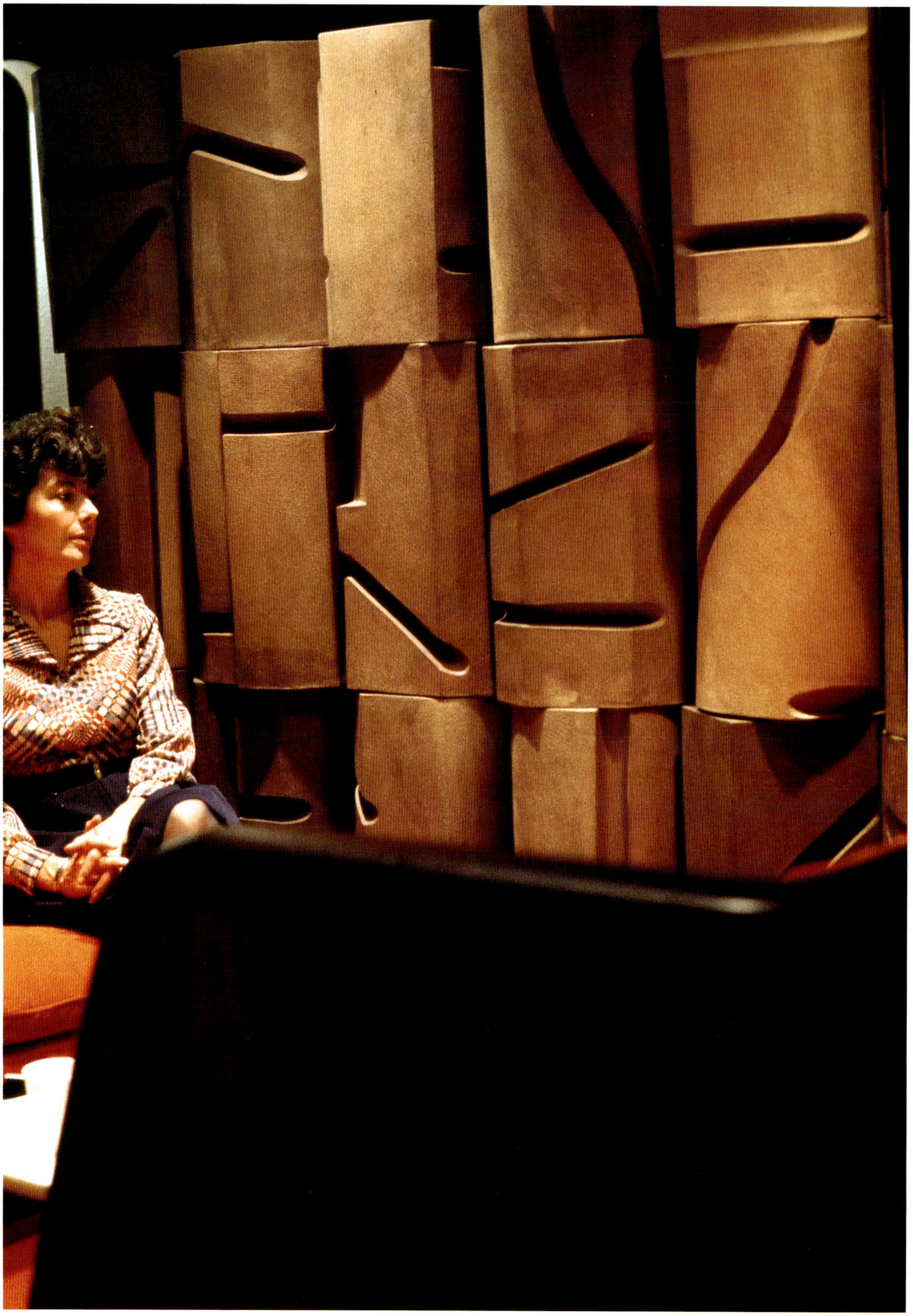

Photo: William Daley

Fig. 258. *Manhattan Canyon Wall*, 1972
Catherine Daley sits in front of the unglazed stoneware wall.
Commission for Fairfield Maxwell Ltd, New York City;
Wilke Davis Mitchell, architects. Collection of the
Houston Museum of Fine Art

Manhattan Canyon Wall, 1972

I am pleased that *Manhattan Canyon Wall* is back among us and will be part of your Art Alliance 20th-Century Masterpieces Exhibition and Catalog. Making the wall in New Mexico with Cate helping all the way was a great joy.

When my friend and former student Barry Wilke called in mid-October 1971 inviting me to make another wall for him, we were already living in Albuquerque for two months. The project he described was to be for a ship leasing corporation's New York City, Park Avenue office. On December 6th, after many calls and exchanges of sketches, we were all pleased with the plans, and I signed a contract with the design firm Wilke Davis Mitchell Associates. We decided together that it would be a ceramic modular wall, which had to be viewed from two sides and visually separate a reception area from a small conference room. We decided it should be six feet high, ten feet long and be made up of nine rows of stoneware modules, stacked three high to a row. Each module would be fastened by being slid over a square aluminum pipe, flanged to the floor.

I began production by building wood, metal, and masonite matched troughs, which, when lined with sheets of clay and joined standing together, formed hollow geometric modules. Some were circular in cross section creating cylinders, others were hexagonal, pentagonal, rectangular, or diamond-shaped. All together, I made a pair of troughs to make in clay the front and back of six related columnar polygons. When the forms were filled, joined, and removed from the troughs, their surfaces were incised with half-round and right-angle tubular lineal patterns across the faces of the twenty-four modules. The outer faces of each of the units were diagonally raked with a wood rasp and the inward creases were smoothed with Teflon burnishers. The outer surfaces felt like cut limestone and the inner creases tempted the mind to finger trace their smooth calligraphs.

I agreed to complete the work in three months and install the piece in New York in early April. The finished wall and I would be flown back east Air Express. Flying with my art and journeying the quick way was a real switch, because we camped our way out to Albuquerque the slow way as a vacationing family. We had arrived in August to allow me to start the fall semester as a guest ceramics teacher at the University of New Mexico. We liked camping, so we kept right on doing it. In the early fall weekends, we used our four-person tent to sleep over in places like Chaco Canyon, Mesa Verde, and Canyon De Chilley. When we weren't camping over, we made day trips from Albuquerque to all the inhabited Pueblos and ancient adobe churches in towns like Truchias and Chiemayo.

When we began actually building the wall in the early winter, Cate and I employed two senior ceramic students, Allen Newman and Kevin Wynkoop, to work on weekends and after classes to help us work rolling clay slabs, filling troughs with clay, cutting tops and bottoms and putting the filled troughs out to partially dry before joining. Kevin and Allen helped me with loading and firing kilns, building the interior aluminum fastening fixtures, and, finally, packing for trucking the crates to the Albuquerque airport.

I fired the unglazed modules four at a time to cone nine in heavy reduction to the right browns for their final oak surroundings. The variety of tones from varying the reduction over the eight kiln loads really worked out well. I felt selecting and arranging the final twenty-four modules was the most satisfying part of the whole project. Cate, Kevin, Allen, and I spent glorious hours and days in the bright New Mexico spring sun working in the outdoor kiln yard playing "Musical Modules." We had a grand time negotiating all the possibilities and finally making everything come together.

During the months of doing parts and pieces, I came to gradually realize that my geometric minimalist "back east pot moves" really were "up" to shaping the feeling of timeless residence I experienced in the fall, drawing enchanted canyons, adobe sacred dwellings, and roofless Spanish stone ruins. Some days working in the studio alone, I felt like an astral traveler revisiting familiar spaces and places traveled in pots in another time and scale.

I felt that pots like *Reptilian Chiclet* and your *D Form* in my 1968 show at the Art Alliance were channels and mappings of places indexed and radiused to connect the fall sketches and the clay modules under our hands. It was a wonderful confirmation of the feelings I had designing, after driving through the striking countryside, remembering the red walls of Jemez Canyon, mesa skylines, and destinations like Taos pueblo.

At the end of making the parts and assembling the wall in the kiln yard, rotating modules, and stepping back to see light and shadow fix form, I felt that my moves had been worked out ages before by my ancient hosts. I feel richly gifted to be an initiate invited there to make mud stand up, turn corners, and be my salient for a Manhattan canyon wall.

Thanks very much for retrieving it.

Bill Daley

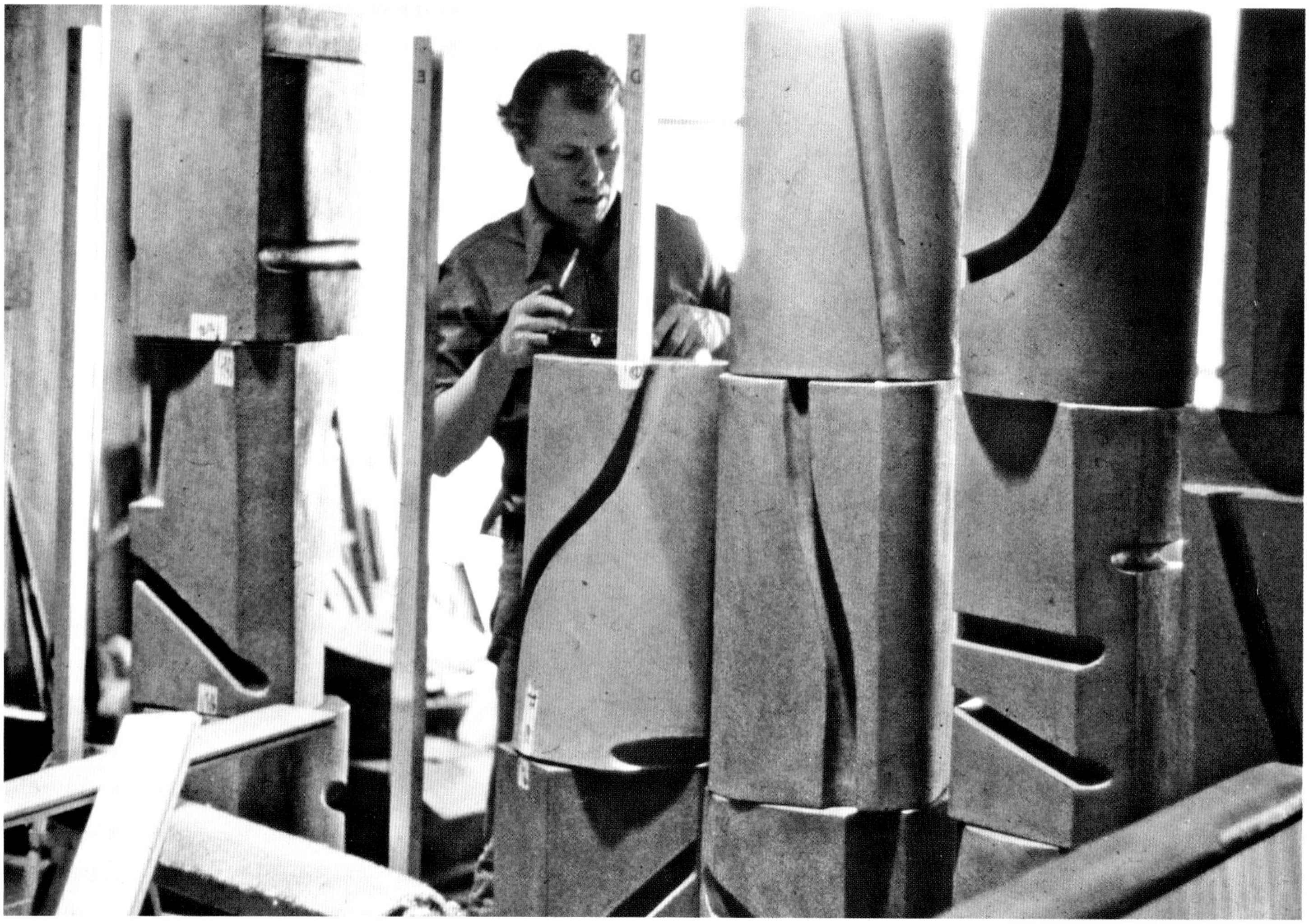

Photo: Joseph O'mara

Fig. 259. Trial installation of the *Manhattan Canyon Wall* in the New Mexico studio. It was a commission for Fairfield Maxwell Limited, New York; Wilke Davis Mitchell, architects

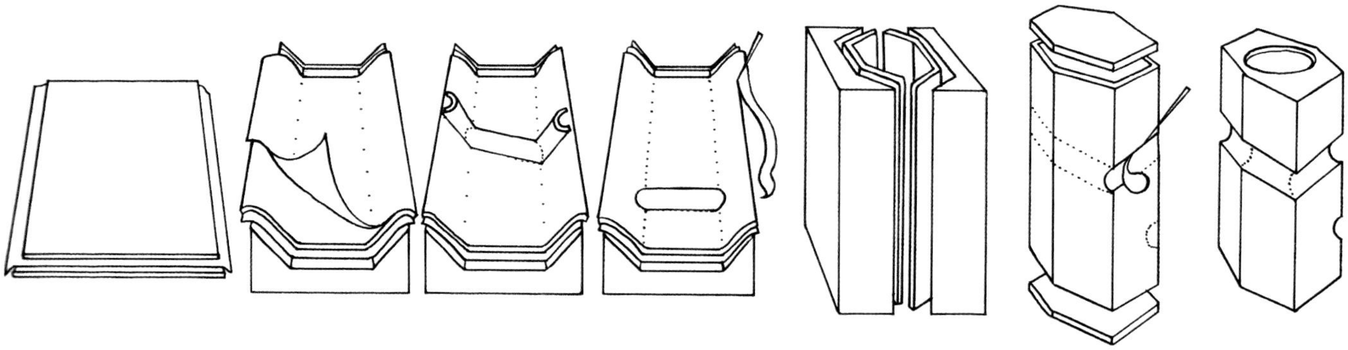

Fig. 260. Drawing for construction process of placing clay sheets into trough molds to create modules – Fairfield Maxwell Ltd, commission

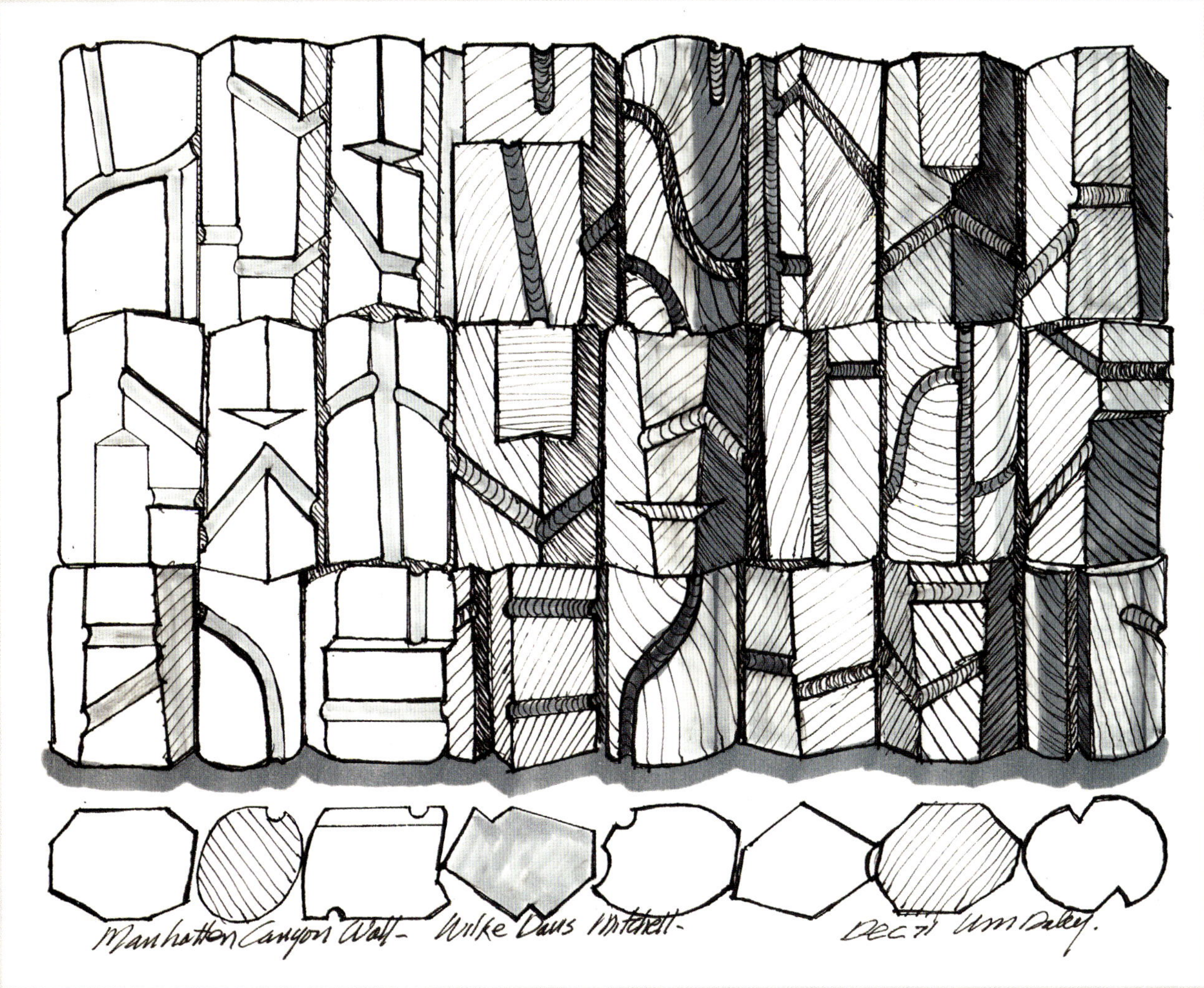

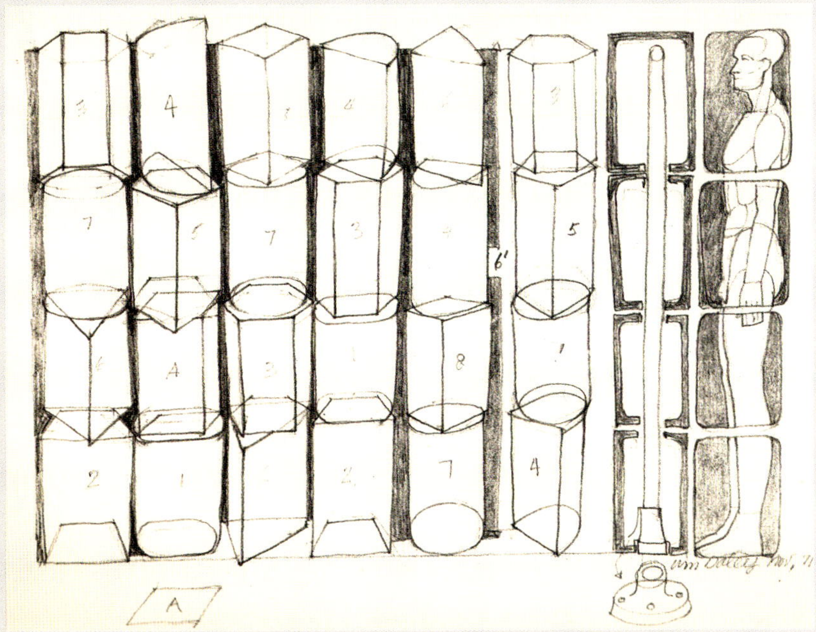

Fig. 261, above. Schematic drawing of *Manhattan Canyon Wall*, 1972

Fig. 262, left. Sketch for stacking modules

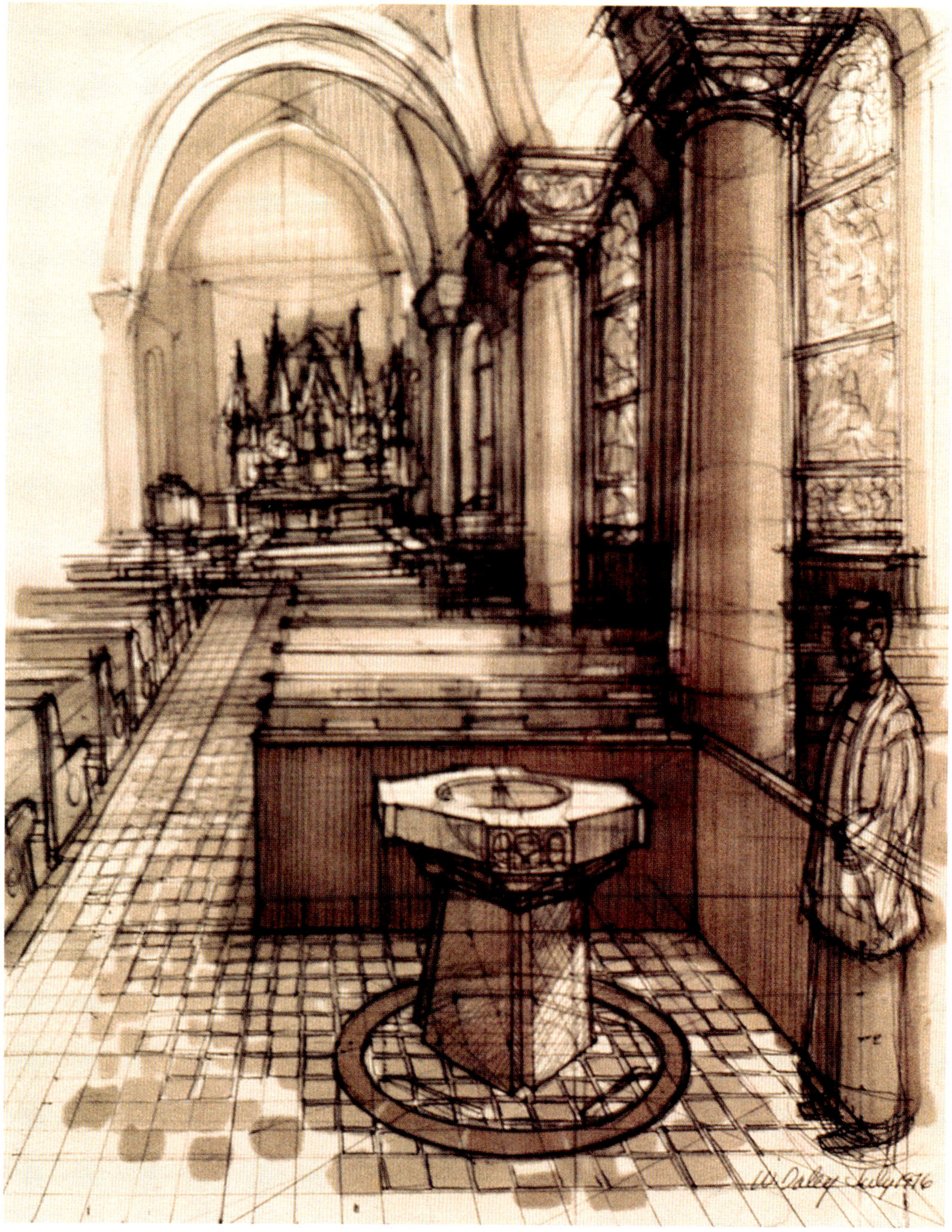

1976

Our Lady of Rosary Church Baptismal Font

Fig. 263. Drawing for *Baptismal Font,* 1976
Our Lady of the Rosary Church
63rd Street, Philadelphia
Commissioned by Bishop McDevitt
for the 41st International Eucharistic Congress,
34 x 36 x 36 inches

Fig. 264, opposite. *Baptismal Font* in the sanctuary

Photo: Amanda Schwab

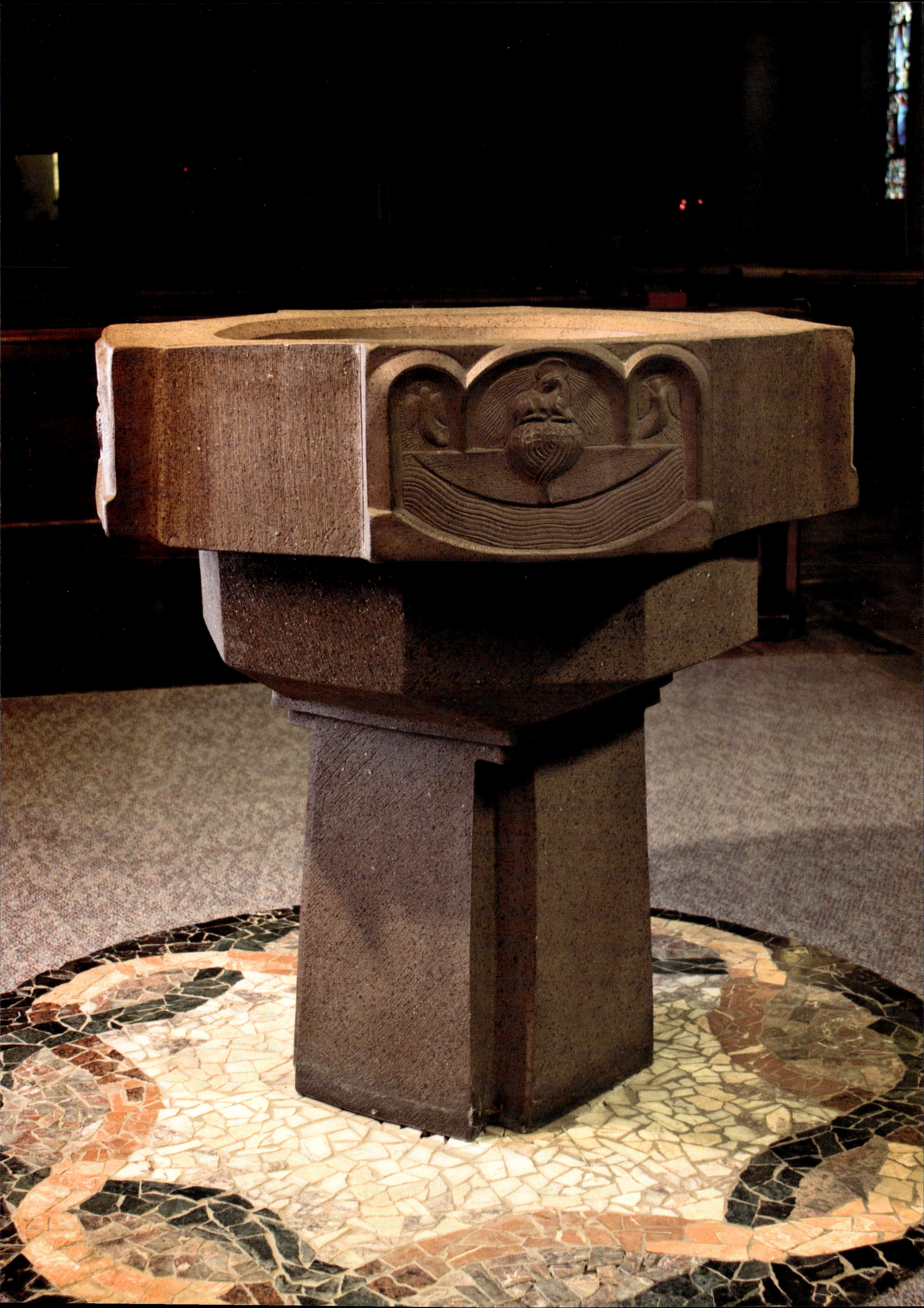

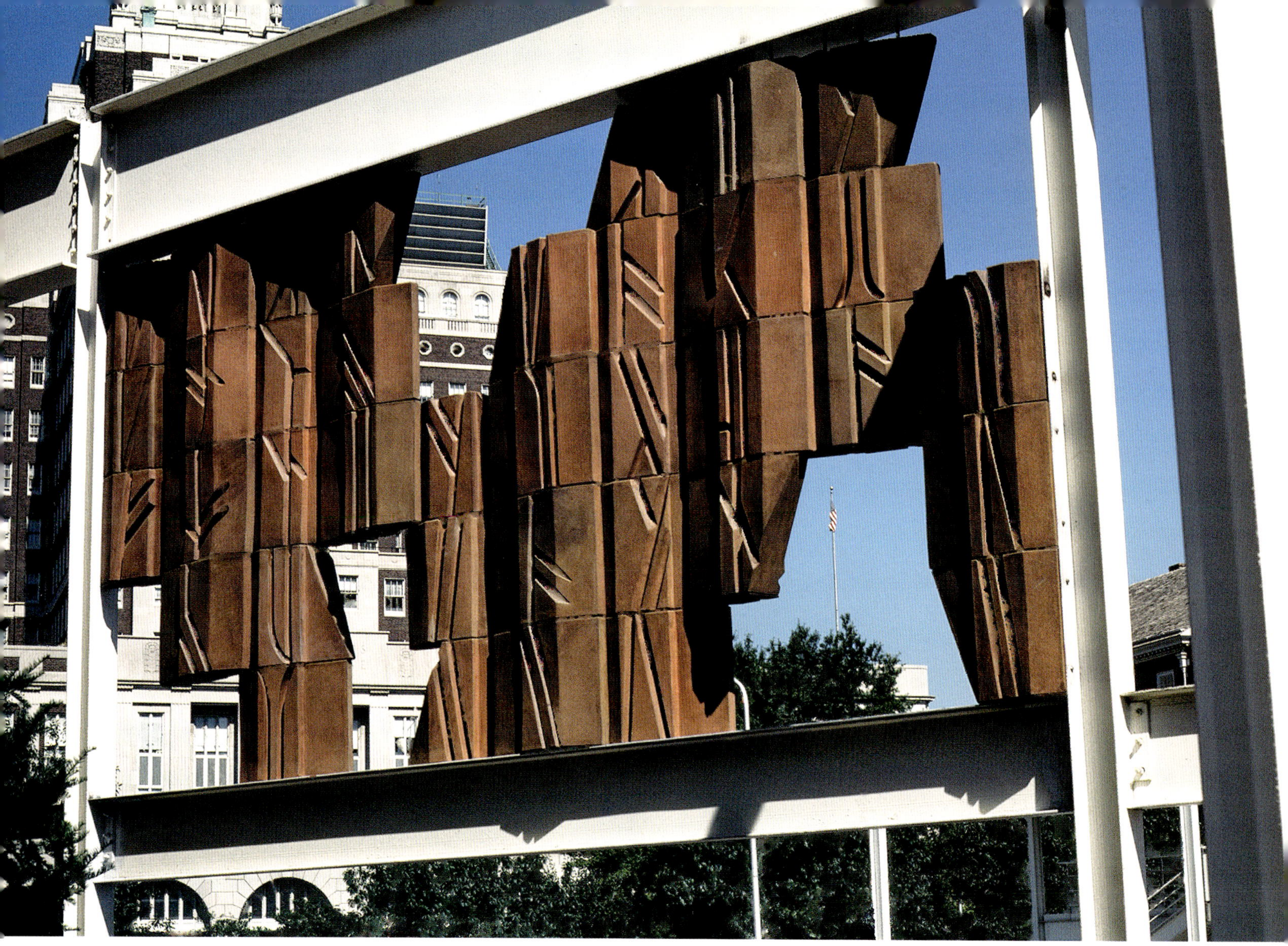

Fig. 266. *Cellular Permutation*, 1978 ceramic screen commission for Ritz Theater, Philadelphia; Posel Enterprises, owner

1978

Ritz Theatre Wall

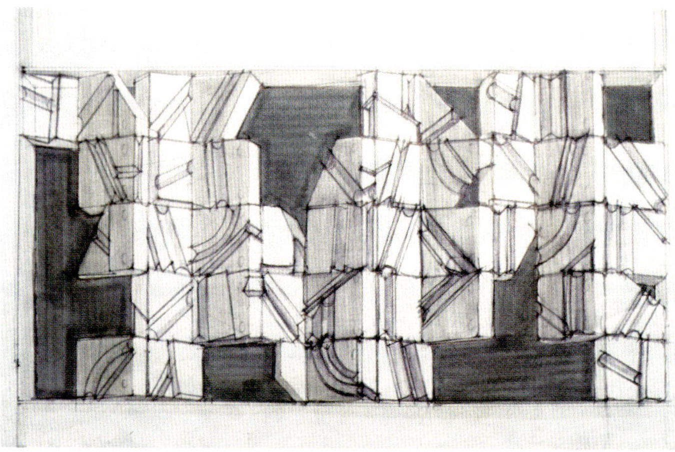

Fig. 265. Ritz wall module sketch installation into steel grid of architecture

Ritz Theatre Thoughts 1.30.09

The Ritz Theatre complex opened in the mid-nineteen-seventies. Now famous, it is Ramon Posel's signature contribution to the ongoing redevelopment of Philadelphia's colonial core – "Old City."

On its site at the juncture of Walnut and Second Streets, the Ritz Theater complex's architect, Robert Getty, wisely used the modern convention of exposing its structural steel construction. Doing so, he processed two large I-beams horizontally across the upper level of the façade. Supported on twenty-foot centers, they form a series of bold white frames. They dramatically relate modernist gestures to the past/future harmonics of "Old City's" Georgian salmon brick and white wood architectural surroundings.

Photos: John Carlano

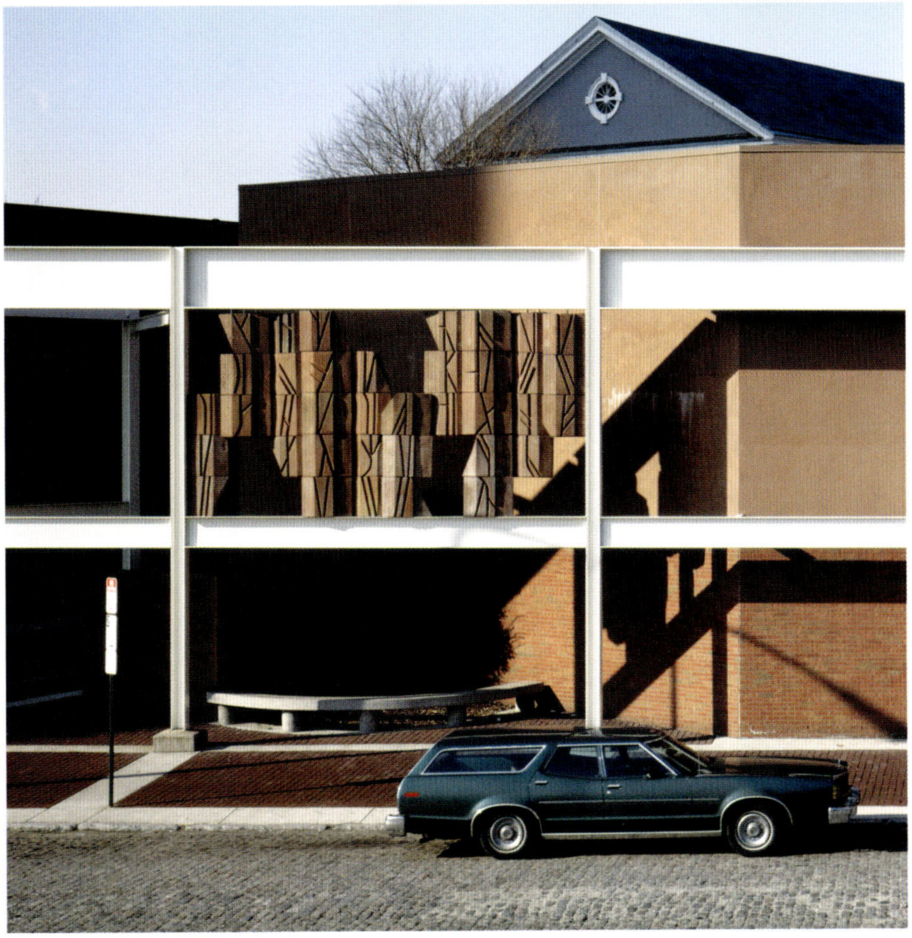

Fig. 268. Ceramic screen shown in second level of steel structure of Ritz Theater building

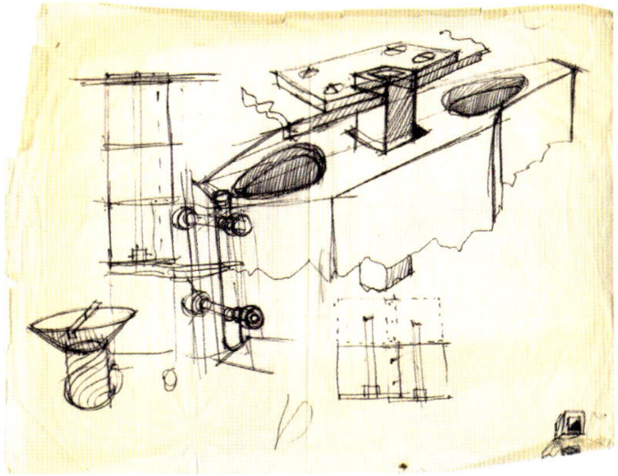

Fig. 269. Drawing of fastenings to attach clay modules

When Ray Posel and Robert Getty chose me to create *Cellular Permutations*, using clay as the connecting material for a sculpture in one of the frames, Cate, my spouse and partner, and I knew joy! Ray and I picked the frame with the best sight lines and began the professional and familial conversations to welcome all discerning cinemaphiles to gladly enter and enjoy virtual dwelling in the world's most telling films – a short walk from Independence Hall.

Beginning our journey, we moved from conversation to images, images to feelings, and feelings to materials for how they might help us create a "welcome scenario" to the movies. I began my part by surveying the Ritz with camera and tape measure to make a line drawing scaled to our chosen aerial frame. Xeroxing the drawing started me working in all three directions. All the "whats, whys, and hows" were provisional and the exhilarating subject for discourse, familial and professional, while reviewing my skeins of sketches and moving toward consensus. When we reached it, we selected the best of the "possibles" and I made a scaled wood frame to hold Bristol board cut outs. Folded and arranged in place as mini sketch models, they helped us reach certainty.

Developing our final choice, I designed thirty-seven clay modules, large hollow bricks. I thought of them together as a two-sided, open, serial screen of aerial coded tablets bearing calligraphies, signifying the geometry of story. The modules overlapped to the internally link together, hanging or stacked, creating telling, patterned spaces, *Cellular Permutations*. They are the gnomon for the sun's and the night's moving light becoming the Ritz's invitation to cinema.

— William Daley

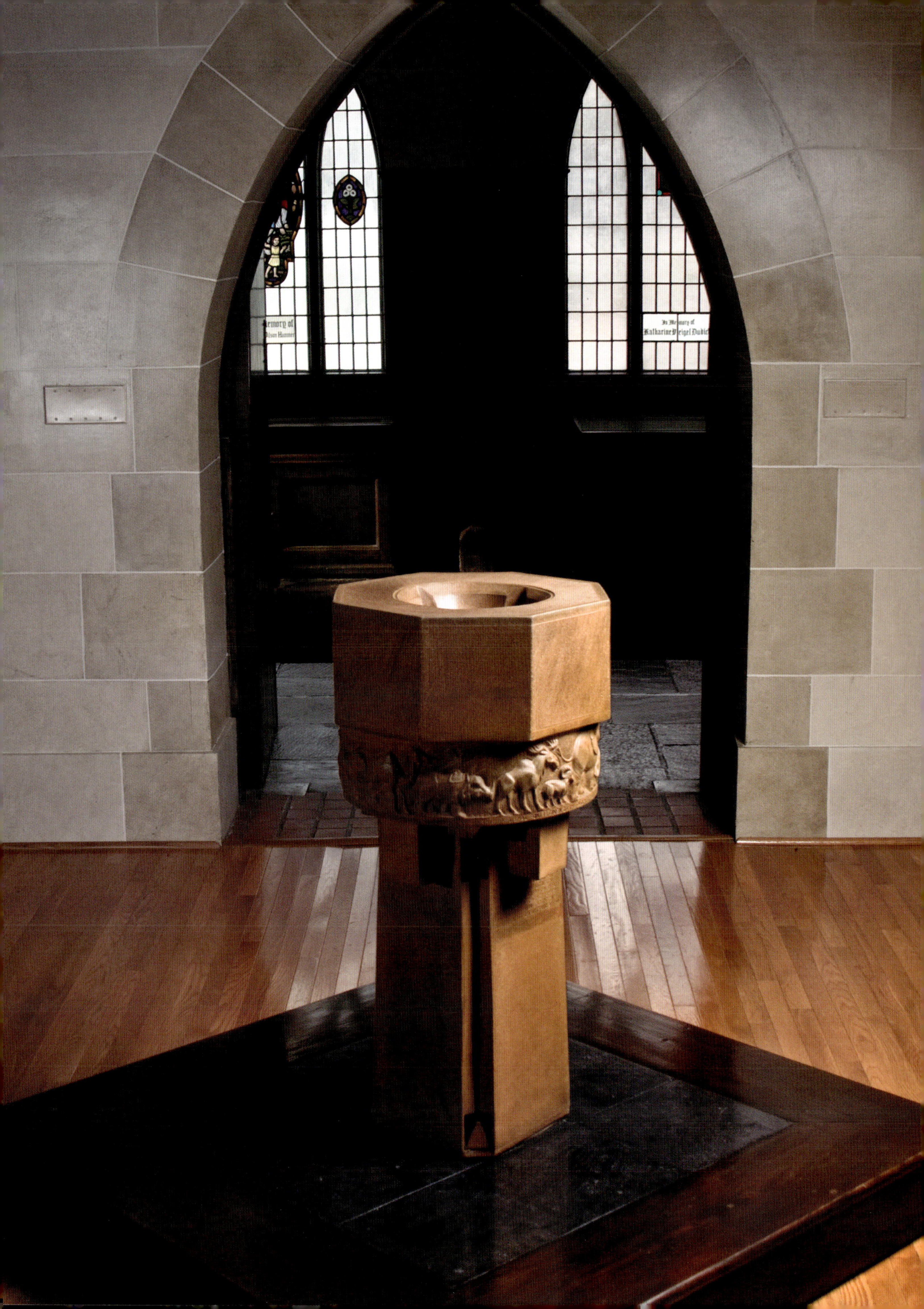
In Memory of

Baptismal Font
St. Paul's Episcopal Church

I made this baptismal font in 1989 for the Reverend Denver Hart, the rector of St. Paul's from 1983 to 1990. The people of St. Paul's made it in tribute to the memory of Dr. Akarue Amachi and his life of service among us.

The font is forty inches high and twenty inches wide at the top. It was made with slabs of stoneware clay in three segments, jointed together, dried slowly and fired in a gas kiln for twenty-four hours.

Any vessel can be used in baptism; it is the water and the words that are sacramental. Historically believers have made objects to give form to the spirit of sacred events. Such liturgical objects always have traditional symbolic references that are reinterpreted in each age of worship. For example, our stained glass windows employ traditional symbols in the style of the day at the beginning of this past century, through the visual voice of the glass master, Louis Comfort Tiffany.

As the maker of this font, I have also worked to link the past to the present by using ancient symbols in new ways. For example the octagon is used in fonts to symbolize the Eighth Day or the first day of paradise, and eternal rather than terrestrial time. In this font the octagon contains a hemispheric water bowl (half of the world) with a cross incised in it symbolizing the four directions and Christ at the center. The octagon is resting on a circular frieze of animals alluding to the loading of the Ark and the first baptism, the flood and the cleansing of the world by water. A cross is holding up the ring of animals, the octagon and the bowl as a buttress on each side of the font attached diamond shaped pedestal or base. There is a tunnel that passes through the base and the buttressing cross to symbolize passage, the journey from one state of consciousness to another; from being outside to inside or going from being spiritually undirected to being directed; such passage can also evoke "coming into" community. The passage and the ring of animal life are not traditional iconography for baptismal fonts but they fit well as powerful symbols of our journey quest for eternal love and our hope for spiritual continuance.

The ring of animals, modeled in relief also honors Akarue and symbols his life as a holy man, a dedicated veterinarian and keeper of souls, a man who dedicated his days to healing God's creatures. By the time the children in the church school are baptizing their children many of the animals in the frieze will be extinct unless we become, like Akarue, more successful guardians of new life in the ancient tradition of Noah.

I chose to make the font of clay because I am a ceramic artist. I also chose clay because it is a powerful symbolic substance. It is just mud, common earth, yet like the butterfly it is capable of striking changes. Changes from eroded rock to dust to mud with water then back to rock by being baked in fire. It is a transformational material, one that dramatizes the cycle of the ordinary becoming exceptional, like sacred water purifying a believer in the sacrament of baptisms.

I made the font as an offering of hope for all of us at St. Paul's and it is fitting that it has pressed into its base the names of our best possibilities… the children baptized by Rev. Denver Hart from 1983 to 1990. They symbolize the only part of eternal time we can effect, the present and the near future through our children and ourselves. Chickwere, Ahamefula, Udochi, and Ndidi Amachi have their names sealed in the font in honor of their father, as he continues through them. God's living promise.

— William Daley

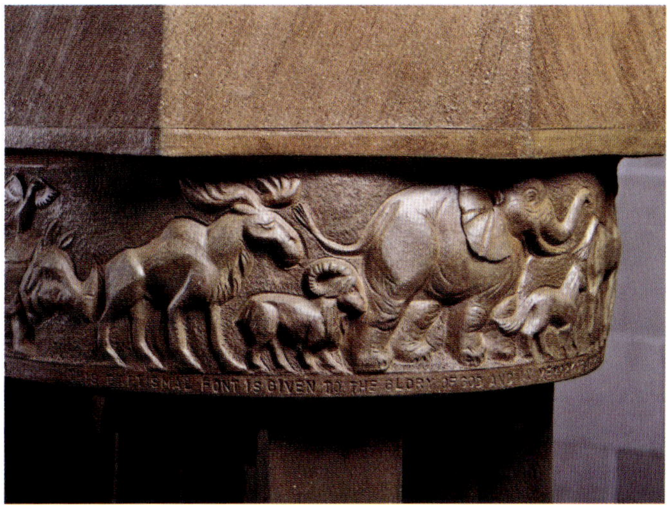

Fig. 270, opposite. *Baptismal Font,* 1989
St. Paul's Episcopal Church
Elkins Park, PA
Stoneware unglazed
41.5 x 27 inches
Dedicated in memory of Akarue Amachi

Fig. 271, right. Detail of animal procession around the font

Photos: Amanda Williams

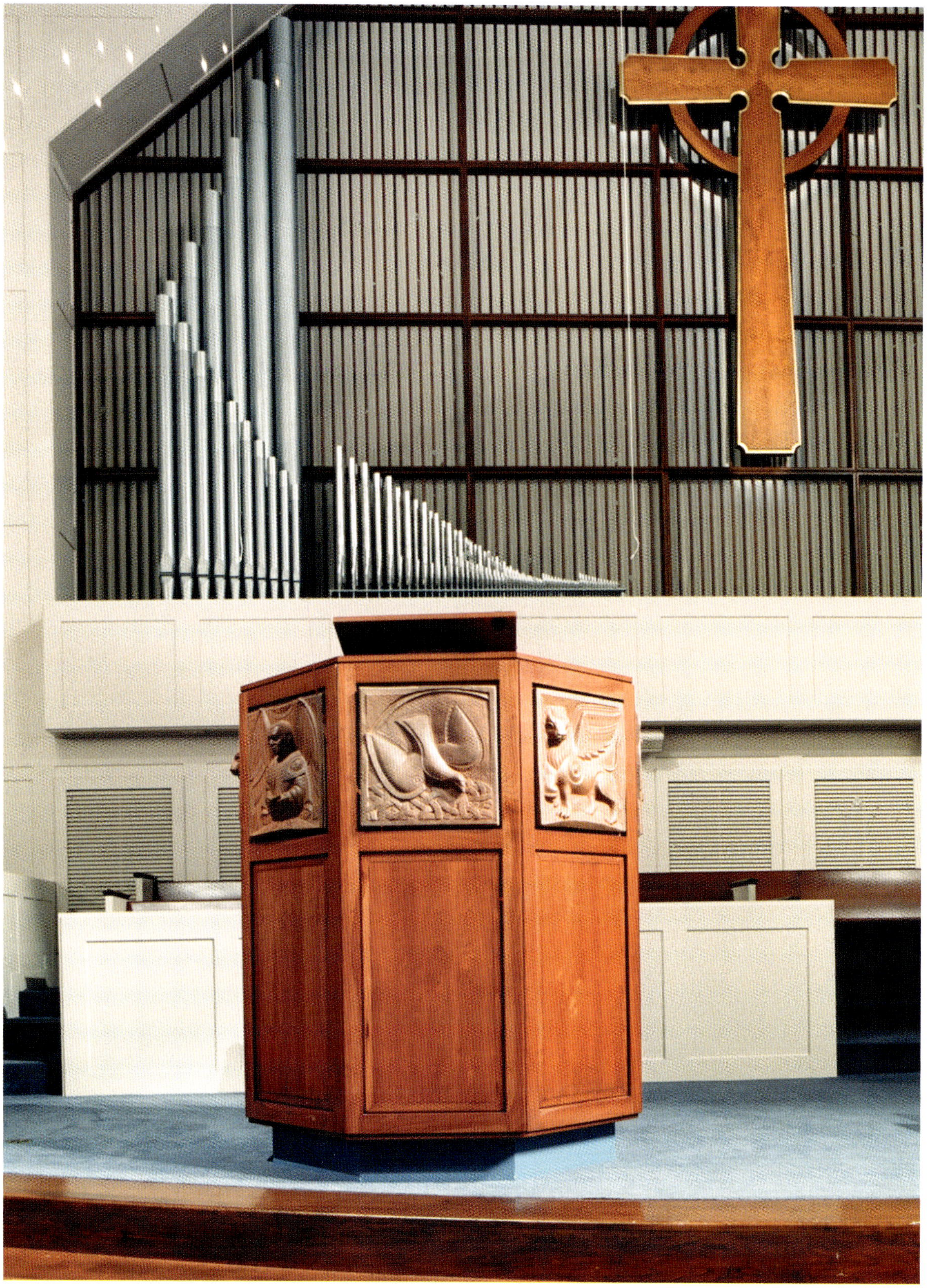

Photos: Don Pearce

Fig. 272. *Pulpit Tiles*, 1995.
Unglazed stoneware
Part of lectern for the sanctuary interior of
Westminster Presbyterian Church
West Chester, PA
Samuel Crothers Associates, Architects
(currently Daley + Jalboot Architects)

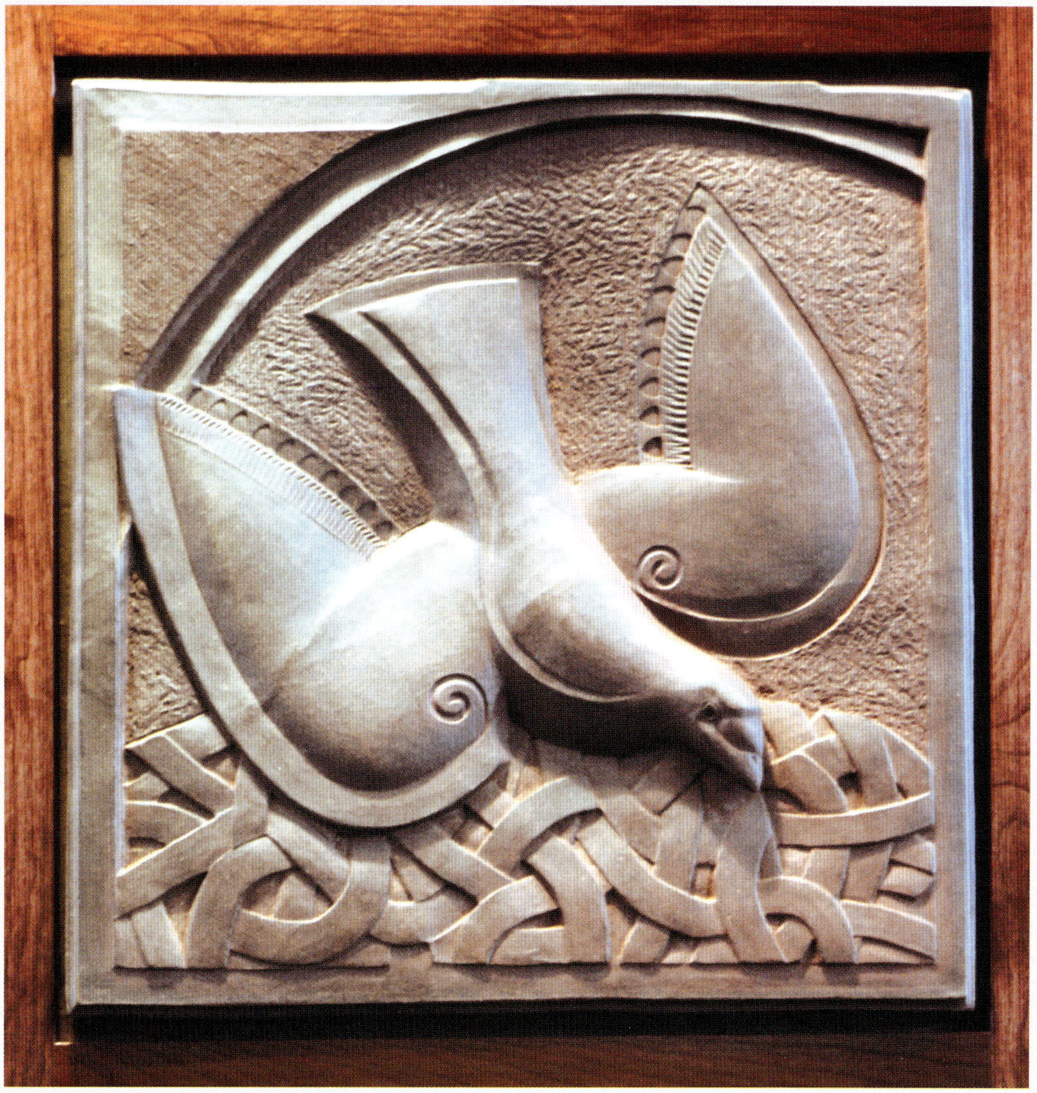

Fig. 273. Detail of *Pulpit Tile* representing the Holy Spirit

1995

Westminster Presbyterian Church

Westminster Presbyterian Church, 1995
West Chester, PA
Lecture: Talk on Art, Worship and Chancel Furniture
Tuesday Oct. 3, 1995

The American philosopher Suzanne Langer states, "all artists are formulators of symbols. Symbols inform our 'life of feeling.'" I like the idea of artists being defined by form words. Given a choice I would choose to call us "transformers"—people who work to "change one form of energy into another" and give feelings their ultimate form. This is the artist's special task. When we succeed we give "an outward visible form to an inward spiritual grace" and become the agents of sacrament.

— William Daley

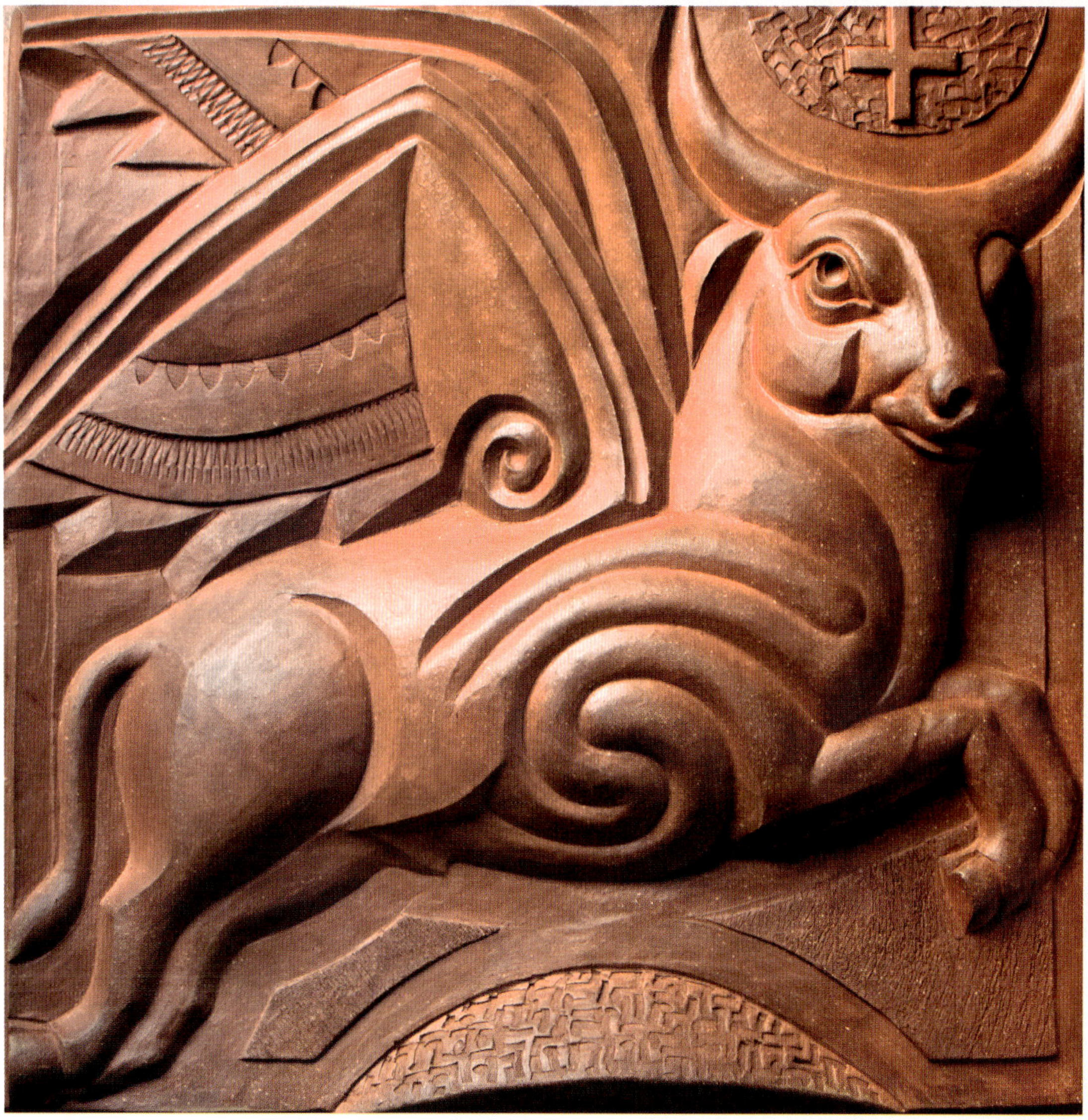

Photo: Amanda Schwab

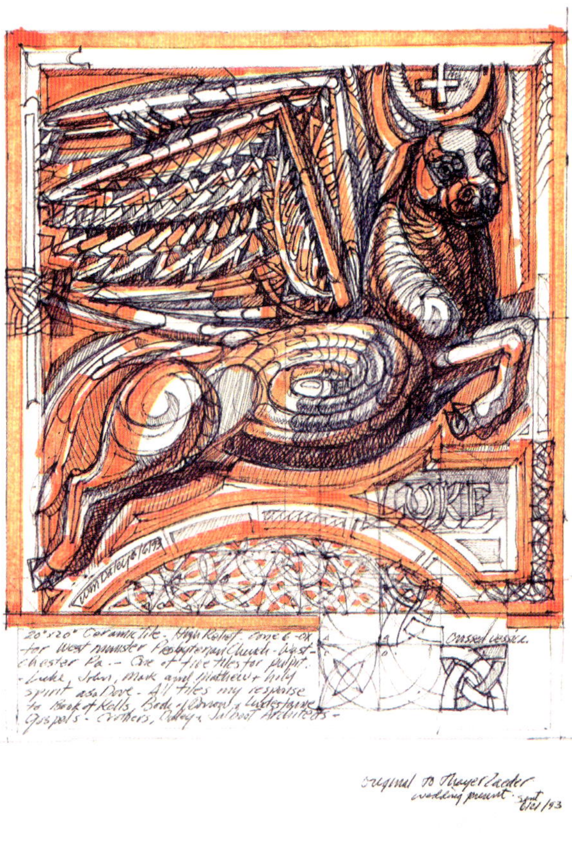

Fig. 274, above. Sample of *St. Luke Pulpit Tile*, inspired by the *Book of Kells*, commission for sanctuary interior of Westminster Presbyterian Church, West Chester, PA
Crothers, Daley, Jalboot Architects.
Collection of Thomas and Donna Weaverling Daley

Fig. 275, left. Drawing of *St. Luke Pulpit Tile*

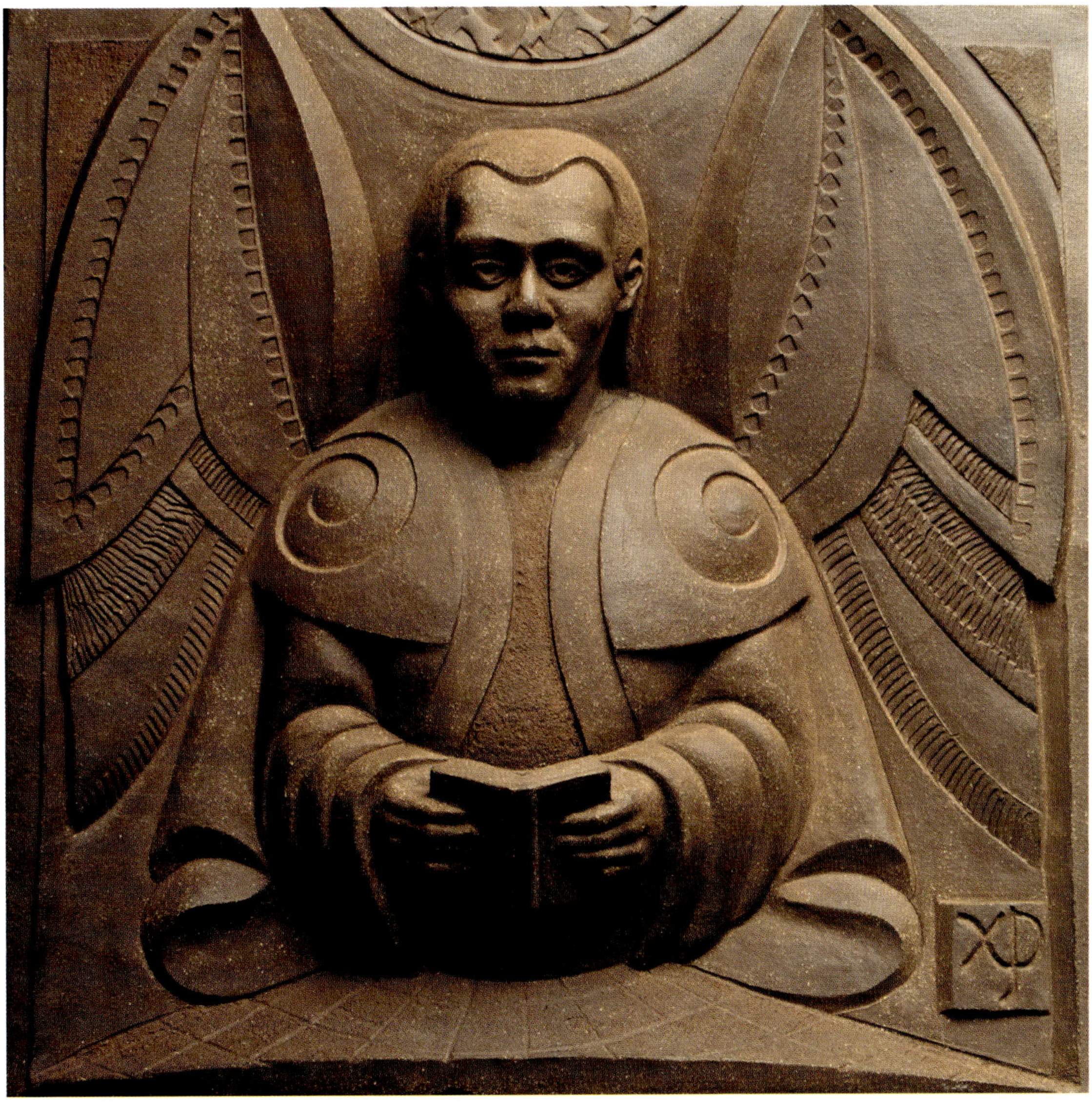

Photo: Amanda Schwab

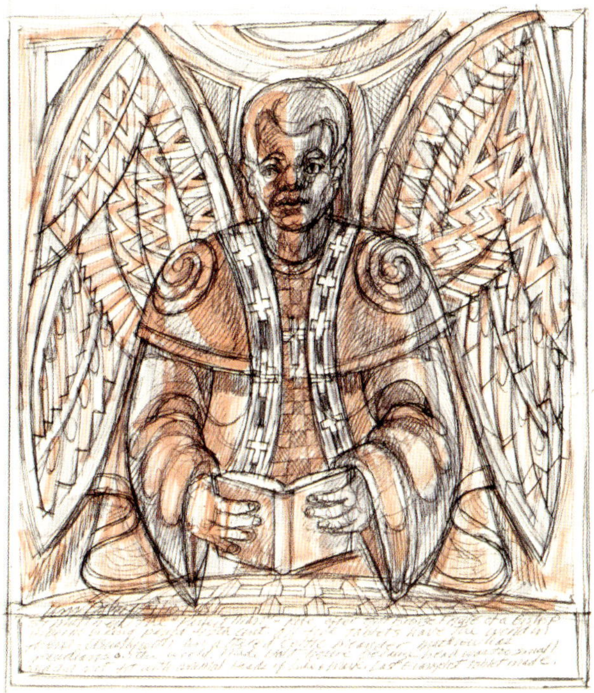

Fig. 276, above. Sample of *St. Matthew Pulpit Tile*
Westminster Presbyterian Church,
Collection of Barbara and George Schwab

Fig. 277, left. Drawing of *St. Matthew Pulpit Tile*

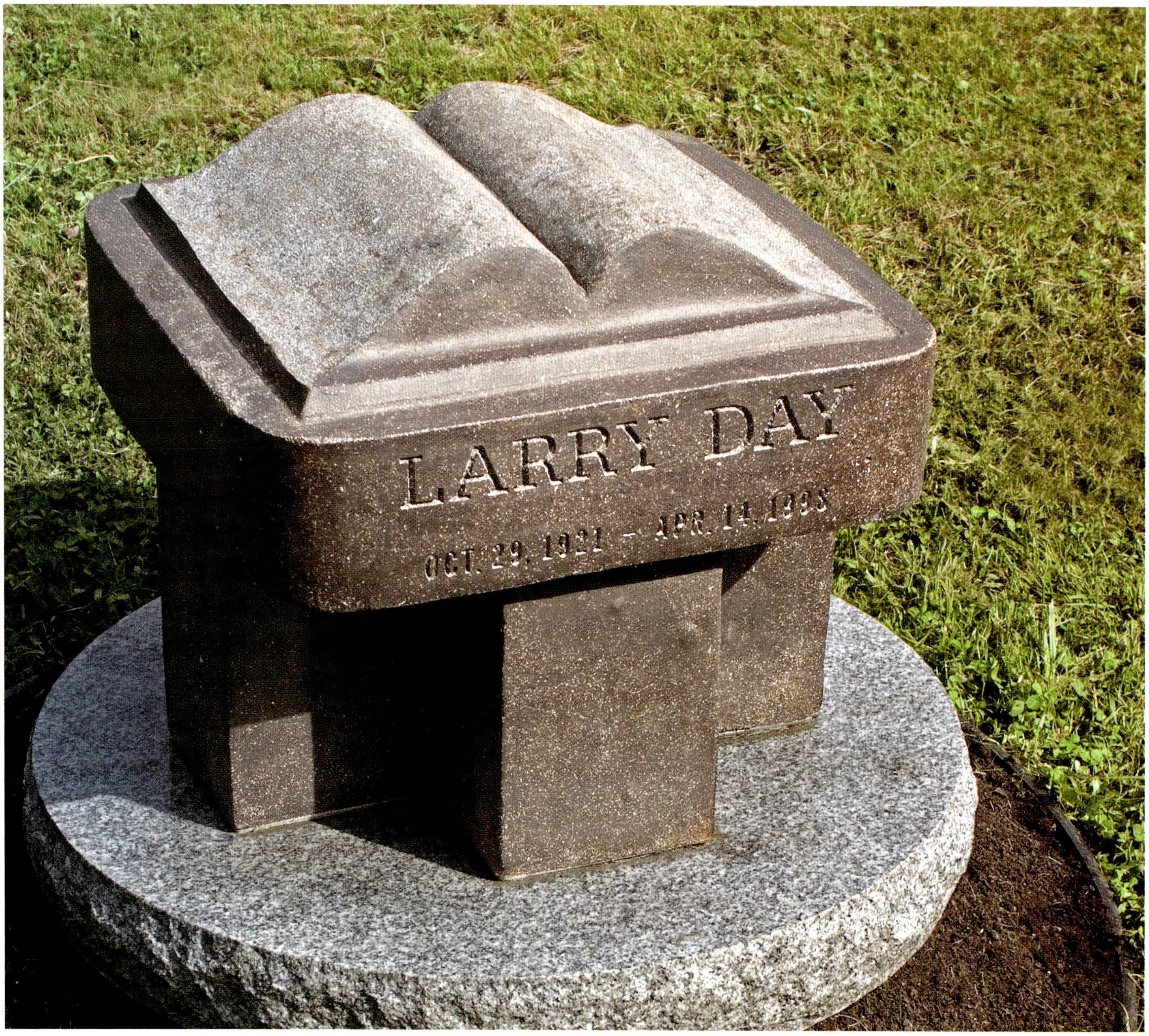

2011

Larry Day Memorial Grave Stone

Fig. 278. *Larry Day Memorial Grave Stone*
Unglazed stoneware
St. Paul's Episcopal Church graveyard,
Elkins Park, PA

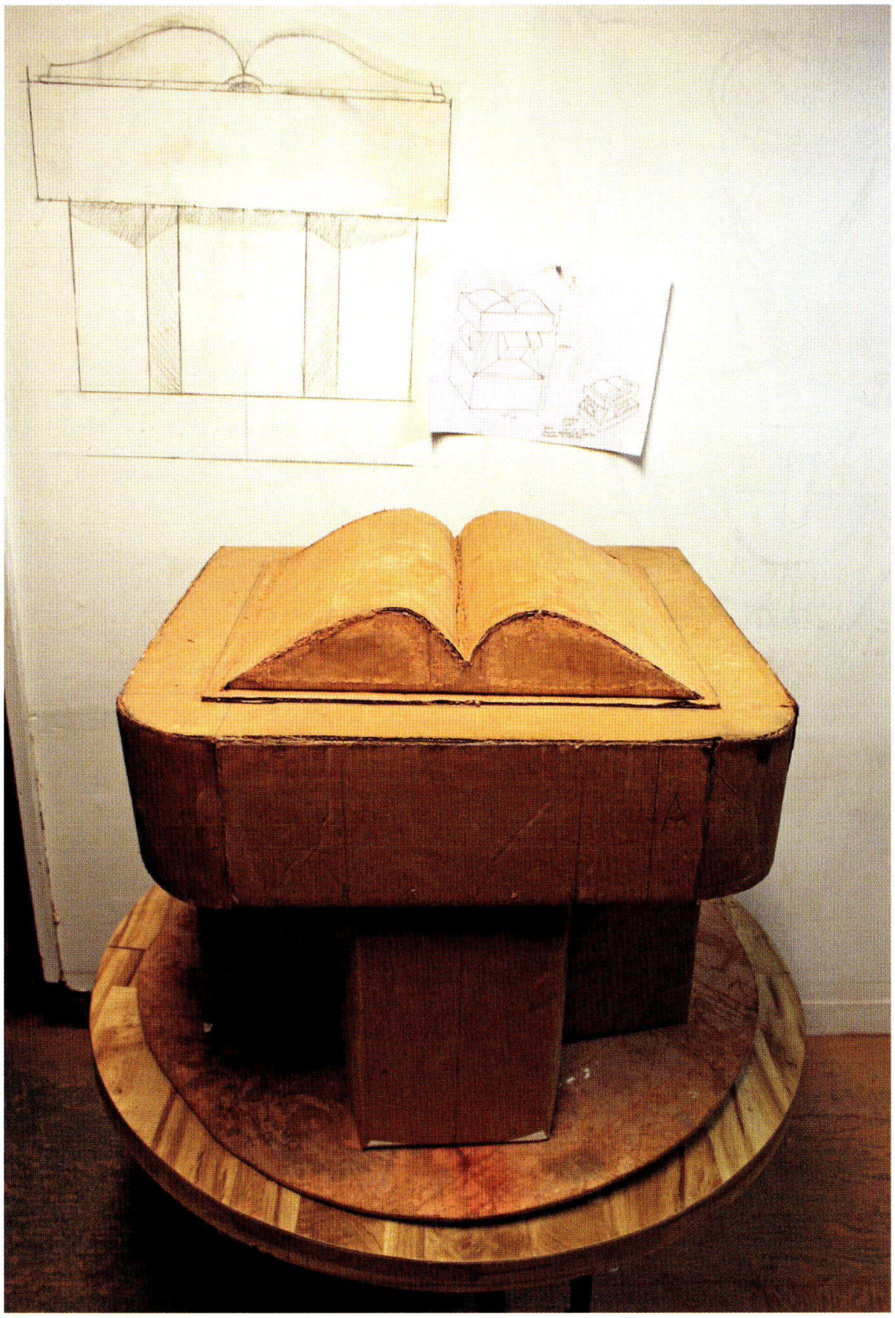

Photos: Charlotte Daley

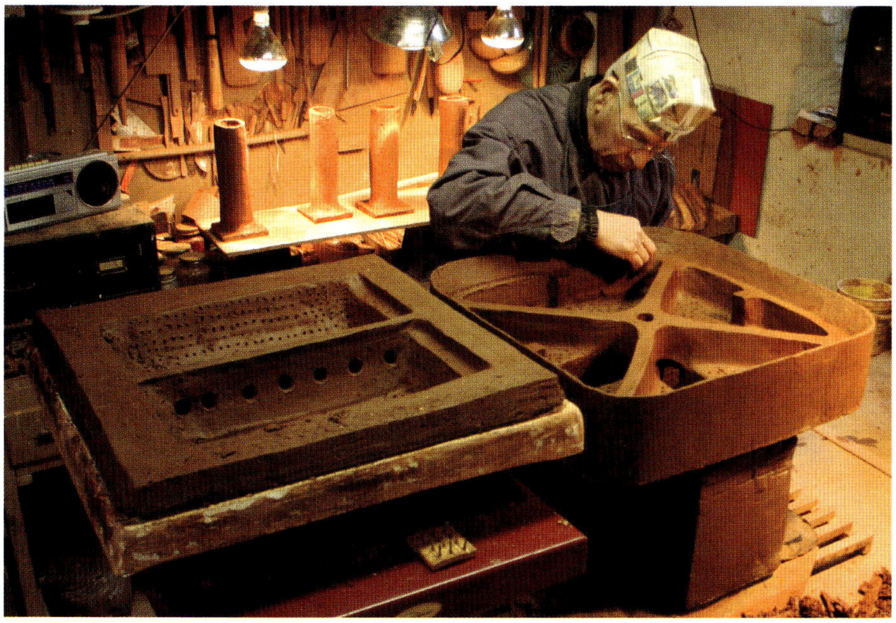

Fig. 279, above. Cardboard model in the basement studio on Ashbourne Road in Elkins Park, PA

Fig. 280, left. Bill constructing the grave stone in his studio

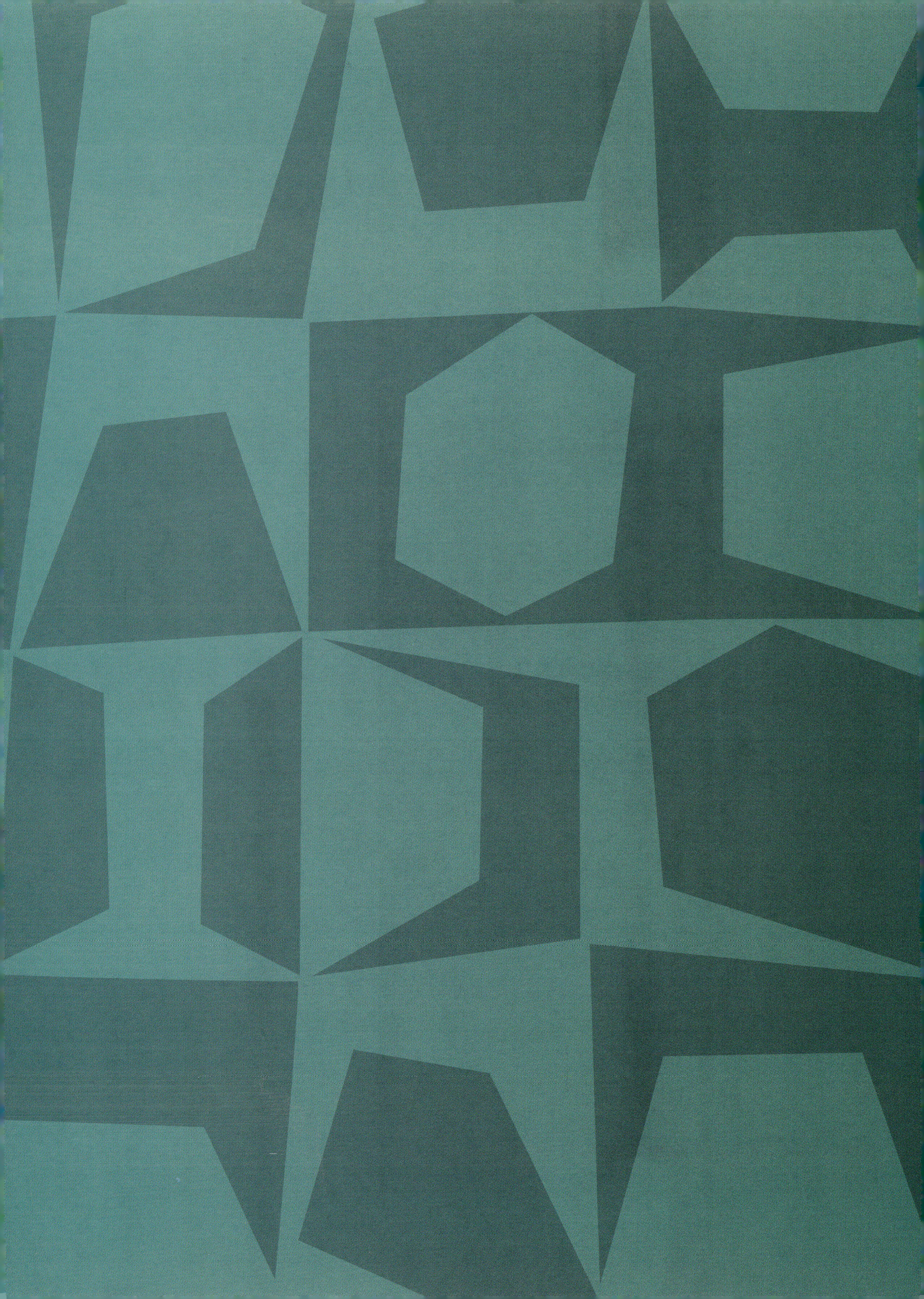

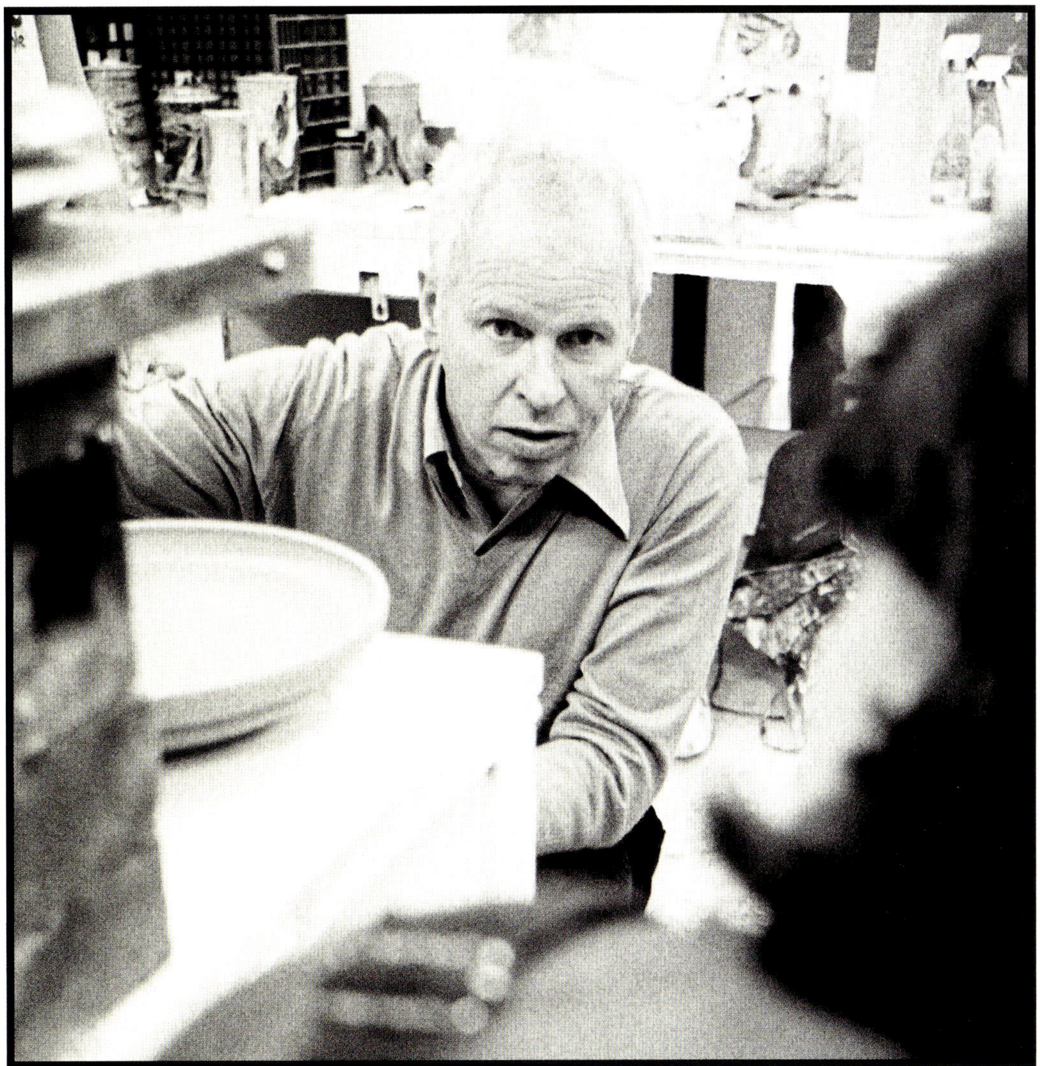

Photo: John Carlano

TEACHING

On the Art of Teaching Art

Stephen Tarantal

"I am a teacher who makes pots and things rather than a craftsman or artist who teaches."
William P. Daley, from a 1978 letter to the America Institute of Architects

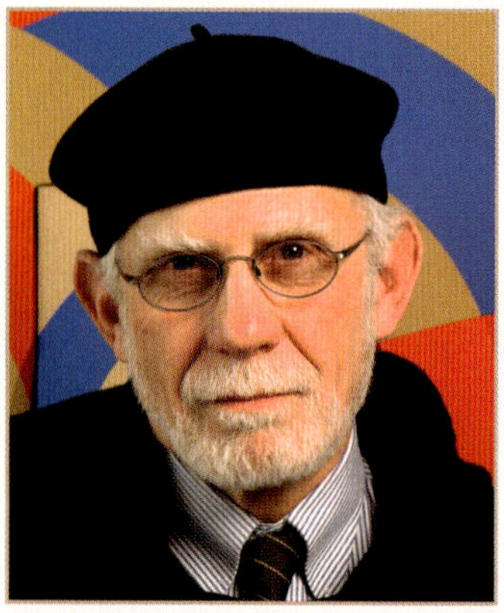
Photo: Victor Rodriguez

Stephen Tarantal

Bill Daley has spent over fifty years affiliated with the art and design school community located at Broad and Pine Streets in Philadelphia. He arrived at this Center City intersection as a young instructor in 1957 at the Philadelphia Museum School of Art (PMSA). He retired as a highly regarded and honored university professor from the Philadelphia College of Art (PCA) in 1990, and he continues to be enthusiastically engaged as a Professor Emeritus at the University of the Arts (UArts).

Shortly after the school received university status in 1987, Bill humorously suggested that our next institutional name change could be the Empire of the Arts! Bill, like all talented jesters, poignantly communicates kernels of truth, wisdom, understanding, irony, and seriousness with humor, which can be a wonderfully effective teaching strategy.

Since he walked through the red painted front door of historic Hamilton Hall, Bill has served (and counseled) multiple presidents and deans, he has been a trusted and valued colleague to faculty from all departments, and an influential teacher and mentor to generations of students—especially those in Crafts/Ceramics and Industrial Design, which he chaired and championed at different times.

"Those of us who want our students to accept jeopardy by going beyond should be involved in seeking jeopardy for ourselves as teachers..." — William P. Daley

Since retiring from the University's full-time teaching ranks, Bill continues to be invited to numerous art programs, professional conferences, and commencements across the U.S. and abroad, to share his insights and reflections about his work, and educating nascent artists and designers. Topics range from focused observations on the process of form-making and artistic discovery, to broad philosophical, spiritual, and critical ideas manifested most recently in his "Hero Slides," and the idea of "Firstness"[1]—a term Bill appropriated for select man-made and natural objects and images, "cosmic icons," that transcend the ordinary and transform our world by their inspired visual presence, rightness, and conceptual uniqueness. His writings and presentations, which encapsulate his ideas about teaching and making visual form, have been accepted into the Smithsonian Archives of American Art. As dean, I have access to faculty files, so I reread a number of his presentations that he routinely sent me over the years, and quickly came to the conclusion that a selection of quotes from Bill's papers would provide the best insight to Bill's dedication to teaching.

Bill's presentations bridge a broad range of passionate interests and activities that include a continuum of six decades of focused, hands-on, experiential learning and discovery with clay—the surprises and disappointments of building large pots, and a prolific breadth of critical readings on art, religion, history, science, philosophy, etc., which have informed and inspired his studio work and teaching—which is always about working on the edge between success and failure. "Those of us who want our students to accept jeopardy by going beyond should be involved in seeking jeopardy for ourselves as teachers. Like students, we must learn to live with our failures and not fall in love for too long with our successes."

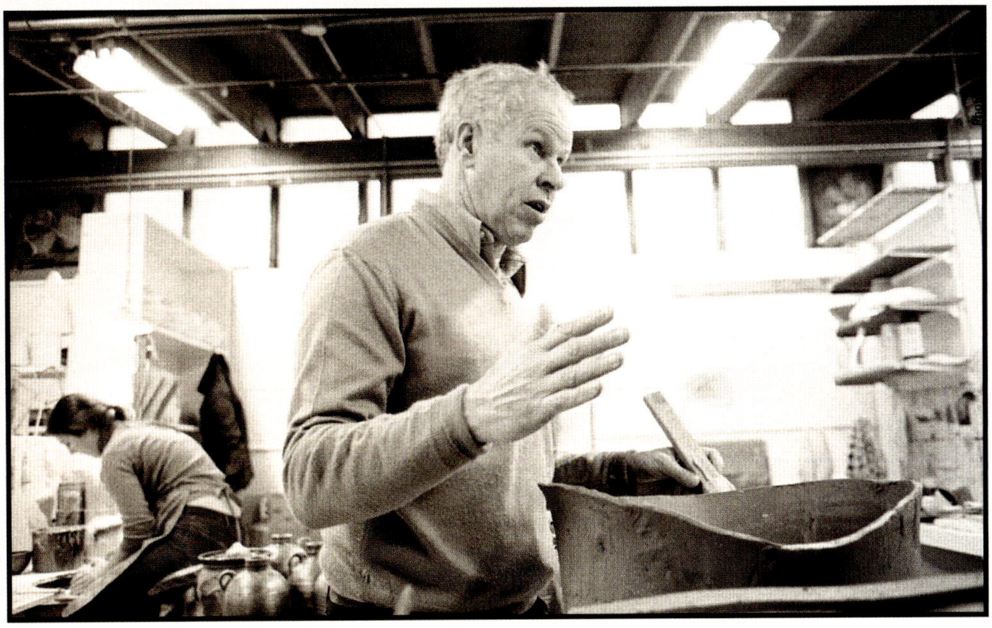

"I have come to believe that the art of teaching art requires one to perform with assurance while dimly perceiving a paradox which makes certainty short-lived..." — William Daley

Bill engages his audience, whether it is one-on-one with a student or in front of hundreds of conference attendees, with his special ability to connect disparate ideas into a fresh new matrix, with a captivating story-telling delivery —no doubt the benefit of a genetic Irish gift. Some of his presentation titles include: "Notes on Sources: A Presentation;" "Thoughts on a Geometry of Presence;" "Canted Closing Remarks of an Octogenarian Mud Man;" "On Drawing;" and "A Form Maker's Perception on 'The Case For Clay In Secondary Education.'"

Bill's public musings typically run the emotional and intellectual gamut, with seriously studied insights, provocative proclamations, self-deprecating humor, spiritual awakenings, and usually a surprising visual prop—often in the shape of a hand-made hat that he carefully unfolds and dons. His unpredictable delivery is always underpinned with important didactic messages, and always enthusiastically received. Bill is extremely energetic and passionate about a great range of ideas and topics, especially about teaching nascent artists and designers. The following paragraph written by Bill in a 1963 article[2], titled "William P. Daley artist and teacher," provides an early perspective on his teaching philosophy.

I have come to believe that the art of teaching art requires one to perform with assurance while dimly perceiving a paradox which makes certainty short-lived. This paradox is the need for all creative people to put themselves in jeopardy by striving to evolve a form language, which will communicate something they do not yet know. The creative person is concerned with transcendence, or 'going beyond what is given or presented by experience.' This basic characteristic of an artist or teacher presents a contradiction in teaching, which I find difficult to resolve. The contradiction is that the artist needs to know so much, which is both necessary and useless. His knowledge of process and technique and his understanding of history and his time are essential. This knowledge helps him to define himself and clarify what he will not and cannot do. Sometime it helps to establish an eclectic beginning point, but it does not declare what he will do because what he already knows is not worth doing. In a sense, he must know a great deal which must be forgotten in order to begin. The teacher then must teach what has to be known while encouraging the student to transcend his experience." [2]

The raison d'être of an art and design school is the sharing of knowledge, passion, joy, failure, instincts, and insights between and among faculty and students, faculty colleagues, and student peers. The act and participation of inquiry, understanding, and mutual trust and respect between a teacher and student is where the intersection of education occurs. Between the mid 1950s and mid-1980s, in residence at the Philadelphia College of Art (PCA), was a particularly dynamic community of art and design faculty who held strong philosophical and pedagogical opinions about the process of art and design thinking and making, each focusing on what an art and design education should be during the last half of the 20th century. *"To me embodying thought through making is the definition of form. Being informed is the purpose of making. Making something you don't know yet is the way to practice all the moves which makes us quick and acutely responsive to our best possibilities."*[3]

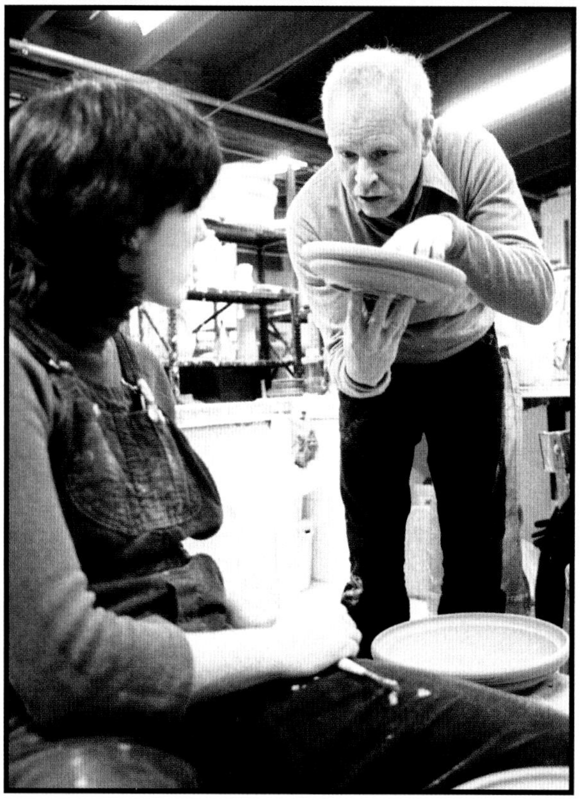

Photo: John Carlano

Bill in the pot shop at the University of the Arts with crafts student Amy Freeman

discipline-centered major programs new departments were added—most notably Liberal Arts and Fine Arts majors—expanding the breadth of the new college. Full-time faculty were hired on tenure-track positions. A faculty chair, who was enabled to shape their program, headed each major department. These were conditions that provided a strong sense of faculty stewardship, leadership, and ownership. Bill flourished in this new environment and was instrumental in the formation and realization of the Industrial Design, Ceramics, and Foundation programs. I do not know any other faculty member at the College who chaired two different programs—especially spanning both design and fine arts departments.

"By becoming masters of form artists are among the first to discover the salient exchanges that reveal the connections between temperament, probability and the shape of the future..."
—William Daley

When Bill started teaching at the school, it was just embarking on plans for major curricular changes from its more vocational diploma beginnings and museum affiliation to an independent, comprehensive degree-granting College of Art and Design. This would be a significant change in the institution's educational mission and identity, since it was founded as the Pennsylvania Museum and School of Industrial Art in 1876. That Museum would eventually become the Philadelphia Museum of Art, which parted ways with the school in 1964.

The anticipation of these two significant changes—of becoming a college and separating from the Museum — resulted in major public discussions and debates during the late 1950s. Bill has always been present at the center of the ongoing educational debate, contributing strongly held persuasive beliefs and positions, so it was not surprising that he would be selected (one of six faculty) to speak at a school-wide faculty forum on *"The Integration of the Arts."*

The school, known as *"the Industrial,"* and the *"Museum School"* during the first half of the 20th century, was beginning to broaden its academic mission. Most importantly, it was decided that for the second half of the 20th century the education of the artist, designer, and teacher must be based on the broadest possible foundation in the humanities. This seminal moment in the evolution of the institution was in sync with Bill's sensibility as an artist, designer, and educator. The college was reorganized around

Effective and memorable faculty like Bill share their passion, knowledge, joy, instincts, and insights when they guide students in their process of artistic discovery. They view teaching as a noble calling, and they take the responsibility of teaching very seriously. Bill often talks about teaching as being a privilege. Teaching and learning are inextricably linked to our personal and professional development—especially for those of us who profess. Faculty must simultaneously learn from their charges about their aspirations and goals, while they artfully steer them toward meaningful experiences. In so doing, we continually fluctuate between being a responsible student and an effective teacher. Bill is constantly curious; this *"octogenarian mud man"* continues to be a vibrant and vigorous studio artist, teacher, and role model. Recently the Center for Teaching and Learning at the University of the Arts invited Bill, along with a few other senior professors and their former students, to be on a panel at a faculty symposium on teaching and learning. Bill and his former student, Alec Karros (who now teaches at UArts), represent the continual cycle of teaching and learning in the arts. Formerly in a student/teacher relationship, Bill and Alec are colleagues, and while Bill's seniority is clearly evident, there is ongoing mutual respect and a continuing penchant for learning from each other. *"I have enjoyed all the satisfactions of sharing the days there [at PCA] as a promising older student becoming an artist. It has been a special pleasure doing so in the changing company of younger and younger teachers."*

Bill's teaching extends to community building. He is always engaged, not only in his specific professional field as an artist, but he is supportive of colleagues in other areas of the College. His involvement includes supporting the larger arts and education communities, most notably advocating and teaching ceramics in K-12. Educationally, artistically, intellectually, emotionally, and perhaps even psychically, Bill is the consummate professional.

"To become a true professional, all the skills I mentioned in the first loop are of little or no real use unless they are grounded by your inner life to nature and humanity. This is the core of the second loop and assures that as professionals you are nurtured by a sense of wonder for how grand we are. Such conceit has to be coupled with deep regard for yourself, all people and all life. I am certain that one reason all good artists are directed from inside as professionals is because of their long study in art schools. They have learned to seek through discovery of self in community. By becoming masters of form artists are among the first to discover the salient exchanges that reveal the connections between temperament, probability and the shape of the future. In locating these points of transformation, they also dramatically create the new signs, symbols and images that shape our emerging consciousness."[4]

In my forty years at the Philadelphia College of Art (now the University of the Arts) my engagement with Bill is unparalleled in the degree to which I have been enriched, enlivened and engaged by the magnitude of his personal gifts of heart, mind, and spirit. In 1991, Bill received the College Art Association's Distinguished Teaching Award. His acceptance speech concluded with these final remarks, which clearly describes his affection for his teaching and learning community:

"Finally, I express my gratitude to the University of the Arts, more precisely to its Philadelphia College of Art. It is a contagious place, peopled with selective obsessive's becoming artists. It holds a community of avid individuals in all the stages of seeking their way. Artists working to give spirit telling form. They are my models and like them I have become skillful in the exchanges of transformation."

Stephen Tarantal, Dean
College of Art and Design
University of the Arts

October 2009

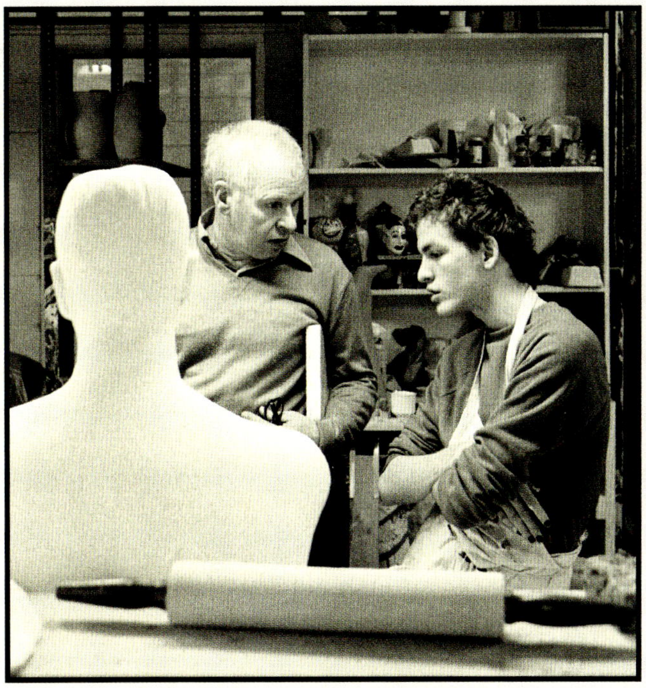

Photo: John Carlano

Student John Detazio

Stephen Tarantal is currently the founding Dean of the College of Arts and Creative Enterprises at Zayed University in the United Arab Emirates with campuses in Dubai and Abu Dhabi. Professor Tarantal taught at the University of the Arts (formerly Philadelphia College of Art) in the Illustration Department, and served as Dean of the College of Art and Design for 24 years. His studio work is directed toward fine art exhibitions and communication design commissions.

1. *Photometric Researches*, 1878, Charles Sanders Peirce – credited as the founder of Semiology, the study of signs and symbols and their connections to cognition and experience. Firstness, Secondness, and Thirdness is best described in an essay titled, "One, Two, Three: Fundamental Categories of Thought and Nature, A Guess at the Riddle".
2. School Arts, Vol. 62, No. 7, March 1963
3. January 28, 1988, A Form Maker's Perceptions On The Case For Clay In Secondary Education
4. 1991, Thoughts for Spring Graduation – Nova Scotia College of Art and Design

William Daley: Teacher Extraordinaire

Marian Pritchard

Bill never demonstrated how to make a specific piece nor did he build his own vessels at school. He would show you a basic technique and let you figure out how to apply it to your work.
— Marian Pritchard

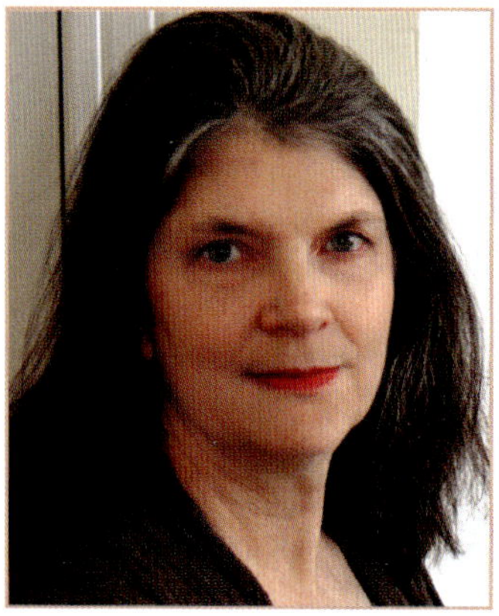
Photo: Charlotte Daley

Marian Pritchard

One word—Bill—describes the heart of the ceramics program at a school variously called the Philadelphia College of Art, the Philadelphia Colleges of the Arts, and the University of the Arts for the years between 1967 and 1990. William Daley first came to the school in 1957 to teach industrial design classes and left in 1960 to teach at State University of New York, Fredonia. He returned in 1963 to revitalize the Industrial Design Department with the understanding that in three or four years he could help develop the Dimensional Design Program, started by Richard Reinhardt and William Parry, who had left to teach in Alfred, New York.

Bill strengthened industrial design and moved to chair the Dimensional Design Program, which he renamed the Craft Department. He included fibers and added glass instruction to the disciplines of ceramics, woodworking, and metals, while continuing the drawing component, which Parry and Reinhardt had initiated.

Students majored in craft, not in a specific material area of study, such as ceramics. Bill wanted students to learn the common factors among the various material disciplines to stimulate a more open attitude to their own work. He also encouraged the faculty to work across their own disciplines by collaborating on student assignments and by rotating the position of department chair. Faculty, such as Richard Reinhardt, who taught both metals and industrial design, and Petras Vaskys in sculpture, ceramics, and industrial design, enhanced this open approach.

Bill tells of building the ceramics department physically as well as philosophically. The school, founded as the Pennsylvania Museum and School of Industrial Art and later known as the Philadelphia Museum and School of Industrial Art, had offered ceramics as part of its vocational mission. In the 1960s, Bill, with some colleagues, built new infrastructure, including kilns, wheels, storage and work areas. The practicality of rebuilding the facilities worked in tandem with the curricular development toward a broadly based approach to ceramics.

When I attended PCA from 1979-1982, the ceramics program provided a strong foundation for working with ceramic materials and processes, focused attention on aesthetics, especially form, introduced students to many artistic achievements from prehistoric to contemporary times, but most importantly encouraged each student to develop work of personal significance. In emphasizing the latter Bill once remarked, "Everyone learned all of it together. The core thing was getting to the bottom of self through working."[1] I can best address the strength and breadth of Bill's teaching by describing my experiences as a student in the ceramic and craft programs that Bill primarily shaped. As he often used the term "extraordinaire" to praise someone, I have applied the term to him. I don't doubt that other graduates would concur with much of my praise for Bill despite their differing experiences. Although focusing on Bill's teaching in school, I know he impacted many individuals, even those he didn't formally teach.

The ceramics program, built around core courses of hand-building, wheel-throwing, and plaster workshop during the sophomore year, evolved into more independent projects junior year, and culminated in a thesis of one's own choosing senior year.

My sophomore year, Bill and Mark Burns taught hand-building, and Rudi Staffel taught wheel-throwing. Petras Vaskys, a.k.a. "Plaster Master," taught plaster workshop and was a wonderful combination of technical knowledge and intuitive thinking. Larry Donahue, technical assistant, was the go-to man when faculty were unavailable.

Because of the major in Craft, a student took two other classes sophomore year in either metals, fibers, glass, or wood.

Bill never demonstrated how to make a specific piece nor did he build his own vessels at school. He would show you a basic technique and let you figure out how to apply it to your work. I vividly remember his hard slab demonstration in which he reinforced the joints with scraps from the slabs used in making the piece. His reasoning about preventing cracks and assuring similar shrinkage rates seemed so rational. When I overheard Bill and Petras discuss whether using slurry or water was more effective in keeping the joint together, I realized that experts could disagree on their preferred techniques. They did not exclude students in their discussions of opposing views, and you quickly learned that you could absorb the basics, then do what works for you.

All of us were shocked one day when Bill sat down at the wheel as a substitute teacher in the wheel-throwing class. Nobody knew the master in hand-building could throw! In no time, he set up a contest to see if anyone could make a bigger thrown pot than he could! It turned out to be a tough task, but the laughter and joy of the day turned into a celebration for making things in clay, a theme that surfaced often when Bill was nearby.

Bill and Richard Reinhardt brought all the craft majors together sophomore year in a class called "Ideation Studies," which emphasized the importance of exploring ideas through drawing. Bill often spoke of the advantage of conceiving one's idea for a piece on paper because making variations in clay could be so labor intensive and time consuming. Bill and Richard, sharing an interest in industrial design, taught us about x, y, and z planes, orthographic projection, and plan, elevation, and perspective views. Through the years we learned how much Bill loved to draw. He sketched his ideas about inside and outside in ballpoint pen in various permutations on a sheet of 8 1/2 x 11 inch typing paper. Bill dramatically gave away one of these gems by number-picking from a hat. We all wanted it, complete with Bill's personal interpretation of plan, elevation, and perspective views! He probably wore a pressman's cap, which he often made out of newspaper for special occasions. We know that Bill's lifelong interest in drawing continues whether on 8 1/2 x 11 inch paper, full scale on his studio wall as a guide to building his work, or as large, imaginative renderings based on completed pieces. In teaching Bill practiced what he espoused, by indirect influence as well as by assignment and program.

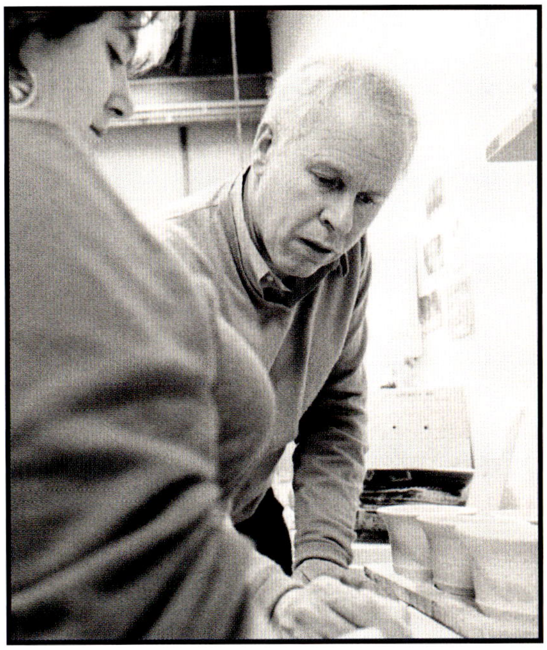

Photo: John Carlano

Student Marjorie Levy

"You all need to work more, you're not working hard enough, what's going on?" Well, what was going on, we figured out, was Bill was working more than his usual amount of constantly because of his upcoming show!
— Marian Pritchard

Junior year started out with "Wake-up with Bill and Petras." At 8:30 a.m. sharp, in a room seemingly less dark after a bit of caffeine, Bill showed us images from his extensive slide collection. He did most of the talking, and Petras laughed a lot, as we did also. Bill's musings often conveyed the essence of a piece by comparing it with other things. I cannot begin to capture the language and imagery, but a comment might have been, "This looks like two big frogs' bellies back to back with flying buttresses coming off the sides." We were used to abstractions and oblique references. Our first sophomore hand-building assignment had been to make "a bump." A what? "A bump. Any kind of bump you want." We'll try, Bill. During our half hour of wake-up we saw different techniques, clays, firing methods, etc., but mostly I remember an emphasis on form and an exposure to different artists, objects, and architectonic structures. The underlying message was, look at all these wondrous possibilities, which take in all of life.

After "Wake-up with Bill and Petras" we would work on assignments given to all the junior ceramic students. Petras, Bill, Liz Stewart, and Rob Forbes worked with us at various times over two days of class time. Teachers talked with individuals as each worked, and group critiques were supportive, but honest in their criticisms. The thrust of junior year seemed to be to wean us from assignment to personal development of our own work. Making a piece based on a historical object or completing a production project had educational validity, but by the second half of the year we had to address our own interests in independent projects. Faculty encouraged us to explore various technologies and forms for our development.

The exploration could be wide or narrow for different concerns. It did not matter. You just worked with several guides looking over your shoulder. Writing out our thoughts for our senior thesis by the end of junior year ultimately forced us to focus our intentions and interests.

Junior year was also the year Rob Forbes taught an informative glaze calculation class. We knew Bill appreciated but rejected the admired aesthetic of "Daddy Binns" and the Chinese Sung Dynasty.[2] Rudi Staffel's approach, when asked sophomore year about a glaze recipe, was to take off all numbers to the right of the decimal point, as most of the class cheered. But glaze calc was part of the program. We knew Bill had certain preferences and didn't use glaze himself, but we knew he valued its educational validity.

Senior year pushed us professionally, and all craft seniors attended a class in Studio Practices to help with the mechanics of being an artist. As part of the course each student had to visually present his or her thesis to the other senior majors. In the pot shop, each student edged toward the senior show with the ever-present sticks and carrots from five faculty – Bill, Petras, Rob, Liz, and a wonderful new teacher, Jeanne Jaffe. Bill's strengths included form analysis and an ability to articulate his response to what you were making as a way to help you improve your work and clarify your intent. Individual delight and angst waxed and waned as each of us tried to resolve work. Even Bill's buoyant nature could occasionally flag, as happened once junior year. "You all need to work more, you're not working hard enough, what's going on?" Well, what was going on, we figured out, was Bill was working more than his usual amount of constantly because of his upcoming show! His words and example propelled us forward. By senior year, we were intently pursuing the ideas we had chosen to resolve, trying to make something of quality, and keeping on course. It did not matter who was hand-building or throwing, making sculpture or pots, adding materials other than clay, opening or closing up space, or concentrating on form or surface, or both. We worked.

As end of school deadlines approached, our group of 12 fragmented a bit as some students pulled ahead, others lagged behind, and various personalities asserted themselves as workspace cubbies overflowed. Bill, the consummate steward, kept the whole afloat with his encouragement, constructive criticism, humor, and tireless leadership. To edge us more closely to senior shows, he gathered us all together for a pep talk. "Spring is beautiful this year (and it is in Philadelphia), the weather's gorgeous, but don't let up. Do your best... and for goodness sake's don't fall in love till the semester's over!" He wanted us to keep pushing our capabilities, and we understood that he supported his messages by example. He coupled a serious investigation in his own work with focused energy, persistent effort, and discipline, but most importantly with an ethic of caring for what he made. What you did mattered. What he did in his own work mattered. Not letting up mattered. The results mattered, but how one sustains one's work with intelligence, desire, capability, and intent really mattered.

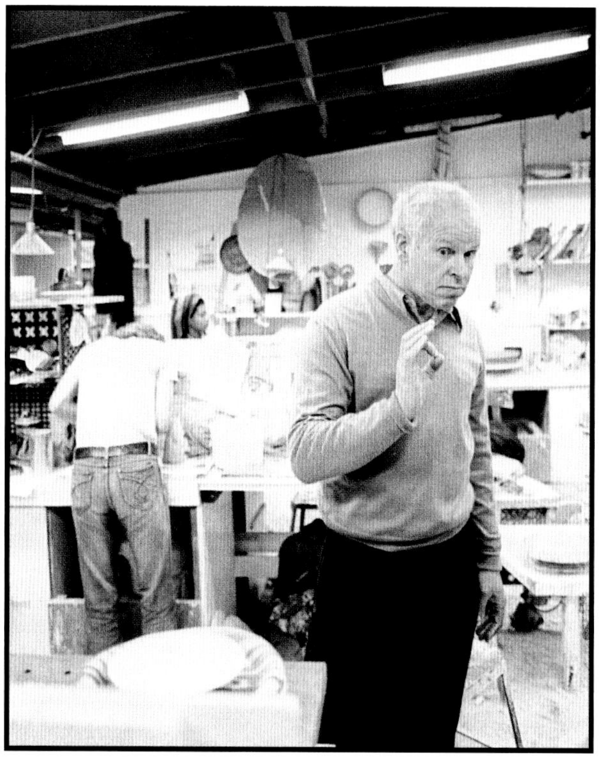
Photo: John Carlano

"Spring is beautiful this year (and it is in Philadelphia), the weather's gorgeous, but don't let up. Do your best... and for goodness sake's don't fall in love till the semester's over!" —William P. Daley

The teachers guided a student to convey his or her ideas in clay. Adopting an already defined philosophical bias, mimicking faculty work, following a trend set by other artists, borrowing without transforming an idea into your own was implicitly discouraged. I do not remember any insistence on following craft, borrowing from fine art, or debating the merits of craft vs. fine art. How you developed your own interests and quality in your work was paramount. Faculty members had various strengths in low- or high-fire, sculpture or function, planned or intuitive processes, concrete or abstract thinking. As they rotated into my cubby, I found that what one teacher said one day could be the complete opposite of what another said two days later. The conflicting suggestions turned out to be a truly positive experience because one felt compelled to reason out or leap into one's own decisions. By the end of senior year I didn't want cubby interruption despite the good intentions. I no longer tried this and that energetically but trusted my instincts more fully. In hindsight my education had concluded successfully. I had the tools to investigate, to focus, to follow through, and to resolve. I had a better understanding of my own aesthetic preferences. I learned that fumbling intuitively could complement rigorous analytical thinking and vice versa. The program had truly enriched me.

Bill and Petras were the cornerstones of the program. The fact that they were such close friends and caring people set a familial atmosphere for all. We went on field trips, had ceramic potlucks, parties, and dances. We had long "bonding" crits. By now, the spring was passing, most of us had fallen in love despite Bill's warning, and we were ready to graduate. When we gorged on our final potluck extravaganza, Bill wore his pressman's hat, and I think we each made and wore one too in solidarity and celebration.

Bill mentions how his Columbia University Teachers College graduate education, especially the work of John Dewey and Herbert Read, helped to form his pedagogy. His hard work and enjoyment in teaching ceramics undoubtedly accounts for most of his success. Bill was the catalyst for an amazing ceramics and craft program. From constructing work spaces, kilns, and wheels, hiring competent and varied faculty, to refining an innovative curriculum over the years, Bill brought his ebullience to a physical and psychologically open environment, called the Pot Shop. With self-effacing wit, he used his broad knowledge judiciously to the benefit of the individual. To this day, when he meets former students or friends, he continues an educational dialogue, sharing his love of making, looking, and reading, and discussing others' concerns while encouraging and praising their contributions.

In school there was an undercurrent of the limitless that surrounded Bill. Being near Bill was being near an embodiment of artistic achievement and celebratory vitality. His pots were audacious and unique, his attitude was open, his exhilaration in learning new things always evident. He taught not only through demonstration and dialogue, but also by exemplifying a path taken by one good artist and one energetic individual. The program provided a wonderful education, but more importantly it introduced one to an enriching professional life in the ceramic arts. He continues to inspire others through his pots and through his joy for life in the arts. More than a teacher extraordinaire, an artist extraordinaire, or a wonderful man, he is a manifestation of fully realized human potential in itself.

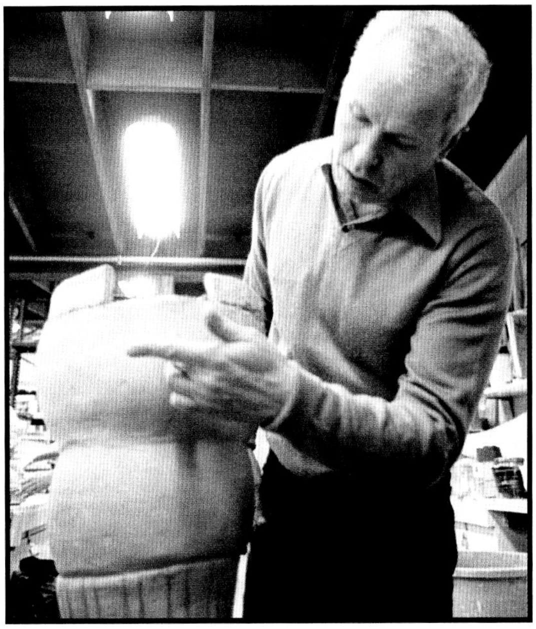

Photo: John Carlano

Marian Pritchard is a Philadelphia ceramic artist who is represented in private and public collections, including the Museum of Fine Arts, Boston. Her exhibitions include major juried shows at the Cleveland Museum of Art and the Philadelphia Museum of Art. She has taught craft history at several city art schools and ceramics at Community College of Philadelphia.

1. Information in this article, especially on the ceramic and craft program before I was a student, was gleaned from phone conversations with Bill Daley on 5/10/2010 and 1/18/2011.
2. Charles Fergus Binns, Founding Director of the New York State School of Clay-Working and Ceramics in Alfred, New York, helped define a pottery aesthetic in the 1950s which valued a simple high-fired pot with a beautiful glaze, inspired mostly by Chinese Sung pottery. (http://ceramics museum.alfred.edu and Lecture by Val Cushing, 4/24/2010, Sub Octo Gallery, Philadelphia.)

Civic Dimensions in Postwar American Ceramic Education

Ezra Shales

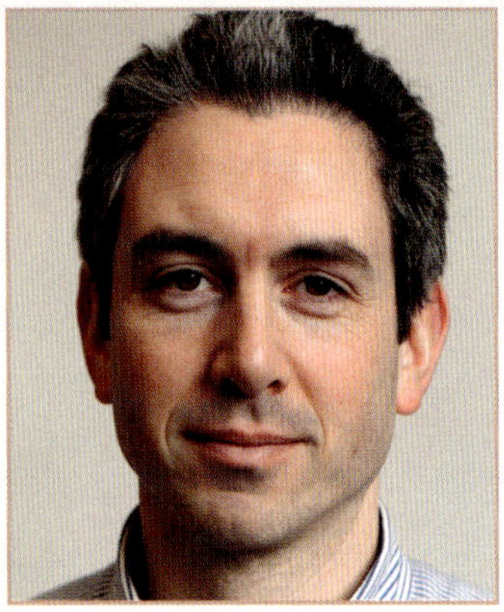

Ezra Shales

Immediately after World War II, William Daley and William Parry shed their fatigues to bound into higher education with immense optimism, valuing school for its "unique ability to serve the talent of the individual."[1] They could have uttered these words of Louis Kahn, architect and fellow Philadelphian, for they entered teaching "Not just to impart knowledge but to serve the talent of the individual." If Kahn asked a brick what it wanted to become, Daley and Parry asked clay what it might imagine. They intended to be artists scaling the heights of culture and also "thinking hands" plumbing the local earth. If that mix of piety and gusto seems contradictory, such was the wellspring of optimism and pragmatism with which G.I.'s reinvented American academia.

Daley and Parry met and collaborated intensely between 1957 and 1963, when their tenures overlapped at the Philadelphia Museum College of Art (now University of the Arts). Their devotion to clay as an expressive medium does not fit into usual story lines of total artistic liberation that permeate postwar art history. Ceramics is often situated as an alternative artistic discourse, with M.C. Richards's *Centering* (1962) insisting on a retreat from society and institutions into soulful-development on the wheel, and Peter Voulkos and the west coast blowing apart the constrictions of tradition. But Daley and Parry had a sense of art as service, and, perhaps because of their time in the service, they saw this in ethical, not utilitarian, terms. "Every time you get freedom, you must remember the other half of it is responsibility," Daley wrote.[2] "The good students use this responsibility to do something good."

Daley and Parry shared admiration for Paul Tillich's *Courage To Be* (1952), and the theological text is a lens through which to perceive clayworking as an avenue toward self-realization. Tillich's observation that "courage must be considered ontologically in order to be understood ethically" they adapted to clay: the *material* encounter between hand and clay was about being and becoming.[3] Their sense of "mindful possibilities" arced across art, craft, and design, even as many perceived these curricula as divergent. They were not seeing art in terms of Kantian philosophical lines of disinterestedness in social terms, what Tillich called a "ground of being." In this brief essay, I hope to make room for a quieter pedagogy of pragmatism, social engagement, and material commitment espoused by these two artists, with the belief that their great integration of possibilities holds equal allure and relevance today.

To their students, they communicated a commitment to self-determination and a faith in the possibilities of self-realization. It was their focus on possibility more than probability that their students remember. Dewey's injunction in *Art as Experience* (1934) that "intelligent moral action leads to self-realization" laid a foundation for their search for the unprecedented in art.[4] In 1963, Daley asked students to probe the enigmatic issue of freedom in artistic inquiry: "how much predictable performance is necessary? When is it necessary to practice procedure? How much exploration and experimentation, or how much search and how much research?"[5] The differentiation between search and research is pivotal: Parry and Daley shared the conviction that self-guided inquiry comprised a *bona fide* art education.

The two are best seen as cooperative contrarians. Material exploration has always sat uneasily between the humanist and vocational poles of academe. Although the rise of unconventional American pedagogical exploration is usually traced to the Bauhaus, Louis Kahn credited his modernist grasp of materiality and autonomy to his high school education in Philadelphia, where the distinction between humanist inquiry and manual training was self-evident.[6] He studied at the Public Industrial Art School (1880-1916) which championed a Progressive holistic process of "thinking through hands." Students in the

"Every time you get freedom, you must remember the other half of it is responsibility. The good students use this responsibility to do something good." — William P. Daley

Northeast Manual Training School (established in 1890) built ingenious mechanical contrivances in preparation for manufacturing work real and imaginary. They mastered lathes and precision milling, and archival images of student work document personalized slide rules. In Girard College (established in 1848 for "poor, white, male orphans"), handicraft built character. Parry attended Girard, and he and Daley experienced enough varieties of art education to value an enlightened individualist approach and to foster the intersection of aesthetic, emotional and intellectual concerns. Like Kahn, they brought a lack of pretension into higher education, in part because of their grasp of primary-school art pedagogy and their wartime service.

Having pole-vaulted from thinking about their parents' lives as housepainters, bobbin-tenders, and coal miners, they embraced the academy's intellectual and professional arenas while still valuing manual agency. Labor was a necessary part of life. Daley's letters to Parry suggest that repairing clapboards and painting the porch were moments when transcendent thoughts were on the move as he whistled, remembered his father's recitals of poetry from memory, and ducked about in his restless mind. Once they left the Depression years of their childhood, they still made operable things and thought of a world in terms of mindful production. Mud was appealing to build social relations in the world. Ceramics and glass practitioners have the distinct habit of living with their work, eating out of their handicraft. Hanging planters were both expressions of lifestyle and avenues for mindful invention. Making a range of things, they were supported by Dewey's "refusal of any sharp distinction between the fine arts and the arts of use, the insistence on ordinariness of aesthetic experience, the emphasis on community and democratic participation and the stress on the cultural dimensions of democracy and the organic relationship between culture and social environment."[7] In 1953, Daley took his students in Iowa to retrieve clay but took a detour to watch an "old-timer" hand-cutting leather hard clay sewage pipe into complex elbows and Ts: this was art made artlessly.

Seven years older, Parry experienced both pre-war vocational training and more liberal post-war schooling. He studied ceramic engineering first at Alfred University between 1938 and 1941 before serving as an Army officer in chemical warfare. Alfred was a preeminent technical school. After the war he returned to study "industrial arts" and "industrial design" there.[8] Alfred was a unique situation because it sought to teach glaze calculation and designing for mass industry as well as hand building and wheel work. Unlike later graduates who might have elected to study solely art or design, Parry applied his technical wizardry to investigating self-glazing clay bodies. He worked on a glaze for Eva Zeisel in the 1940s and holds a patent for whiteware. Parry's teacher and friend, Charles Harder celebrated this juggling of science, art and industry. In contrast, Daley began as a painter at the Massachusetts College of Art, finding clay through a teacher, Charles Abbott.

As teachers, Herbert Read's *Education through Art* informed their fear of school as "moral despotism," "the tendency of one man to make another just like himself."[9] They neither built aesthetic dogma nor groomed acolytes. When schools achieve a perceptible style, they are often considered successful, but these pragmatic humanists prized unorthodox inquiry. No similar-looking ceramicists emerged from their classes. Both had ambitious and successful students, all of whom eagerly shower praise upon their teachers. Daley's students remember him on fire, gesturing excitedly and leaving their ambitions stoked, their eyes set on a course of action. Parry's recall his gentle demeanor that instilled confidence.

They were motivated by the belief that art would be essential to build a new postwar world. In 1949, Alan Kaprow, who later invented Happenings, read *Art as Experience* while an art student and valued Dewey's aim to make art less estranged from everyday life. "Art not separate from experience," Kaprow wrote in the margins, "what is an authentic experience?... environment is a process of interaction."[10] Kaprow began to work outside conventions, to assemble blocks of ice and watch them melt. Daley and Parry embraced everyday life by digging clay and finding an enduring metaphorical and physical engagement. While we expect clay to be soft under our thumbs, its resilience is as awesome. Imagine Daley and Parry in an abandoned pit in Burlington or Elm, New Jersey, shoveling pounds of resistant clay into their truck. Their earthwork was both a progressive and recuperative act. The clay beds dated back to the Erie Canal, America's first wave of industrialization. The digging was a liberation, an expression of self-reliance. "Marvelously full of iron pyrites" recalls Daley, describing that mud bank with a sense of the earthly joy that comes from the doing, the making, the being, and the sharing. Material investigation blurred art into life; clay took hold of them and they embraced its struggle.

In 1964, Parry lent Daley a copy of Henri Focillon's *La Vie des Formes* (1934), and they both found his "Praise of Hands" to integrate matter and spirit and articulate their own sense of the catalytic dimensions of material exploration and haptic learning.

Sight slips over the surface of the universe. The hand knows that an object has physical bulk, that it is smooth or rough, that it is not soldered to heaven or earth from which it appears to be inseparable.... Surface, volume, density and weight are not optical phenomena. The sense of touch fills nature with mysterious forces.[11]

— Henri Focillon, *La Vie des Formes* (1934)

The two ceramicists manually inhabited Celtic interlace and geometric structures as material-spiritual intersections. Both Daley and Parry made students feel their way through numeric sequences and materials simultaneously. Plaster was about flipping insides and outsides, not about mindless replication. As a teacher in the Industrial Design Department, Daley had a reputation for giving "far out" problems that demanded designers "think like artists": make "nine extraordinary examples of exaggeration."[12]

Visiting the Swiss artist and designer Max Bill's exhibition at the Philadelphia Art Alliance, they derived immense satisfaction from each other's company as they puzzled over the metaphysical meanings of the tension between interior and exterior forms. Revelatory discussions ensued about ways they might define structure in terms of tubes instead of walls. Tubes, like pots, have indivisible interiors and exteriors. To a twenty-first century eye, the Swiss sculptures might appear characteristic of formalist abstraction, but the salient point is that Daley and Parry perceived a common-traveler. The idea of a permeable membrane connected the bronzes to their own searches.

Parry's 1967 "Manifesto" (following page) – which Daley treasured – argued that "Success has to be looked upon as partially possible, rather than as total and confirmable." A high regard for experimentation and self-expression motivated their work as artists and educators. The "Manifesto" condemned the "beautiful employment of craft for recognized ends... that is dependent upon example and prototype" because it was "research," not "searching." He characterized "the search" as finding "fresh means to support a partially sensed idea without being stifled by what one does not yet know or cannot do with skill."

Perhaps it is apt to wonder if such civic materialism has a distinct vernacular identity in American modernism. Yes, Bauhauslers brought much to America. But Louis Kahn often demonstrated drawing exercises to build ambidexterity that he had first encountered at the Public Industrial Art School in 1912-14. Turning his back to his audience with chalk in each hand and inscribing spiral forms, he tried to communicate what drawing could be when a somatic and psychic journey. Bill Daley and Bill Parry undertook a similar endeavor to relate social, emotional, and aesthetic dimensions, to engage the body, hand, and mind deliberately with material. They considered pots to have a theological gravity and "psychic utility."[13] One of Parry's former colleagues at Alfred still treasures a gift from him, one of his scroll drawings over eighty feet long that delineates the curvature of the earth's surface. It speaks of the bigger context in which we live, that the permeable membrane on which we walk is both an interior and exterior. A year ago, I watched Daley invite a group of college students to drum on his amphora in a gallery and witnessed his pleasure in making the inside and outside vibrate and reverberate, and I thought that was how he taught. The sight recalled his admission that he lost a job teaching Sunday school decades earlier because his similar kinesthetic acts were considered too unruly for a sanctuary. As teachers, they had high-minded expectations that their students build a better world, not just make art. As sculptors they built metaphysical and literal cisterns to drink from, spoons to ladle out the mind's deepest thoughts. Life-affirmative construction was an imperative and materiality a demonstrable ethic. Joy was a clay coil entreated to gain structural integrity through a rhythmic kneading, and standing up on its own legs.

Ezra Shales, Associate Professor at Massachusetts College of Art and Design, focuses on the intersection of design, craft, and art. His book, *Made in Newark* (Rutgers University Press, 2010), explores handicraft as an anchor of regional identity in Progressive-era New Jersey. He has a Ph.D. from the Bard Graduate Center and an M.F.A. from Hunter College and is currently working on a book about communal drinking vessels in American history.

1. Joseph A. Burton, "The Aesthetic Education of Louis I. Kahn, 1912-1924," *Perspecta*, Vol. 28 (1997), p. 209.
2. "Comments on the Color and Design Class" *NOUS* (April 25 1966).
3. Paul Tillich, *Courage To Be* (New Haven: Yale University Press, 1952), p. 1.
4. Andrew Feffer, *Chicago Pragmatists and American Progressivism* (1993), p. 20.
5. "Artist + Teacher" *School Arts Magazine* 3 (1963)
6. Burton,"The Aesthetic Education of Louis I. Kahn."
7. Andrew Hemingway, *Artists on the Left* (New Haven: Yale University Press, 2002), p. 152.
8. Melvin H. Bernstein, *Art and Design at Alfred* (Cranbury, NJ: Associated University Presses, 1986), v121.
9. Herbert Read, *Education through Art* 2nd ed. (1943; New York: Pantheon, 1945), p. 259. On Daley's interest in Read see mss March 21, 1993, submitted to American Craft for its 50th Anniversary issue, "The Sure Companionship of Books."
10. Allan Kaprow, *Essays on the Blurring of Art and Life*, ed. Jeff Kelley (1993; rev ed. Berkeley: University of California Press, 2003).
11. Henri Focillon, *The Life of Forms in Art*, trans. C.B. Hogan and G. Kubler (London: Zone Books, 1989), pp. 162-163.
12. Marlise Mason, "Comments on the Color and Design Class" *NOUS* (April 25 1966).
13. William Daley, Juror's Statement "Strictly Functional Pottery National 1994," typed mss.

1967

A Manifesto

William Parry

My concern is for the purely expressive use of materials and with the question of motivations in making sculpture.

While the process demands experience and control in the use of clay as well as other materials, it is separate from the functional problem-solving necessary for the making of well-designed objects. The very skills and analytical patterns of thought required for the success of the latter are encumbrances in; the search to find the self-available means to materialize ideas in a form statement that is essentially expressive.

One has to proceed to find fresh means to support a partially sensed idea without being stifled by what one does not yet know or cannot do with skill.

The beautiful employment of craft for recognized ends is something else and while it is greatly to be admired, it is irrelevant and must be set aside in favor of the other. The beautiful employment of craft is dependent upon example and prototype – this search has none. The one is a continuum; the other is not.

Where education is involved, a method as obscure as this has to be protected and part of that protection has to be exposure to points of view equally strongly held in opposition in other subject areas where they occur naturally. Success has to be looked upon as partially possible, rather than as total and confirmable. There is no proof of it. When judgments of craft insert themselves, it is due to lack of success in the expressive force of the work, and thus it can be seen that the most compelling works of expression are artless.

William D. Parry
1967

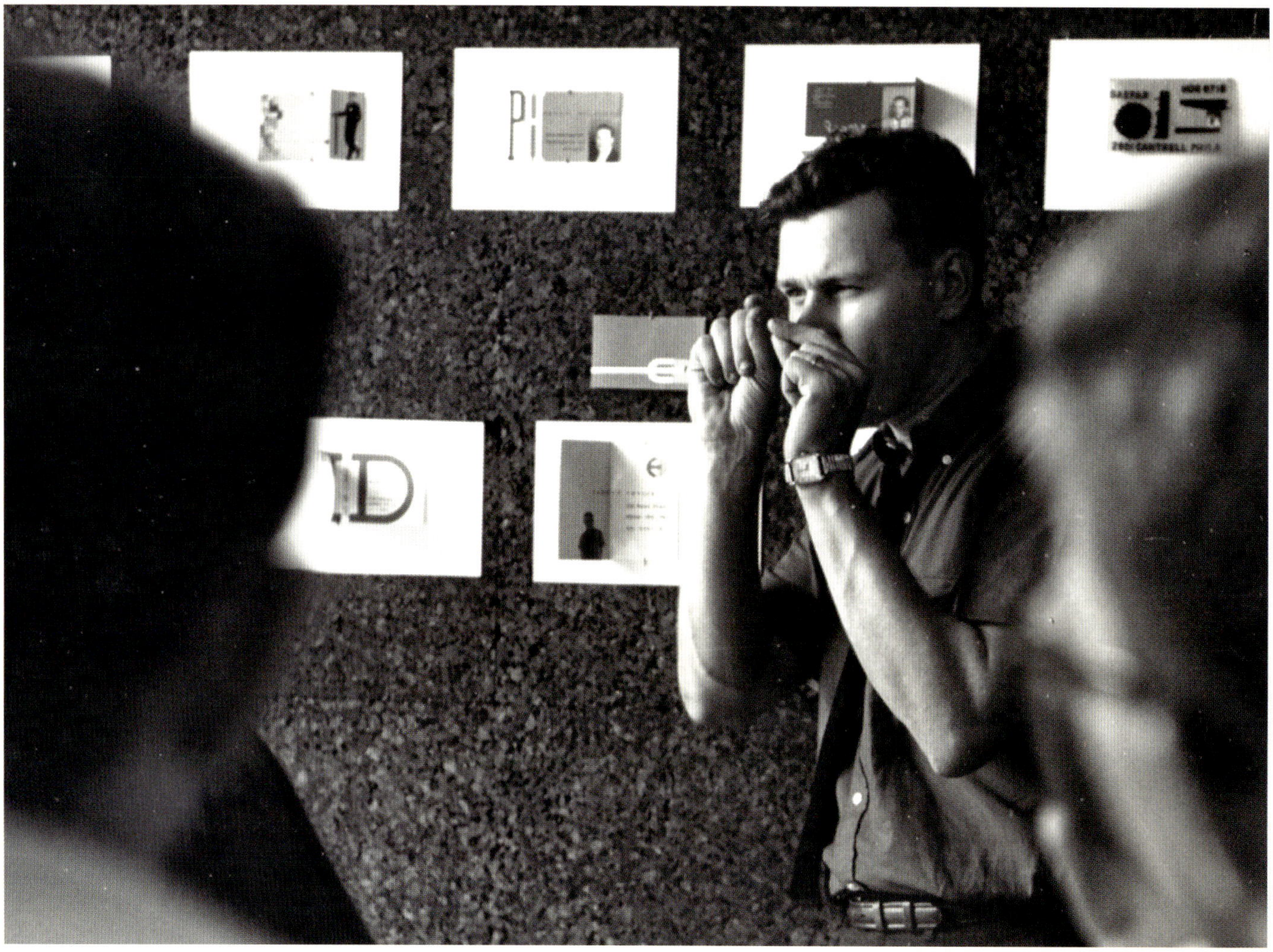

Photos: Murray Weiss

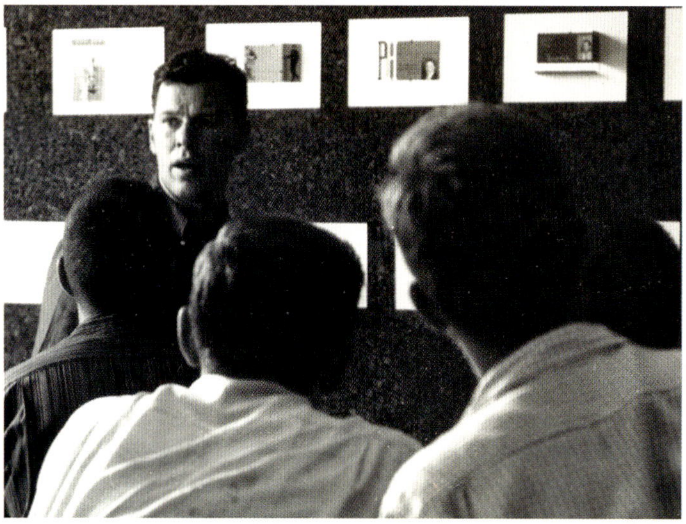

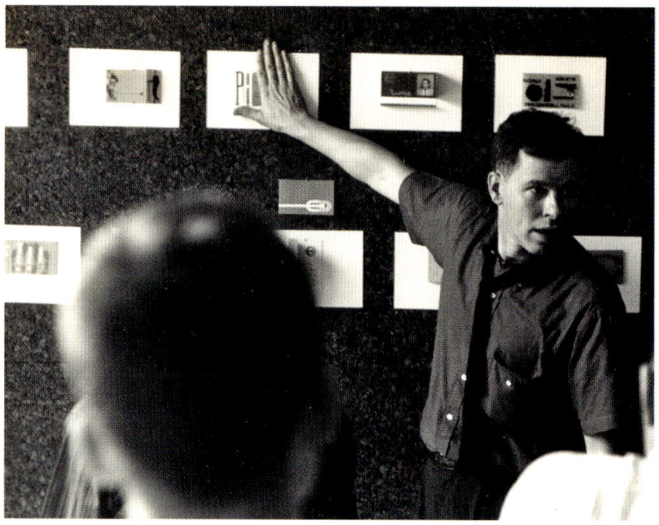

Bill Daley teaching in industrial Design Department,
Philadelphia College of Art (University of the Arts)

Artist + Teacher

William P. Daley, 1963
editor John Cataldo

I have been teaching long enough to find myself supporting positions which I criticized ten years ago. More often, I find that the issues seem less clear and my responses more tentative and tolerant. I have come to believe that the art of teaching art requires one to perform with assurance while dimly perceiving a paradox which makes certainty short-lived.

This paradox is the need for all creative people to put themselves in jeopardy by striving to evolve a form language which will communicate something they do not yet know. The creative person is concerned with transcendence, or "going beyond what is given or presented by experience." This basic characteristic of an artist or teacher presents a contradiction in teaching which I find difficult to resolve. The contradiction is that the artist needs to know so much which is both necessary and useless. His knowledge of process and technique and his understanding of history and his time are essential. This knowledge helps him to define himself and clarify what he will not or cannot do. Sometimes it helps to establish an eclectic beginning point, but it does not declare what he will do because what he already knows is not worth doing. In a sense, he must know a great deal which must be forgotten in order to begin. The teacher then must teach what has to be known while encouraging the student to transcend his experience.

In our efforts to establish this equilibrium, we are faced with questions of how much, how long, and when? For example, how much predictable performance is necessary? When is it necessary to practice procedure? How much exploration and experimentation, or how much search and how much research? In short, how much order and how much disorder—with its consequent risk for the teacher and the student?

Ascribing values to these questions is an individual matter which varies with each student and teacher. I do now know the answers, but I do have a bias which I honor in working toward a tenuous equilibrium. I try to seek the degree of risk which permits me enough confidence to proceed while I do not know what I am doing. As a teacher, I attempt to develop an awareness of this position as I work with my students. This attitude rests on a person's opinion of himself. He must have a capability in which he can believe. Such a capability is the result of successfully resolving the conditions of previous risks. It is this believed capability that gives a person the confidence to proceed without knowing the outcome of his endeavors. Successfully resolving the conditions of previous risks seems to me to be largely dependent on the ability to be speculative and to seek alternative options in evolving a workable strategy or means of proceeding. We have all witnessed the disastrous results of developing a believed capability by a series of ready-make triumphs when a person is called upon to perform without a precedent. This often occurs when skill and process become ends and eliminate speculation by developing the false idea that there is a right way to do everything.

I attempt to develop the student's ability to see possibilities by making my requests as open as I can to permit the greatest number of strategies. Two which worked well in a design class were "surprise me" and "make me laugh." This involves risk for the teacher because he is forced to rationalize the student's intent on the spot, which increases the opportunities to be wrong in the same ratio that each student faced in qualifying the open problem. I attempt to evolve with the student as many variations and alternative strategies as possible after he has concluded his work.

I attempt to develop the student's ability to see possibilities by making my requests as open as I can to permit the greatest number of strategies. Two which worked well in a design class were "surprise me" and "make me laugh." — William Daley

If a student is not capable of performing in an open situation, I make my requests more specific by increasing the number of "givens" until the student feels confident enough with the limitations to proceed. When a student's resolution of a problem is inadequate for technical reasons, I attempt to work out with him technical procedures which are in keeping with the intention of his work. The development of skills takes time. I find it is the easiest necessity to have the student understand and accept. I also find it is the part of process which can be criticized easily without damaging his belief in his capability. Mastery to me is the clear resolution of an idea. I attempt to emphasize the imaginative quality of the idea and expect its clear exploitation in time.

Those of us who want our students to accept jeopardy by going beyond should be involved in seeking jeopardy for ourselves as teachers. Like students, we must learn to live with our failures and not fall in love for too long with our successes.

Thoughts for Spring Graduation

William Daley

Many thanks to President Ian Clark for inviting me to speak to you at this graduation. It's my first time in the role of the "Emeritus, Old One with words," Actually I feel like the graduates, joyously nervous. I say this having had two graduations from art school: my first was in 1950 and my second last year from an art school as marvelous as this one.

As a co-conspirator, teacher, and artist over that time, I have heard forty graduation addresses. I really know why Emeritus means, "to serve out one's term." Quite often these speeches were variations on the "Travail to Triumph" story. The ones I liked best usually began with recollections of walking barefoot five miles to school. Maybe in a blizzard carrying a great wet oil painting. (hope for laughter)

It pleases me to hear you laugh; it makes me less fearful and expresses more powerfully than words that this occasion is first and foremost a celebration. Your laughter crystallizes here the dreams, love, and labor of the graduates, their mentors and families. This is a time for savoring the pleasures of shared thanks and approval. It is a time to acknowledge genius in passage, graduates, teachers, parents, administrators, and speechmakers. I am honored to confirm your laughter and congratulate all assembled here for your parts in the creation of this glorious day.

As an Insider from Outside, I am impressed by the education you have started here. Art schools are one of the two things I know about. I know you are well equipped to begin the careers you have been preparing for. You can use tools well, organize tasks, research the past, manipulate precedent, solve problems, and clearly present your findings. In the other half of that loop you can compare yourself with your peers, present yourself with assurance, make contracts, keep accounts, and work to prosper.

I have a high respect for competence in becoming a true professional. But I would like to describe another loop inside the one I have just described. This second loop as inner circles which arts education has special capacities and traditions for manifesting. To me it's the thing that makes us special, and I believe, provocatively useful for the future. If the first moment in art is to be alive and in love with consciousness, the second has to be to prove it. I believe artists are among those committed to making this connection. Making it by discovering the shape of themselves and finding their unique correspondence with everything in their lives, their homes, their country, the world, and the cosmos. My name for seeking this connection is to be a person in search of significant form. Significant form is always moving toward creating community and artists become masters of these exchanges.

To become a true professional, all the skills I mentioned in the first loop are of little or no real use unless they are grounded by your inner life to nature and humanity. This is the core of the second loop and assures that as professionals you are nurtured by a sense of wonder for how grand we are. Such a conceit has to be coupled with a deep regard for yourself, all people, and all life. I am certain that one reason all good artists are directed from inside as professionals is because of their long study in art schools. There they have learned to seek through materials the discovery of self in community. By becoming masters of form, artists are among the first to discover the salient exchanges that reveal the connections between their temperament, probability, and the shape of the future. In locating these points of transformation, they also dramatically create the new signs, symbols, and images that shape our emerging consciousness.

Artists of all kinds, poets, musicians, painters, and potters, evoke our communal consciousness by looking inward to fix through form, spirit and matter. The artist's mind is directed by practice to know how to manipulate the quickening power of light, sound, touch and time to recognize through material the promise in possibilities. In this process, good artists become aware of unthought connections toward richer affirmations for all of us. Artists become masters at adjusting stuff, materials, tools, processes, and structures to shape feeling, intuitions, findings, and dreams to create what I call significant form. As artists we live by inhabiting *form* words like informer, performer, transformer, conformer—all the form words that together elaborate and describe becoming informed. We as artists are primary models and experts in all skills that form words imply. We know all about how to adjust the parts to become wholes. We know how to recognize the cracks that become openings. We know through making that what is vital is always efficient, economical, and intelligent. All intelligent artists know that most often the way becomes the destination. We grow to know that expedience is loss and shortcuts are illusions. This litany could become a book. Let me just say that *artists are the keepers* of becoming connections. They are the exemplars empowered to amplify the exchanges needed to make our children's livable future.

Somehow there never is a good time to graduate. As artists we are experts in knowing the fear of not knowing what will happen next. We are also trustful about our abilities to work it out even in "hard times." As a society we are capable of anything. This is not a time to be apocalyptic but a quick list of our impending "hard times" might go this way: crack, AIDS, racism, sexism, endangered species, global warming, pollution, nuclear waste, hydrocarbons, non-biodegradable garbage. A final shameful descriptor of our dysfunctions as a technocratic society is our recent triumph in turning annihilation into a Nintendo game. A television mega-series hosted by a general in creased combat fatigues trimmed with yellow ribbons.

I believe the recognition that we are lost is clear. I believe that the ways to build a real "New World Order" are known. I believe we, as artists are the vanguard called to amplify the means and teach the connections. Connections that develop the communal exchanges can assure a livable future. A future in which we are globally informed; a future in which professional action emerges from the inside loop and integrates our social and spiritual capacities to survive and prosper.

In conclusion, let me tell you a story. When I was thirteen I asked my parents if I could sail in the Hudson River with my friend and fellow paperboy, Bill Shields. He had just finished building a very small sailboat in his cellar. My parents feared the river; our neighbor's husband had drowned in it. After two nights of Irish parental night talk, my mother gave the answer. She said, "Billy, if you are born to be hung, you will never be drowned, so sail." It is fifty and more years since that permission. At the time, I was jubilant that it meant YES. I am still thinking about what else it means. It has been in my mind when unknown changes and corrosive fears were part of my "hard times." Every crisis has precipitated useful meanings from this enigmatic dark and celebratory "Irish" permission. I know I am on the edge of one of those "Travail to Triumph" stories, so, as the "Invited Old One with Words," I give this permission to you as a present. It's not much, but it's the best I can do to help you work through despair, to hope, and go for it. Perhaps you will remember it at a later time and it will be useful. I am still finding it so. I have bounced it off what YES means; it has been the Plimsol for lots of speculation about alternatives. Now I spend a lot of my time thinking about what it has to do with probability and what the odds are for living forever.

In closing, let me read the beginning of the Aztec poem I have given you. It says what I have been trying to say another way, much more beautifully than I ever could.

The artist: disciple, abundant, multiple, restless.
The true artist: capable, practicing, skillful;
Maintains dialogue with his heart,
meets things with his mind.
The true artist: draws out all from his heart,
Works with delight, makes things with calm,
with sagacity,
Works like a true Toltec, composes his objects,
works dexterously, Invents;
Arranges materials, adorns them,
makes them adjust. *

It will please me if I am around when your genius has its turn. Enjoy the days, save that soaped Brillo pad and remember, "If you are born to be hung, you will never be drowned."

* Aztec Poem: English version by Denise Ledertov, *Shaking the Pumpkin*, edited by Jerome Rothenberg, Doubleday & Company, Inc.

Thoughts on "Craft and Learning"

William Daley

This session on "craft and learning" is really exciting for me because we have here diverse faculty representing different kinds of education and experience. In Max Frohlich, we have an octogenarian master metal smith who graduated from a Swiss art academy before World War II. In Lino Tagliapietra we have a master glassblower from Murano who learned all his moves in Italian factories from the age of twelve. Two "old art school" products are represented in Audrey Walker and myself—she English, I American. The rest of the faculty represents present college and university (M.F.A.) practice at its best: Judith Salomon, Wendy Maruyama, Margaret Sahlstrand, and Susan Wilchins.

We are all players who make up the history of craft education—the Apprenticeship, the Academy, the Art School, and the University. The first four of us have lived through a gradual metamorphosis in crafts education, which began shortly after World War II. Now we see the granting of degrees instead of diplomas, and the combining of vocational, professional, and academic aspects in arts curricula. What really is wonderful is that places like Haystack, and all the art centers and continuing education programs in colleges and universities, didn't even exist when we four were making our starts.

When I think of the cooperative studios and residency programs, places like the Clay Studio and Lill Street that have sprung up over the past ten years, I know that the crafts are in the middle of a grand transformation. A new kind of craft and learning has emerged, one that has resulted in a range of approaches that have kept the best of professional education. We are seeing a shift from the rather narrow vocational/professional/elitist focus to a sort of democratizing self-expression. Access to learning has broadened immensely, and our old ways of doing things have been changed in the process. We now have greater mastery over materials; we can make them behave in new and exciting ways; and we have more master crafts people than ever before in history.

There are many reasons for these changes and combinings. I suppose most of the changes have to do with our historic commitment to public education and a deep belief in the need for a free people to form spirit through art. What a fascinating thing it would be for the founders of the Liberty Tadd School in Philadelphia, for example, if they were able to travel from 1900 to the present to sit in on this session. They would marvel at the ways all levels of knowing have been advanced since the first public school program dedicated to learning through "thinking" hands was formed. I suspect they would be most surprised by the informality of present day teaching, and by the changed role of the teacher from that of authority and master of the "right way"

I suppose most of the changes have to do with our historic commitment to public education and a deep belief in the need for a free people to form spirit through art. — William Daley

to that of accomplished friend, one with information to share about the many ways of doing. The best teachers in craft are those who focus on shared possibilities, rather than the "right answers".

Our time travelers would experience another revelation about access and choice for individual students. As a student in 1946, I made two choices: One was to go to art school; the second was to choose a major—everything after that was someone else's idea, from how to sharpen a pencil to what was beautiful and why. I am convinced that seeing ten students in a Haystack studio all successfully using different media and techniques of their choice for totally different expressive reasons would be enormously satisfying for those founders.

Attending this session's evening slide presentation would provide another revelation. For me, seeing the slides of work by Audrey Walker's students from the experimental fibers program that she helped found at London's Goldsmith College was thrilling. That extraordinary work really came out of open and experimental ways of teaching craft that are very much a part of how the best teaching takes place at Haystack. The thirty-five millimeter slide and the photographic copy process have made possible profound visual exchanges in our personal vision. Our founders would discover a time when shared visual thought is taking precedence over written information. I sometimes wonder how the videocassette recorder and the television camera will reshape our access to, and the sharing of, personal work in the future. I imagine that, for those who come after us, it might be as great a change as is looking back to sepia photos of masterworks mounted on cards for us. (I never saw a slide of my teachers' work, or that of anyone else as a student back in the '40s.) Thinking about other technologies today, most teachers are teaching the use of tools, processes, and techniques that did not exist when they themselves were students. Cultural exchanges of information about the how, the what, and the why of making things will continue to make the exchanges in teaching a more mutual enterprise.

In some ways we are all time travelers in relation to the future. For me, this first year of my retirement is like a year of Januarys—I keep looking back and ahead to try and understand the present, and to wonder about the future. Seeing how vital craft and learning are across this country is exhilarating; I can see that we are in the middle of a marvelous arts education revolution. The crafts in programs at most levels have made grand strides, giving shape to one of my old dreams—that everyone can become an artist. I feel certain it's possible; actually getting it to happen will be different than anything I can imagine. I know there will be obstacles. At times, I worry about what they might be, and sometimes I fear the most ground may be lost if we over-enjoy our success. For instance, I wonder if arts education, particularly learning through crafts, is as broad and accessible as it should be. If it is expansive enough to prosper in the future. Does crafts education have its hand out far enough to serve?

True, we have a super system for educating professional craftsmen and those who are already interested. While we have given our energy to building a new kind of college/university professional craft education that is coupled with life-long education in exciting ways, we have not expended the same energy on primary and secondary education. What are we doing for the young ones—children in our public school system—who don't know how rewarding it is to "make" things? For the most part their programs are bad, often little more than copying and imitative drawing, but rarely "building" through craft media.

What are we doing for the young ones—children in our public school system—who don't know how rewarding it is to "make" things? For the most part their programs are bad, often little more than copying and imitative drawing, but rarely "building" through craft media. — William Daley

This year I helped jury a regional scholastic high school art competition, and later I saw the national show. It was ninety-eight percent drawing, painting, and graphics, and just two percent sculpture and crafts. That two percent included ceramics, jewelry, some wood carving, and, with a few exceptions, it all was far below standards. For thirty years, I have observed the same thing reviewing student portfolios at art school career days. So I think our first priority should be to find ways to improve what is offered in junior and senior high school programs. What once was the hands-on approach to teaching industrial arts, using real materials for vocational reasons, now should become a function of the total arts curriculum—working real materials to develop cognitive skills through making, conceptual thinking and problem solving. I am certain that creating objects for personal use with primary materials—clay, wood, fibers, metals, etc.—will encourage problem solving by teaching students to understand the nature of being responsible for making something from beginning to end, and will serve as a link between creative cognitive and social development. One way to make this happen might be to have every school that educates professional craftsmen to include volunteer teaching service in public schools as part of their curricula, along with all their seminars on marketing and portfolio preparation. Upper level teachers should be used as lecturers and workshop leaders for high school students on their turf. Art centers in communities could also serve as the means for connecting artists with these students.

Plans to add more history, appreciation, and aesthetics to visual arts education curricula generally are a clear indicator that crafts are not being emphasized. For the most part, craft and learning is not being employed to confirm the ways that "thinking hands" can mediate the cognitive skills of modeling for young people in real situations. All theoretical endeavors only become "real" as they acquire form. Form is more than idea; it is idea tempered by possibilities and limits—the limits of material, structure and the passion of avid participants. The syntax of verbal, written, and graphic means to inform ideas all have temporal and spatial constraints that often make them schematic and confounding.

I worry that, as we have more and more highly abstract ways of knowing reality, we will place a lesser value on the primary ones. I think this is beginning to happen in some schools where arts programs are only seen as frills and tight budgets are using limited funds for computers and science programs. The opposite must happen if we are to advance.

As we learn to use computers and their modeling potential we will need to confirm our judgment by developing discernment and value through making. None of these cognitive abilities should be left to the science of robotics. The purpose of art is to enhance consciousness by urging individuals to give form to spirit. This means that in living we all become artists—masters at adjusting stuff, materials, tools, processes, and structures to shape feelings, intuitions, facts, findings, and dreams. By becoming artists we will know how to recognize what is imminent. We will then know how to adjust parts to become wholes. We will know through cycles of making the ways to recognize what is authentic, intelligent, and essential in community.

Many of the future ecological and global problems we face will diminish only if we reinforce primary ways of doing things, growing things, and attending to our inner and outer near space while maintaining strong connections to the earth, its materials, and our traditions of making.

Phillips Exeter Academy's Lamont Gallery Presents "William Daley – Selected Works: 1967–2010"

Sara Zela

Phillips Exeter Academy presented an exhibit by William Daley, a leading figure in American studio ceramics. The exhibit entitled *"William Daley – Selected Works: 1967–2010"* included a selection of hand-built, unglazed stoneware vessels and drawings. An artist's reception was held at the gallery on Friday, September 16, [2011] 6:30-8 p.m., and a gallery talk took place on Saturday, September 17 at 10 a.m. Daley attended these events and was available to answer questions about his work. The Lamont Gallery is in the Frederick R. Mayer Art Center on Tan Lane.

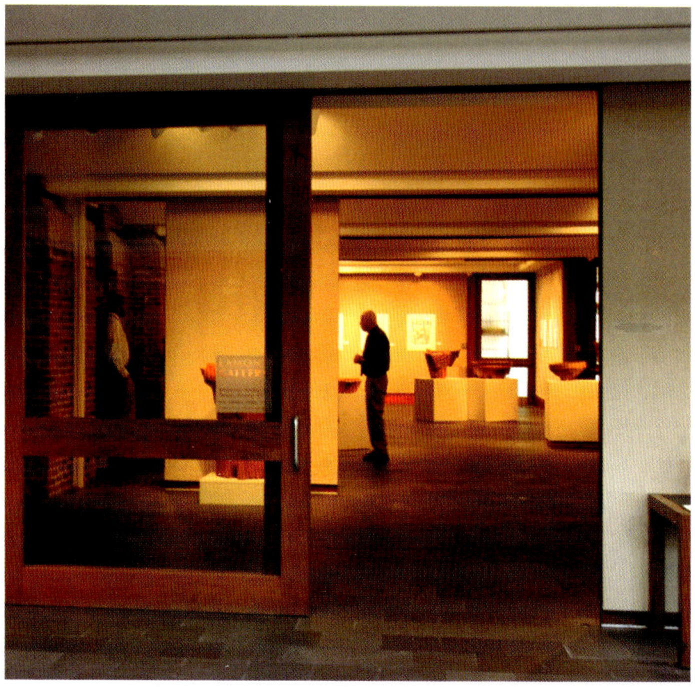

Photo: Sara Zela, Lamont Gallery, Phillips Exeter Academy

Ceramics Artist William Daley Brings Wonder to the Classroom

Nichole Pellaton

Daley playfully invited students to touch his ceramic pieces on display in Lamont Gallery September 26, 2011. Artist William Daley is no ordinary Exeter visitor. He wears, and makes, paper hats. He greets strangers on the path, asking questions about what they do, where they come from, their interests. And he teaches students to approach ceramics with the wide-open eyes, hands, and mind of a child discovering clay for the first time.

This 85-year-old creator of large, highly architectural ceramics works was in his element as he challenged students to explore his pieces with their hands—to drum on them, to reach inside and feel the textures, to interact with the artwork in ways they wouldn't have otherwise dared. The students, after a few moments of surprise, started exploring his sculptures inch-by-inch, feeling the intricate, unglazed surfaces.

Daley came to campus for the opening of his exhibition, *"William Daley – Selected Works: 1967 – 2010."* He met with several ceramics and drawing classes in Lamont Gallery, where students loved his playful approach. "One of the great things about Exeter is its tendency to introduce students to truly interesting people," says Andrew '12, a ceramics student. "Each time I spoke with Mr. Daley, I felt as if I were analyzing a novel—in each joke he made, in each turn of phrase, there was a striking idea. He seems to love giving people bait and making them take a closer look at everything he says, peppering into his lectures references that don't seem to make sense at first, but that ultimately sum up a profound observation about humanity or, as he puts it, 'the ineffable.' He's very clearly an educator at heart."

The gallery opening reception and talk—held in Mayer Auditorium—was packed. Afterwards, Daley and visitors returned to the gallery to make paper hats from newsprint. Daley has been making paper hats since he first started dating his wife of over 60 years, Cate. Cate's father, a newspaper printer, challenged Daley to prove his marriage-worthiness by folding paper hats, so the story goes, and the artist has never stopped folding.

"Bill has amazing energy and passion," says Sara Zela, Lamont Gallery manager. "His vast knowledge and experience, and his willingness to engage with everyone he encounters, has made his visit here an incredible experience for our students, visitors, faculty and staff."

Daley was accompanied by his wife, with whom he works, and his son, Thomas Daley, an architect. During the visit, Thomas Daley met with architecture students and gave a presentation about his architecture firm, his philosophy, and a collaborative project undertaken with his father at the Northern Clay Center.

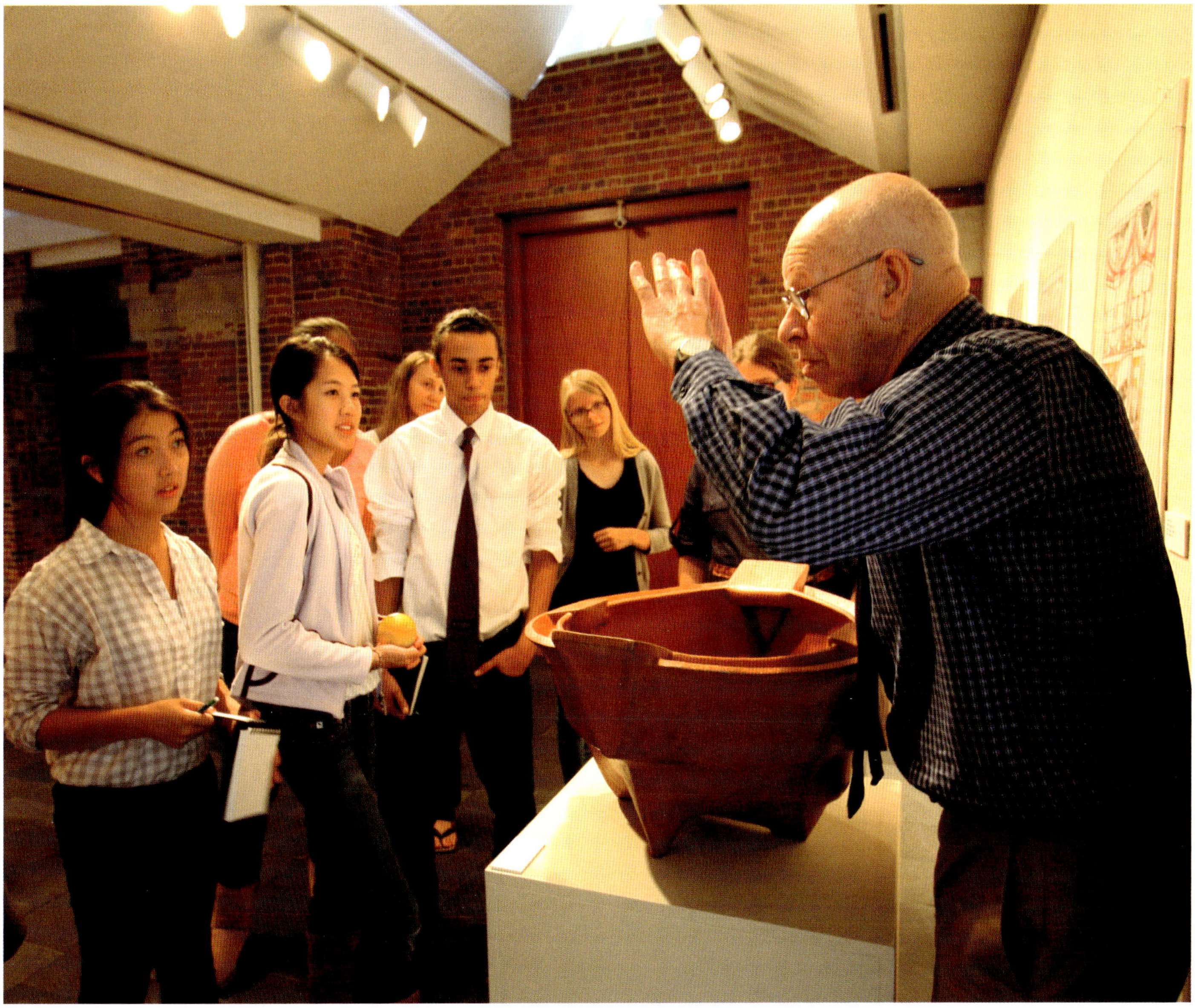

Photo: Sara Zela, Lamont Gallery, Phillips Exeter Academy

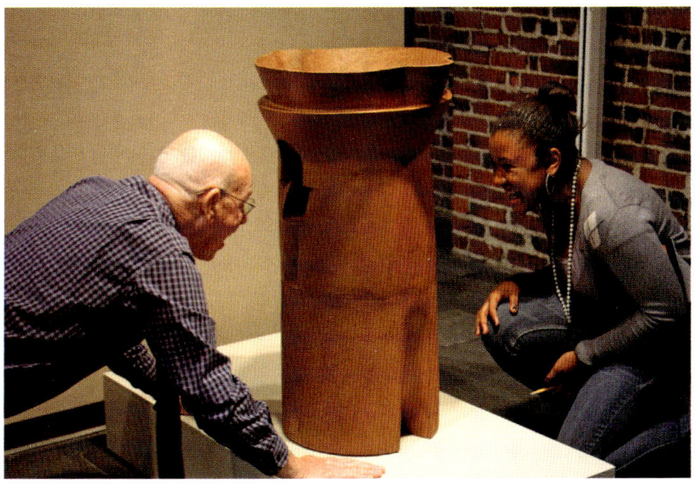

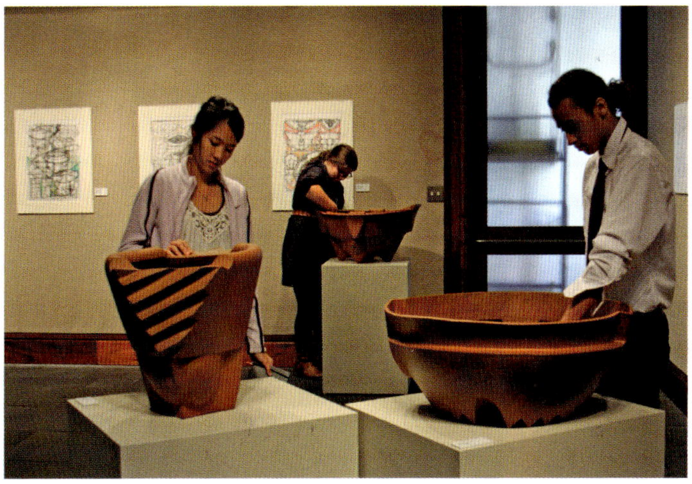

Photos: Nicole Pellaton, Phillips Exeter Academy

Selected Students of Bill's

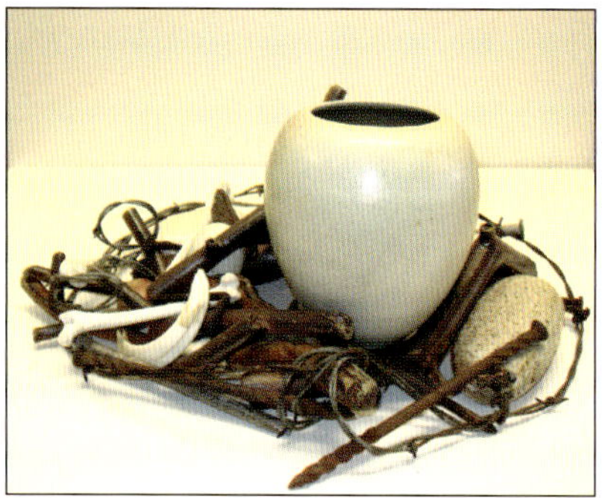
Photo: Doug Weissman

Karen Aumann | Associate Professor
Art Department, Community College of Philadelphia
Crafts Student 1969 to 1973

As freshmen at the Philadelphia College of Art, my roommate and I persuaded Bill Daley to sign us into his elective ceramics class, which, as freshman, we were not supposed to do. After that, I quickly dropped my Art Education courses and spent my time in the "Pot Shop". Who knew you could major in pottery making at College?

As a teacher of Ceramics and Design at Community College of Philadelphia, Bill's words of wisdom are never far from my mind as I attempt to inspire my students. I especially remember how Bill, then "middle aged", liked to talk about how he was looking forward to being in his dotage when he could then say anything he liked without worry. Funny, since I never felt he was ever holding back much. He once told me that the handle I had just applied to my very first mug looked like a snake that got run over by a truck.

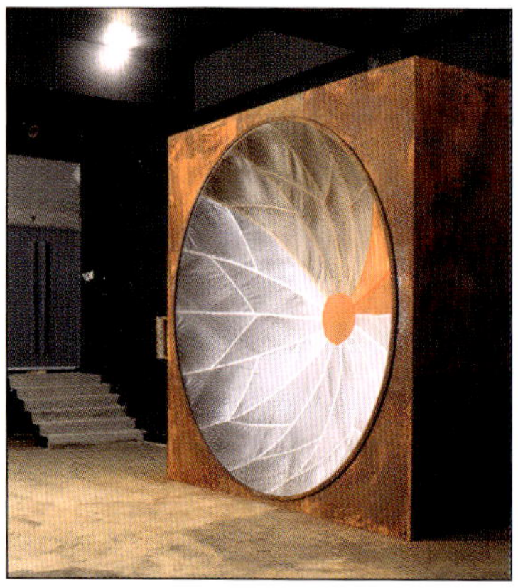

Jin Hwan Bae | Professor, Korea National University of Arts, South Korea

Jin met Bill when he attended Bill's two-week workshop at Banff. Jin then came to UArts as a resident. In the ceramics department, studying with Bill, Petras Vaskys, Liz Stewart, and Rob Forbes. After his stint as a student, Jin taught "Korean Ceramics for Americans" at night, in the Continuing Education Program.

In addition to his teaching career, Jin's work has been exhibited in more than 10 solo and group exhibitions at galleries like the To-Art Space in Seoul, Korea; the Triennale for Form & Import in Frankfurt, Germany; the Seoul Museum of Art; and more.

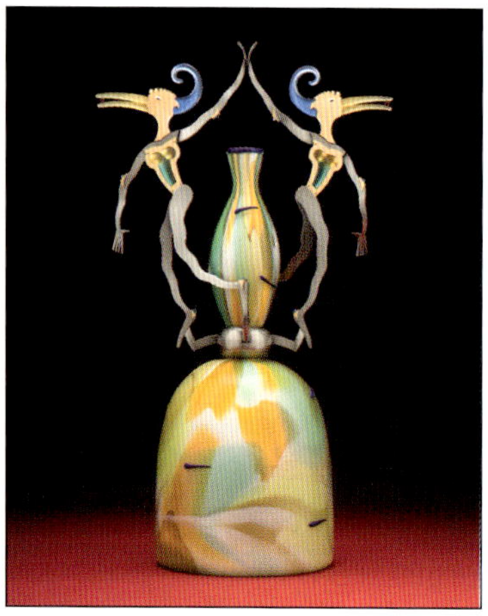
Photo: Bill Truslow

Dan Dailey | Professor Massachusetts College of Art. Boston
Crafts Student 1964 to 1969

Bill taught Dan from his freshman to senior year. They shared a common last name, and Dan grew up in Elkins Park, where Bill lived. Dan was the epitome of someone who thinks and lived "outside the box," according to Bill. He was remarkable in every way—bending rules, stretching boundaries, and problem solving in a manner that defied convention. After graduation, Bill hired Dan to work on several large architectural commissions, including the South African Airlines ceramic and bronze wall.

Today, Dan Dailey works primarily in his Kensington, New Hampshire, studio. He is a professor at Massachusetts College of Art where he founded the Glass Department in 1973. Dailey is an independent artist/designer and has received fellowships from both the Massachusetts Council on the Arts and the National Endowment for the Arts. Since 1971, Dailey has participated in over 300 juried and invitational exhibits, including a major retrospective at the Renwick Gallery of the Smithsonian Institution. His work is represented in 48 museum and public collections in the United States, Europe, Australia and Japan.

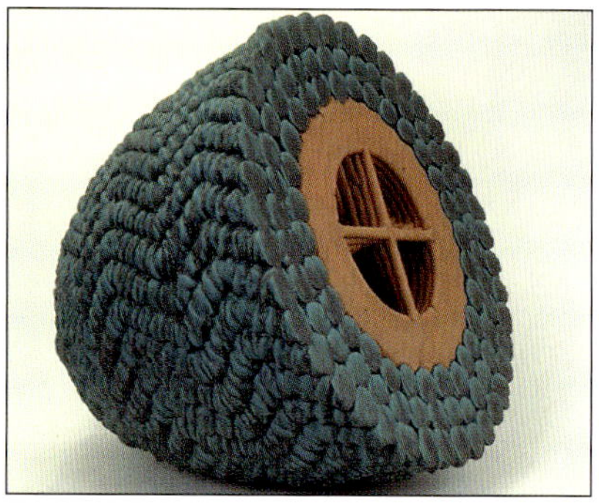

Graham Marks | Master of Acupuncture
Alfred, New York
Crafts Student 1968 to 1974

When Graham was looking for a ceramics program, he came across a poster for a solo show of Bill Daley's at the Philadelphia Art Alliance. "I want to study with that man," he thought. He did, and received his BFA from The Philadelphia College of Art and his MFA from Alfred University.

Bill recalls Graham as being inspirational in spirit.

Graham taught ceramics for 10 years, first at Kansas State University and later at the School for American Craftsman at Rochester Institute of Technology. He also spent several years as the Head of Ceramics at Cranbrook Academy of Art. His work is in numerous private and public collections including the Detroit Institute of the Arts, Cranbrook Academy of Art Museum, Dienst Beeldende Kunst, Hertogenbosch, Holland, the Museum of Arts and Design, New York City, The Everson Museum of Art and the Renwick Gallery of the Smithsonian Institution. Since 1995, Graham has been practicing acupuncture and operates a private practice in Alfred, New York.

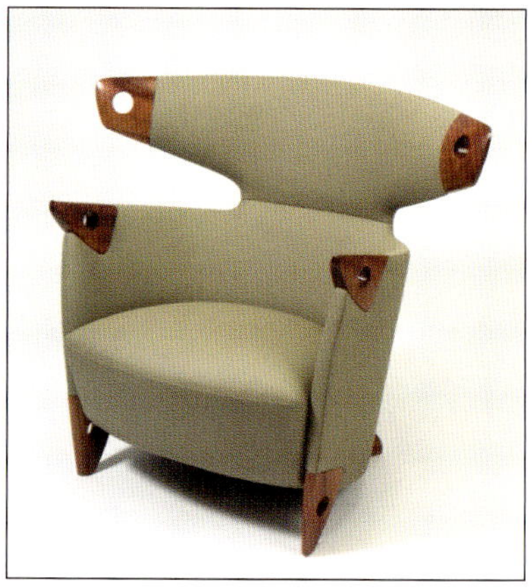
Photo: Eric Gould

Alphonse Mattia | Professor, Rhode Island School of Design, Crafts Student

Alphonse has received fellowships from Massachusetts Artists Foundation and the National Endowment for the Arts, and is known as one of the country's leading educators and studio furniture makers. His work is in many private collections and permanent collections, including The Museum of Art and Design, The Philadelphia Museum of

Arts, and the Museum of Fine Arts Boston, The Yale University Art Gallery and The Phillip Morris Corporation.

Mattia dismisses the dichotomy between functional and nonfunctional objects. He prefers instead the concept of "usefulness"—something that conveys meaning and emotion along with utility.

Photo: Courtesy of Lutron Corporation

Noel Mayo | Owner and President
Noel Mayo Associates
Industrial Design Student 1956 to 1960

Noel Mayo, owner and president of Noel Mayo Associates, Inc. (NMA), the first African-American industrial design firm in the United States, is also the first African-American Ohio Eminent Scholar in Art and Design Technology at The Ohio State University (OSU) where he teaches undergraduate and graduate students.

Noel serves on the boards of The University of the Arts and the Free Library of Philadelphia.

His firm's completed design projects include the Philadelphia Airport's Communication Center interiors, highway approach, and exit signage, Philadelphia Convention Center lighting fixtures, and street lighting fixtures for Pennsylvania Avenue, in Washington, D.C. with Raymond Grenald Associates.

During his 40-year relationship with Lutron Electronics Company, Inc., the world leader in lighting controls, Mayo designed all of their products, including the Vierti dimmer design, which won the 2009 Red Dot Award.

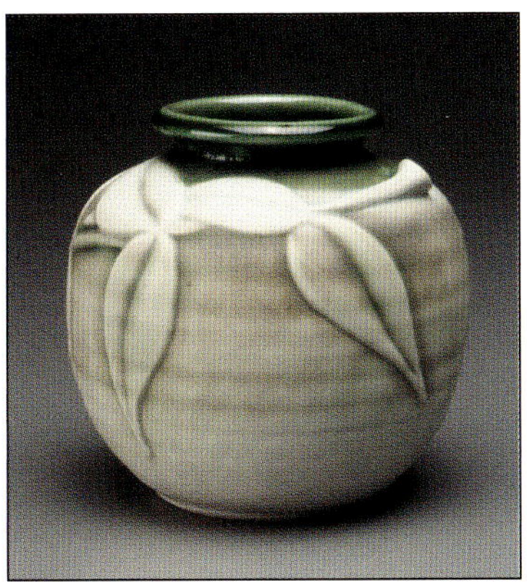

Kathryn Narrow
Crafts Student 1966 to 1969

After trying a variety of liberal arts majors, Kathryn transferred to Philadelphia College of Art and, after just one course with Bill, switched her major once more from Industrial Design to Ceramics. A major influence in Kathryn's life, she credits Bill's "enthusiasm for play" as the spark that inspired the creation of the Claymobile program, which she founded while working at the Clay Studio as Education Director. Claymobile is a mobile ceramics studio that partners with schools, community organizations, and social service agencies to bring hands-on art education to under-served populations.

Narrow creates carved and altered porcelain pottery that is both functional and decorative, influenced by techniques and styles of Korean, Chinese, and Japanese ceramics as well as early American art pottery. Her aim is to produce intimate work that can be held in the hands and which, by its use, brings beauty to the user's environment.

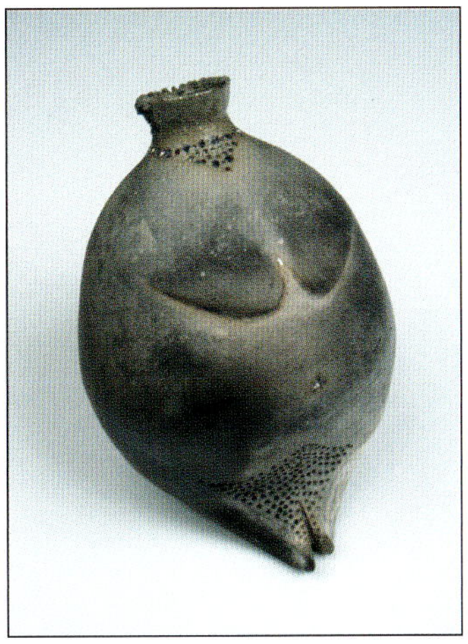
Photo: Jarvis Grant

Winnie Owens-Hart | Professor
Howard University
Crafts Student 1967 to 1971

As a student of Bill at the Philadelphia College of Art, Winnie was most influenced by seeing his work installed in public places. Though her most prominent memories of Bill are his famous paper hats, seeing his work reminded her that he was also a serious artist.

Throughout her career, Winnie has researched contemporary and traditional African ceramics, African-American clay workers, and the ceramic techniques of aboriginal cultures. Prior to teaching at Howard University, Winnie taught ceramics in Nigeria. She has authored numerous publications including *Ceramics: From Africa to America* and *Revivals!*

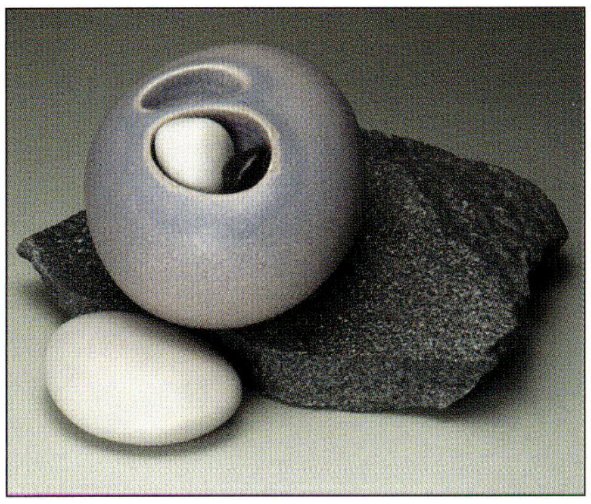
Photo: Marian Pritchard

Marian Pritchard |
Crafts Student 1979 to 1982

Marian Pritchard is a Philadelphia ceramic artist and teacher. She recently showed her hand-built sculptures of glazed ceramic stones in two shows connected with NCECA 2010. Her porcelain vases are in various collections, including the Museum of Fine Arts, Boston, and she has participated in major juried shows at the Cleveland Museum of Art and Philadelphia Museum of Art. In 1991, she worked in Wales as a British American Arts Fellow. She has taught art education at the University of Massachusetts, ceramics at Swarthmore College, craft history and contemporary crafts at the University of the Arts, and contemporary crafts at Moore College of Art and Design, where she also co-taught a summer course in Italy. As acting director of the Clay Studio in 1983, she helped coordinate their first national show. She has taught ceramics part-time at Community College of Philadelphia since 1986. She earned an A.B. Magna cum Laude in art history from Wheaton College, Massachusetts, an M. A. T. in art education from the University of Massachusetts, as well as a B.F.A. from the Philadelphia Colleges of the Arts.

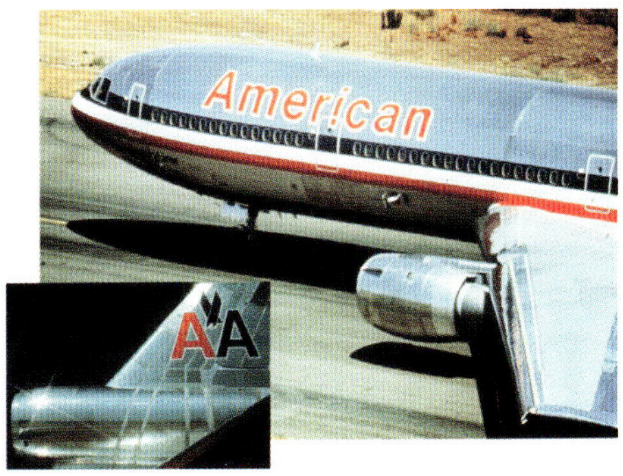

Photo: archive of Henry Dreyfuss

James Ryan | Consultant
Industrial Design Student 1957 to 1960

Jim Ryan studied Industrial Design with Bill Daley at the Philadelphia College of Art during Bill's earliest years teaching there. Jim credits Bill with having a very strong influence on him.

Jim developed the American Airlines fleet markings and logo in 1969, including the mark of the 2 "A's" and the eagle, and they are still in use more than 40 years later, unmodified. Jim joined the design-consulting firm Henry Dreyfuss Associates in 1966, and eventually became a partner before retiring in 2002.

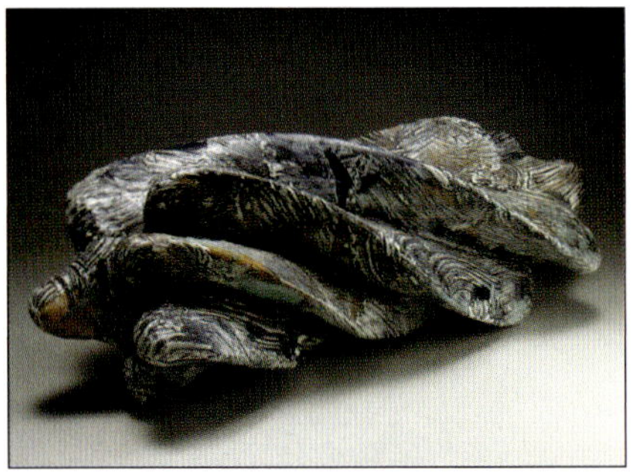

Photo: John Stephenson

John Stephenson | Artist and Educator

John Stephenson has been an artist and educator in Ann Arbor, Michigan, for over thirty-five years. He began his career teaching ceramics at the University of Michigan in 1959 after receiving his MFA from Cranbrook Academy of Art. While at Michigan he received the Catherine B. Heller Distinguished Professorship, an honorary Doctorate of Arts from Grand Valley State in Grand Rapids, and even served as interim Dean of the School of Art.

John has exhibited both nationally and internationally since the beginning of his career. His work is included in collections at Alfred University, Ohio State University, Parish Museum of Art, Portland Museum of Art, Cranbrook Museum of Art, Eastern Michigan University, Detroit Institute of Art, Museum of Contemporary Crafts in New York City, Faenza Museum (Italy), Museum of Decorative Arts (Prague), Keramion Museum, (Germany), Foshan/Nanfeng Museum (PR China), Benaki Museum (Athens), Ichon World Ceramic Center (Korea) and many corporate and private collections in the United States.

Selected Colleagues of Bill's from University of the Arts

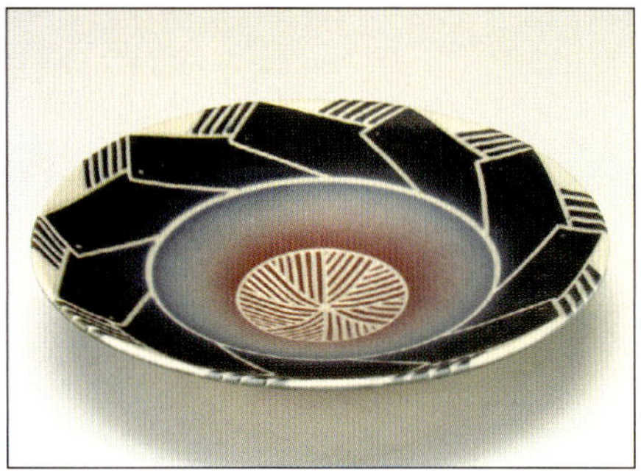

Photo: Wayne Bates

Wayne Bates | Ceramic Artist, Wayne Bates Studio

Bates has lived and worked at his studio in Murray, Kentucky, since 1978, where he makes functional sgraffito decorated bowls, platters, and vases. His work is shown at fairs such as the Baltimore Winter Market, American Artisan Festival, Kentucky Crafted: The Market, and at numerous galleries and shops across the country.

Wayne taught ceramics at the University of the Arts in Philadelphia from 1970 until 1978 and co-chaired the Crafts Department with Bill Daley during that time. Wayne has taught at Haystack Mountain School for Crafts and Arrowmont School for Crafts. For three years he was the ceramic consultant to reactivate The Moravian Pottery and Tile Works in Doylestown, Pennsylvania. The Tile Works makes and sells museum quality reproductions of Mercer Tiles and conducts workshops in tile making.

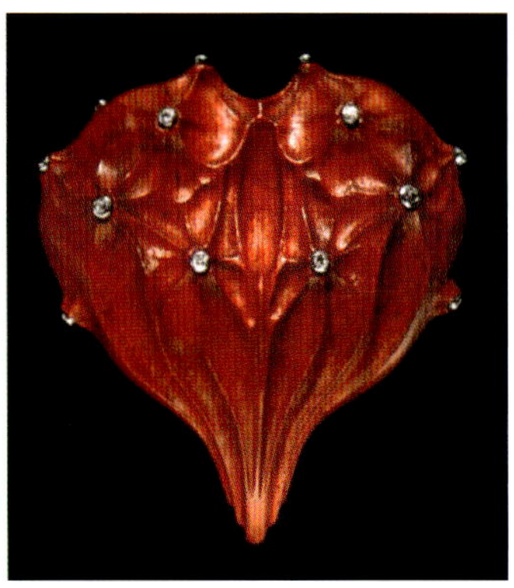

Photo: Jack Ramsdale

Sharon Church | Professor, Crafts, University of the Arts, Philadelphia

Sharon grew up in Wilmington, Delaware, studied art at Skidmore College, and received her M.F.A. from the Rochester Institute of Technology. After a brief stint of teaching, she launched a limited production jewelry business and received a National Endowment for the Arts Individual Craftsman's Grant.

But life changed, radically, when she was hired by Dick Reinhardt to teach Jewelry and Metalsmithing at the Philadelphia College of Art. Mentored by Dick and his great friend, Bill Daley, she learned that teaching and making studio art are intimately entwined and deeply joyous. She eventually became a professor at what is now the University of the Arts, and Sharon's carved jewelry forms have come to reside in collections at the Philadelphia Museum of Art, the Helen Williams Drutt Collection of the Museum of Fine Arts, Houston, the National Gallery of Australia in Canberra, and the Museum of Art and Design, New York.

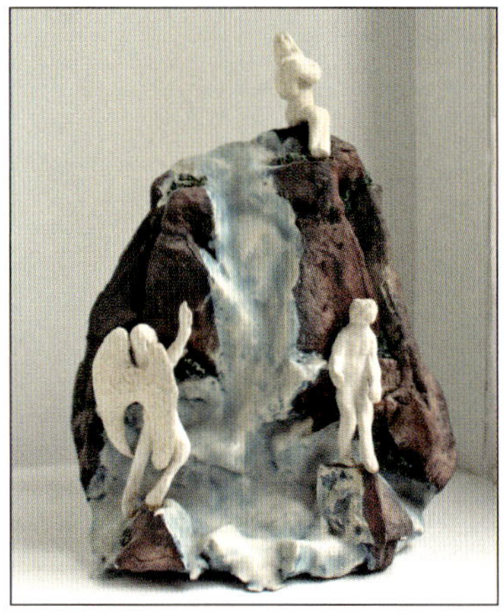

Photo: Larry Donahue

Larry Donahue | Adjunct Professor, Crafts, University of the Arts, Philadelphia

Larry is a master mold-maker, continuing in the tradition of his teacher, Petras Vaskys. His unique use of materials and processes inspired Bill Daley to remark, "He is a master vessel maker who invigorates students toward innovation and risk taking."

In addition to teaching at UArts, Bucks County Community College, and the Clay Studio, Larry has also worked as a mold maker and ceramist at Moravian Pottery and Tile Works in Doylestown, Pennsylvania. His ceramic work has been exhibited nationally and internationally at locations including the American Craft Museum in New York, Philadelphia's University of the Arts, Professional Institute for Educators in Philadelphia, and Sydney, Australia. His cast vases have been sold through *Martha Stewart Living* magazine.

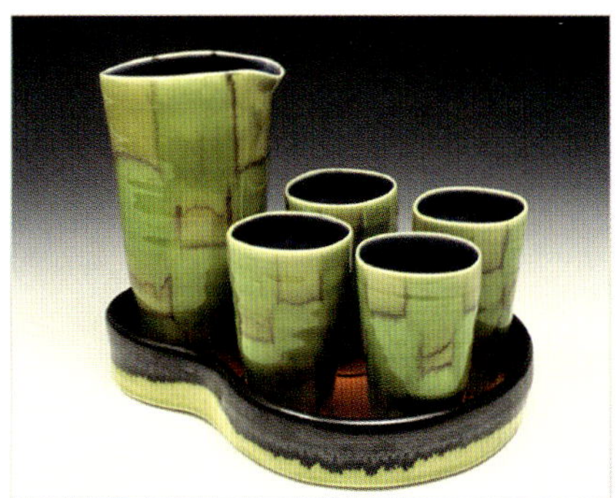

Photo: Alec Karros

Alec Karros | Adjunct Assistant Professor, Crafts, University of the Arts, Philadelphia

Alec came to UArts as a junior, transferring from the University of Illinois. He was an accomplished thrower, setting a demanding pace for his fellow students.

Bill remembers Alec as one who ignited everyone he came in contact with – students and faculty alike – a quality that has translated into his teaching as well.

Alec grew up working in his father's kitchen, and his initial aesthetic sensibilities were shaped by the preparation and presentation of food. Studying pottery in art school was a natural extension of this experience. Karros continues to explore the relationship between necessity, invention, and aesthetics, producing pots that function effortlessly, revealing their elegance and uniqueness through daily use.

Karros has taught at State University of New York (Oswego), the University of Colorado (Boulder), and currently teaches at the University of the Arts. In addition to his academic work, he creates one-of-a-kind porcelain ware at his Mountainville, New Jersey, studio.

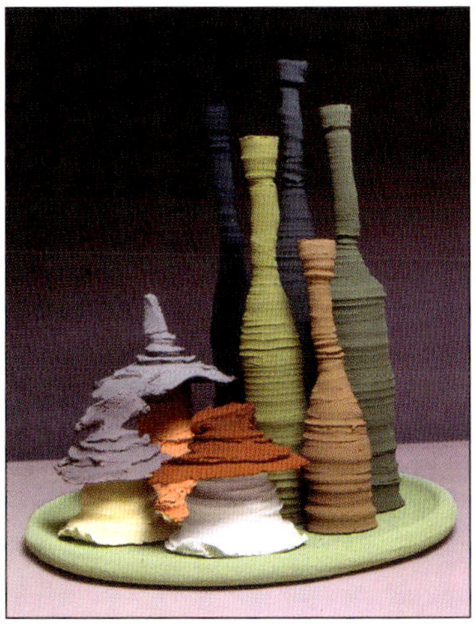

Photo: Joshua Schreier

James Makins | Professor, Crafts, University of the Arts, Philadelphia

James Makins, a ceramic artist originally from Johnstown, Pennsylvania, is the primary innovator of working in the plastic kinetic quality of throwing porcelain vessels. Like Bill Daley, he majored in both ceramics and art education (Makins at UArts, Daley at Mass Art) and was the assistant of Byron Temple in New Hope, Pennsylvania. Jim went to Cranbrook after graduating from UArts and was mentored there by Richard Devore.

James's work has been widely exhibited at galleries and museums including the Philadelphia Museum of Art, Los Angeles County Art Museum, American Craft Museum, Renwick Gallery (Smithsonian), Izushi Art Museum and Shigaraki Ceramic Cultural Museum in Japan, Museum Boymans-van Buenigen in the Netherlands. He has been awarded two National Endowment for the Art Awards, Professional Fulbright Fellowship to Japan, Japan Foundation Fellowship,Grand Prize, Izushi International Porcelain Trienniale, Silver Medal, Faenza International Ceramic Exhibition and three New York State Council for the Arts Fellowships.

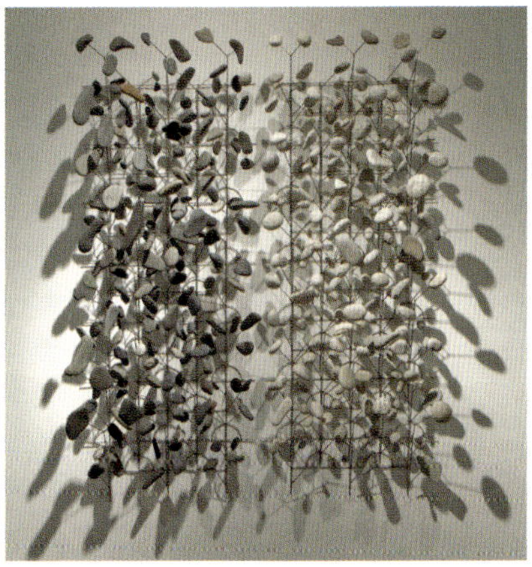

Photo: Jack Ramsdale

Warren Seelig | Professor, Crafts, University of the Arts, Philadelphia

Warren Seelig lives and works in Rockland, Maine. He holds the rank of distinguished visiting professor in the Craft/Fibers program at the University of the Arts in Philadelphia, where he teaches, curates and writes on various subjects related to fiber, textiles and material studies. He received a BS from the Philadelphia College of Textiles & Science and an MFA from the Cranbrook Academy of Art. Seelig has twice received individual fellowships from the National Endowment for the Arts and three fellowships from the Pennsylvania Council on the Arts. His work has been included in more than fifty major museum exhibitions in the United states, Europe, Japan and Korea with many solo and group exhibitions world wide Seelig was elected a Fellow of the American Crafts Council and was a member of its board.

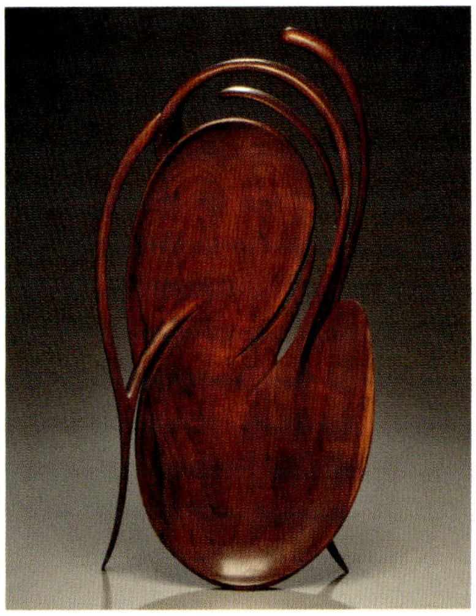

Photo: John Carlano

Richard Stetser | Professor Emeritus, University of the Arts
Student 1955 -1960

Dick Stetser's career includes forty-plus years teaching at the Philadelphia College of Art/University of the Arts. He has taught in both the Sculpture and Foundation Departments, and has chaired both during his career. Stetser also taught at the American College in Paris for several years.

Dick is a graduate of the Dimensional Design Department of PCA under Bill Parry, and upon graduation was hired to be the Shop Master. He then began teaching. When the school became a college, Dick was named, along with colleague Leonard Lehrer, as co-chair of the Freshman Department. Stetser taught 3D, and Lehrer 2D Design. In the 1960s Stetser taught with Bill, and Renzetti in 3D Design classes.

Stetser's honors include three Venture Fund Awards from UArts, and a National juried show first prize from the Wharton Esherick Museum. His work has been exhibited in numerous galleries, and he has completed many private and public commissions throughout the Philadelphia area.

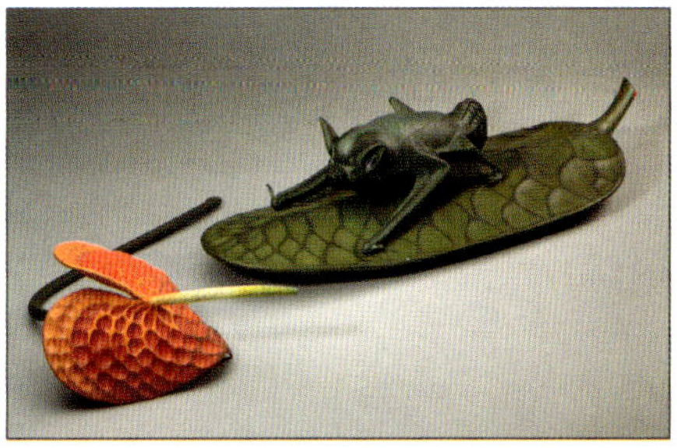

Photo: Ken Yanoviak

Lizbeth Stewart | Professor, Crafts, University of the Arts, Philadelphia

Liz was hired to teach at the Philadelphia College of Art in 1981, and accepted the position with much trepidation. Working with Bill Daley and Petras Vaskys allowed Liz to learn by observation, and, with their encouragement and support, she learned that a studio practice could be an example to students without overshadowing one's ability to teach and to nurture the expressive nature of emerging student artists. Bill describes her as "an extraordinary teacher who calls on her students to pursue their own aspirations as artists."

A native Philadelphian, Stewart attended school in Bryn Quelled, Bucks County, and graduated with a B.F.A. from Moore College of Art in 1971. In 1976, she was awarded a Craftsmen's Fellowship from the National Endowment for the Arts. Stewart was Artist in Residence at The Fabric Workshop in 1981. Her work is represented in numerous private collections throughout the United States and in the following permanent collections: Philadelphia Museum of Art; Campbell Museum, Camden, New Jersey; Joseph H. Hirshhorn Collection, Washington, D.C.; Lannan Foundation, Palm Beach, Florida.

Teaching Colleagues and Mentors

William P. Daley, 2012
editor Charlotte Daley

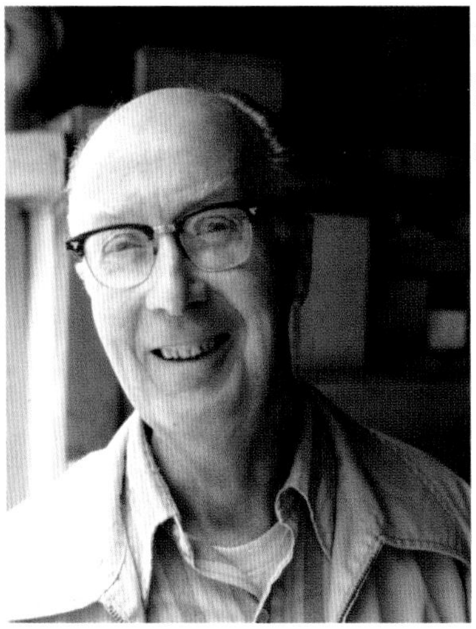
Photo: William Daley

Charles Abbott
Teacher from 1911 to 1986

Charles Abbott | In September 1948, Charles Abbott drove to Boston from his Pottery in South Berwick, Maine, to teach a ceramics day-class at the Massachusetts School of Art. Catherine and I were in that class, and, when the day ended, I knew from then on that working clay was to be part of my Art Life.

We were not alone; he spoke with spirited enthusiasm preparing clay and throwing lidded bowls on the wheel. The sense of his telling story—studying Architectural Engineering at Syracuse University, working as a draftsman during the depression, and changing course to become a studio potter—spoke to our hopes as returning veterans.

As the semesters processed, we came to realize how rich a trove his attentive practice and oblique council really was. The gaps in his story surfaced as we took our turn on the wheel, learning in-between, to coil, slab build, load kilns, formulate glazes, and hear more of his stories.

He made his liberation move by enrolling to study Asian Art History at the University of Michigan. At the end of the first semester studying Islamic and Chinese pottery, he heard of a teacher at Cranbrook Academy of Art who was "actually making pots."After a visit he enrolled to become one of Maija Grotell's first students. Three terms later, he returned to Maine, in love with mud, to build his studio and earn his living for a decade by making pottery "people want to use."

C.A., known to us as "Mr. Abbott,"was a persuasive combination of engaging opposites. One lobe was that of a Techno, pragmatic, problem-solving Yankee. The other domain was of a romantic Emersonian New Englander profoundly smitten by Chinese culture. A gentle follower of the Tao at Walden Pond, he sang the praises of Sung dynasty pots and shared his dual beliefs with ironic wit to our anti-authoritarian pleasure. A secure traditionalist, he was open to "all of it," from science to metaphysics. His examples, as a practiced Janus, opened enticing doors for us nascent artists to worlds promising clay paths for avid spirits.

To cite one of his bifold Techno/Thoreauen strategies "that we could do it all," he brought in Apple Ash from his studio wood stove to glaze our early earthenware pots nearly for free. He then whetted our techno, "free at last" appetites to scrounge and scrape rust, strip insulated copper wire, pulverize green, brown, and clear beer and soda bottles to flux clay and color our ash glazes. It all augmented our limited studio supplies. It did even more for our self-esteem, savoring the gift of possible self-sustainability. The true free part was sharing his expansive understanding of the material chemistries of fire and clay becoming glass.

He also often talked about the "symmetry of wonder" celebrating the contours of Sung vases and their limpid celadon glazes. To see them, C.A. arranged a very privileged class trip to the closed chambers of the Boston Museum of Fine Arts to experience the porcelains that Croomaswami had collected for the museum. Since we had no slides then, Mr. A. would carry in pieces from his home collection, providing special occasions to hear his thoughts on why he admired them. Our favorites were those of his teacher Maija Grotell and the large bravura bowls of his friend in New Hampshire, Mary Schrirer. From time to time, he slyly offered rare treats by putting in sequestered view a few of his latest small (3.5") vases, the alchemic splendors on which he tested his quest for the Grail—Crystalline Glazes with Cobalt, Zinc and Rutile—"His Search-Research" for clandestine discovery by the curious obsessives among us.

His myriad letters to me from 1952 to 1986 brimmed with useful formulae, much valued technical information, cogent critique, teaching counsel, and telling acerbic commentary. His most extraordinary offering was the detailed, step-by-step directions for successfully firing a large, newly acquired, gas kiln.

It was a first time for me, and I still marvel that I succeeded. The students were delighted by the results of our first gas firing. I was very grateful that I only singed my eyebrows with a blowback when igniting the first of four burners.

Writing some years later, querying my mentor, I asked: "Do you know any one who thinks of Shang Bronzes as useful objects?" He quickly replied "I do!—I find Shang Bronzes powerfully compelling with the knowledge of meaning. I am intellectually awed by the profoundness of their content. When you can express your beliefs as to the power and glory of your world in a vessel you can drink from—You will have arrived!"

At this writing, I am hopeful that libations from my next "Vesica Cisterne" in your honor will please you. I drink in grateful tribute for your grand elliptic teachings illuminating seven generations of "Mass Art" ceramicists questing for their Grail's arrival.

Teaching Trajectory

My first teaching job was at Iowa State Teachers College. Later I also taught at the State University of New York at New Paltz and at S.U.N.Y, Fredonia, the University of New Mexico and the University of the Arts, which, as my second home, I still call the Philadelphia College of Art.

For over forty years, these five have offered me places to practice teaching, learn to make telling pots, and become a believable teacher of makers. It all happened venturing in community with avid colleagues and students aspiring to be Artists. We had a grand time together exploring the boundaries of our inspired ignorance in classes, over coffee, in studios, or wherever two or more obsessives gathered. We often advanced together while making things—Building kilns, workspaces, curricula, and exhibitions, and sharing our own work in progress. Citing all the participant's names at these five Eden's is not possible here so let me offer the following placeholders.

As a first-employed teacher at the then Iowa State Teachers College, my principal mentors were the painters David Delafield and Paul Smith, the jeweler Robert Von Neumann, the printmaker David Driesbach, and the then student, John Stevenson. At the State University of New York at New Paltz, I count most all the colleagues I left town with after the "Great Dismissal," (a group of teachers were dismissed and two of us resigned in protest). First among them were the painters Richard Fraenkel, Jules Olitsky, and Robert Newman, the photographer Robert Forth, and printmaker Alfred Maurice. All were, for me, inspiring vibrant models as Artist/Teachers.

After resigning from New Paltz in 1957, being invited by Joseph Carreiro and William Parry to teach at the Philadelphia Museum School of Art after the Exodus was a great gift. Professing there for more than three decades in four programs, as they evolved from an industrial art school, to a college of art, and now the University of the Arts, privileged me to teach while venturing to become an artist.

As a nascent artist and faculty colleague, among mega masters and their students, sharing their practice was revelatory. Witnessing their moves from ideation to realization, I learned so much about my own "doing." Helping to assist students work through their cycles of transformation to successful outcomes confirmed my own trials. In studio time, the gift exchanges offered exhilarating openings to learn where body, material, media, and mind became a cored site to incubate compelling Form.

Praising these master-makers, teachers, and their students, from PMSA to PCA to U of Arts over three decades would take a book in itself. Let me pay homage to five for the many, and cite my early closest colleagues at PMSA: Joseph Carreiro, William Parry, Richard Reinhardt, Aurelius Renzetti, and Petras Vaskys. All together, they were five points of light, illuminators for us all, modeling the quest for our artist selves.

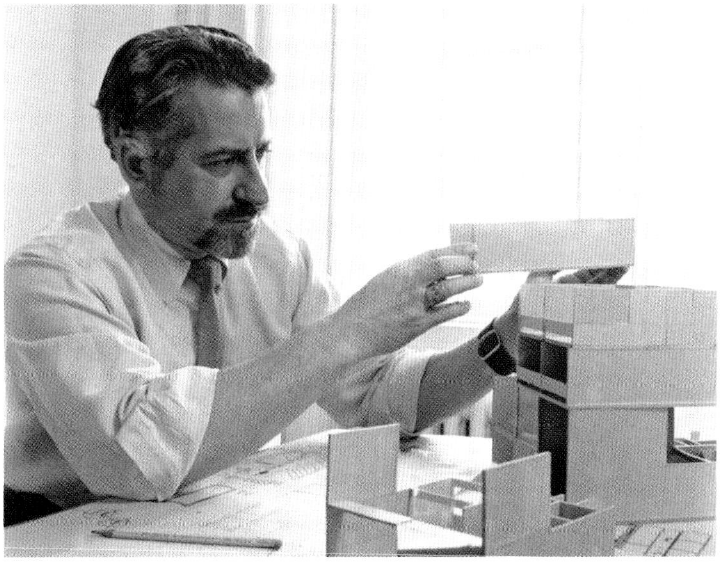

Photo: Cornell University Archives

Joseph Carreiro
1920 to 1978

Joseph Carreiro | Descendent of Portuguese cabinet makers and a World War II U.S. Navy Yard painter of war ships, for over six years, Joe majored in two programs, Painting and Art Education, at the Massachusetts School of Art. He also managed the school's student operated scholarship art supply store. He attended graduate school at Harvard University, taught design at Cornell University, and chaired the Philadelphia College of Art Industrial Design program. He then returned to Cornell, in 1965, as an internationally renowned educator/ designer and was the Professor/ Chair of Cornell's Design and Environmental Analysis program until his passing in 1978.

Joseph Carreiro, Catherine Stennes, and I were classmates at "Mass Art" class of 1950. As three of ten seniors, we each taught venturesome, free Saturday classes for talented high school students from Boston area public schools. With Charles Quillen, Joe and I also made our senior thesis together—a working tubular flexible telescoping pole and panel system for presenting student artwork in public places. It was the beginning of our close collaborations as teachers, designers, artists, and friends. As a successful professional practitioner and founder of Carreiro Design Associates, Joe invited me to be an associate and work with him on the Brussels and Seattle World Fairs, international trade fairs, exhibitions, and interior designs by night, while teaching drawing and design to Industrial and Dimensional Design students by day. Together these gifted occasions became the genesis for later projects carried out with student assistants in my own studio.

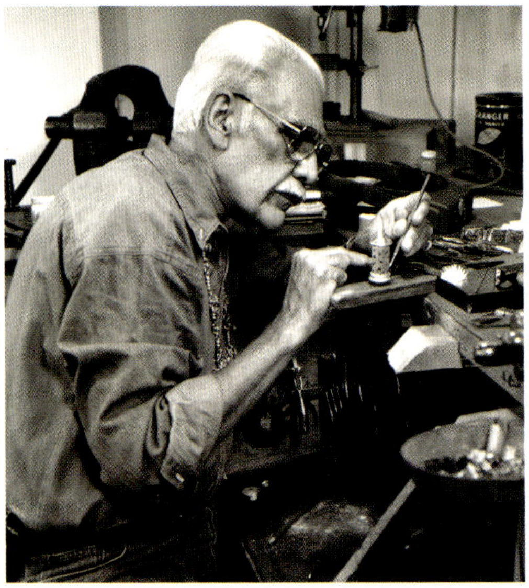

Photo: William Daley

Richard Reinhardt
1921 to 1998

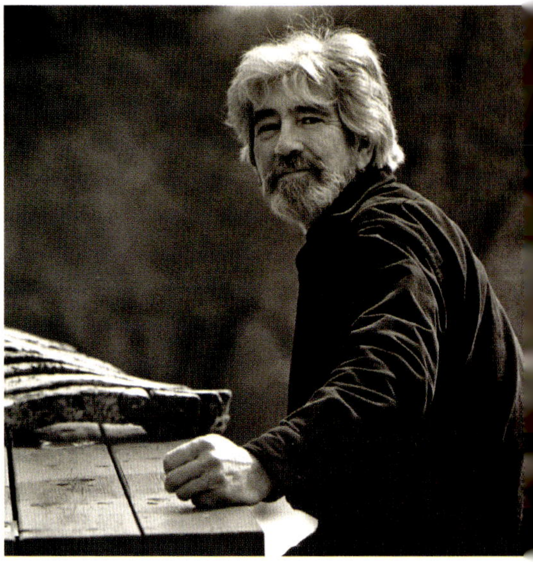

Photo: Brian Oglesby

William Parry
1918 to 2004

Richard Reinhardt | Elmer, Reinhardt's father, a master tinsmith and Philadelphia public school industrial arts teacher, nurtured Richard's early precocity as "a maker of things." Awarded a Board of Education scholarship to attend Saturday classes at the then Philadelphia Museum School of Industrial Arts, Richards began his long distinguished career at the school as an educator, master silversmith, jeweler, and furniture designer/craftsman.

Returning from the Marine Corps in 1946 to complete his studies on the GI Bill, he continued his career as a nascent silversmith at PMSIA, studying with Douglas Gilchrist and Virginia Cute Curtain. In 1948, Curtain and Reinhardt, as teachers, studied with Baron Eric Von Fleming, Sweden's "Royal Smith." In the 1960s, he successfully helped establish new parameters for educating Dimensional/ Industrial Designers. Later, Carreiro and Reinhardt together were invited to help establish a Dimensional/Industrial Design program in Japan, modeled on Philadelphia Museum School of Art's curricular transformations to become the Philadelphia College of Art. During his long career, he taught with his mentor Renzetti and colleagues Carreiro and Parry, re-informing craft traditions.

Reinhardt executed many commissions. These ranged from corporate wood conference tables to forged silver, liturgical serving tureens, platters, ewers, wedding crowns, and processional regalia. His masterful, forged silver and rosewood mace for the University of the Arts, designed and constructed around an 1876 medal struck for the Philadelphia Bicentennial and founding of the Museum School. This mace is carried by the Marshall in graduation processions celebrating the new graduates passage to the future. I felt close to its maker when I was the retiring Marshall in 1990.

During the 1980s in Philadelphia, Jane Korman at the Swan Gallery, Helen Drutt at the Drutt Gallery, and the Quadrant Gallery in Chestnut Hill, Massachusetts, presented solo exhibitions of Reinhardt's signature forged silver bracelets, torques, buckles, and silver chains. In 1998, a splendid, retrospective exhibition and catalog, "Full Circle: A Legacy of Metal Work," at the University of the Arts Rosenwald-Wolf Gallery, honored Reinhardt's genius as an artist and educator. In 1949, as an emerging silversmith, his work was exhibited at the Metropolitan Museum. Now his work is in the principal collections of the Boston Museum of Fine Arts, the Philadelphia Museum of Art, The Smithsonian Renwick Museum, and the renowned jewelry collection of the Museum of the Rhode Island School of Art. He served what is now the University of the Arts with distinction from 1947 to 1986, professing in all the offices from instructor to professor and department chair of Industrial Design and the Craft Programs to the Dean of the University of the Arts. The recipient of a Doctorate Honoris Causa from his beloved alma mater, he retired serving with distinction over three decades and died in 1998.

William Parry | Son of a Welsh Pennsylvania coal miner, Parry attended the renowned Philadelphia Girard School as a young boy. After his graduation and short employment at a Manhattan Bank, he began his merit studies at the New York State School of Ceramics at Alfred. He elected to major in Charles Harder's Ceramic Design and Engineering Program, which was interrupted by his enlistment in the army, where he served as an Army Lieutenant in Europe. After the war, he returned to Alfred to complete his studies on the G.I. Bill. He and his family then moved to California, employed by the Riverside China Company as a ceramic designer and formulator of glazes. They returned east for Parry to teach at the Philadelphia Museum School of Art as its principal ceramist and to co-chair the emerging Dimensional Design Program with Reinhardt. As returning veterans after the carnage, they believed a way to promise the future was by educating individuals—maker/craftsmen/designer/ artists—who aspire to transform energies through material to help vibrantly form our deepest, equitable, needed, communal, lived-for aspirations.

William Parry, known by most as Bill, was a prescient artist/teacher. His gifts for nurturing insights, evolving boundaries, and engendering self trust in our sought for ventures called us all to wager risk with avidity. His own studio practice mirrored his core professed beliefs. From his tests and trials to unthought of "what ifs", we found often "possibility becoming" as persuasive productive outcomes. His teaching was a major way for media to morph thought to evocative form. For me, his professed gift has become a much-practiced preference.

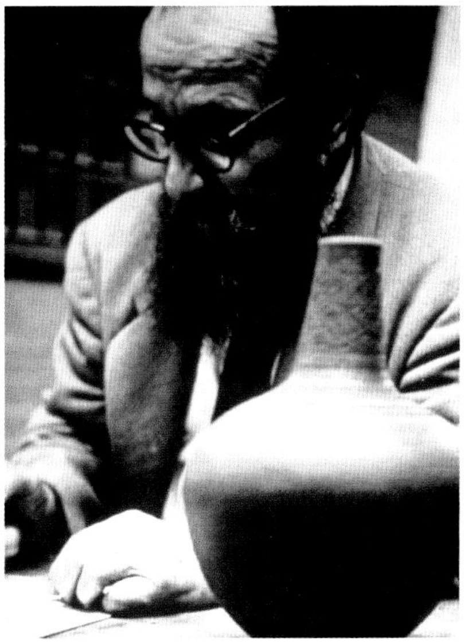

Marcus Aurellus Renzetti
1896 to 1975

His techno, serendipitous venturing-becoming sculptures are many. His test play with soft sheets of clay that stood alone on end with the least support became seemingly weightless avians. From clay-slab raptors and doves to butterflies and moths on point, they conjured gravity. Related sets of trials and tests supported clay sheets horizontally with legs from their corners; it was based on the under-structures of large tropical Lilly Pads. Another variation was based on tree leaves and their stem structures after insects had eaten out the spaces between the veins. Both nature searches were supported by parallel searches of coffered industrial concrete floors and ceilings becoming tables, trays, tiles and pierced screens.

A final example that pleased me was a stoneware, self-glazing clay body to behave like Egyptian Paste. He began by dipping K26 soft firebricks in alkaline copper solutions. When they came from the kiln, they were brilliant, iridescent turquoise-stained bricks. When fired stacked together, the touching surfaces fused. Later, this experiment led him to pre-cut, dip, and fuse-join elements that would become the stuff of his iridescent Stack House Sculpture. In 1986, Bill presented his findings in a solo exhibition at the Harstook Gallery at New York's Greenwich House. Reinhardt, Vaskys, and I attended, enjoying a vigorous conversation on the train home about Parry's smart moves.

William Parry was not about connecting ALL the dots; with the spaces in between, called "what if," for himself and his students, he answered to en-form them in the present through his seminal work, shaping shared wonder until his death in 2004.

Marcus Aurelius Renzetti | Born in Abruzzi, Italy in 1896, he moved to Philadelphia as a young man where he attended the famed experimental "Liberty Tadd School," officially the Public Industrial Art School of Philadelphia. In his after-school time, he inhabited the Fleisher Free School, founded by Samuel Fleisher. In 1916-17 and 1922-23, he studied sculpture at the Pennsylvania Academy of the Fine Arts. In 1922, Samuel Fleisher, impressed by Aurelius's precocity, awarded him a scholarship to study in Italy. Returning to Philadelphia, he opened his studio doing portrait commissions and industrial design projects. In the 1930s, he began teaching evening sculpture classes at the Fleisher Free School and day classes at Museum School—a regimen he kept for over four decades, teaching figure modeling, ceramic sculpture, industrial design, ceramics, basic science and engineering, three dimensional design, and freshman form study. At various times he chaired the Sculpture Program and Dimensional Design Program as one of its three founders.

Renzetti mentored all of us, students and faculty alike, with engaging wit and generous, confirming, tacit/spoken wisdom. He nurtured, the same way, primal materials—clay, fiber, glass, plaster, metal, wood, and stone—to embody touch forming content. He was in love with "material making," and his body, mind, and spirit were worked by his telling hands in action as a carver, constructor, caster, forger, smith, and deft lover of tools and their ways. His keen senses and haptic body intelligence were wondrously tuned to a deep intuitive awareness of nature's elegant, primal, logarithmic symmetries. We all felt privileged when offered an opportunity to be his acolyte, teaching with him. Witnessing his demonstrations with graced equilibrium, forming fitness, making a waste mold, putting a faux patina on a plaster portrait head, throwing a huge bowl on the potter's wheel, or forging and sharpening a tool was a deep learning joy. Whatever the media or its form M.A.R. was the Mega Hand/Mind/Meister.

Honored as "Master/Sculptor/Teacher" by Philadelphia's da Vinci Society, M.A. Renzetti retired in 1973 as Distinguished Professor Emeritus at the Philadelphia College of Art and Design. He was much beloved and revered by Philadelphia's art community and generations of students and colleagues at PCA, Fleisher, PAFA and Tyler. He died in the home he beautifully rebuilt, in Arden, Delaware,in 1975.

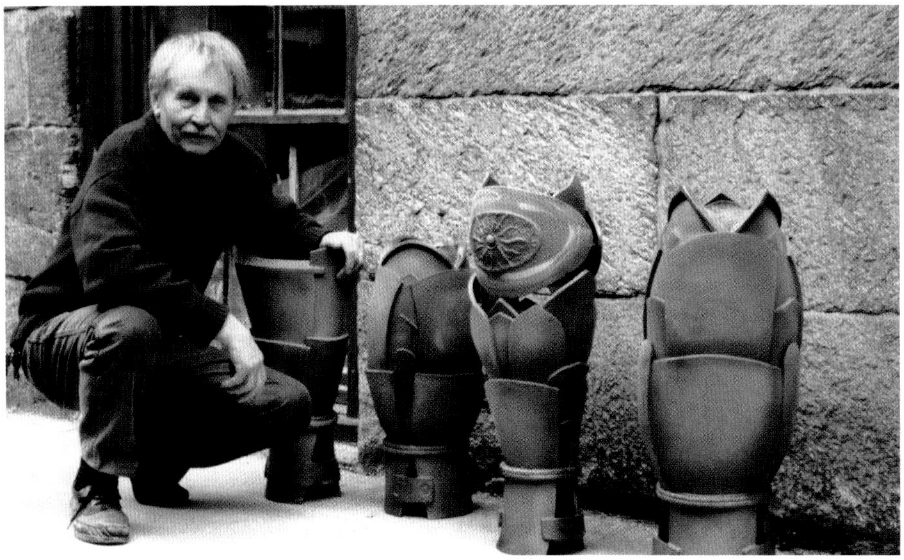

Petras Vaskys
1902 to 2001

Petras Vaskys | Vaskys was born in Mosedis, Lithuania, the son of a tailor/subsistence farmer and part-time traveling bookseller. With his sisters and brother, he grew up assisting his parents producing food and tailoring clothes. While attending town school and hands-on home schooling, Petras's precocity earned him a place at the Kounas Art Academy and the opportunity to study by working at the Kounas Botanical Gardens. His ceramic and botanical studies were curtailed by Lithuania's dual occupations and his conscription to agricultural forced labor during World War II. He survived the bombing of Dresden and later escaped to Italy. He then continued his education in 1944, living in a Monastery and studying sculpture at the Academy of Fine Arts in Florence. He completed his study in 1948 at the Academy of Fine Arts in Rome.

Beginning in 1948, at the Biennnalle-Nazionale De'Arte Sacra, Novara, Italy, Petras enjoyed a continuos parallel ceramic life as a sculptor/vessel maker.

In 1957, Petras and I both came to work at the Philadelphia Museum School. Petras, came to teach form study and plaster mold making with Aurelius Renzetti. They both taught ceramics with William Parry as a trio, prospering the clay studio's ceramic majors and elective students. Petras was the primary kiln-keeper, teaching all hands how to stack and fire the gas kilns. When George Roselli became our first "super tech," he took over the firing supervision, material, and maintenance logistics, as Petras spent more and more time teaching.

Vaskys maintained a very successful plaster-casting studio in Cornwall Heights, Pennsylvania, designing specialty cast industrial prototypes and complex plastic pieced anatomical medical demonstration models. His life-scale hand, skin, muscle, ligatures, bones, and nails, all in color, as removable parts, were a prize-winning wonder. He was also awarded a patent for his successful design and development of a silicone breast prosthesis. In 1963, Petras was commissioned to create the President tableau figures for the American Wax Museum at the Bourse in Philadelphia.

Represented by Philadelphia's Rosenfeld Gallery, he exhibited in the Ceramic Nationals at the Everson Museum in Syracuse, New York, and the invited exhibition Craft'67 at the Museum of the Philadelphia Civic Center. In 2010, at the Philadelphia Museum of Art he was part of "Interactions in Clay," a grand survey of Philadelphia's past half-century of Ceramic worthies. Cate and I enjoy being worship-full "keepers" of his work in our home collection, along with most of the artists cited in this writing.

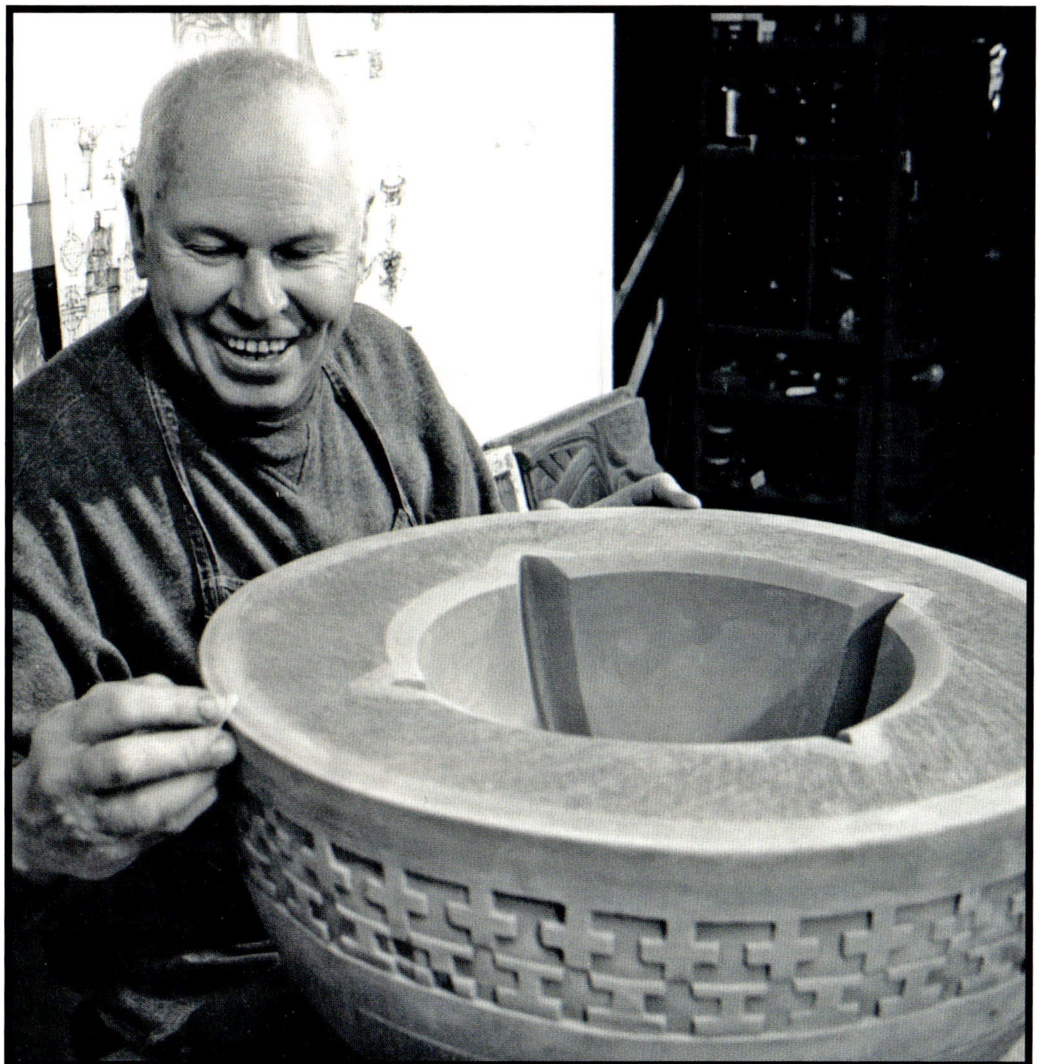

Photo: Bennett Lorber, MD, MACP

APPENDICES

Fig. 281. Bill Daley in his studio constructing the baptismal font for the Westminster Presbyterian Church, 1995

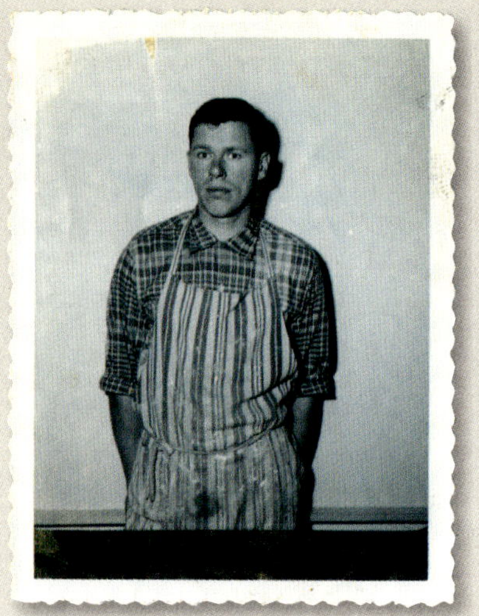

1951, Joins faculty of University of Northern Iowa, Cedar Falls, as instructor in ceramics and sculpture

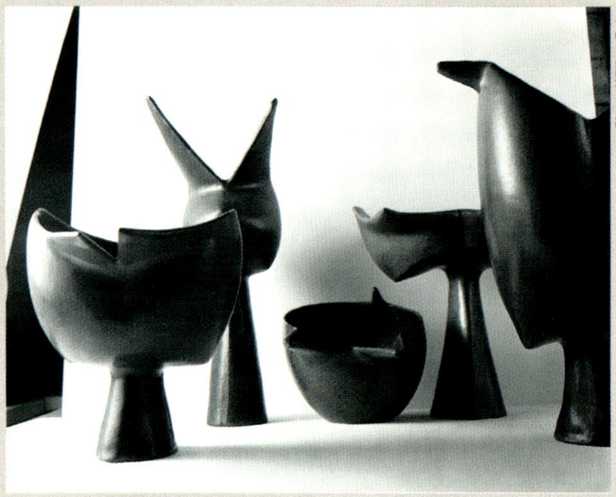

Photo: William Daley

1956, *Pottery by William Daley,* New York University, New York

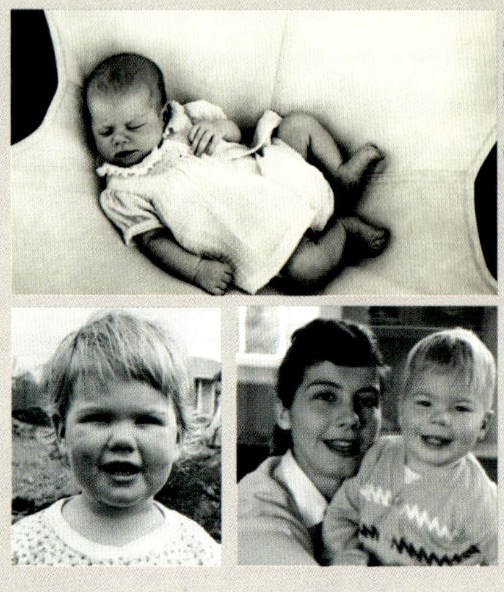

Photos: William Daley

Daughters Barbara, Charlotte, wife Catherine, and son Thomas.

Selected Chronology

1950

Receives Bachelor of Science from Massachusetts College of Arts, Boston, MA

Marries Catherine Anne Stennes September 2, 1950

1951

Receives Master of Arts in Education from Columbia University Teacher's College, New York, NY

Joins faculty of University of Northern Iowa, Cedar Falls, as instructor in ceramics and sculpture

1952

Iowa Artists 4th Annual Competitive Exhibition, Des Moines Art Center, Des Moines, IA

Seventh National Decorative Arts and Ceramic Exhibition, Wichita Art Association, Wichita, KA

Daughter Barbara is born

1953

Fiber, Clay, Metal 1953: Second Annual Craft Competition, St. Paul Gallery and School of Art, St. Paul, MI

5th Annual Iowa Artists Competitive Exhibition, Des Moines Art Center, Des Moines, IA

1954

Daughter Charlotte is born

Joins faculty of State University of New York College at New Paltz, as assistant professor teaching ceramics, sculpture and design

6th Annual Iowa Artists Show, Des Moines Art Center, Des Moines, IA (receives award)

Young Americans 1954, America House, New York, NY

1955

New York Society of Ceramic Arts: 61st Annual Exhibition, Cooper Union Museum for the Arts of Decoration, New York, NY

Young Americans 1955, American House, New York, NY

1956

Son Thomas is born on Bill's birthday

Pottery by William Daley, New York University, New York, NY

1957

Joins faculty of Philadelphia College of Art as associate professor teaching industrial design and coordinating sophomore three-dimensional program

Fifth Annual Miami National Ceramic Exhibition, University of Miami, Miami, FL (receives honorable mention)

1958

Becomes associate, Carreiro Design Associates, Philadelphia, PA (through 1960)

Featured on inaugural edition of *Adventures in the Hand Arts* – NBC educational series hosted by Shari Lewis for Girl Scouts of America

1959

Fountain, ceramic commission for John Pickett, Falls Church, VA

Phoenix, clay and wood sculpture commission for Robert John Company, Philadelphia, PA; Carreiro Design Associates, architect

William Daley, Pace Gallery, Boston, MA

1960

Becomes affiliate, Charles Broudy Architect, Philadelphia (through 1963)

Columbine, ceramic and wood sculpture commission for Herman Wilke, PA

William Daley, William Parry, Marcus Aurelius Renzetti, Lisa Langley, Philadelphia Art Alliance, Philadelphia, PA

1962

Abacus, ceramic, wood, and bronze screen commission for IBM Pavilion, Seattle World's Fair; Carreiro Design Associates, architects

Third Philadelphia Arts Festival, Philadelphia Museum of Art and Pennsylvania Academy of Fine Arts, Philadelphia, PA

22nd Ceramic National Exhibition, Everson Museum of Art, Syracuse, NY, Philadelphia Museum of Art and Pennsylvania Academy of Fine Arts, Philadelphia, PA

1963

Becomes affiliate, William Sklaroff Design Associates, Philadelphia, PA

Joins faculty of State University of New York College (SUNY) at Fredonia, as associate professor teaching ceramics, sculpture, and design

Screens, stoneware and teak screens, commission for Penn Center Inn's Singing Waters Restaurant, Philadelphia, William Sklaroff Design, architects

Centurion Marker, copper sheet and pipe fountain commission, Falls Church, VA: Pickett Siessto Hook, architects

1964

20th Annual Invitational Ceramics Exhibition, Lang Art Building, Scripps College, Claremont, CA

Rejoins faculty of Philadelphia College of Art (PCA) as associate professor coordinating sophomore professional program for graphic design, illustration, and textile design

Diana (1964-65) enters permanent collection of St. Lawrence University, Canton, NY

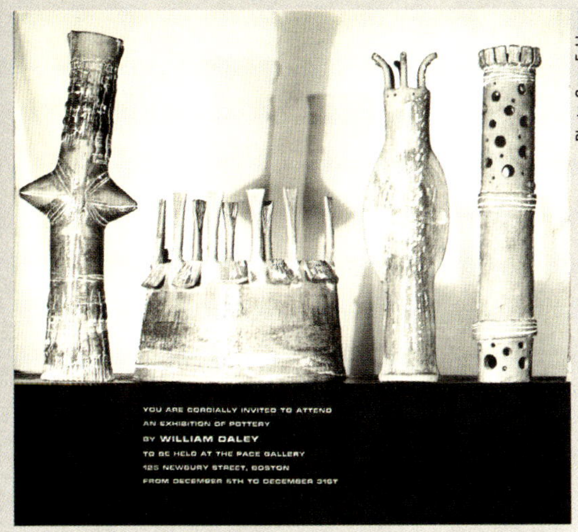

Photo: Gene Feldman

1959, Exhibit poster
Pace Gallery, Boston

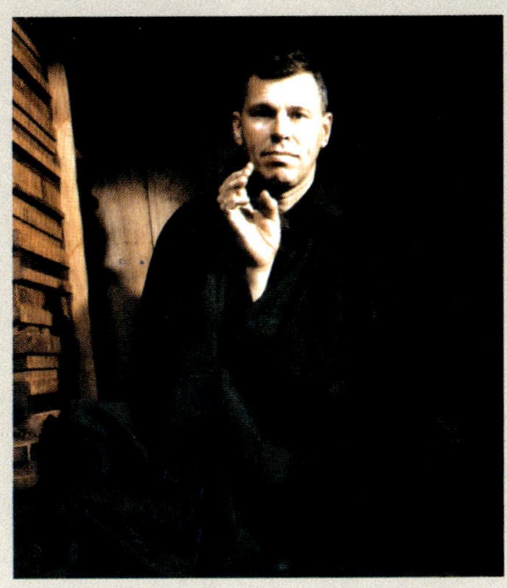

Photo: William Parry

1963, Joins faculty of State University of New York College (SUNY) at Fredonia

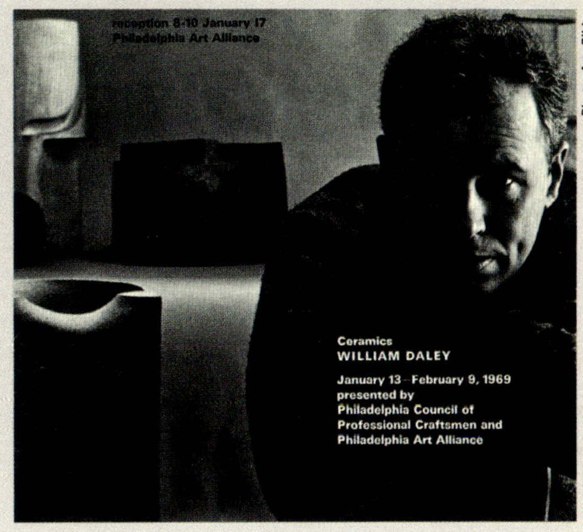

Photo: Ave Pildas

1969, *William Daley*, Philadelphia Art Alliance, sponsored by Philadelphia Council of Professional Craftsmen

Phoenix: Memorial to Lilian Gaubedian, brazed copper, relief commission for SUNY, Fredonia, NY

William Daley: Sculpture, Ceramics, Drawings, SUNY, Fredonia, NY

1966

Becomes chair of the Dimensional Design (Crafts Department), Philadelphia College of the Arts

Transformation, welded bronze sculpture commission ARCO Chemical Company, Philadelphia, PA

An Exhibition of Sculpture of William Daley, Pennsylvania State College, State College, PA

24th Ceramic National Exhibition, Everson Museum of Art, Syracuse, NY

William Daley, Samuel Maitlin, Jerome Johnson, Gallery 1015, Wyncote, PA

1967

Joins board of directors, Philadelphia Council of Professional Craftsmen (through 1971)

Craftsmen '67: First Regional Exhibition by the Museum of the Philadelphia Civic Center, Philadelphia, PA

First Survey of Contemporary American Crafts, University Art Museum, University of Texas at Austin, TX

William Daley/William Parry, Philadelphia Art Alliance, Philadelphia, PA

1968

Large Planter No. 1 (Composite Form) (1977) and *Large Planter No. 2 (Chiclet 1)* (1967), Philadelphia Art Alliance, Philadelphia, PA

Burning Bush, cupra nickel sculpture, commission for Herman Wilke, Philadelphia, PA

20th Anniversary Exhibition: Philadelphia Chapter Artists Equity, Museum of the Philadelphia Civic Center, Philadelphia, PA

25th Ceramic National Exhibition, Everson Museum of Art, Syracuse, NY (traveled)

1969

Joins board of directors, National Council on Education in the Ceramic Arts (through 1970)

Untitled, nickle-plated copper sculpture, commission for Holloway Realty Company, Philadelphia; Vincent Kling Associates, architects

Philadelphia Council of Professional Crafts, Philadelphia College of Art, Philadelphia, PA

William Daley, Philadelphia Art Alliance, sponsored by Philadelphia Council of Professional Craftsmen

1970

Becomes member of American Craft Council

Crafts 1970, Institute of Contemporary Art

Craftsmen '70, Museum of the Philadelphia Civic Center, Philadelphia, PA

Making It – 1 for 1. Franklin Institute Philadelphia, PA

Flight, commission for South African Airways, New York, NY

Peace Exhibition, Philadelphia Museum of Art, Philadelphia, PA

Lectures on panel with Robert Venturi and George Nelson, Collab 1971, Philadelphia Museum of Art, Philadelphia, PA

1971

Helix, bronze sculpture commission for Germantown Friends School, Philadelphia, PA

1972

Wall commission for Fairfield Maxwell Limited, New York, NY, Wilke Davis Mitchell, architects

1973

Becomes Chair of the Industrial Design Department, Philadelphia College of Art

Craftsmen '73, Museum of the Philadelphia Civic Center, Philadelphia, PA

Invitational Craft Exhibit, University of New Mexico Art Museum, Albuquerque, NM

1974

William Daley, Helen Drutt Gallery Philadelphia, PA

1975

Architectonic Volume II (Stadium Pot) enters the permanent collection of Central Iowa Art Association, Ceramic Study Collection, Marshalltown, IA

Soup Tureen enters permanent collection of Campbell Museum, Camden, NJ

Floor Pot (vessel) enters permanent collection of Philadelphia Museum of Art, Philadelphia, PA

Contemporary Clay: Ten Approaches, Dartmouth College, Hanover, NH (traveled)

Exhibition of Liturgical Arts, Museum of the Philadelphia Civic Center (organized by 41st International Eucharistic Congress)

Philadelphia: Three Centuries of American Art, Philadelphia Museum of Art

William Daley, Helen Drutt Gallery, Philadelphia, PA

1977

Contemporary Crafts, Marian Locks Gallery in association with Helen Drutt Gallery, Philadelphia, PA

The Philadelphia Craft Show 1977 Guide, Philadelphia Museum of Art

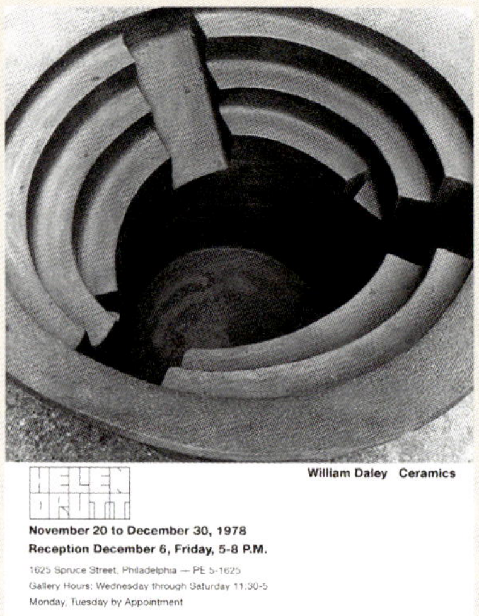

1974 William Daley at the Helen Drutt Gallery show announcement

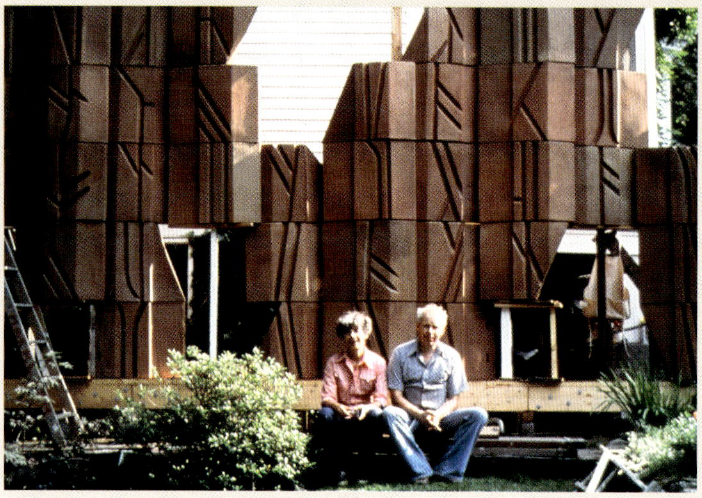

1978 trial assembly in backyard of house in Elkins Park of *Cellular Permutation* ceramic screen, commission for Ritz Theater, Philadelphia

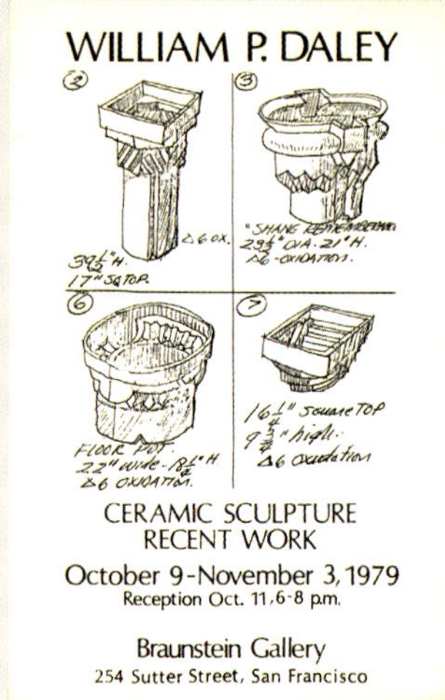

1979, William Daley: Ceramic Sculpture/Recent Work, **Braunstein Gallery, San Francisco**

1978

Cellular Permutation, ceramic screen, commission for Ritz Theater, Philadelphia, PA

Artist Equity Triennial Exhibition, Museum of the Philadelphia Civic Center, Philadelphia, PA

1979

A Century of Ceramics in the United States, 1878-1978, Everson Museum of Art, Syracuse, NY (traveled)

Clay Fiber Metal: An Invited Exhibition of Pennsylvania Crafts, Beaver College, Glenside, PA (currently Arcadia University)

Invitational Clay/Fiber/Metal Exhibition, Gallery of the College of Visual and Performing Arts, Southeastern Massachusetts College of Art, Boston, MA

William Daley: Ceramic Sculpture/Recent Work, Braunstein Gallery, San Francisco

1980

Untitled (Stadium Pot), 1979, enters the permanent collection of St. Louis Art Museum, MO

Receives award for Outstanding Achievement in the Visual Arts, Massachusetts College of Art, Boston, MA

The Contemporary American Potter: New Vessels, University of Northern Iowa, IA

Philadelphia: Clay 1980, Tyler School of Art, Temple University, Philadelphia, PA

Robert L. Pfannebecker Collection: A Selection of Contemporary American Crafts, Moore College of Art, Philadelphia, PA

William Daley, Helen Drutt Gallery, Philadelphia, PA

1981

American Potters I, Hadler/Rodriguez Gallery, New York, NY

American Potters II, Garth Clark Gallery, Los Angeles, LA

Centering on Contemporary American Clay: American Ceramics from the Joan Mannheimer Collection, University of Iowa Museum of Art, Iowa City, IA

Explorations: Ceramic Seminar '81, Nickle Art Museum, Calgary, Alberta, Canada

Fifth National Crafts Invitation Exhibition, Art Gallery, Skidmore College, Saratoga Springs, NY

25th Anniversary of the American Craft Museum Art Gallery, Skidmore College, Saratoga Springs, NY

William Daley, Exhibit A Gallery, Chicago

**Broad Spectrum: Artists who teach at the Philadelphia College of Art*, Allentown Art Museum, PA (traveled)

1982

William Daley: Selected Works, 1954-1982, Carreiro Gallery, Massachusetts College of Art, Boston, MA

1983

Danced Place (1982) enters permanent collection of Dienst Beeldende Kunst, Kruithuis, s'Hertogenbosch, The Netherlands

Becomes Honorary Life Member, National Council on Ceramic Education in the Arts

American Clay Artists: Philadelphia '83, Philadelphia, PA

Betty Woodman/Elsa Rady/William Daley, Douglas Drake Gallery, Kansas City, MO

Ceramic Echoes: Historical References in Contemporary Ceramics, Nelson Atkins Museum of Art, Kansas City, MO

Clay in New Mexico 1983, University of New Mexico, Albuquerque, NM

Who's Afraid of American Pottery?, Dienst Beeldende Kunst, Kruithuis, Stedelijk Museum, Hertogenbosch, The Netherlands (traveled)

William Daley, Helen Drutt Gallery, Philadelphia, PA

1984

Clay Vessels: Works by Ten Modern Masters, Palo Alto Cultural Center, CA

Multiplicity in Clay Metal Fiber, American Craft Museum, New York, NY

1985

Ceramic Art of the '70's: A Review of a Decade, Exhibit A Gallery, Chicago, IL

William Daley: Large Scale Handbuilding, Pewabic Pottery, Detroit, MI

Pottery Questions II, Broward Community College, Fort Lauderdale, FL. (traveled)

Surface/Function/Shape: Selections from the Earl Millard Collection, Southern Illinois University, Edwardsville, IL

William Daley, Garth Clark Gallery, New York, NY

William Daley: Ten Vessels, New York State College of Ceramics at Alfred University, Alfred, NY (traveled)

William Daley: Vessels, Braunstein Gallery, San Francisco, CA

1986

To Josiah W. (1980) enters permanent collection of Victoria & Albert, London, UK (gift of Mr. & Mrs. John W. Pickett)

Speaker, *American Potters Today: 5th International Ceramics Symposium* of the Institute for Ceramic History, Victoria & Albert Museum, London, UK

Participant in symposium and catalog, *Architecture of a Vessel: A Symposium*, School for American Craftsmen, Rochester Institute of Technology, Rochester, NY

Big Containment: Large Scale Ceramic Vessels by Contemporary American Ceramists, Northern Illinois University, DeKalb, IL

Contemporary Crafts: A Concept in Flux, Society for the Arts in Crafts, Pittsburg, PA

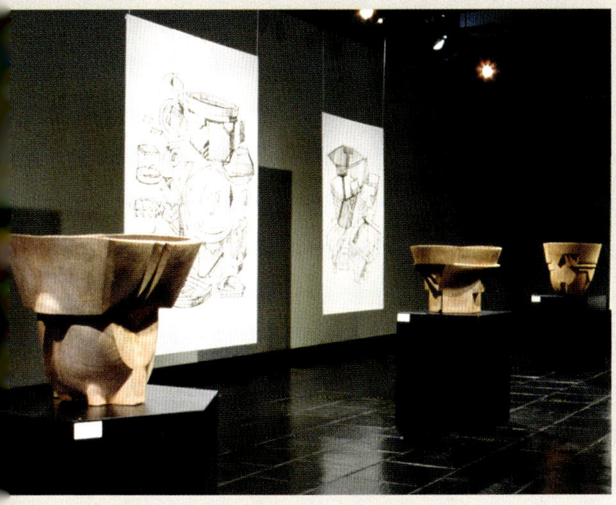

1985, *William Daley: Ten Vessels,* New York State College of Ceramics at Alfred University, Alfred, NY (traveled)

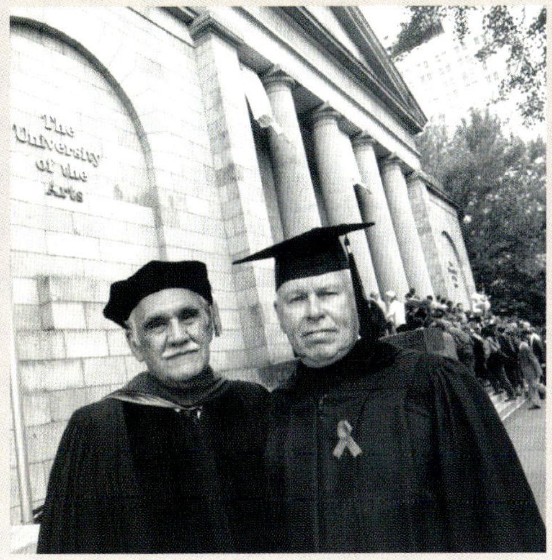

Photo: John Carlano

1990, William Daley becomes Professor Emeritus, University of the Arts, pictured with Richard Reinhart

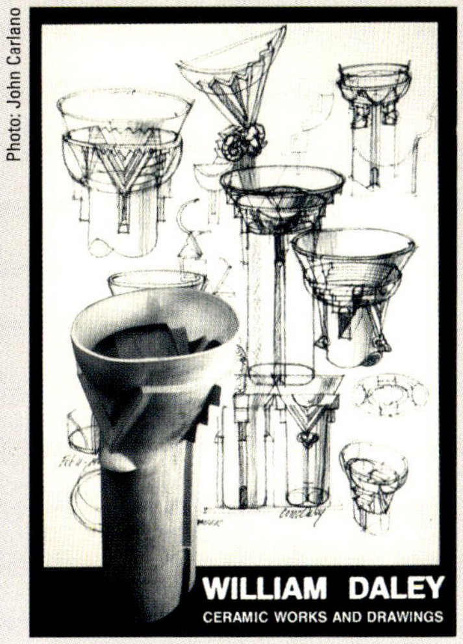

1991 Receives Art Matters Lifetime Achievement Award for Excellence in the Arts

Craft Today: Poetry of the Physical, American Craft Museum, New York, NY (traveled)

1987

Apsed Toras (1987) enters permanent collection of National Museum of Contemporary Art, Seoul, Korea (gift of the artist)

American Ceramics Now: The 27th Ceramic National Exhibition, Everson Museum of Art, Syracuse, NY (traveled)

Ceramic Artists: Works on Paper, Syracuse University, Syracuse, NY

Contemporary American Ceramics, National Museum of Contemporary Art, Seoul, Korea

1988

Oval Chamber (1987) enters permanent collection of Renwick Gallery of the National Museum of American Art, Smithsonian Institute, Washington, DC

Clay in the East IV: Ceramic Conference and Exhibition, The University of the Arts, Philadelphia, PA

Elected to College of Fellows, American Crafts Council

East-West Ceramics Exhibition, Korean Culture and the Arts Foundation, XXIV Olympics, Seoul, Korea

Solo Exhibition William Daley, Helen Drutt Gallery, Philadelphia, PA

1989

Baptismal Font, ceramic, commission for St. Paul's Episcopal Church, Elkins Park, PA, in memory of Akarue Amachi

Becomes first University Distinguished Professor, University of the Arts, Philadelphia, PA

American Clay Artists 1989, William Daley, Patterson Sims, and Ken Vavrek, Port of History Museum, Philadelphia, PA

Artful Objects: Recent American Crafts, Fort Wayne Museum of Art, Fort Wayne, IN

**Craft Today USA,* Musée des Arts Decoratifs, Paris, France. (travels to 15 countries until 1992)

Kansas City Collects Contemporary Ceramics, Nelson Atkins Museum of Art, Kansas, MO

2nd International Ceramics Competition, Festival Grounds, Mino, Japan

1990

Joined Haystack Mountain School of Crafts Board of Trustees (1990-1999)

Bowmans Rests (1989) enters the permanent collection of Los Angeles County Museum of Art, Los Angeles, CA (gift of Gwen Laurie Smits)

Conic Loop (1989) enters the permanent collection of Arkansas Arts Center, Little Rock, AR

Silla's Well (1988) and *Bifold Cloister* (1990) enter permanent collection of National Museum of Contemporary Art, Seoul, Korea

Taos Revisited (1989) enters permanent collection of Everson Museum of Art, Syracuse, NY

Becomes Professor Emeritus, University of the Arts, Philadelphia, PA

1991

Conical Procession (1982) enters the permanent collection of American Craft Museum, New York, NY (gift of Jane and Leonard Korman)

Floor Pot (1979) enters the permanent collection of Boca Raton Museum of Art, Boca Raton, FL (gift of Jane and Leonard Korman)

Receives Distinguished Teaching of Art Award, College Art Association

Receives *Art Matters* Lifetime Achievement Award for Excellence in the Arts

Functional Ceramics 1991, Wayne Center for the Arts, Wooster, OH

Scale/Detail 1991, Habitat/Shaw Gallery, Farmington Hills, MI

1992

Greened Oval Squared (1992) enters the permanent collection of Arrow International Inc., Reading, PA

1993

Vesica III (1992) enters the collection of Shin Sang Ho

Receives the Distinguished Alumni Award and Exhibition, Teachers College, Columbia University, New York, NY

**William Daley: Ceramic Works and Drawings,* text by Ricard Torchia and Matthew Drutt, The Levy Gallery for the Arts in Philadelphia, Moore College of Art and Design, Philadelphia. (traveled to The Renwick Gallery - 1994, Washington D.C., The Arkansas Arts Center - 1994, Little Rock, and The American Craft Museum, New York - 1996)

Becomes a Trustee to the American Craft Council. (1993-1996)

**Collectors Annual: The Art of Craft,* Boca Raton Museum of Art, Boca Raton, FL

1994

**The Collector's Eye: Contemporary Ceramics, American, Canadian and British,* The Koffler Gallery, Koffler Centre of the Arts, North York, Ontario, Canada

Receives the degree of Doctor of Fine Arts honoris causa, University of the Arts, Philadelphia, PA

Receives the Distinguished Alumni Award, Massachusetts College of Art, Boston, MA

**Clay Studio 20th Anniversary Celebration Exhibition,* The Philadelphia Maritime Museum, Philadelphia, PA

Entries with * asterisks denote titles of exhibitions with catalogs

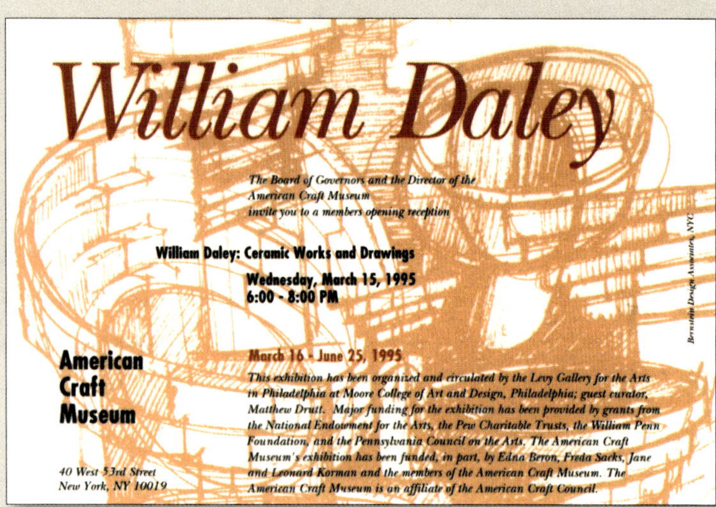

1995 *William Daley: Ceramic Works and Drawings* exhibit at the American Craft Museum, New York, NY

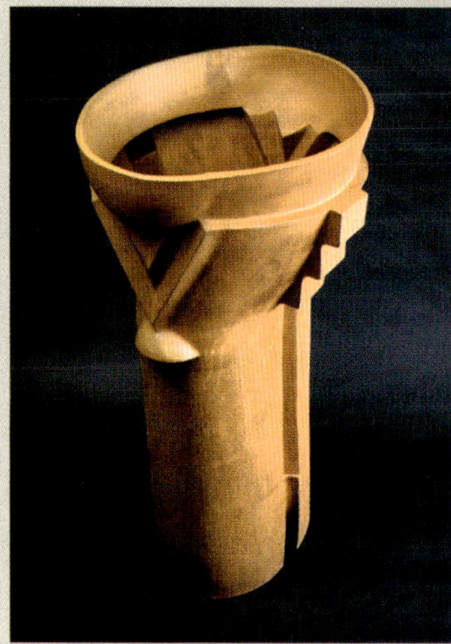
2001 USA Clay, *Oval Chamber*, 1986, The Renwick Gallery, Smithsonian American Art Museum, Washington DC

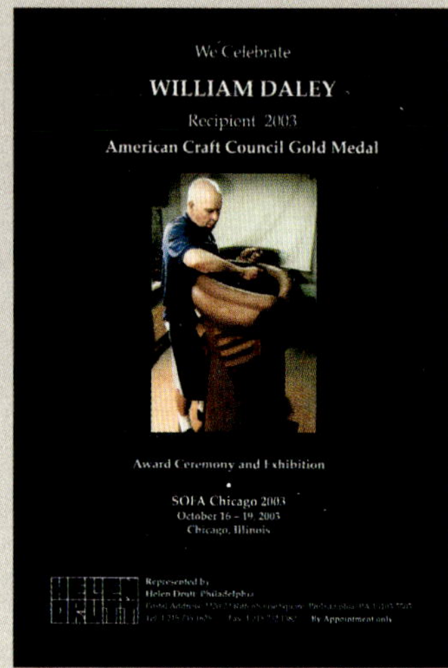

2003 Recipient of the American Craft Council Gold Medal for consummate craftsmanship, SOFA. Chicago, IL

1995

Keynote speaker, *25th Annual Glass Art Society Conference*, Asheville, NC.

William Daley: Ceramic Works and Drawings exhibit at the American Craft Museum, New York, NY

Baptismal Font, Pulpit Tiles, Lectern, commission for sanctuary interior of Westminster Presbyterian Church, West Chester, PA; Crothers Daley Jalboot Architects.

Lecture on *Art, Worship, and Chancel Furniture* at Westminster Presbyterian Church, West Chester, PA

Abbot's Space (1984) enters permanent collection of Mint Museum of Craft and Design, Charlotte, NC (gift of Allan Chasanoff)

1996

Untitled, stoneware and copper wall commission for South African Airlines (1970) enters permanent collection of Arrow International Inc., Reading, PA

1997

**Celebrating American Craft 1975 – 1995*, text by Bodil Busk Laursen, Susie Elson and Paul J. Smith, The Danish Museum of Decorative Art, Copenhagen, Denmark

1998

**Contemporary Ceramics: Selections from The Metropolitan Museum of Art*, The Metropolitan Museum of Art, New York, NY

Prides Crossing (1985) and *Squared Up* (1985) enter permanent collection of The Metropolitan Museum of Art, New York, NY

1999

Burning Bush & Descending Dove, chapel tiles commission for chapel interior of Westminster Presbyterian Church, West Chester, PA; Crothers Daley Jalboot Architects

**Choice from America: Modern American Ceramics*, text by Arthur C. Danto and Janet Koplos, The Kruithuis Museums Collection, Hertogenbosch, The Netherlands

Columnar Form (1978) enters permanent collection of The Kruithuis Museums Collection, Hertogenbosch, The Netherlands

Regis Master Series: Honors, Lecture, and Exhibition, Northern Clay Center, Minneapolis, MN.

2000

**Color and Fire: Defining Moments in Studio Ceramics, 1950 – 2000*, Los Angeles County Museum of Art, Los Angeles, CA. (traveled)

Allan Chasanoff Ceramic Collection, Mint Museum of Craft and Design, Charlotte, NC

Participant in the Senior Artists Initiative, Philadelphia, PA

Represents The Haystack Mountain School of Crafts, *SOFA Seventh Annual International Exposition of Sculpture Objects & Functional Art*, Chicago, Navy Pier, Chicago, IL

2001

**Studio Days 2001: A View of Contemporary Ceramics*, Chester Springs Studio, Chester Springs, PA

**Centennial Arts & Crafts Exhibition: Celebrating Contemporary Crafts 2001*, Providence Art Club, Providence, RI

USA Clay, The Renwick Gallery, Smithsonian American Art Museum, Washington DC

**World Contemporary Ceramics, World Ceramic Exposition 2001*, Korea, Ichon World Ceramic Center, Ichon, Korea

William Daley: Ceramic Vessels and Drawing From His Personal Collection, A lecture and exhibition, Keystone College, La Plume, PA

Senior Artists Initiative Selected Works, Woodmere Art Museum, Philadelphia, PA

2002

Poetics of Clay: An International Perspective, The Philadelphia Art Alliance, Philadelphia, PA (traveled)

Participant in *Digital Dialogues, Technology And The Hand: A Studio-Based Symposium*, Haystack Mountain School of Crafts, Deer Isle, ME

RF's Passing Place (2001) acquired by The Rhode Island School of Design Museum.

2003

Haystack Vesica (2003) enters the collection of Candice Groot

Orphic Vesica (2003) enters the permanent collection of Arizona State University Art Museum, Tempe, AZ

**Great Pots: Contemporary Ceramics From Function To Fantasy*, The Newark Museum, Newark, NJ

Conic Chamber (1986) enters the permanent collection of The Newark Museum, Newark, NJ

Prior Oval (1986) enters the permanent collection of Minneapolis Institute of Arts, Minneapolis, MN

Recipient of the American Craft Council Gold Medal for consummate craftsmanship, SOFA, Chicago, IL

*Honorary Exhibitor *SOFA Tenth Annual International Exposition of Sculpture Objects & Functional Art*, Chicago, Navy Pier, Chicago, IL

*Represents Helen Drutt: Philadelphia, *SOFA Sixth Annual International Exposition of Sculpture Objects & Functional Art New York*, Seventh Regiment Armory, New York, NY

Manhattan Canyon Wall, stoneware wall commission for Fairfield Maxwell Limited (1971) enters permanent collection of The Museum of Fine Arts, Houston, TX (gift of Marlin Miller Jr. and Helen Williams Drutt English). Companion drawings given by William and Catherine Daley

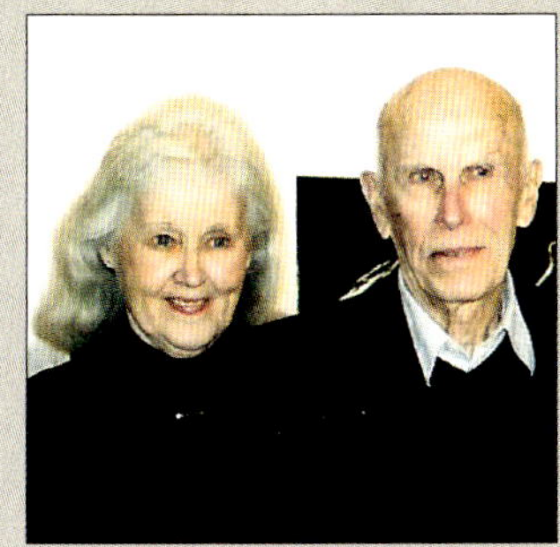

Photo: Charlotte Daley

Bill's sister Alice Daley LaPier and his brother Joseph O'Mara at Bill's 80th birthday celebration

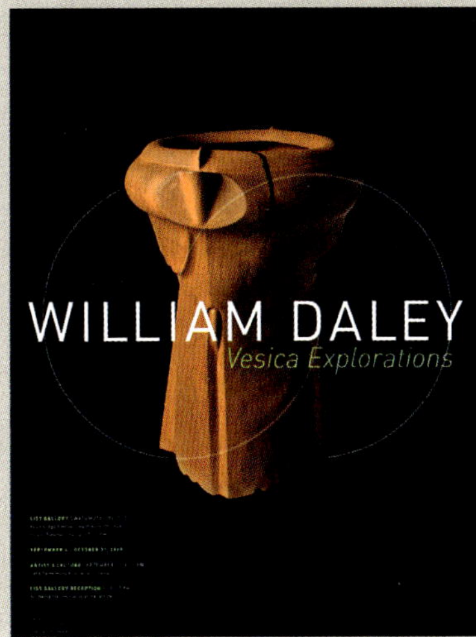

2009 Poster for *Vesica Exploration* exhibition, List Gallery Swarthmore College, Swarthmore, PA Poster designed by George Andrew Schwab

Awarded a Pew Fellowship in the Arts, October, 2010 Clay Artist

Floor Pot (1973) enters permanent collection of The Museum of Fine Arts, Houston, TX (gift of Jenna Cooper)

2004

Lecture *Where Matter and Spirit Are Met*, Church of St. Martin-In-The-Fields, Philadelphia, PA

Panelist for *Craft Education: A 75th Anniversary View from Penland School of Crafts, SOFA Eleventh Annual International Exposition of Sculpture Objects & Functional Art,* Chicago, Navy Pier, Chicago, IL

Panelist for *Craft Education: A 75th Anniversary View from Penland School of Crafts, SOFA Seventh Annual International Exposition of Sculpture Objects & Functional Art,* Chicago, Seventh Regiment Armory, New York, NY

Guest lecturer *Ceramics in Arts Education,* a workshop and panel discussion with Wayne Bates, Hood College, Fredrick, MD

Becomes an Honorary Trustee, Haystack Mountain School of Crafts, Deer Isle, ME

Receives the James Renwick Alliance Distinguished Educator Award, The Renwick Gallery of the Smithsonian American Art Museum, Washington, DC

Commissions *Daley's Arc*, a concerto for clarinet and piano composed by Arne Running

Phoenix Vesica (2004) enters the collection of Alfred and Rica Fenaughty

2005

Keynote speaker *39th Annual Conference of the National Council on Education* for the Ceramic Arts, Baltimore, MD

*Represents The Museum of Fine Arts, Houston *Eighth Annual International Exposition of Sculpture Objects & Functional Art*, Seventh Regiment Armory, New York, NY

Vesica Monday (2005) enters the collection of Marlin and Ginger Miller

2006

Keynote speaker *2006 International Architectural Ceramic Exhibition: New Way of Architectural Ceramics*, Clayarch Gimhae Museum, Gimhae, Korea.

Oracular Vesica (2006), *Nuptual Vesica* (2006), *Processing Vesica* (2005), *Extierus Vesica* (2004), *Johns Passing* (1995), *B's Place* (1995), enter the permanent collection of the Clayarch Gimhae Museum, Gimhae, Korea (*Oracular Vesica* gift of William P. Daley)

Suite of ten drawings, *Vesica Vessels* (2006), enters the permanent collection of the Clayarch Gimhae Museum, Gimhae, Korea

2007

Receives the Watershed Inaugural Legends Award, Watershed Center for the Ceramic Arts, New Castle, ME

Crossed Vesica (2007) enters the collection of Alfred and Rica Fenaughty

Award and exhibition held at SOFA 2007 Chicago, Chicago, IL

Faced Fore (1981) enters the permanent collection of The Philadelphia Museum of Art, Philadelphia, PA

2008

Cyclops Eye Cup Revisited (2007) enters the collection of Marlin and Ginger Miller

Vesica Vesica II (2008) enters the collection of Alfred and Rica Fenaughty. (gift of William and Catherine Daley)

Architecture and Ceramics: Material. Structure. Vision an exhibit and lecture with son Thomas Daley, Architect, at the University of Minnesota College of Design, Minneapolis, MN

2009

**Vesica Explorations Exhibition*, List Gallery Swarthmore College, Swarthmore, PA in October 2009

2010

Daley and Daley A collaboration, March 29, 2010, Philadelphia Center for Architecture, Philadelphia, PA

Awarded a Pew Fellowship in the Arts, October, 2010 Clay Artist

SOFA Chicago Nov 5-7, 2010 Perimeter Gallery, Chicago, IL

2011

Permanent collection University of Minnesota Architecture and Ceramics, *A material for the Ages*, Minneapolis, MN

William Daley Selected Works 1967-2010 Phillips Exeter Academy Exeter, NH

William Daley at SOFA Chicago 2011, Chicago, IL

Entries with * asterisks denote titles of exhibitions with catalogs

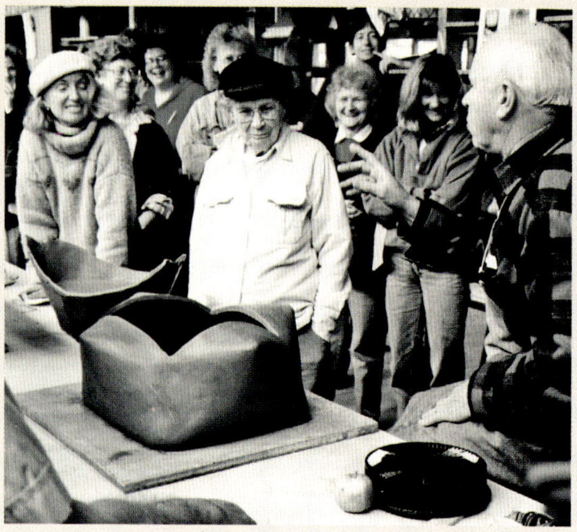

1989 Workshop at Haystack School of Crafts, Deer Island, Maine

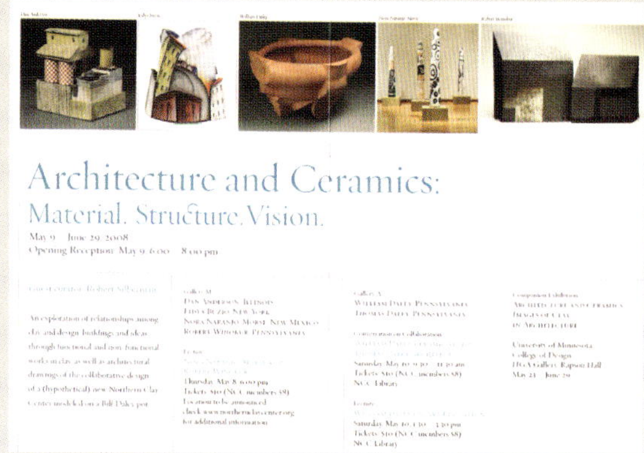

2008 *Architecture and Ceramics: Material. Structure. Vision,* an exhibit and lecture with son Thomas Daley, Architect, at the University of Minnesota College of Design.

Workshops

1958

NBC's (National Broadcasting Company Inc.) Adventuring in the Hand Arts: The Hands of Man Adventuring in Pottery, RCA Building, Radio City, New York, NY

1961

General Motors Styling Staff Summer Program, General Motors Technical Center, Warren, MI

1966

Multimedia presentation with Natalie Charkow, South Senior High School, Great Neck, NY

1968

Girl Scouts of the United States of America's World of Arts: Raku Workshop, Pleasantville, NY

1970

Ceramic Workshop, Waldorf College, Forest City, IA

1971

The Inter-Society Committee for 20th Century Decorative Arts and Design: Colab '71, Philadelphia Museum of Art, Philadelphia, PA

1973

Summer Session Course: Ceramic Sculpture Workshop, California College of Arts and Crafts, Oakland, CA

1979

Artists and Commissions, Purdue University Department of Creative Arts, West Lafayette, Indiana

1979

Summer Session: An Approach to Design for Craftsmen, The Banff Centre School of Fine Arts, Banff, Alberta, Canada

1982

Workshop, Emily Carr College of Art, Vancouver, British Columbia

1983

Artist in the Blossom Festival School – Art: Summer Program, Kent State University, Kent, OH

Pottery Workshop, Luther College, Decorah, Iowa

1984

Large Scale Hand Building, Napa Valley College, Napa, CA

Summer Workshops 1984: William Daley, Anderson Ranch Arts Center, Aspen, CO

Workshop, Lecture, Studio Evaluation, School of Art, Kutztown State College, Kutztown, PA

1985

Ceramic Workshop, Long Beach Island Foundation of the Arts and Sciences, Loveladies, NJ

Visiting Artist Workshop, Chautauqua Institution, Chautauqua, NY

Visiting Artist Workshop, Nova Scotia College of Art and Design, Halifax, Nova Scotia, Canada

1986

William Daley: Lecture and Workshop,Greenwich House Pottery, New York, NY

Fall Education Series: William Daley, Lillstreet Gallery, Chicago, IL

1987

Many Minds and Hands: Bill Daley Visiting Artist, Alfred University, Alfred, NY

Pots in a Box, University of South Carolina, Columbus, SC

1988

Hands On Workshop: William Daley and Lisbeth Stewart, Mountain House Craft Center, Lancaster, PA

Visiting Artist: William Daley, Hartwick College, Oneonta, NY

1989

Workshop, The Mercersberg Academy, Mercerberg, PA

Hands On Workshop, The University of Manitoba School of Art, Winnipeg, Manitoba

First Session: Clay: William Daley, Haystack Mountain School of Crafts, Deer Isle, ME

Handbuilding Workshop: Bill Daley, Mainline Center of the Arts, Haverford, PA

Clayfolk Workshop, Southern Oregon State College, Ashland, OR

1990

Artist-in-Residence: Ceramic Workshop, Cheltenham Art Center, Cheltenham, PA

Visiting Artist: William Daley, Southern Illinois University at Carbondale, Carbondale, IL

Workshop, University of Miami, Department of Art and Art History, Coral Gables, FL

Visiting Artist Workshop, Chautauqua Institution, Chautauqua, NY

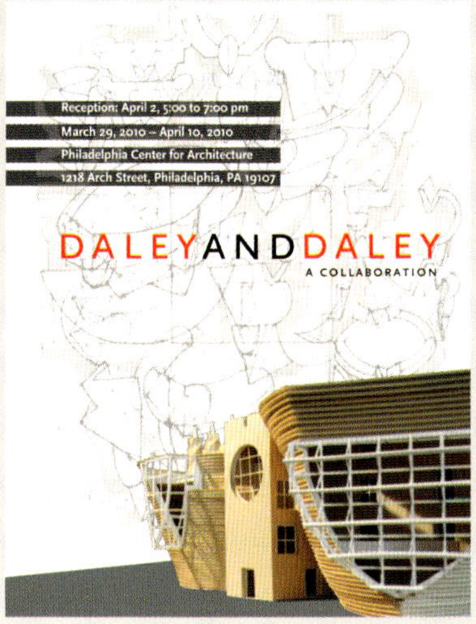

2010 *Daley and Daley—A collaboration,* exhibition at the Philadelphia Center for Architecture, Philadelphia, PA

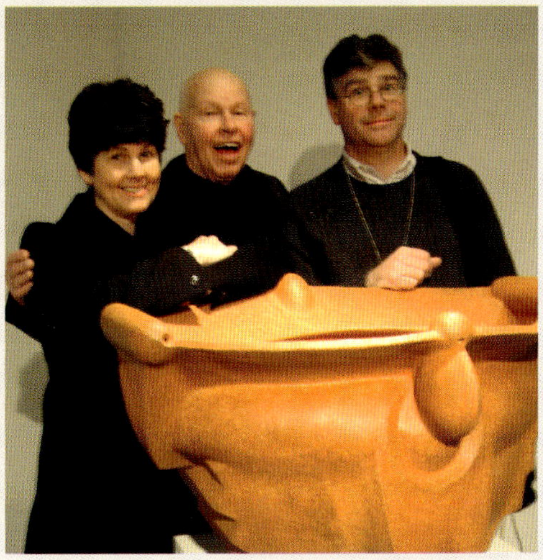

Daughter Charlotte, Bill, and son Thomas at SOFA 2010

Hands On Workshop, Arvada Center for the Arts and Humanities, Arvada, CO

William Daley Workshop, Whitemarsh Community Art Center, Lafayette Hill, PA

Bill Daley, Radcliffe Ceramics Studio, Boston, MA

1991

Functional Ceramics Workshop, Wayne Center for the Arts, Wooster, OH

Large Scale Handbuilding, Converse College, Spartanburg, SC

Pots from Preforms, Peter's Valley Craft Center, Layton, NJ

Summer Session: Large Scale Hand Building, Penland School, Penland, NC

Visiting Artists/ Scholars Program: Bill Daley, Northern Illinois University, DeKalb, IL

1992

Handbuilding Ceramics, Touchstone Center for Crafts, Uniontown, PA

1995

Tar Paper to Clay: Large Scale Slab Building, University of Oregon Department of Fine and Applied Arts, Eugene, OR

1998

Tar Paper to Clay: Large Scale Slab Building, Baylor University Department of Art, Waco, TX

Tar Paper to Clay: Large Scale Slab Building, Souderton High School, Souderton, PA

1999

Neshaminy Art Teachers Workshop, Neshaminy High School, Neshaminy, PA

Pottstown Kid's Fest, Directed by Chester Springs Studio, Pottstown, PA

Summer Art Institute for Teachers, Abington Art Center, Abington, PA

Professional Studio/Lecture Series: Slab-building with Freshman), Massachusetts College of Art, Boston, MA

2000

Art and Mud: A course for teachers, Anderson Ranch Arts Center, Snowmass Village, CO

The New England Workshop, Haystack Mountain School of Crafts, Deer Isle, ME

Tar Paper to Clay: Large Scale Slab Building, Cleveland Institute of Art, Cleveland, OH

Tar Paper to Clay: Large Scale Slab Building, La Salle College High School, Wyndmoor, PA

2001

Tar Paper to Clay: Large Scale Slab Building (participants were high school art teachers and students), Northern Clay Center, Minneapolis, MN

Ceramics: a course for teachers, Anderson Ranch Arts Center, Snowmass Village, CO

Pots From Pre-forms, Peter's Valley Craft Education Center, Layton, NJ

Fall Clay Intensive: Up-Scale Handbuilding, Truro Center for the Arts: Castle Hill, Truro, MA

Tar Paper to Clay: Large Scale Slab Building, Juniata College, Huntington, PA

2002

Tar Paper to Clay: Large Scale Slab Building, National College of Art & Design, Dublin and University of Ulster School of Art, Belfast, Ireland

Tar Paper to Clay: Large Scale Slab Building, Maine College of Art, Portland, ME

2003

Clay in Mind Workshop: high school teachers and their students, Northern Burlington County Regional High School, Columbus, NJ

Studio Based Learning, Haystack Mountain School of Crafts, Deer Isle, ME

2005

Upscale Handbuilding, Archie Bray Foundation for the Ceramic Arts, Helena, MT

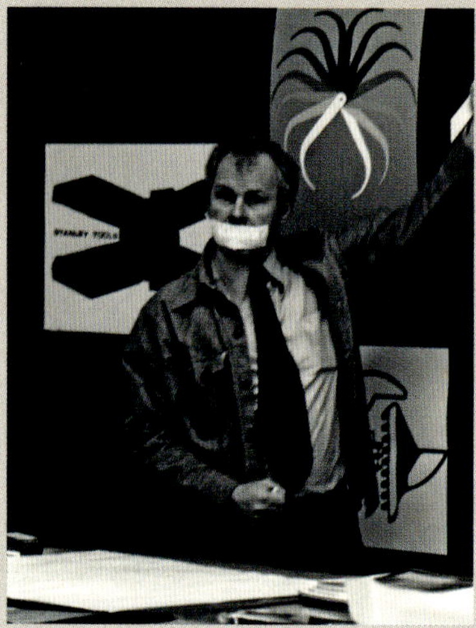

Photo: Murray Weiss

Bill encourages the Industrial Design students to talk about their work

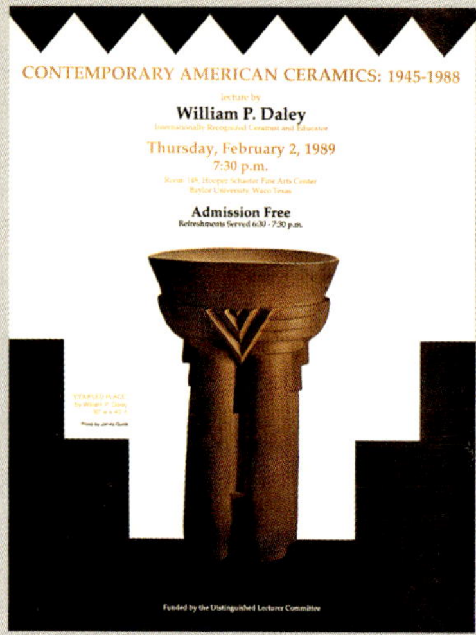

1989 Announcement for Lecture at Baylor University, Waco, TX

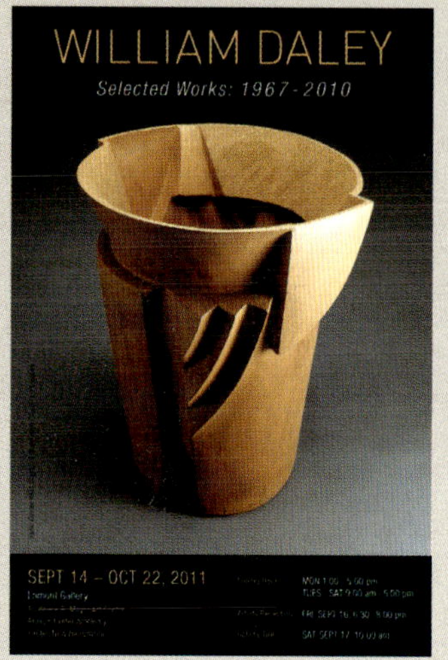

2011 Announcement for Phillips Exeter Academy Exhibit and Lecture

Lectures

1974

Design, (Architecture) and Art Symposium, University of Cincinnati, Cincinnati, OH

1976

Lecture, Montgomery County Community College, Blue Bell, PA

1977

The School District of Philadelphia's Board of Education and The Philadelphia Art Alliance: Artist to Artist Program, Philadelphia Art Alliance, Philadelphia, PA

Lecture for Department of Industrial Design Professional Practices Course, Ohio State University, Columbus, OH

Lecture to Freshman Foundation Class, Alfred University, Alfred, NY

Jam Craft Media Day and Exhibition Art in Crafts: From Function to Fantasy, Jamaica Art Center, Jamaica, NY

Arts Council: Contemporary Crafts Considered, YM/YWHA, Philadelphia, PA

Artist-in-the-Schools Program: Monumental Clay Works, William Allen High School, Allentown, PA

1978

William Daley Ceramic Works, Cleveland Institute of Art, Cleveland, OH

William Daley Ceramic Works, Ceramics Department, Rhode Island School of Design, Providence, RI

1979

William Daley: Slide Lecture, Fine Arts Center, University of Rhode Island, Kingston, RI

NCECA Supermud Conference: Women in Clay, The Pennsylvania State University, University Park, PA

Ceramics in Interior Spaces, Stein Auditorium, Drexel University, Philadelphia, PA

Through the Ages (with asides to Indian ancestors, present friends, and others), Purdue University Department of Creative Arts, West Lafayette, IN

1980

The Contemporary American Potter: William Daley, Lecture, Washington State University Museum of Art, Pullman, WA

NCECA 1980: Three Ceramic Artists, William Daley, Viola Frey and Jun Kuneko, The University of Michigan, Ann Arbor, MI

1981

Ceramics Seminar '81: Lecture, The Nickle Arts Museum, Calgary, Alberta

The Philadelphia Museum of Art Celebration of Crafts: Ceramics: Philadelphia/USA, Van Pelt Auditorium, The Philadelphia Museum of Art, Philadelphia, PA

Visiting Artist Lecture Series Fall '81, State University of New York Fredonia College, Fredonia, NY

1982

Ceramist Lecture, The University of Texas at Austin, Austin, TX

William Daley at Swarthmore College, Swarthmore College, Swarthmore, PA

1983

NCECA 1983: The Geometry of Residence, Georgia State University, Atlanta, TX (Included "Honorary Membership Award)

1984

NCECA 1984: Presentation of the Apple Award, George Sherman Union, Boston University, Boston, MA

1985

Architecture Interaction of Art and Technology in Architecture: William Daley, Temple University, College of Engineering and Architecture, Philadelphia, PA

Talk / Slides / Discussion: William Daley & Garth Clark, Kansas City Art Institute, Kansas City, Kansas

The Clay Studio Lecture Series: Bill Daley, The Clay Studio, Philadelphia, PA

1986

5th International Ceramics Symposium of the Institute for Ceramic History: William Daley, Victoria and Albert Museum, London, UK

Architecture of the Vessel Symposium: Thoughts on a Geometry of Presence, Rochester Institute of Technology, Rochester, NY

1986

Hands On Workshop, Connecticut Clay Artists, Newtown, CT

Panel Discussion Architectural Ceramics: Michael Rubin, Rudy Autio, William Daley, Ruth Duckworth, George Woodman, Chicago International New Art Forms Exposition, Chicago, IL

The American Craft Council Presents: Art/Culture/Future: American Craft '86: Education Panel, Oakland Museum, Oakland, CA

1987

Visual Artists Series: Slide Lecture, State University of New York, Plattsburg, NY

NEA Visiting Artists Forum Series: William Daley, University of Arkansas, College of Arts and Sciences, Fayetteville, AK

International Creators: William Daley, The Craft Studio at Harbourfront, Toronto, Ontario, Canada

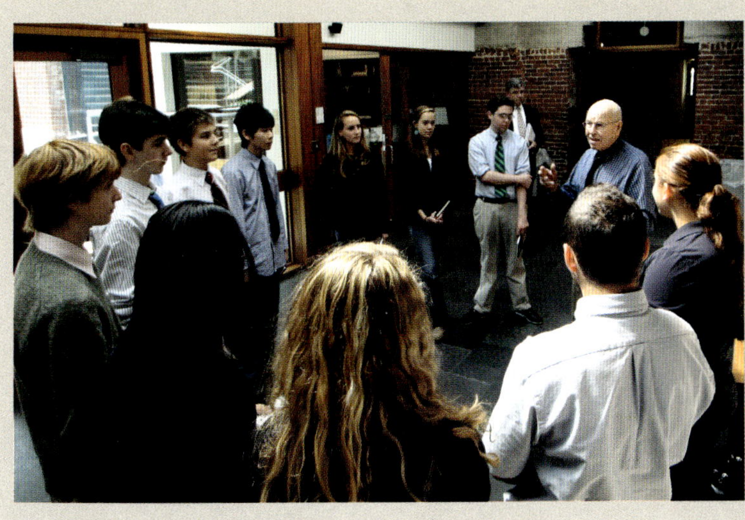

2011 Bill talks to the high school students at Phillips Exeter Academy about his work

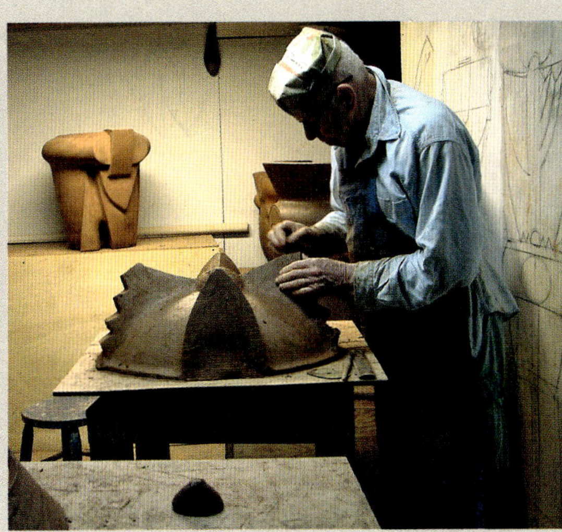

2005 Bill working in his basement studio

1988

The Case for Clay in Secondary Art Education: Perception-Past and Present: A Form Makes Thoughts, New York University, New York, NY

Art Workshop: William Daley – Camden County College, Camden, NJ

1989

Contemporary American Ceramics: 1945-1988, Baylor University Department of Fine Arts, Waco, TX

William Daley Ceramic Artist: Slide Lecture on Ten Years of Work, Princeton University, Princeton, NJ

Art Forum: William Daley Mud Now, Montclair State College, Upper Montclair, NJ

The Marjorie Heilman Visiting Artist Lecture Series: William P. Daley Ceramicist, Kirby Lecture Hall, Swarthmore College, Swarthmore, PA

1990

Guest Lecturer, The Banff Centre School of Fine Arts, Banff, Alberta, Canada

Spring Forum: William Daley, James Renwick Alliance, Bethesda, MD

1991

Guest Lecturer for class Earth-Water-Fire: Ceramic Art 3600 BC to the Present, John Hopkins University, School of Continuing Studies, Baltimore, MD

Presenter: Conversations '91 The Inside of Art, SUNY The College at New Palz, New Paltz, NY

1995

Keynote speaker, 25th Annual Glass Art Society Conference, Asheville, NC

Lecture on Art, Worship, and Chancel Furniture at Westminster Presbyterian Church, West Chester, PA

2004

"Where Matter and Spirit Are Met," Church of St. Martin-In-The-Fields, Philadelphia, PA

2005

Keynote speaker 39th Annual Conference of the National Council on Education for the Ceramic Arts, Baltimore, MD

2009

"Vesica Explorations" lecture, Swarthmore College, Swarthmore, PA

2011

"William Daley Selected Works 1967–2010" lecture, Phillips Exeter Academy, Exeter, NH

"Reflections of an American Craft Council Gold Medalist: A Conversation with William Daley and Janet Koplos" at SOFA 2011, Chicago, IL

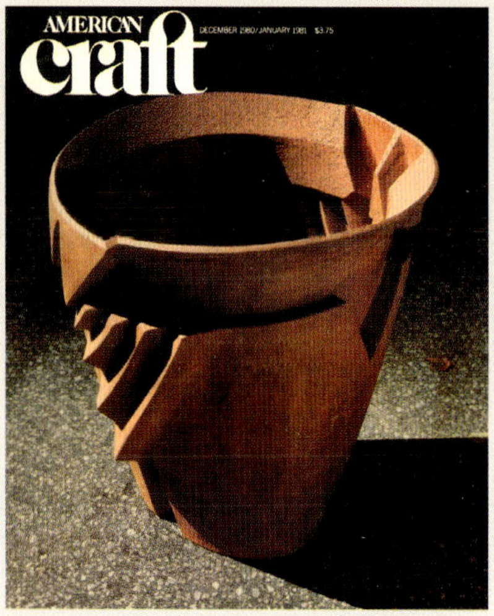

1980 *American Craft* cover

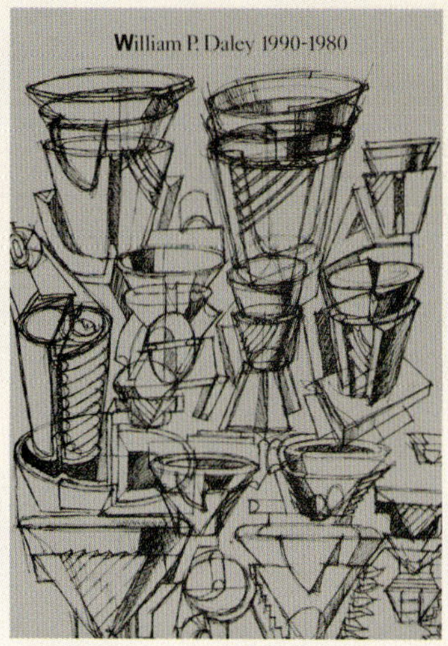

William Daley 1990–1980
University of the Arts, Philadelphia

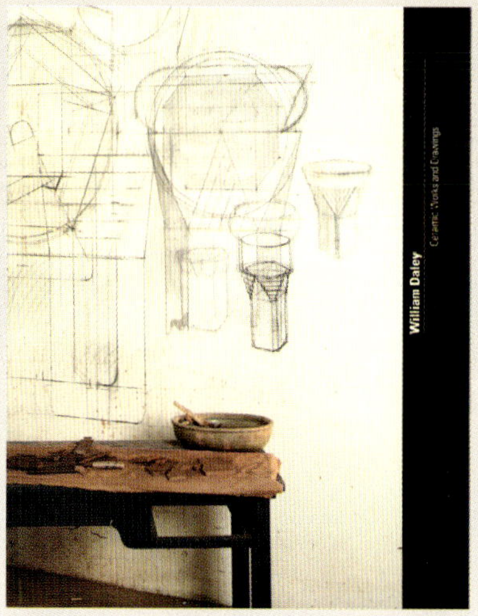

William Daley: Ceramic Works and Drawings,
Moore College of Art and Design, Philadelphia

Bibliography

Periodicals

Anderson, John. "New Accent on Fifth Avenue." *Interiors* (October 1968)

___. "The Non-Exotic Ticket Office." *Interiors* (May 1971): 108-11

"Basic Study on Architectural Ceramics." *Clay In Art International Yearbook* (2006-2007): 150-151

Bodine, Sarah, and Michael Dunas. "On View: Philadelphia." *New Art Examiner* (June/Summer 1992): 26-38

Borum, Jenifer P. "William Daley American Craft Museum." *Art Forum International* (February 1996): 87

Clark, Garth. "American Ceramics." *Crafts* (May/June 1986): 40-47

Daley, William. "Sand-Mold Ceramics." *Ceramics Monthly* (1953): 24

___. "Artist and Teacher." *School Arts*, no. 3 (1963): 13-15

___. "Helen Drutt Gallery, Nov. 8 – Dec. 2, 1978." *Goldsmith Journal* (February 1974): 4

___. "Celebration of Clay." *American Craft* (August/September 1979): 2-11

___. "Notes on Sources: A Presentation." *NCECA Journal* (1, 1980): 20-21

___. "On Drawing." *American Ceramics* (1, Winter 1982): 38-41

___. "Apple Award: Charles Abbott. Presented by William Daley." *NCECA Journal* (April 1984): 6-8

___. "Form Makes Thought: Perceptions of a Form-Maker." *Studio Potter* (June 1988): 27

___. "Comment: Permission." *Ceramics Monthly* (November 1991): 86

___. "Thoughts on Craft and Learning." Haystack Mountain School of Crafts (1991): 4-6

___. "Viewpoints: Issues in Art Education." *American Craft* (May/June 1992): 77

___. "Looking Backward." Studio Potter (June 1992): 1-14

___. "Exchanges of the Spirit." *Digital Dialogues 2002: Technology and the Hand* (Fourteenth Ed. ©2004 Haystack Mountain School of Crafts): 9-10

___. "Pennsylvania Passages Part One: Bill Daley." *The Studio Potter* (Volume 28, Number 2, June 2000): 83-84

Daley, William, et al. "Statement: Visual Artists Fellowships Crafts Panel, June 15-19, 1992." *New Arts Examiner* (September 1992): 1-2

Davis, Spencer L."William Daley: Ceramics and Drawing." *Ceramics Monthly* (September 1994): 57-60

Doubet, Ward. "William Daley: Facts of Faith." *American Ceramics* (2, 1990): 16-25

Drutt-English, Helen W. "Poetics of Clay: An International Perspective." *Kerameiki Techni: International Ceramic Art Review* (Issue 39, December 2001): 55-59

Drutt, Matthew. "William Daley Ceramic Works and Drawings." *Ceramics Art and Perception* (Issue 19, 1995): 10-18

Ellison, Robert A., Jr. "William P. Daley." *American Craft* (August/September 1990) 70-71

"Gold Medal: William Daley." *American Craft* (October/November 2003): 58-59

Hays, Joanne. "Talents Outside the Mainstream." *Artweek* (January 28, 1985): 3

Hunt, William. "Bill Daley: Architectonic Vessels." *Ceramics Monthly* (February 1980): 29-31

Jepson, Barbara. "American Ceramists." *Town and Country* (May 1981): 205-8

Kau, Yi-Wen. "Bill Daley." *Ceramic Art* (Issue 1, Autumn 1993): 44-49 (Korean)

Kaye, Ellen. "Putting Down Roots." *Discover Magazine, Philadelphia Inquirer* (May 30, 1976): 12-13

Koplos, Janet. "William P. Daley." *New Art Examiner* (June 1981): 17

Levy, Marge. "NCECA Honorary Membership Awards: William Daley." *NCECA Journal* (1983): 9-10

Mahoney, William. "Looking/Learning: Shang Play." *School Arts* (December 1985): 23-26

Mallet, John. "Pottery As Art Not Craft?" *Arts Review* (June 1989): 328-29

Mason, Marlise. "Mr. Daley." *Nous* (April 28, 1966)

McTwigan, Michael. "Duality in Clay: William Daley." *American Craft* (December 1980/January 1981): 30-33

Moran, Lois. Ed. "Arrow International – 5 Commissions." *American Craft* (April/ May 1997): 44-47.

Olitski, Jules. "Pots for a Show." *Ceramics Monthly* (December 1956): 17, 32

Saline, Carol. "If It Felt So Good, How Come I Feel So Bad?" *Philadelphia Magazine* (November 1975): 122-27.

Schreckengost, Viktor. "In Conversation: Viktor Schreckengost/William Daley." *American Craft* (June/ July 1997): 44-49, 70-72

"Sculpture by William Daley." City Art: The Fine Arts Program (1978): 6

Sewell, Darrel, et al. Philadelphia Museum of Art Bulletin: *Contemporary American Crafts* (Fall 1991): 56

Slemmer, Leslie. "William Daley." *Craft Horizons* (June 1975): 20

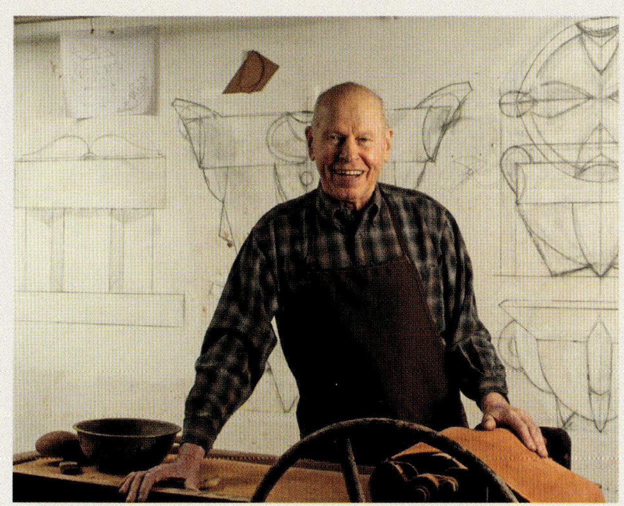

2010 Awarded a Pew Fellowship in the Arts, October, 2010 Clay Artist

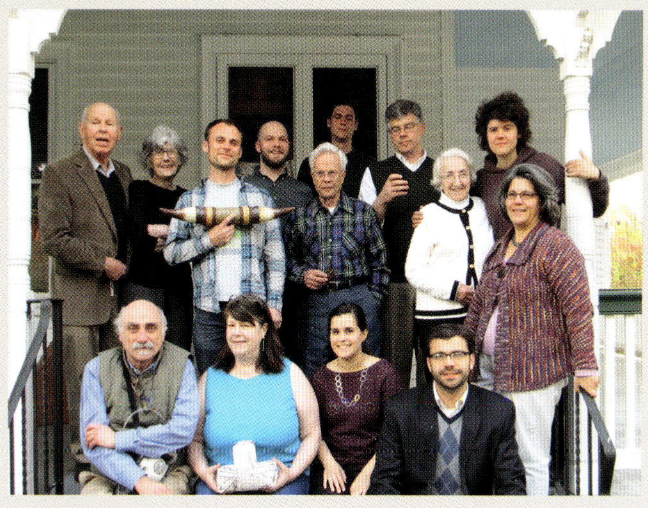

Photo: John Carlano

2012 Easter – family and friends on the front porch at the Daley House

Staal, Gert. "Pottenbakken knustel-imago ontstegen." *Volkskurant* (February 2, 1983)

"Transformations." ARCO (March/April 1967)

Wasserman, Burt. "The Unique Tureen Collection of Campbell Museum Reflects Singular Aesthetics in Functional Vessels." *Art Craft Magazine* (October/November 1980): 6-13

Watson, Oliver. "American Potters Today." *Ceramic Review* (May/June 1986): 24-28

Wieder, Lawrence. "Helen Drutt Gallery." *American Craft* (August/September 1979): 30-31

"William Daley." *Ceramica* (1988): 42-45

Williams, Gerry. "Nine Potters from Philadelphia." *The Studio Potter* (Volume 2, Number 1, Summer 1973): 11-18

Winokur, Paula. "William Daley." *Craft Horizons* (August 1976): 57

Wright, David G. "Red Clays for Mid-Range Oxidation." *Ceramics Monthly* (March 1999): 70-72

Books

Bach, Penny Balkin. *Public Art in Philadelphia*, Philadelphia: Temple University Press, 1992

Borrmann, Gottfried. *Keramik der Welt*, Dusseldorf, Germany: Kunst + Handwerk, 1984

Ceramic Study Collection, The. Marshalltown, Iowa: Central Iowa Art Association, 1979

Clark, Garth. *American Potters.* New York: Watson-Guptill, 1981

___. American Ceramics: 876 to the Present. New York: Abbeville Press, 1987

Daley, William. "Inside/Outside." In *Conversations 1991: The Inside of Art.* New Paltz, NY: State University of New York, New Paltz Metals Department, 1991

Dormer, Peter. *The New Ceramics: Trends + Traditions.* London: Thames & Hudson, 1986

Levin, Elaine. *The History of American Ceramics.* New York: Harry N. Abrams, 1988

Lord, Joan and Liza Kirwin. *Artists In Their Studios: Images from the Smithsonian's Archives of American Art*, New York: Collins Design, 2007

Margetts, Martina. *International Crafts.* London: Thames & Hudson, 1991

Nelson, Glenn. *Ceramics: A Potter's Handbook*, 5th ed. New York: Holt, Rinehart and Winston, 1984

Philadelphia's Treasures in Bronze and Stone. New York: Walker Publishing Co., 1976

Ramljak, Suzanne, *Crafting A Legacy: Contemporary American Crafts in the Philadelphia Museum of Art*, Philadelphia: Philadelphia Museum of Art, 2002

Sang Ho, Shin. *Contemporary Ceramics – A Step Toward the Future.* Seoul, Korea: Hongik University Ceramic Research Institute, 1993

Speight, Charlotte. *Hands in Clay: An Introduction to Ceramics*, 2nd ed. Mountain View, Calif.: Mayfield Publishing Co., 1989

Wilson, Janet, Skilled Work: *American Craft in the Renwick Gallery*, Washington: Smithsonian Institution Press, 1998.

"2400" Collection, The. Washington, DC: U.S. News and World Report, 1984

Photo: ©Victoria and Albert Museum

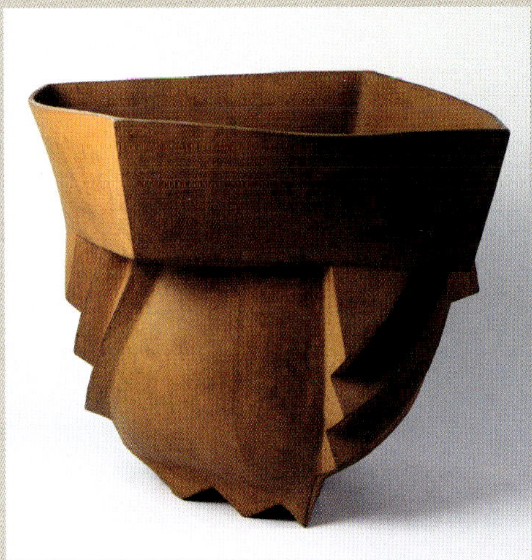

1980 *To Josiah W* in the collection of the Victoria and Albert Museum, London, UK

Photo: Amanda Schwab

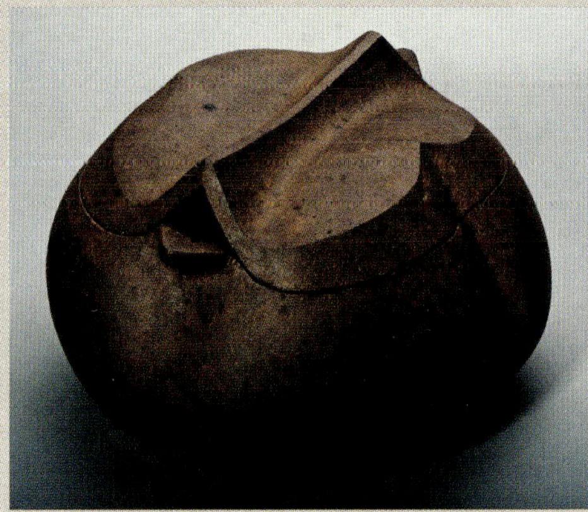

1975 *Soup Tureen* enters permanent collection of Campbell Museum, Camden, NJ

Museums

Arkansas Art Center
Little Rock, Arkansas

Boca Raton Museum of Art
Boca Raton, FL

Campbell Museum,
Camden, NJ

Carnegie Museum of Art
Pittsburgh, PA

Central Iowa Art Association
Marshalltown, IA

Charles A. Wustum Museum of Fine Arts
Racine, WI

Clayarch Gimhae Museum
Gimhae, Korea

Des Moines Art Center
Des Moines, Iowa

Everson Museum of Art
Syracuse, NY

Los Angeles County Museum
Los Angeles, CA

Metropolitan Museum of Art
New York, NY

Mint Museum, The
Charlotte, NC

Minneapolis Museum of Art
Minneapolis, MN

Museum of Arts and Design
New York, NY

National Museum of Contemporary Art
Seoul, Korea

Philadelphia Museum of Art
Philadelphia, PA

Rhode Island School of Design,
Museum of Art
Providence, Rhode Island

Reading Public Museum
Reading, PA

Renwick Gallery of the National Museum
of American Art, Smithsonian Institution,
Washington, DC

Saint Lawrence University Museum
Canton, NY

Saint Louis Art Museum
St. Louis, MO

Stedelijk Museum
Amsterdam, The Netherlands

Victoria and Albert Museum
London, UK

author
Ruth Fine

contributors
William Daley
Helen W. Drutt English
Janet Koplos
Marian Pritchard
Ezra Shales
Stephen Tarantal

editor
Charlotte Daley

editor/designer
Barbara Daley Schwab

editing consultant
Sara Jane Daley

artist representative
Thomas W. Daley

chronology
Mathew Drutt, 1950-1993

chronology/design consultant
George Andrew Schwab, 1994-2012

transcription
George M. Schwab

major photography
John Carlano
James Quaile
Amanda Schwab